Fundamental Optical Models

Fundamental Optical Models addresses the theory of several fundamental optical problems and provides insights into converting theory into computer software. The book brings together a variety of essential theories and modeling tools – some only found in very old research papers – into a single text and helps researchers to understand the theories needed to develop their own software that can be customized for different projects. The book's nine chapters cover the optics of multilayer thin films, planar waveguides, multilayer conformal thin films on diffraction gratings, the method of lines for modeling more general waveguides, cylindrical waveguides (i.e., optical fibers) with multilayers, fields in the vicinity of the focus of a lens, "near field to far field" transforms, and scattered field finite difference time domain modeling. To aid learning, the book includes a large number of examples and helpful hints in converting theory into computer algorithms.

This book is suitable for undergraduate and graduate students studying electromagnetism and optics, as well as industrial researchers.

Key Features:

- Addresses the critical gap between optical theory and practical implementation.
- Brings together a variety of fundamental optical theories and modeling tools that are often scattered across very old research papers and difficult-to-find sources.
- Includes numerous examples and helpful hints from a distinguished industrial physicist.

Fundamental Optical Models

William A. Challener

CRC Press
Taylor & Francis Group
Boca Raton London New York

CRC Press is an imprint of the
Taylor & Francis Group, an **informa** business

Designed cover image: William A. Challener

First edition published 2026
by CRC Press
2385 NW Executive Center Drive, Suite 320, Boca Raton FL 33431

and by CRC Press
4 Park Square, Milton Park, Abingdon, Oxon, OX14 4RN

CRC Press is an imprint of Taylor & Francis Group, LLC

ISBN: 978-1-041-14836-4 (hbk)
ISBN: 978-1-041-19936-6 (pbk)
ISBN: 978-1-003-71423-1 (ebk)

DOI: 10.1201/9781003714231

Typeset in Times
by Apex CoVantage, LLC

Contents

Preface

Over the course of my career, I have often found it necessary and/or convenient to develop my own optical modeling software. Obviously, there are many very powerful commercial programs available to make calculations similar to many of those described in this book and to solve even much more complex problems, so why would one even think of doing this? In my case, there were several reasons. Commercial software is often prohibitively expensive, sometimes requiring an annual license renewal that is a large fraction of the original purchase price. Commercial software is often so complex that it can take several years to master. Even then, technical support is frequently required for learning the steps to perform some new type of calculation, and the vendors often offer courses for their customers to learn just the basics of using their software.

In addition, the commercial software may calculate only one part of a research problem. With a personal, customized program, the user can string together different types of calculations to provide the final solution to a more complicated optical device. For example, if one desires to efficiently couple the electric fields from a focused beam of radially polarized light into a nano-antenna of arbitrary design, it is highly unlikely that there is any commercial software which combines complex lens focusing calculations with numerical nano-antenna calculations.

Perhaps the most important reason for writing my own software, however, has been the lack of commercial software able to solve particular optical research problems. Commercial software is primarily useful for the engineer who is optimizing the design of well-known optical structures such as grating couplers into ridge waveguides. If the type of structure is already routinely used in commercial products, then it is likely that there is commercial software for modeling it. If the optical device is sufficiently novel, however, as is often the case in research settings then there may be no useful commercial software. For example, I have looked in vain for thin film software that can solve for the optical properties of a multilayer film stack that includes a magneto-optic or liquid crystal layer. Another example is the energy coupling from a near field transducer located at the focus of a planar solid immersion mirror fed by waveguide modes that are out-of-phase. Such a model was used to design and optimize the near field transducer for heat assisted magnetic recording, a hard disc drive storage technology that is now commercially available.

On the other hand, it is also true that it is "expensive," that is, time-consuming, to write and debug one's own software and so there should be a good reason for doing so. In this book I have attempted to provide some helpful hints here and there for converting a physics equation into a working computer algorithm. For those who are just interested in solving the types of problems considered herein, there is also low-cost software that was used to generate the figures in this book available at Junonia Photonics.

I should also mention that I have tried to present multiple examples of calculations for each chapter, some from the technical literature and some from my own experience, as an aid for those who would develop their own software. Of course, I am far more familiar with the problems on which I have personally worked and which have been the original motivation for developing these models and software, so I hope I will be forgiven for including some of my own research examples and undoubtedly leaving out many other more worthy examples.

I would like to thank many colleagues with whom I have worked over the years on a variety of interesting projects for their help and suggestions. Dr. Amit Itagi receives a special mention for his collaboration on the development of the scattered field FDTD algorithm that could model tightly focused beams of laser light. Many times, when stuck on a bug in my software, he would come to the rescue. I have greatly missed that collaboration when writing this book and having to redevelop the FDTD software! Dr. Chubing Peng also deserves special mention for our electromagnetic physics discussions and for his amazing laboratory skills that generated, among other things, the spectacular scanning near field microscopy image in Figure 4.24(b). The author also benefited greatly in better understanding magneto-optics, plasmonics, and optical data storage from discussions and interaction with Prof. Masud Mansuripur. Professors Ed Schlesinger, Dan Stancil, and Jim Bain were important collaborators in the HAMR project and first exposed the author to solid immersion lenses and FDTD modeling. Drs. Sabarni Palit and Jeff White each suggested specific optical problems that became valuable examples to extend and exercise the modeling software. Of course, I am also indebted to the many scientists who have published their work over the years that provided the basis for this book. There are perhaps a few new ideas herein, but by far most of the theory and examples in this book were first developed and described, sometimes many decades ago, by others from whom I have been privileged to learn.

Finally, I must especially thank my wife, Helen, who has been extremely patient over the years that I have been occupied writing this book, understanding that it has been one of my most important life projects.

William A. Challener,
Glenville, NY

About the Author

William A. Challener, PhD, is an industrial physicist. After completing his doctorate in physics from the University of California at Berkeley in 1983, he was a postdoctoral fellow at the Los Alamos National Laboratory for two years studying the far infrared properties of various materials. Following that, he joined 3M to conduct research in rewritable optical data storage where he worked for eleven years. In the Imation spin-off from 3M, he continued his research for four years in the application of surface plasmon physics to novel gas and biosensors. Transferring to the Seagate research center, he worked for nine years on perhaps his most commercially successful endeavor where he invented the lollipop near field transducer and the planar solid immersion lens that led to the first demonstration of heat-assisted magnetic recording. He was awarded the International Storage Industry Consortium technical achievement award in 2007 and was elected a fellow of the American Physical Society in 2013. He completed his research career at General Electric, working on a variety of projects involving fiber optics, optical sensors, and quantum optics, and retired after ten years during the covid pandemic. He has received 94 US patents and authored or co-authored about 60 journal articles and a text on surface plasmon physics. He currently develops optical modeling software for his company, Junonia Photonics, and is a part-time consultant.

1 Introduction

Eight different optical models are derived in this book with a stand-alone chapter for each. The equations have also been implemented in software to calculate the plots. Deriving the equations can often be easier than developing the computer code to implement those equations. Even software math libraries for special functions may be found to have certain limitations. For example, the Bessel functions in Mie theory sometimes require precision far beyond that available from a software library. Throughout the book, therefore, there are hints at software development that go beyond just the raw equations. Also, many examples are worked out in each chapter with specific numerical results as an aid for those who are developing their own software.

Chapter 2 presents a technique for modeling a multilayer thin film stack that incorporates materials that are characterized by dielectric tensors. For example, magneto-optical (MO) data storage disks are composed of a four-layer thin film stack. The MO layer has a diagonal permittivity tensor in the cylindrical (θ, ϕ, z) coordinate system, or permittivity tensor with off-diagonal components in the Cartesian coordinate system. As a result, it affects right- and left-circularly polarized light differently. It generates a differential phase shift and absorption between the two circular polarizations. By adjusting the thicknesses of the four layers, a disk can be fabricated to meet the disk drive industry standards in which the overall reflectivity is ~20% and the incident linearly polarized light is rotated by ~±1° left or right depending on the magnetization state of the recorded data bit.

Of course, optical disks are not simply a multilayer thin film deposited onto a flat substrate. Rewritable optical disks are mastered with a spiral tracking groove extending from the inside to the outer edge of the disk with dimensions comparable to the wavelength of the incident laser light. In other words, the multilayer thin films are deposited onto essentially a diffraction grating. The grooves cause the incident focused laser beam to diffract into a 0 and ±1st order cones of light. Light from part of the ±1st orders and all of the zero order is captured by the focusing lens. Because there is a phase shift between the orders, there is also interference in the reflected beam at the tracking detector. The ±1st orders interfere differently with the zero order if the beam is not exactly centered on the track. Using differential detection with a split tracking detector, an error signal can be generated when the focused beam is not precisely centered on the "land" between grooves.

There are two well-known techniques for solving for the diffracted fields from a periodically grooved surfaced coated with thin films: the Chandezon technique and the rigorous coupled-wave analysis technique. For applications using sinusoidal or trapezoidal-shaped grooves, the Chandezon technique is ideal and is described in the second chapter. The rigorous coupled-wave technique is more suitable for rectangular gratings but is not included in this book.

Of course, there are many other applications of diffraction gratings, and often the gratings must be coated to optimize their function. Over a century ago, Robert Wood noticed strange "absorption lines" produced in the zero-order reflectance for certain angles from his metal-coated gratings. These lines were extremely narrow and could almost entirely extinguish the zero-order reflected beam at certain "resonance" angles. Ugo Fano explained the effect in terms of a surface plasmon excitation. A surface plasmon is a collective oscillation of electron charge coupled to the electromagnetic field that propagates across a surface bounded by a dielectric and a metal. The evanescent field extends into the dielectric above the surface and is extremely sensitive to changes in the refractive index at the surface. This effect is employed as an extremely sensitive sensor for surface biomolecular binding interactions.

Another area in which diffraction gratings are important is for light coupling into thin film waveguides. Modern grating couplers make use of advanced designs with curved and chirped gratings which is far beyond the scope of this book. Such gratings must be designed and optimized through

DOI: 10.1201/9781003714231-1

numerical techniques such as finite element or finite difference. However, a feel for at least some of the design issues can be obtained even from the semi-analytical Chandezon technique by making use of a focused beam of light on the grating that limits the illuminated area of the grating to a finite distance. The focused beam is composed of an infinite sum of plane waves and can thereby be approximated by the Chandezon solver. The electric field cross section at the edge of the spot in the grating can then be calculated. An overlap integral for this field with the optical waveguide modes then completes the estimate of coupling efficiency into a waveguide mode. Of course, this calculation also requires the electric field cross-section calculation of the planar waveguide.

Planar waveguides are perhaps the simplest topic in this book but certainly a very important one. Understanding the different transverse electric (TE) and transverse magnetic (TM) planar waveguide modes is a useful first step towards understanding the more technologically important modes in fiber optics or strip or rib waveguides in photonic integrated circuits (PICs). A planar waveguide shaped like a parabola was used in the initial demonstration of heat-assisted magnetic recording (HAMR). The electric field at the focus of the waveguide can be accurately approximated by summations of planar waveguide modes incident over the effective numerical aperture of the parabola. The mode index of planar waveguides can be made quite large by utilizing a high index core material. Moreover, the incidence angle of the marginal rays in a parabolic mirror approaches 90°. As a result, it is possible to focus light to an extremely small, diffraction-limited, spot. Using 413 nm laser light, a so-called "solid immersion mirror" or SIM was found to focus light to a FWHM spot of only ~90 nm! This planar optical element may not yet have been exploited to the level it deserves.

PICs are becoming key components of modern technology, and critical parts of each PIC are the waveguides that carry the light between components on the chip. Therefore, a technique for modeling rectangular, rib, and other waveguide cross sections is important. Unfortunately, it is not possible to model these waveguides analytically except as an approximation which is generally not sufficiently accurate for commercial devices. Commercial waveguide solvers are quite sophisticated and expensive. The method-of-lines technique popularized by Reinhold Pregla and his colleagues handles a wide variety of waveguides including curved waveguides and waveguides designed in cylindrical coordinate systems. The chapter in this book is merely an introduction to that technique.

Fiber optic waveguides are at least as important in modern technology as strip or rib waveguides in PICs. Depending on the core and cladding materials and their radii, a single mode or a wide variety of modes can be supported in an optical fiber. The chapter in this book derives the usual fiber optic equations for core and cladding but extends the usual equations to handle multiple cladding layers. This enables one to solve for unusual fiber designs including a fiber with a rather strange radial index profile designed to carry Airy-like mode in one of its a higher order modes. Getting light into and out of an optical fiber is also critical. Therefore, some discussion is provided for overlap calculations to estimate coupling efficiency of a focused spot into a fiber. Often the end of a fiber is cleaved at a small angle to mitigate back reflections. This cleave also has the effect of slightly tilting the direction of the emitted beam. The tilt of the beam can be determined by a "near field to far field" transform described in this chapter, another useful technique to understand for general applications. This chapter, however, does not develop the technique for modeling photonic crystal fibers, anti-resonant hollow core fibers, or other newer fiber designs that require numerical modeling techniques.

The magneto-optic disk drive focuses light onto the disk through a very high numerical aperture objective in order to illuminate a very small spot and thereby increase the storage density or number of bits per unit area. An optical model of thin films on a planar surface will not adequately model this situation. Neither will a model of the diffracted fields from incident plane waves. One needs to model a *focused* beam of light onto the disk substrate. The brute force approach is to recognize that a focused spot is generated by summing an infinite number of plane waves over a range of polar and azimuthal angles with the appropriate intensity apodization. Although it is a lot of work to develop such a model, this was in fact accomplished and used to design some novel magneto-optic and phase change disk structures. On the other hand, the electric fields near the focus of a high numerical aperture lens are often encountered in the research environment and a general-purpose method for calculating these fields is needed. A vector field calculation is desired to account for the effects of polarization. The Richards-Wolf technique satisfies this need quite well but does not seem to be well-known. Furthermore, it lends itself to straightforward calculation of the fields in the focal plane of radially and azimuthally polarized incident waves on the lens.

The final chapter describes the scattered field finite difference time domain (FDTD) technique for modeling light absorption, scattering, and transmission from small objects or periodic arrays of objects. It is a numerical technique that can handle light scattering from arbitrarily shaped objects. As such it is a very powerful calculation tool, though its main drawback is the time required to setup the calculation as well as for it to run. The most common FDTD technique is called the total field FDTD technique. The equations are much simpler for total field FDTD than scattered field FDTD, but the latter has certain advantages that made it indispensable for modeling the near field transducer (NFT) used in HAMR. In particular, the solid immersion waveguide mirror previously mentioned generates a large electric field in a very small focus at the location of the NFT. This couples efficiently into the "lollipop" NFT that in turn couples the energy into a spot in the recording medium that is about two orders of magnitude smaller than the diffraction limit. As far as the author is aware, the total field FDTD technique cannot be used to model this device, and certainly the author was unable to find any commercial FDTD codes that could model the absorbed and scattered light of a transducer at the focus of what is effectively a lens with a numerical aperture > 0.9. Again, this chapter covers all the basics of scattered field FDTD, but it is merely a drop in the bucket of the technical literature for FDTD. The interested researcher should find this chapter a helpful introduction to scattered field FDTD to enable further studies into this useful technique if so desired.

A portion of this book covers some useful tips for software programming. All of the figures have been calculated by programs implementing the derived equations and are available on the Junonia Photonics website at relatively low cost. Deriving an equation is not the same as programming a computer to calculate the results of that equation. Often the available precision of the computer or programming language is a hard limit on the types of calculations that can be performed. In a few of these chapters, double precision (64-bit) is required, and in a couple of chapters, quadruple precision is necessary for some of the calculations. When Gustav Mie published his theory of light scattering from small spherical particles, he apparently worked all of his examples by hand. This was an amazing mathematical feat. However, it limited him to extremely small metallic particles. With modern computers it is of course now possible to quickly calculate the scattering from much larger particles. The holy grail of Mie theory is, perhaps, calculating the rainbow from millimeter-sized water droplets. This is not trivial, even with modern computers. It can be accomplished, however, on a modern PC with a 64-bit operating system using quadruple precision. The algorithm for doing this was published more recently, but does not appear to be widely known.

Finally, it should be said that models are usually developed to better understand how a system works or to design and optimize a system. As such, models are only useful in as much as they accurately reflect reality. A good model is one which includes all the relevant physics but not much more. The models in this book all rely on classical electromagnetic theory, specifically the linear Maxwell's equations. Nonlinear and quantum effects are not included. If such effects are important to a system being studied, then of course the model must include the more complex physics.

2 Multilayer Thin Films

2.1 INTRODUCTION

One of the most common and important optical models is that for plane waves incident upon multiple layers of thin films on a planar substrate. Generally, one desires to know the reflected and transmitted intensities and perhaps fields as a function of angle of incidence, wavelength, and polarization. Usually the thin films are described by a scalar refractive index which varies with wavelength. Sometimes the films must be represented by a dielectric or refractive index tensor. Optically active materials, which change the polarization of light as it passes through them, must be represented by dielectric tensors. This includes electro-optic, magneto-optic, birefringent, and chiral materials. It is also often the case that one desires to design a multilayer film stack to achieve certain properties such as high or low reflectance or transmittance, or perhaps to determine the unknown refractive index of a thin film based on its measured optical properties. In this case, a computer program that searches a range of thin film thicknesses to optimize a specific figure of merit can be quite useful.

There are at least two different approaches to calculation of the optical properties of thin film stacks. The transfer matrix method employs 4 × 4 matrices describing the transfer of the electric and magnetic field components [1–3] of the electromagnetic waves through each thin film. A series of matrix multiplications then calculates the reflected and transmitted fields from the film stack as well as the fields within each layer using the standard interface boundary conditions. The transfer matrix method can fail for thick films in which there are evanescent waves such as occurs in absorbing films or with total internal reflection. In such a case, some modes exhibit exponential decay and others exponential growth. When the modes are added together at certain points in the calculation, the number of significant figures of precision for the sum may exceed that available from the computer and critical information is lost, thereby corrupting the result. A 4 × 4 matrix multiplication method based on a scattering matrix approach has been developed to solve this problem [4].

A different transfer matrix technique that can handle dielectric tensors and that only requires 2 × 2 matrix multiplication has also been developed [5]. The author has used this technique in a variety of thin film projects such as the development and optimization of magneto-optic and phase change recording media, optical biosensors, anti-reflection coatings, etc. and found that it does not seem to suffer from the same thick film/evanescent wave problems of the 4 × 4 transfer matrix method. Nevertheless, an implementation of the algorithm which avoids such possible loss of precision is discussed in this chapter.

2.2 MATRIX EQUATIONS

2.2.1 Wave Equation

In a medium without free charges or currents, Maxwell's equations are

$$\nabla \cdot \boldsymbol{E} = 0 \tag{2.1}$$

$$\nabla \cdot \boldsymbol{B} = 0 \tag{2.2}$$

$$\nabla \times \boldsymbol{E} = -\frac{\partial \boldsymbol{B}}{\partial t} \tag{2.3}$$

$$\nabla \times \boldsymbol{H} = \frac{\partial \boldsymbol{D}}{\partial t} \tag{2.4}$$

DOI: 10.1201/9781003714231-2

where $\boldsymbol{E}$ is the electric field, $\boldsymbol{D}$ is the displacement field, $\boldsymbol{H}$ is the magnetic field, $\boldsymbol{B}$ is the magnetic induction and t is time. Bold type indicates a vector. The constitutive relations generally appropriate for media at optical frequencies are

$$\boldsymbol{D} = \epsilon_0 \tilde{\epsilon} \boldsymbol{E} \tag{2.5}$$

and

$$\boldsymbol{B} = \mu_0 \boldsymbol{H} \tag{2.6}$$

where ϵ_0 is the vacuum permittivity, μ_0 is the vacuum permeability, and $\tilde{\epsilon}$ is the dielectric tensor of the medium. The tilde over the $\tilde{\epsilon}$ indicates that it is in general a tensor quantity. More often, however, a thin film is represented by a scalar dielectric constant or refractive index. The permittivity of a medium is the product of its dielectric constant and the permittivity of free space, ε_0. It is perhaps somewhat confusing that the dielectric *constant* is actually a function of wavelength for every dense material.

The magnetic fields may be eliminated from (2.3)–(2.6) to give the standard wave equation,

$$\nabla \times (\nabla \times \boldsymbol{E}) - k_0^2 \, \tilde{\epsilon} \, \boldsymbol{E} = 0. \tag{2.7}$$

In this chapter, the time dependence for the electromagnetic waves is chosen to be $e^{-i\omega t}$ and the relation between the angular frequency and the vacuum wavevector is

$$k_0 = \sqrt{\epsilon_0 \mu_0} \, \omega. \tag{2.8}$$

The electric and magnetic field vectors of a plane wave are

$$\boldsymbol{E}(r,t) = E_0 \exp[i(\boldsymbol{k} \cdot r\text{-}\omega t)] \tag{2.9}$$

and

$$\boldsymbol{H}(r,t) = H_0 \exp[i(\boldsymbol{k} \cdot r\text{-}\omega t)], \tag{2.10}$$

respectively. For an isotropic medium, the wavevector is

$$\boldsymbol{k} = \left(\frac{2\pi n}{\lambda}\right) \hat{\boldsymbol{k}} = n k_0 \, \hat{\boldsymbol{k}} \tag{2.11}$$

where λ is the free space wavelength and n is the scalar refractive index of the medium. The caret symbol is used to denote a unit vector.

2.2.2 Film Stack Geometry

The film stack geometry is pictured in Figure 2.1. Layer 0 is an isotropic incident medium with refractive index n_0. There are N thin film layers, each with a thickness d_n and dielectric tensor $\tilde{\varepsilon}_n$. The transmitting medium is layer $N + 1$ with (complex) refractive index n_{N+1}. The z-axis is perpendicular to the thin film layers, and $z = 0$ occurs at the top surface. The incident beam comes from the top and is propagating in the $-z$ direction.

The incident plane wave in general has wavevector components along all three axes. The polar and azimuthal angles of incidence are θ and φ, respectively. If the electric field of the plane wave

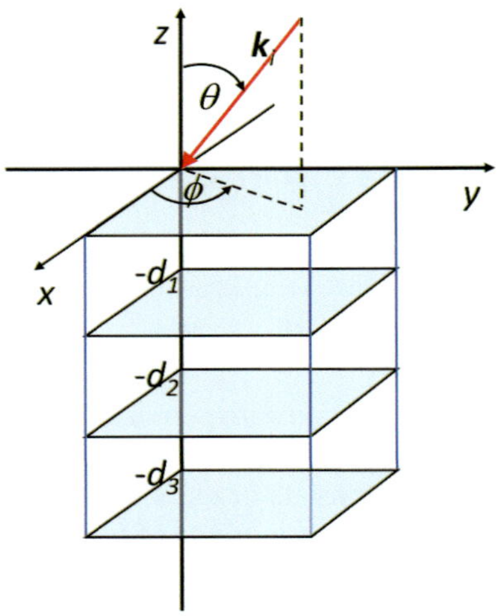

FIGURE 2.1 Thin film geometry consisting of a multilayer thin film stack. The z direction is normal to the films. Each layer has thickness d_n and dielectric tensor $\tilde{\varepsilon}_n$. The incident beam has wavevector $\boldsymbol{k}_i$, is propagating in the $-z$ direction with polar and azimuthal angles of incidence θ and φ, respectively.

lies in the plane of incidence, this is the transverse magnetic (TM) or p-polarization. If the electric field is orthogonal to the plane of incidence, this is the transverse electric (TE) or s-polarization.

Phase matching of the waves at each interface requires that the k_x and k_y components of the wavevector be unchanged throughout the film stack (Snell's Law) and are solely determined by the wavevector of the incident plane wave. Therefore, for all layers,

$$k_x = -k_i \sin\theta \cos\phi, \tag{2.12}$$

and

$$k_y = -k_i \sin\theta \sin\phi, \tag{2.13}$$

where the incident wavevector is

$$k_i = n_0 \left(\frac{2\pi}{\lambda} \right) = n_0 k_0. \tag{2.14}$$

2.2.3 k_z Wavevectors and Electric Field Components

The k_z component of the wavevector in a given thin film layer will vary throughout the film stack. In the most general case, there may be four different values for k_z within each layer that satisfy the wave equation. These four wavevectors are denoted by k_{zi} where i is the index of the wavevector. Including the incident and transmitting media, there are thus $4(N + 2)$ wavevectors to be determined for the complete film stack.

(2.7) can be separated into three scalar equations,

$$\left(-\tilde{k}_y^2 - \tilde{k}_z^2 + \epsilon_{xx}\right) E_x + \left(\tilde{k}_x \tilde{k}_y + \epsilon_{xy}\right) E_y + \left(\tilde{k}_x \tilde{k}_z + \epsilon_{xz}\right) E_z = 0, \tag{2.15}$$

$$\left(\tilde{k}_x \tilde{k}_y + \epsilon_{yx}\right) E_x + \left(-\tilde{k}_z^2 - \tilde{k}_x^2 + \epsilon_{yy}\right) E_y + \left(\tilde{k}_y \tilde{k}_z + \epsilon_{yz}\right) E_z = 0, \tag{2.16}$$

and

$$\left(\tilde{k}_x\tilde{k}_z+\epsilon_{zx}\right)E_x+\left(\tilde{k}_y\tilde{k}_z+\epsilon_{zy}\right)E_y+\left(-\tilde{k}_x^2-\tilde{k}_y^2+\epsilon_{zz}\right)E_z=0. \tag{2.17}$$

The tilde over each wavevector component in (2.15)–(2.17) indicates that it has been normalized by the vacuum wavevector, k_0. The four k_z wavevectors must each solve these three equations simultaneously. This is true if the determinant of the coefficient matrix is zero.

$$\left|\tilde{\boldsymbol{m}}_{1-4}\right|=\begin{vmatrix}-\tilde{k}_y^2-\tilde{k}_{z,1-4}^2+\epsilon_{xx} & \tilde{k}_x\tilde{k}_y+\epsilon_{xy} & \tilde{k}_x\tilde{k}_{z,1-4}+\epsilon_{xz}\\ \tilde{k}_x\tilde{k}_y+\epsilon_{yx} & -\tilde{k}_{z,1-4}^2-\tilde{k}_x^2+\epsilon_{yy} & \tilde{k}_y\tilde{k}_{z,1-4}+\epsilon_{yz}\\ \tilde{k}_x\tilde{k}_{z,1-4}+\epsilon_{zx} & \tilde{k}_y\tilde{k}_{z,1-4}+\epsilon_{zy} & -\tilde{k}_x^2-\tilde{k}_y^2+\epsilon_{zz}\end{vmatrix}=0. \tag{2.18}$$

(2.18) is a quartic equation in k_z,

$$A\tilde{k}_z^4+B\tilde{k}_z^3+C\tilde{k}_z^2+D\tilde{k}_z+E=0 \tag{2.19}$$

with coefficients

$$A=\epsilon_{zz}, \tag{2.20}$$

$$B=\tilde{k}_x\left(\epsilon_{xz}+\epsilon_{zx}\right)+\tilde{k}_y\left(\epsilon_{zy}+\epsilon_{yz}\right), \tag{2.21}$$

$$C=\tilde{k}_x^2\left(\epsilon_{xx}+\epsilon_{zz}\right)+\tilde{k}_y^2\left(\epsilon_{yy}+\epsilon_{zz}\right)+\tilde{k}_x\tilde{k}_y\left(\epsilon_{xy}+\epsilon_{yx}\right)-\left(\epsilon_{xx}+\epsilon_{yy}\right)\epsilon_{zz}+\epsilon_{xz}\epsilon_{zx}+\epsilon_{yz}\epsilon_{zy}, \tag{2.22}$$

$$\begin{aligned}D=&\left(\tilde{k}_x^2+\tilde{k}_y^2\right)\left[\tilde{k}_x\left(\epsilon_{xz}+\epsilon_{zx}\right)+\tilde{k}_y\left(\epsilon_{yz}+\epsilon_{zy}\right)\right]\\&+\tilde{k}_x\left[\epsilon_{xy}\epsilon_{yz}+\epsilon_{yx}\epsilon_{zy}-\epsilon_{yy}\left(\epsilon_{zx}+\epsilon_{xz}\right)\right]+\tilde{k}_y\left[\epsilon_{xz}\epsilon_{yx}+\epsilon_{xy}\epsilon_{zx}-\epsilon_{xx}\left(\epsilon_{yz}+\epsilon_{zy}\right)\right],\end{aligned} \tag{2.23}$$

and

$$\begin{aligned}E=&\left(\tilde{k}_x^2+\tilde{k}_y^2\right)\left[\tilde{k}_x^2\epsilon_{xx}+\tilde{k}_y^2\epsilon_{yy}+\tilde{k}_x\tilde{k}_y\left(\epsilon_{xy}+\epsilon_{yx}\right)+\epsilon_{xy}\epsilon_{yx}-\epsilon_{xx}\epsilon_{yy}\right]+\tilde{k}_x^2\left(\epsilon_{xz}\epsilon_{zx}-\epsilon_{xx}\epsilon_{zz}\right)\\&+\tilde{k}_y^2\left(\epsilon_{yz}\epsilon_{zy}-\epsilon_{yy}\epsilon_{zz}\right)+\tilde{k}_x\tilde{k}_y\left(\epsilon_{yz}\epsilon_{zx}+\epsilon_{xz}\epsilon_{zy}-\epsilon_{xy}\epsilon_{zz}-\epsilon_{yx}\epsilon_{zz}\right)\\&+\left(\epsilon_{xy}\epsilon_{yz}-\epsilon_{xz}\epsilon_{yy}\right)\epsilon_{zx}+\left(\epsilon_{xz}\epsilon_{yx}-\epsilon_{xx}\epsilon_{yz}\right)\epsilon_{zy}+\left(\epsilon_{xx}\epsilon_{yy}-\epsilon_{xy}\epsilon_{yx}\right)\epsilon_{zz}.\end{aligned} \tag{2.24}$$

These coefficients are expressed in a somewhat different form in Eq. (20) of [5]. Although there is an analytical procedure for solving for the complex roots of a quartic equation, limited machine precision makes it difficult to implement this algorithm in practice in a manner that always generates accurate results for the roots. When developing a computer program, it is often much easier to use a polynomial solver such as DZPOCC from the IMSL standard numerical library.

When the fields are propagating within isotropic, non-absorbing media, all four wavevectors are real numbers and have the same magnitude. Two of the wavevectors are negative, representing transmitted waves with orthogonal polarizations propagating in the $-z$ direction. The other two wavevectors are positive, representing reflected waves with orthogonal polarizations propagating in the $+z$ direction. Of course, the k_z wavevector components may also be complex numbers or pure imaginary numbers. The latter situation occurs, for example, for the evanescent waves caused by total internal reflection. Following the notation of [5], k_{z1} and k_{z2} are defined as the "downward" wavevectors in the direction of the incident beam, and k_{z3} and k_{z4} as the "upward" wavevectors in the direction of the reflected beam. Each wavevector corresponds to a specific relationship between the various field components. The downward wavevectors, k_{z1} and k_{z2}, correspond to fields that have orthogonal polarizations ($\boldsymbol{E}_1 \cdot \boldsymbol{E}_2 = 0$). Likewise, the upward wavevectors, k_{z3} and k_{z4}, correspond to fields that have orthogonal polarizations ($\boldsymbol{E}_3 \cdot \boldsymbol{E}_4 = 0$).

[5] teaches that k_{z1} and k_{z2} should be selected for the lower half of the complex plane (i.e., propagating downward), and k_{z3} and k_{z4} should be in the upper half (i.e., propagating upward). This statement is almost correct. If the wavevectors are complex numbers, then indeed k_{z1} and k_{z2} should be selected from the four wavevector solutions as the two in the lower half plane with negative *imaginary* parts, which are not necessarily *propagating* modes, but are modes which decay exponentially in the downward direction. Likewise, the remaining two wavevectors should have positive imaginary parts so that they decay exponentially in the upward direction. However, sometimes two, and often all four, of the wavevectors are real numbers and cannot be allocated by this rule. In that case, initially the smallest unassigned wavevectors, usually negative but not always, are assigned to the remaining unassigned k_{z1} and k_{z2} wavevectors, and the largest are initially assigned to any unassigned k_{z3} and k_{z4} wavevectors. These assignments may be changed upon further analysis as will be discussed.

Often the B and D coefficients in (2.19) for the quartic equation are zero. In that case, the four wavevectors are

$$k_z = \pm\sqrt{\frac{-C \pm \sqrt{C^2 - 4AE}}{2A}}. \tag{2.25}$$

Following [5], for the k_{z1} and k_{z3} wavevectors, the E_y and E_z field components are specified as functions of E_x,

$$E_{y1,3} = a_{1,3} E_{x1,3}, \tag{2.26}$$

and

$$E_{z1,3} = b_{1,3} E_{x1,3}. \tag{2.27}$$

The E_x and E_z field components for the k_{z2} and k_{z4} wavevectors are specified as functions of E_y,

$$E_{x2,4} = a_{2,4} E_{y2,4}, \tag{2.28}$$

and

$$E_{z2,4} = b_{2,4} E_{y2,4}. \tag{2.29}$$

The coefficients, a_i and b_i, must be determined for each of the four wavevectors in each of the $N + 2$ media. There are many special situations in which some care must be taken. For example, according to (2.17) for normal incidence ($k_x = k_y = 0$), within a material with no off-diagonal elements in its dielectric tensor, $E_z = 0$. Hence, under these circumstances, the coefficients $b = 0$. The four wavevectors are $\tilde{k}_{z1,3} = \pm\sqrt{\epsilon_{xx}}$, and $\tilde{k}_{z2,4} = \pm\sqrt{\epsilon_{yy}}$, and $a = 0$.

Frequently, the incident beam lies in either the xz plane or the yz plane. In this case, the quartic equation (2.19) can be separated into two quadratic equations. In particular, when the incident beam lies in the xz plane and there are at least four off-diagonal elements of the dielectric tensor for a given layer that are zero, then

$$\epsilon = \begin{pmatrix} \epsilon_{xx} & 0 & \epsilon_{xz} \\ 0 & \epsilon_{yy} & 0 \\ \epsilon_{zx} & 0 & \epsilon_{zz} \end{pmatrix}. \tag{2.30}$$

This situation will occur for any material described by a scalar refractive index, any birefringent material for which the principal axes are aligned to the Cartesian coordinate system, a polar magneto-optic (MO) or electro-optic (EO) material, or a liquid crystal in which the nematic director lies in the xz plane. The s-polarized beam just has an electric field component along the y-axis and the upward and downward k_{z2} and k_{z4} wavevectors for this beam are simply

$$k_{z2,z4} = \pm\sqrt{n_m^2 - n_i^2 \sin^2\theta_i} = \pm\sqrt{\epsilon_{yy} - k_x^2} \tag{2.31}$$

where $n_m = \sqrt{\epsilon_{yy}}$, n_i is the refractive index of the incident medium, and θ_i is the polar angle of incidence. This result can be easily verified by substituting (2.31) into (2.20)–(2.24) and setting $\epsilon_{xy} = \epsilon_{yx} = \epsilon_{yz} = \epsilon_{zy} = k_y = 0$. These two roots are for the E_y electric field component, which in turn require $a_2 = b_2 = a_4 = b_4 = 0$ because for these two wavevectors there are no E_x or E_z field components. If k_{z2} and k_{z4} are complex numbers, then the value for k_{z2} with a negative imaginary part corresponding to evanescent decay in the downward direction is selected. If they are both real numbers, then the negative root for k_{z2} corresponding to propagation in the downward direction is selected.

In the general case, (2.18) may be written as three separate equations,

$$m_{11}E_x + m_{12}E_y + m_{13}E_z = 0, \tag{2.32}$$

$$m_{21}E_x + m_{22}E_y + m_{23}E_z = 0, \tag{2.33}$$

and

$$m_{31}E_x + m_{32}E_y + m_{33}E_z = 0. \tag{2.34}$$

The four values of k_z each make these three equations linearly dependent. Therefore, only two of these equations are required to solve for the a and b coefficients for each wavevector. The following procedure can be followed to determine the coefficients.

2.3 *a* AND *b* COEFFICIENTS FOR k_{z1} AND k_{z2} WAVEVECTORS

To determine the a_1, b_1, a_2 and b_2 coefficients for the k_{z1} and k_{z2} wavevectors, each of the three pairs of equations (2.32)–(2.34) should be considered with each initially assigned value for k_{z1} and k_{z2}. Although in cases of high symmetry, as discussed in the previous paragraph, it is possible to immediately determine the correct values for k_{z1} and k_{z2}, this is not the case in general. Hence, the initial choice for k_{z1} and k_{z2} may be incorrect and need to be switched. For example, sometimes, using k_{z1} it is possible to find values for a_1, b_1, a_2 and b_2, but using k_{z2} it is only possible to find values for a_1, and b_1. In that case, the values for k_{z1} and k_{z2} must be switched.

2.3.1 First and Second Equations, (2.32) and (2.33)

(2.32) and (2.33) are solved for E_y and E_z in terms of E_x, and for E_x and E_z in terms of E_y.

$$E_y = -\frac{(m_{23}m_{11} - m_{13}m_{21})}{(m_{23}m_{12} - m_{13}m_{22})} E_x \equiv a_1 E_x \tag{2.35}$$

$$E_z = -\frac{(m_{12}m_{21} - m_{11}m_{22})}{(m_{12}m_{23} - m_{13}m_{22})} E_x \equiv b_1 E_x. \tag{2.36}$$

$$E_x = -\frac{(m_{23}m_{12} - m_{13}m_{22})}{(m_{23}m_{11} - m_{13}m_{21})} E_y \equiv a_2 E_y \tag{2.37}$$

$$E_z = -\frac{(m_{22}m_{11} - m_{12}m_{21})}{(m_{23}m_{11} - m_{13}m_{21})} E_y \equiv b_2 E_y. \tag{2.38}$$

In (2.35) if the denominator is zero because $m_{13} = m_{23} = 0$, then if $m_{12}\neq 0$ and $m_{22}\neq 0$,

$$E_y = -\left(\frac{m_{21}}{m_{22}}\right) E_x = -\left(\frac{m_{11}}{m_{12}}\right) E_x \equiv a_1 E_x. \tag{2.39}$$

However, b_1 is undefined. In (2.36) if the denominator is zero because $m_{12} = m_{22} = 0$, then if $m_{13}\neq 0$ and $m_{23}\neq 0$,

$$E_z = -\left(\frac{m_{21}}{m_{23}}\right) E_x = -\left(\frac{m_{11}}{m_{13}}\right) E_x \equiv b_1 E_x. \tag{2.40}$$

However, a_1 is undefined. This process is repeated for the other two pairs of equations, (2.32) and (2.34), and (2.33) and (2.34), solving for any values of a_1, a_2, b_1 and b_2. Then, the entire process is repeated for the k_{z2} wavevector. At this point, after considering all three pairs of equations for both k_{z1} and k_{z2}, if the originally assigned k_{z1} provides values for both a_1 and b_1, and the originally assigned k_{z2} provides values for both a_2 and b_2, then all of these parameters are used for the remainder of the calculation. However, if one or more of the parameters a_1, a_2, b_1 and b_2 cannot be determined with the original selection for k_{z1} and k_{z2}, then these two wavevectors are switched so that the original k_{z1} now becomes k_{z2} and the values for a_2 and b_2 that go with that wavevector are used, and likewise k_{z2} becomes k_{z1} along with the values for a_1 and b_1 that were calculated for that wavevector. Finally, this entire process is repeated for the wavevectors k_{z3} and k_{z4}. This obviously is a complex procedure and perhaps there is a simpler way to properly assign wavevectors and find the a and b coefficients, but at least this procedure seems to work.

2.3.2 Magnetic Field Components

The components of the magnetic field must also be determined as functions of the electric field. In each layer n, and for wavevectors k_{z1} and k_{z3}, the magnetic field components are functions of E_x,

$$H_x^{(n)} = c_n E_x^{(n)}, \qquad H_y^{(n)} = d_n E_x^{(n)}, \qquad H_z^{(n)} = e_n E_x^{(n)}. \tag{2.41}$$

For wavevectors k_{z2} and k_{z4}, the magnetic field components are functions of E_y,

$$H_x^{(n)} = c_n E_y^{(n)}, \qquad H_y^{(n)} = d_n E_y^{(n)}, \quad \text{and} \quad H_z^{(n)} = e_n E_y^{(n)}. \tag{2.42}$$

The magnetic field components are related to the electric field components by (2.3),

$$H_x = \frac{1}{Z_0}\left(\tilde{k}_y E_z - \tilde{k}_z E_y\right), \tag{2.43}$$

$$H_y = \frac{1}{Z_0}\left(\tilde{k}_z E_x - \tilde{k}_x E_z\right), \tag{2.44}$$

and

$$H_z = \frac{1}{Z_0}\left(\tilde{k}_x E_y - \tilde{k}_y E_x\right) \tag{2.45}$$

where Z_0 is the vacuum impedance,

$$Z_0 = \sqrt{\frac{\mu_0}{\epsilon_0}}. \tag{2.46}$$

The c, d and e coefficients in (2.41) and (2.42) can be determined for each of the four wavevectors within each of the $N + 2$ different media using (2.35)–(2.45).

2.3.3 2 × 2 Matrices for Each Layer and Boundary Conditions

2 × 2 matrices are now defined for each layer n,

$$\tilde{\boldsymbol{A}}_{12}^{(n)} \equiv \begin{bmatrix} 1 & a_2^{(n)} \\ a_1^{(n)} & 1 \end{bmatrix}, \qquad \tilde{\boldsymbol{A}}_{34}^{n} \equiv \begin{bmatrix} 1 & a_4^{(n)} \\ a_3^{(n)} & 1 \end{bmatrix}, \tag{2.47}$$

and

$$\tilde{\boldsymbol{B}}_{12}^{(n)} \equiv \begin{bmatrix} c_1^{(n)} & c_2^{(n)} \\ d_1^{(n)} & d_2^{(n)} \end{bmatrix}, \qquad \tilde{\boldsymbol{B}}_{34}^{(n)} \equiv \begin{bmatrix} c_3^{(n)} & c_4^{(n)} \\ d_3^{(n)} & d_4^{(n)} \end{bmatrix}. \tag{2.48}$$

The boundary conditions at the interface between layer $n - 1$ and layer n, which ensure continuity of the tangential components of the electric and magnetic fields, are

$$\tilde{\boldsymbol{A}}_{12}^{(n-1)} \begin{bmatrix} E_{x1} \\ E_{y2} \end{bmatrix}_{n-1} + \tilde{\boldsymbol{A}}_{34}^{(n-1)} \begin{bmatrix} E_{x3} \\ E_{y4} \end{bmatrix}_{n-1} = \tilde{\boldsymbol{A}}_{12}^{(n)} \tilde{\boldsymbol{P}}_{12}^{(n)} \begin{bmatrix} E_{x1} \\ E_{y2} \end{bmatrix}_{n} + \tilde{\boldsymbol{A}}_{34}^{(n)} \tilde{\boldsymbol{P}}_{34}^{(n)} \begin{bmatrix} E_{x3} \\ E_{y4} \end{bmatrix}_{n} \tag{2.49}$$

and

$$\tilde{\boldsymbol{B}}_{12}^{(n-1)} \begin{bmatrix} E_{x1} \\ E_{y2} \end{bmatrix}_{n-1} + \tilde{\boldsymbol{B}}_{34}^{(n-1)} \begin{bmatrix} E_{x3} \\ E_{y4} \end{bmatrix}_{n-1} = \tilde{\boldsymbol{B}}_{12}^{(n)} \tilde{\boldsymbol{P}}_{12}^{(n)} \begin{bmatrix} E_{x1} \\ E_{y2} \end{bmatrix}_{n} + \tilde{\boldsymbol{B}}_{34}^{(n)} \tilde{\boldsymbol{P}}_{34}^{(n)} \begin{bmatrix} E_{x3} \\ E_{y4} \end{bmatrix}_{n} \tag{2.50}$$

where $\tilde{\boldsymbol{P}}_{12}^{(n)}$ and $\tilde{\boldsymbol{P}}_{34}^{(n)}$ are phase shift matrices for layer n,

$$\tilde{\boldsymbol{P}}_{12}^{(n)} \equiv \begin{bmatrix} e^{ik_{z1}^{(n)}d_n} & 0 \\ 0 & e^{ik_{z2}^{(n)}d_n} \end{bmatrix} \quad \text{and} \quad \tilde{\boldsymbol{P}}_{34}^{(n)} \equiv \begin{bmatrix} e^{ik_{z3}^{(n)}d_n} & 0 \\ 0 & e^{ik_{z4}^{(n)}d_n} \end{bmatrix} \tag{2.51}$$

containing the information about the propagation of the light from the bottom surface of layer n up to the top surface of layer n (and bottom surface of layer $n-1$).

Because there are no reflected fields in the bottom transmitting medium, $E_{x3}^{(N+1)} = E_{x4}^{(N+1)} = 0$. The boundary conditions at the bottom interface, therefore, are simply

$$\tilde{\boldsymbol{A}}_{12}^{(N)} \begin{bmatrix} E_{x1} \\ E_{y2} \end{bmatrix}_N + \tilde{\boldsymbol{A}}_{34}^{(N)} \begin{bmatrix} E_{x3} \\ E_{y4} \end{bmatrix}_N = \tilde{\boldsymbol{A}}_{12}^{(N+1)} \begin{bmatrix} E_{x1} \\ E_{y2} \end{bmatrix}_{N+1}, \tag{2.52}$$

and

$$\tilde{\boldsymbol{B}}_{12}^{(N)} \begin{bmatrix} E_{x1} \\ E_{y2} \end{bmatrix}_N + \tilde{\boldsymbol{B}}_{34}^{(N)} \begin{bmatrix} E_{x3} \\ E_{y4} \end{bmatrix}_N = \tilde{\boldsymbol{B}}_{12}^{(N+1)} \begin{bmatrix} E_{x1} \\ E_{y2} \end{bmatrix}_{N+1}. \tag{2.53}$$

2.3.4 Transmitted and Reflected Fields

(2.52) and (2.53) can be solved for the transmitted fields, $E_{x1}^{(N+1)}$ and $E_{y2}^{(N+1)}$, and for the fields of the upward modes in the bottom thin film layer, $E_{x3}^{(N)}$ and $E_{y4}^{(N)}$, in terms of the fields of the downward modes in the bottom thin film layer, $E_{x1}^{(N)}$ and $E_{y2}^{(N)}$,

$$\begin{bmatrix} E_{x1} \\ E_{y2} \end{bmatrix}_{N+1} = \tilde{\boldsymbol{T}}_{xy}^{(N)} \begin{bmatrix} E_{x1} \\ E_{y2} \end{bmatrix}_N, \tag{2.54}$$

and

$$\begin{bmatrix} E_{x3} \\ E_{y4} \end{bmatrix}_N = \tilde{\boldsymbol{R}}_{xy}^{(N)} \begin{bmatrix} E_{x1} \\ E_{y2} \end{bmatrix}_N \tag{2.55}$$

where

$$\tilde{\boldsymbol{T}}_{xy}^{(N)} = \left(\tilde{\boldsymbol{I}}^{(N+1)}\right)^{-1} \tilde{\boldsymbol{J}}^{(N)}, \tag{2.56}$$

$$\tilde{\boldsymbol{I}}^{(N+1)} \equiv \left(\tilde{\boldsymbol{A}}_{34}^{(N)}\right)^{-1} \tilde{\boldsymbol{A}}_{12}^{(N+1)} - \left(\tilde{\boldsymbol{B}}_{34}^{(N)}\right)^{-1} \tilde{\boldsymbol{B}}_{12}^{(N+1)} \tag{2.57}$$

$$\tilde{\boldsymbol{J}}^{(N)} \equiv \left(\tilde{\boldsymbol{A}}_{34}^{(N)}\right)^{-1} \tilde{\boldsymbol{A}}_{12}^{(N)} - \left(\tilde{\boldsymbol{B}}_{34}^{(N)}\right)^{-1} \tilde{\boldsymbol{B}}_{12}^{(N)}, \tag{2.58}$$

and

$$\tilde{\boldsymbol{R}}_{xy}^{(N)} = -\left(\tilde{\boldsymbol{K}}^{(N+1)}\right)^{-1} \tilde{\boldsymbol{L}}^{(N+1)}, \tag{2.59}$$

$$\tilde{\boldsymbol{K}}^{(N+1)} \equiv \left(\tilde{\boldsymbol{A}}_{12}^{(N+1)}\right)^{-1} \tilde{\boldsymbol{A}}_{34}^{(N)} - \left(\tilde{\boldsymbol{B}}_{12}^{(N+1)}\right)^{-1} \tilde{\boldsymbol{B}}_{34}^{(N)}, \tag{2.60}$$

$$\tilde{\boldsymbol{L}}^{(N+1)} \equiv \left(\tilde{\boldsymbol{A}}_{12}^{(N+1)}\right)^{-1} \tilde{\boldsymbol{A}}_{12}^{(N)} - \left(\tilde{\boldsymbol{B}}_{12}^{(N+1)}\right)^{-1} \tilde{\boldsymbol{B}}_{12}^{(N)}. \tag{2.61}$$

There is also a relation between $\tilde{\boldsymbol{T}}_{\text{xy}}^{(N)}$ and $\tilde{\boldsymbol{R}}_{\text{xy}}^{(N)}$,

$$\tilde{\boldsymbol{T}}_{\text{xy}}^{(N)} = \left(\tilde{\boldsymbol{A}}_{12}^{(N+1)}\right)^{-1} \tilde{\boldsymbol{A}}_{34}^{(N)} \left[\left(\tilde{\boldsymbol{A}}_{34}^{(N)}\right)^{-1} \tilde{\boldsymbol{A}}_{12}^{(N)} + \tilde{\boldsymbol{R}}_{\text{xy}}^{(N)}\right]. \tag{2.62}$$

(2.52) and (2.53) for the interface between layers can also be solved for the downward mode fields in layer n in terms of the downward mode fields in layer $n-1$,

$$\begin{bmatrix} E_{\text{x1}} \\ E_{\text{y2}} \end{bmatrix}_n = \tilde{\boldsymbol{T}}_{\text{xy}}^{(n-1)} \begin{bmatrix} E_{\text{x1}} \\ E_{\text{y2}} \end{bmatrix}_{n-1} \tag{2.63}$$

and the upward mode fields in layer $n-1$ in terms of the downward mode fields in layer $n-1$,

$$\begin{bmatrix} E_{\text{x3}} \\ E_{\text{y4}} \end{bmatrix}_{n-1} = \tilde{\boldsymbol{R}}_{\text{xy}}^{(n-1)} \begin{bmatrix} E_{\text{x1}} \\ E_{\text{y2}} \end{bmatrix}_{n-1}. \tag{2.64}$$

With the definitions,

$$\tilde{\boldsymbol{D}}^{(n)} \equiv \left[\left(\tilde{\boldsymbol{A}}_{12}^{(n)}\right)^{-1} \tilde{\boldsymbol{A}}_{34}^{(n)} - \left(\tilde{\boldsymbol{B}}_{12}^{(n)}\right)^{-1} \tilde{\boldsymbol{B}}_{34}^{(n)}\right]^{-1} \left[\left(\tilde{\boldsymbol{A}}_{12}^{(n)}\right)^{-1} \tilde{\boldsymbol{A}}_{12}^{(n-1)} - \left(\tilde{\boldsymbol{B}}_{12}^{(n)}\right)^{-1} \tilde{\boldsymbol{B}}_{12}^{(n-1)}\right], \tag{2.65}$$

$$\tilde{\boldsymbol{F}}^{(n)} \equiv \left[\left(\tilde{\boldsymbol{A}}_{12}^{(n)}\right)^{-1} \tilde{\boldsymbol{A}}_{34}^{(n)} - \left(\tilde{\boldsymbol{B}}_{12}^{(n)}\right)^{-1} \tilde{\boldsymbol{B}}_{34}^{(n)}\right]^{-1} \left[\left(\tilde{\boldsymbol{A}}_{12}^{(n)}\right)^{-1} \tilde{\boldsymbol{A}}_{34}^{(n-1)} - \left(\tilde{\boldsymbol{B}}_{12}^{(n)}\right)^{-1} \tilde{\boldsymbol{B}}_{34}^{(n-1)}\right], \tag{2.66}$$

$$\tilde{\boldsymbol{G}}^{(n)} \equiv \left[\left(\tilde{\boldsymbol{A}}_{34}^{(n)}\right)^{-1} \tilde{\boldsymbol{A}}_{12}^{(n)} - \left(\tilde{\boldsymbol{B}}_{34}^{(n)}\right)^{-1} \tilde{\boldsymbol{B}}_{12}^{(n)}\right]^{-1} \left[\left(\tilde{\boldsymbol{A}}_{34}^{(n)}\right)^{-1} \tilde{\boldsymbol{A}}_{12}^{(n-1)} - \left(\tilde{\boldsymbol{B}}_{34}^{(n)}\right)^{-1} \tilde{\boldsymbol{B}}_{12}^{(n-1)}\right], \tag{2.67}$$

$$\tilde{\boldsymbol{H}}^{(n)} \equiv \left[\left(\tilde{\boldsymbol{A}}_{34}^{(n)}\right)^{-1} \tilde{\boldsymbol{A}}_{12}^{(n)} - \left(\tilde{\boldsymbol{B}}_{34}^{(n)}\right)^{-1} \tilde{\boldsymbol{B}}_{12}^{(n)}\right]^{-1} \left[\left(\tilde{\boldsymbol{A}}_{34}^{(n)}\right)^{-1} \tilde{\boldsymbol{A}}_{34}^{(n-1)} - \left(\tilde{\boldsymbol{B}}_{34}^{(n)}\right)^{-1} \tilde{\boldsymbol{B}}_{34}^{(n-1)}\right], \tag{2.68}$$

and

$$\tilde{\boldsymbol{S}}^{(n)} \equiv \tilde{\boldsymbol{P}}_{34}^{(n)} \tilde{\boldsymbol{R}}_{\text{xy}}^{(n)} \left(\tilde{\boldsymbol{P}}_{12}^{(n)}\right)^{-1}, \tag{2.69}$$

the results are

$$\tilde{\boldsymbol{T}}_{\text{xy}}^{(n-1)} \equiv \left(\tilde{\boldsymbol{P}}_{12}^{(n)}\right)^{-1} \left(\tilde{\boldsymbol{G}}^{(n)} + \tilde{\boldsymbol{H}}^{(n)} \tilde{\boldsymbol{R}}_{\text{xy}}^{(n-1)}\right), \tag{2.70}$$

and

$$\tilde{\boldsymbol{R}}_{\mathrm{xy}}^{(n-1)} \equiv \left(\tilde{\boldsymbol{F}}^{(n)} - \tilde{\boldsymbol{S}}^{(n)}\tilde{\boldsymbol{H}}^{(n)}\right)^{-1}\left(\tilde{\boldsymbol{S}}^{(n)}\tilde{\boldsymbol{G}}^{(n)} - \tilde{\boldsymbol{D}}^{(n)}\right). \tag{2.71}$$

The nonzero elements of ($\tilde{\boldsymbol{P}}_{12}^{(n)}$)$^{-1}$ and $\tilde{\boldsymbol{P}}_{34}^{(n)}$ in (2.69) all have positive exponentials according to (2.51) and including the sign differences that have been selected for the k_z, so there will not be any elements of $\tilde{\boldsymbol{S}}^{(n)}$ that are composed of only exponentially decreasing terms while others are composed of only exponentially increasing terms leading to difficulties maintaining the precision of the computation.

The procedure for computing the fields for the entire film stack can be summarized as follows. The reflectance matrix is calculated from (2.59) for the bottom interface. The reflectance matrices for all other layers are calculated sequentially from (2.71). The transmittance matrix for the bottom interface is calculated from (2.56). The transmittance matrix for all other layers is calculated from (2.70). The transmittance for the whole film stack is obtained by multiplying the transmittance matrices for each layer,

$$\begin{bmatrix} E_{\mathrm{x1}}^{(N+1)} \\ E_{\mathrm{y2}}^{(N+1)} \end{bmatrix} = \tilde{\boldsymbol{T}}_{\mathrm{xy}}^{(N)}\tilde{\boldsymbol{T}}_{\mathrm{xy}}^{(N-1)}...\tilde{\boldsymbol{T}}_{\mathrm{xy}}^{(0)}\begin{bmatrix} E_{\mathrm{x1}} \\ E_{\mathrm{y2}} \end{bmatrix}_0. \tag{2.72}$$

In [5], an equation for the reflectance matrix $\tilde{\boldsymbol{R}}_{\mathrm{xy}}^{(n-1)}$ is derived in a different manner, and the results are

$$\tilde{\boldsymbol{R}}_{\mathrm{xy}}^{(n-1)} = \left(\tilde{\boldsymbol{K}}^{(n)}\tilde{\boldsymbol{A}}_{34}^{(n-1)} - \tilde{\boldsymbol{B}}_{34}^{(n-1)}\right)^{-1}\left(\tilde{\boldsymbol{B}}_{12}^{(n-1)} - \tilde{\boldsymbol{K}}^{(n)}\tilde{\boldsymbol{A}}_{12}^{(n-1)}\right), \tag{2.73}$$

and

$$\tilde{\boldsymbol{K}}^{(n)} \equiv \left(\tilde{\boldsymbol{B}}_{12}^{(n)}\tilde{\boldsymbol{P}}_{12}^{(n)} + \tilde{\boldsymbol{B}}_{34}^{(n)}\tilde{\boldsymbol{P}}_{34}^{(n)}\tilde{\boldsymbol{R}}_{\mathrm{xy}}^{(n)}\right)\left(\tilde{\boldsymbol{A}}_{12}^{(n)}\tilde{\boldsymbol{P}}_{12}^{(n)} + \tilde{\boldsymbol{A}}_{34}^{(n)}\tilde{\boldsymbol{P}}_{34}^{(n)}\tilde{\boldsymbol{R}}_{\mathrm{xy}}^{(n)}\right)^{-1}. \tag{2.74}$$

The equation for the transmittance matrix is also derived differently, resulting in

$$\tilde{\boldsymbol{T}}_{\mathrm{xy}}^{(n-1)} = \left(\tilde{\boldsymbol{A}}_{12}^{(n)}\tilde{\boldsymbol{P}}_{12}^{(n)} + \tilde{\boldsymbol{A}}_{34}^{(n)}\tilde{\boldsymbol{P}}_{34}^{(n)}\tilde{\boldsymbol{R}}_{\mathrm{xy}}^{(n)}\right)^{-1}\left(\tilde{\boldsymbol{A}}_{12}^{(n-1)} + \tilde{\boldsymbol{A}}_{34}^{(n-1)}\tilde{\boldsymbol{R}}_{\mathrm{xy}}^{(n-1)}\right). \tag{2.75}$$

Both $\tilde{\boldsymbol{P}}_{12}^{(n)}$ and $\tilde{\boldsymbol{P}}_{34}^{(n)}$ are present in (2.74) and (2.75). If a layer is very thick and if the wavevectors have an imaginary part, then according to (2.51), $\tilde{\boldsymbol{P}}_{12}^{(n)}$ will have elements that become exponentially large with layer thickness and $\tilde{\boldsymbol{P}}_{34}^{(n)}$ will have elements that become exponentially small with increasing layer thickness. This can potentially give rise to errors in the calculation of $\tilde{\boldsymbol{K}}^{(n)}$ if the machine precision in the calculation is insufficient to handle both the exponentially small and exponentially large terms when they get summed. In particular, the terms involving $\tilde{\boldsymbol{P}}_{34}^{(n)}$ will drop out of (2.74) and (2.75). By (2.73), then, the reflectivity matrix of layer $n - 1$ will have no dependence on the reflectivity matrix of layer n. Similarly, the transmittance matrix of layer $n - 1$ will have no dependence on the reflectivity matrix of layer n. The wavevectors within a given layer can have an imaginary component if the index of refraction is complex or if the electric field is evanescent. In the former case, the electric field decays exponentially through absorption in the layer and so it makes sense that reflection from the bottom interface should not contribute substantially to the reflection from the top surface. On the other hand, if the layer is birefringent and dichroic, then it is possible for the electric field from one polarization to be continuously converted into the other polarization as it propagates through a thick layer.

On the other hand, the equations from [5] are computationally more efficient. Calculating the matrix $\tilde{\boldsymbol{K}}^n$ requires six matrix multiplications and one matrix inversion. Calculating the reflectance matrix $\tilde{\boldsymbol{R}}_{\text{xy}}^{n-1}$ from (2.73) then requires three additional matrix multiplications and another matrix inversion for a total of nine matrix multiplications and two matrix inversions. When these equations were implemented within *Mathematica*©, it was found that calculations took ~27% more time with the new algorithm.

Once the matrices $\tilde{\boldsymbol{R}}_{\text{xy}}^n$ and $\tilde{\boldsymbol{T}}_{\text{xy}}^n$ have been determined for each layer, the electric and magnetic fields are easily found from the incident fields. The reflected fields in the incident medium are obtained from (2.71) for $n = 1$ and the transmitted fields in the bottom medium are obtained from (2.72). It is frequently more useful to calculate reflectance and transmittance coefficients in terms of incident s and p polarizations. s and p polarization components can be converted to Cartesian field components by the matrix equation,

$$\begin{bmatrix} E_x \\ E_y \end{bmatrix} = \begin{bmatrix} \cos\theta\cos\phi & -\sin\phi \\ \cos\theta\sin\phi & \cos\phi \end{bmatrix} \begin{bmatrix} E_p \\ E_s \end{bmatrix} \equiv \tilde{\boldsymbol{W}}_{\text{inc}} \begin{bmatrix} E_p \\ E_s \end{bmatrix}. \tag{2.76}$$

The reflectivity matrix is

$$\begin{bmatrix} E_p \\ E_s \end{bmatrix}_{\text{ref}} = \tilde{\boldsymbol{R}} \begin{bmatrix} E_p \\ E_s \end{bmatrix}_{\text{inc}} \tag{2.77}$$

where

$$\tilde{\boldsymbol{R}} = \left(\tilde{\boldsymbol{W}}_{\text{inc}}\right)^{-1} \tilde{\boldsymbol{R}}_{\text{xy}} \tilde{\boldsymbol{W}}_{\text{inc}}. \tag{2.78}$$

Similarly, we have

$$\begin{bmatrix} E_p \\ E_s \end{bmatrix}_{\text{trn}} = \tilde{\boldsymbol{T}} \begin{bmatrix} E_p \\ E_s \end{bmatrix}_{\text{inc}} \tag{2.79}$$

where

$$\tilde{\boldsymbol{T}} \equiv \left(\tilde{\boldsymbol{W}}_{\text{trn}}\right)^{-1} \tilde{\boldsymbol{T}}_{\text{xy}} \tilde{\boldsymbol{W}}_{\text{inc}} \tag{2.80}$$

and

$$\tilde{\boldsymbol{W}}_{\text{trn}} = \begin{bmatrix} \cos\theta_t\cos\phi & -\sin\phi \\ \cos\theta_t\sin\phi & \cos\phi \end{bmatrix}. \tag{2.81}$$

If the incident plane wave is p-polarized, the total reflectance is

$$R_p = |R_{11}|^2 + |R_{21}|^2. \tag{2.82}$$

If the incident plane wave is s-polarized, the total reflectance is

$$R_s = |R_{12}|^2 + |R_{22}|^2. \tag{2.83}$$

The total transmittance is calculated in a similar manner with the one additional complexity if the refracted angle of the transmitted beam differs from that of the incident angle. In particular, if the refractive index of the transmitting medium is different from that of the incident medium, then the component of the Poynting vector in the z direction is proportional to the absolute value of the electric field amplitude squared, multiplied by $n \cos \theta$, where n is the refractive index of the medium and θ is the polar angle of propagation of the plane wave within the medium. Therefore, if the incident plane wave is p-polarized, the total transmittance is

$$T_p = \left(|T_{11}|^2 + |T_{21}|^2\right)\left(\frac{n_t \cos\theta_t}{n_i \cos\theta_i}\right). \tag{2.84}$$

If the incident plane wave is s-polarized, the total transmittance is

$$T_s = \left(|T_{12}|^2 + |T_{22}|^2\right)\left(\frac{n_t \cos\theta_t}{n_i \cos\theta_i}\right). \tag{2.85}$$

2.3.5 Electric Fields Within Each Layer

Once the reflectance matrix has been found, it is straightforward to calculate the electric fields throughout the film stack. The reflected field in the incident medium is given by (2.77) and the transmitted field by (2.79). For the fields within each layer, (2.63) and (2.64) are applied successively starting with the first layer below the incident medium.

For incident p-polarization of unit amplitude, for k_{z1}, the incident field components are

$$E_{x1}^{(0)} = \cos\theta\cos\phi, \tag{2.86}$$

$$E_{y1}^{(0)} = \cos\theta\sin\phi, \tag{2.87}$$

and

$$E_{z1}^{(0)} = \sin\theta. \tag{2.88}$$

For k_{z2}, they are

$$E_{x2}^{(0)} = E_{y2}^{(0)} = E_{z2}^{(0)} = 0. \tag{2.89}$$

The magnetic field components are found from (2.43)–(2.45) by substituting (2.35)–(2.40).

$$H_{x1}^{(0)} = \frac{1}{Z_0}\left(-\tilde{k}_{z1}^{(0)} a_1^{(0)} + \tilde{k}_y b_1^{(0)}\right) E_{x1}^{(0)} \tag{2.90}$$

$$H_{y1}^{(0)} = \frac{1}{Z_0}\left(+\tilde{k}_{z1}^{(0)} - \tilde{k}_x b_1^{(0)}\right) E_{x1}^{(0)}, \tag{2.91}$$

$$H_{z1}^{(0)} = \frac{1}{Z_0}\left(-\tilde{k}_y + \tilde{k}_x a_1^{(0)}\right) E_{x1}^{(0)}, \tag{2.92}$$

and

$$H_{x2}^{(0)} = H_{y2}^{(0)} = H_{z2}^{(0)} = 0. \tag{2.93}$$

For incident s-polarization of unit amplitude, for k_{z1}, the incident field components are

$$E_{x1}^{(0)} = E_{y1}^{(0)} = E_{z1}^{(0)} = 0 \tag{2.94}$$

and for k_{z2}, they are

$$E_{x2}^{(0)} = -\sin\phi, \tag{2.95}$$

$$E_{y2}^{(0)} = \cos\phi, \tag{2.96}$$

and

$$E_{z2}^{(0)} = 0. \tag{2.97}$$

The remaining field components are again found from (2.43)–(2.45) by substituting (2.35)–(2.40).

$$H_{x1}^{(0)} = H_{y1}^{(0)} = H_{z1}^{(0)} = 0, \tag{2.98}$$

$$H_{x2}^{(0)} = \frac{1}{Z_0}\left(-\tilde{k}_{z2}^{(0)} a_2^{(0)} + \tilde{k}_y b_2^{(0)}\right) E_{y2}^{(0)}, \tag{2.99}$$

$$H_{y2}^{(0)} = \frac{1}{Z_0}\left(+\tilde{k}_{z2}^{(0)} - \tilde{k}_x b_2^{(0)}\right) E_{y2}^{(0)}, \tag{2.100}$$

and

$$H_{z2}^{(0)} = \frac{1}{Z_0}\left(-\tilde{k}_y + \tilde{k}_x a_2^{(0)}\right) E_{y2}^{(0)}. \tag{2.101}$$

For each layer, starting from the incident layer, the reflected fields are calculated corresponding to wavevectors k_{z3} and k_{z4} using (2.64) for each layer j,

$$E_{x3}^{(j)} = R_{11}^{(j)} E_{x1}^{(j)} + R_{12}^{(j)} E_{y2}^{(j)}, \tag{2.102}$$

$$E_{y3}^{(j)} = a_3^{(j)} E_{x3}^{(j)}, \tag{2.103}$$

$$E_{z3}^{(j)} = b_3^{(j)} E_{x3}^{(j)}, \tag{2.104}$$

$$H_{x3}^{(j)} = \frac{1}{Z_0}\left(-\tilde{k}_{z3}^{(j)} a_3^{(j)} + \tilde{k}_y b_3^{(j)}\right) E_{x3}^{(j)}, \tag{2.105}$$

$$H_{y3}^{(j)} = \frac{1}{Z_0}\left(\tilde{k}_{z3}^{(j)} - \tilde{k}_x b_3^{(j)}\right) E_{x3}^{(j)}, \tag{2.106}$$

$$H_{z3}^{(j)} = \frac{1}{Z_0}\left(-\tilde{k}_y + \tilde{k}_x a_3^{(j)}\right)E_{x3}^{(j)}, \tag{2.107}$$

$$E_{y4}^{(j)} = R_{21}^{(j)}E_{x1}^{(j)} + R_{22}^{(j)}E_{y2}^{(j)}, \tag{2.108}$$

$$E_{x4}^{(j)} = a_4^{(j)}E_{y4}^{(j)}, \tag{2.109}$$

$$E_{z4}^{(j)} = b_4^{(j)}E_{y4}^{(j)}, \tag{2.110}$$

$$H_{x4}^{(j)} = \frac{1}{Z_0}\left(-\tilde{k}_{z4}^{(j)} + \tilde{k}_y b_4^{(j)}\right)E_{y4}^{(j)}, \tag{2.111}$$

$$H_{y4}^{(j)} = \frac{1}{Z_0}\left(+\tilde{k}_{z4}^{(j)}a_4^{(j)} - \tilde{k}_x b_4^{(j)}\right)E_{y4}^{(j)}, \tag{2.112}$$

and

$$H_{z4}^{(j)} = \frac{1}{Z_0}\left(-\tilde{k}_y a_4^{(j)} + \tilde{k}_x\right)E_{y4}^{(j)}. \tag{2.113}$$

The transmitted field components at the bottom of layer $j + 1$ are calculated using (2.63) with the fields in layer j.

$$E_{x1}^{(j+1)} = T_{11}^{(j)}E_{x1}^{(j)} + T_{12}^{(j)}E_{y2}^{(j)}, \tag{2.114}$$

$$E_{y1}^{(j+1)} = a_1^{(j+1)}E_{x1}^{(j+1)}, \tag{2.115}$$

$$E_{z1}^{(j+1)} = b_1^{(j+1)}E_{x1}^{(j+1)}, \tag{2.116}$$

$$H_{x1}^{(j+1)} = \left(-\tilde{k}_{z1}^{(j+1)}a_1^{(j+1)} + \tilde{k}_y b_1^{(j+1)}\right)E_{x1}^{(j+1)}, \tag{2.117}$$

$$H_{y1}^{(j+1)} = \left(\tilde{k}_{z1}^{(j+1)} - \tilde{k}_x b_1^{(j+1)}\right)E_{x1}^{(j+1)}, \tag{2.118}$$

$$H_{z1}^{(j+1)} = \left(-\tilde{k}_y + \tilde{k}_x a_1^{(j+1)}\right)E_{x1}^{(j+1)}, \tag{2.119}$$

$$E_{y2}^{(j+1)} = T_{21}^{(j)}E_{x1}^{(j)} + T_{22}^{(j)}E_{y2}^{(j)}, \tag{2.120}$$

$$E_{x2}^{(j+1)} = a_2^{(j+1)} E_{y2}^{(j+1)}, \tag{2.121}$$

$$E_{z2}^{(j+1)} = b_2^{(j+1)} E_{y2}^{(j+1)}, \tag{2.122}$$

$$H_{x2}^{(j+1)} = \left(-\tilde{k}_{z2}^{(j+1)} + \tilde{k}_y b_2^{(j+1)}\right) E_{y2}^{(j+1)}, \tag{2.123}$$

$$H_{y2}^{(j+1)} = \left(+\tilde{k}_{z2}^{(j+1)} a_2^{(j+1)} - \tilde{k}_x b_2^{(j+1)}\right) E_{y2}^{(j+1)}, \tag{2.124}$$

and

$$H_{z2}^{(j+1)} = \left(-\tilde{k}_y a_2^{(j+1)} + \tilde{k}_x\right) E_{y2}^{(j+1)}. \tag{2.125}$$

Finally, there are no "reflected" fields within the transmitting medium so

$$E_{x3}^{(N+1)} = E_{y3}^{(N+1)} = E_{z3}^{(N+1)} = H_{x3}^{(N+1)} = H_{y3}^{(N+1)} = H_{z3}^{(N+1)} = 0, \tag{2.126}$$

and

$$E_{x4}^{(N+1)} = E_{y4}^{(N+1)} = E_{z4}^{(N+1)} = H_{x4}^{(N+1)} = H_{y4}^{(N+1)} = H_{z4}^{(N+1)} = 0. \tag{2.127}$$

The fields within each layer are then given by

$$E_{x,y,z}^{(j)} = E_{x1,y1,z1}^{(j)} e^{ik_{z1}^{(j)}\left(z - z_{\text{bot}}^{(j)}\right)} + E_{x2,y2,z2}^{(j)} e^{ik_{z2}^{(j)}\left(z - z_{\text{bot}}^{(j)}\right)} + E_{x3,y3,z3}^{(j)} e^{ik_{z3}^{(j)}\left(z - z_{\text{bot}}^{(j)}\right)} + E_{x4,y4,z4}^{(j)} e^{ik_{z4}^{(j)}\left(z - z_{\text{bot}}^{(j)}\right)}. \tag{2.128}$$

Converting back into s- and p-polarization is accomplished by inverting (2.76),

$$\begin{bmatrix} E_p^{(j)} \\ E_s^{(j)} \end{bmatrix} = \begin{bmatrix} \dfrac{\cos\phi}{\cos\theta_j} & \dfrac{\sin\phi}{\cos\theta_j} \\ -\sin\phi & \cos\phi \end{bmatrix} \begin{bmatrix} E_x^{(j)} \\ E_y^{(j)} \end{bmatrix}. \tag{2.129}$$

It should be noted that although the azimuthal angle ϕ remains constant throughout the film stack, the polar angle θ depends on the medium index of refraction through Snell's law. Therefore, it is labelled with a subscript j in (2.129).

The Poynting vector is

$$\boldsymbol{S} = \frac{1}{2}\text{Re}\left(\boldsymbol{E} \times \boldsymbol{H}^*\right). \tag{2.130}$$

The z-component of the Poynting vector determines the power/area propagating through the film stack.

2.3.6 Magneto-Optics and Birefringence

When one or more of the thin film layers is magnetic, the thin film stack can exhibit Faraday and Kerr rotation and ellipticity. Linearly polarized incident light is converted upon reflection or transmission into elliptically polarized light and the major axis of the ellipse is rotated from the polarization angle of the incident light. It is straightforward to determine the rotation and ellipticity of the electric field. Given the reflection matrix, $\tilde{\boldsymbol{R}}$, for example, the rotation of the plane of polarization for incident p-polarized light is

$$\vartheta_p = \tan^{-1}\left[\mathrm{Re}\left(\frac{R_{12}}{R_{11}}\right)\right], \tag{2.131}$$

and for s-polarized light is

$$\vartheta_s = \tan^{-1}\left[\mathrm{Re}\left(\frac{R_{21}}{R_{22}}\right)\right]. \tag{2.132}$$

The ellipticity of the reflected polarization for incident p-polarized light is

$$\varepsilon_p = \tan^{-1}\left[\mathrm{Im}\left(\frac{R_{12}}{R_{11}}\right)\right], \tag{2.133}$$

and for incident s-polarized light is

$$\varepsilon_s = \tan^{-1}\left[\mathrm{Im}\left(\frac{R_{21}}{R_{22}}\right)\right]. \tag{2.134}$$

Alternatively, if the p- and s-polarized field amplitudes are known, then for incident p-polarized light,

$$\vartheta_p + i\varepsilon_p = tan^{-1}\left(\frac{E_s}{E_p}\right), \tag{2.135}$$

and for incident s-polarized light,

$$\vartheta_s + i\varepsilon_s = tan^{-1}\left(-\frac{E_p}{E_s}\right). \tag{2.136}$$

The negative sign is inserted to ensure that rotation is measured in the same direction.

2.4 EXAMPLES

2.4.1 Dielectric Thin Film, Light Incident From Vacuum

Consider the simple thin film stack in Figure 2.2(a) consisting of a single dielectric layer with a refractive index of 2 deposited onto a substrate with a refractive index of 1.5 and a thickness of 40 nm. The light is incident from vacuum with a wavelength of 850 nm. The s- and p-polarization reflectance

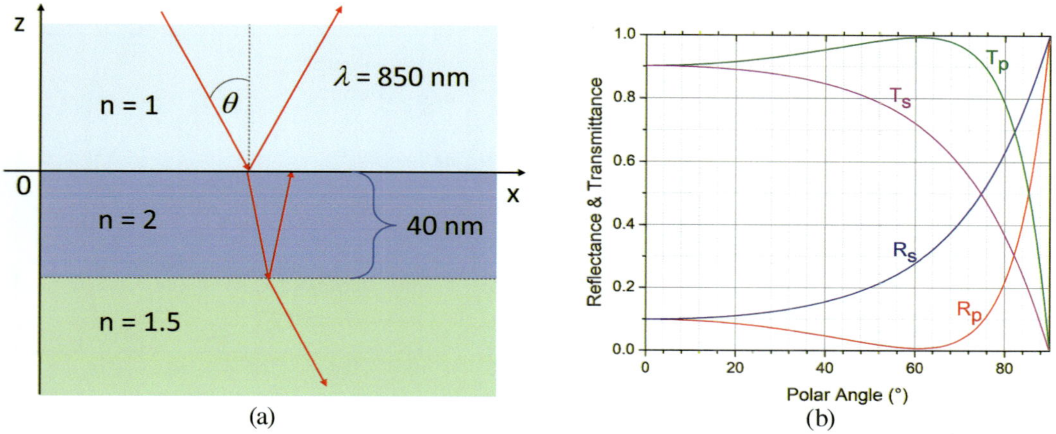

FIGURE 2.2 (a) 40 nm thin film with refractive index 2.0 deposited onto semi-infinite substrate with refractive index 1.5. Incident beam of wavelength 850 nm from vacuum onto film. (b) Reflectance and transmittance for *s*- and *p*-polarizations.

and transmittance is graphed as a function of the polar angle of incidence in Figure 2.2(b). At an angle of 61° the reflectance of the *p*-polarization drops to 0.0067. This is due to the Brewster angle at the surface of the thin film given by

$$\theta_B = tan^{-1}(n) = 63.4° \tag{2.137}$$

The residual reflectance is due to the second film/substrate interface.

2.4.2 Dielectric Thin Film, Light Incident From Substrate

Consider the effect of the previous example when the light is incident from the substrate side as shown in Figure 2.3. In this case, once the angle of incidence exceeds the critical angle, all of the light is reflected. The critical angle is given by

$$\theta_c = sin^{-1}\left(\frac{n_m}{n_i}\right) = 41.6°. \tag{2.138}$$

2.4.3 Surface Plasmons

Surface plasmons are electromagnetic excitations that occur at interfaces between dielectrics and metals or materials with a negative real part of their dielectric constant. Consider the thin film stack in Figure 2.4(a) which is laminated onto the bottom of a prism as shown in Figure 2.4(b).

The prism configuration in Figure 2.4(b) is called the Kretschmann configuration [6] and is frequently used for exciting surface plasmons on planar metal surfaces. The reflectance for both *s*- and *p*-polarizations is plotted in Figure 2.5(a) as a function of the angle of incidence. Figure 2.5(b) is an expanded plot around the surface plasmon resonance. At an angle of 42.8° the *p*-polarized reflectance nearly vanishes. At this angle the energy and momentum conservation laws are satisfied for transferring the incident light energy nearly entirely into the surface plasmon mode that propagates along the gold-air interface. The approximate resonance angle condition is [7]

$$\sin\theta_r \cong \frac{1}{n_i}\sqrt{\frac{\epsilon_d\left|Re(\epsilon_m)\right|}{\left|Re(\epsilon_m)\right| - \epsilon_d}}. \tag{2.139}$$

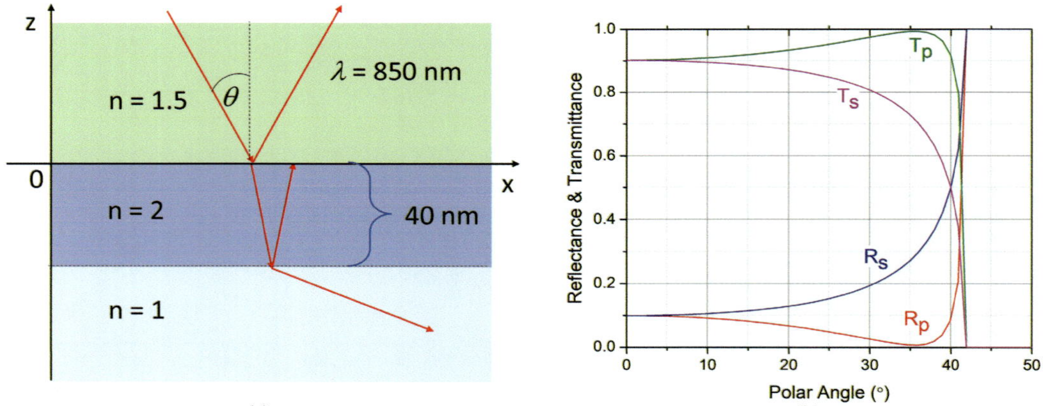

(a) (b)

FIGURE 2.3 (a) 40 nm thin film with refractive index 2.0 deposited onto semi-infinite substrate with refractive index 1.5. The beam is incident at a wavelength of 850 nm through the substrate onto the film. (b) Reflectance and transmittance for *s*- and *p*-polarizations.

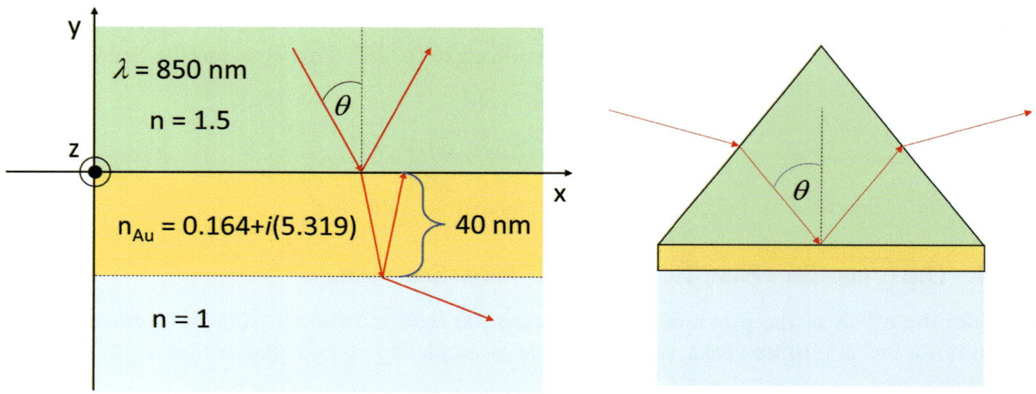

(a) (b)

FIGURE 2.4 Light at a wavelength of 850 nm is incident upon a 40 nm gold film through a glass substrate. In practice, this is often done by using a prism as shown in (b).

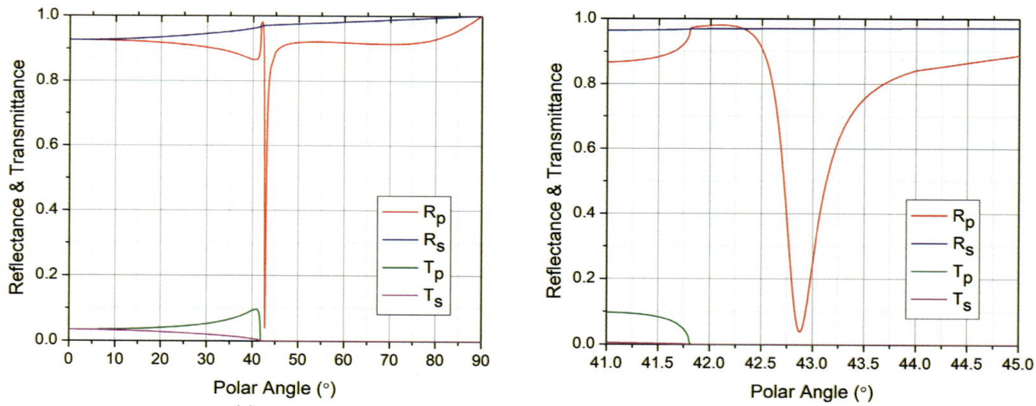

(a) (b)

FIGURE 2.5 Reflectance and transmittance for *s*- and *p*-polarized light as a function of polar angle for the system in Figure 2.4. The angle scale is expanded in (b) around the resonance.

n_i is the refractive index of the incident medium, ε_d is the dielectric constant of the medium below the prism, and $\text{Re}(\varepsilon_m)$ is the real part of the dielectric constant of the metal film. For the parameters in Figure 2.4, the dielectric constant of gold is $-28.26 + i\,(2.745)$ and (2.139) predicts a surface plasmon resonance at an incident angle of 42.7°.

The real part of the electric field is plotted in Figure 2.6(a). For $z > 0$, because the reflectance is so small at the resonance angle, the electric field is approximately that of an incident plane wave of unit amplitude that almost goes to zero at the gold interface. The field then steadily increases through the 40 nm gold film until it reaches the air interface. It then begins an exponential decay into the air that is characteristic of an evanescent mode. The x- and z-components of the Poynting vector are plotted in Figure 2.6(b). S_x exhibits a sharp large rise in amplitude at the gold-air interface characteristic of the energy in the surface plasmon that is propagating along the x-direction at the surface.

A surface plasmon example [8] using a silver film is shown in Figure 2.7 which is incorporated into the liquid crystal example 8 at the end of the chapter. The calculated reflectance at a wavelength of 632.8 nm exhibits a strong resonance at ~68.5° angle of incidence.

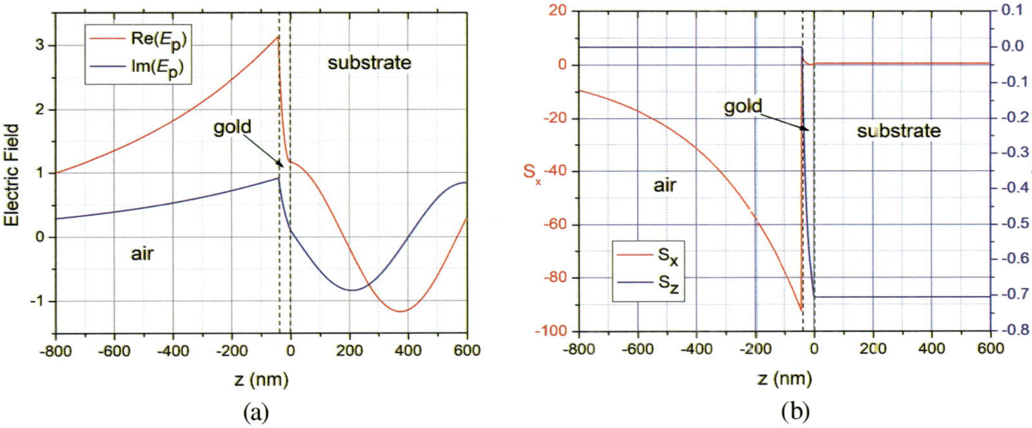

FIGURE 2.6 (a) Real and imaginary parts of the electric field at the surface plasmon resonance $\theta = 42.88°$. (b) x and z components of the Poynting vector.

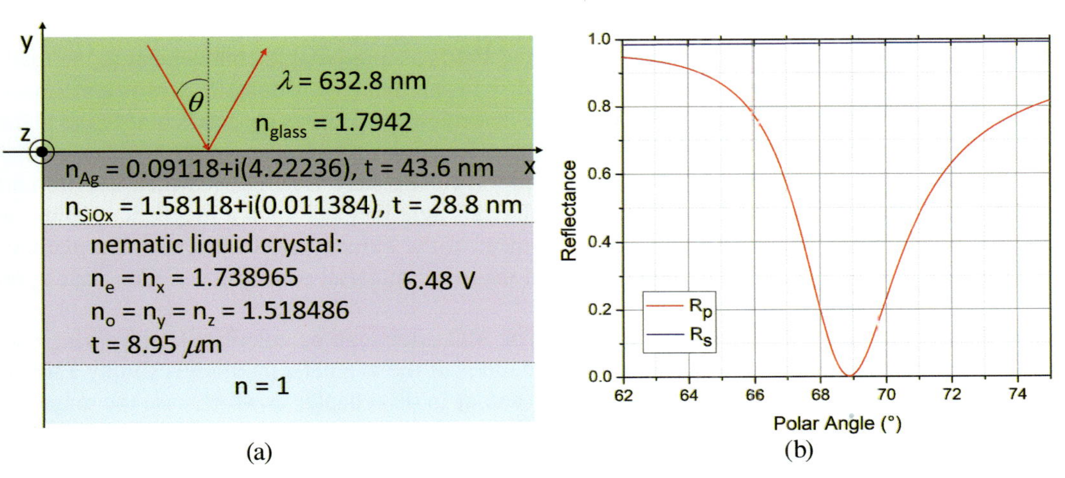

FIGURE 2.7 (a) Surface plasmon resonance for film stack for comparison to Figure 8(a) in [8]. (b) Reflectance as a function of angle of incidence exhibiting the SP resonance at ~68.5°.

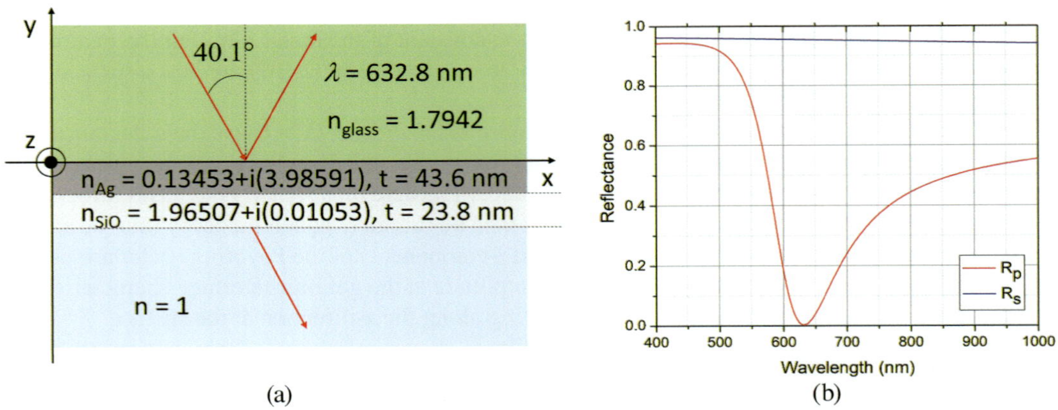

FIGURE 2.8 (a) Film stack for surface plasmon resonance of silver. (b) Wavelength-dependence of resonance.

The surface plasmon resonance can also be calculated as a function of wavelength. The film stack at a wavelength of 632.8 nm is shown in Figure 2.8(a). Using the wavelength-dependent refractive indices for silver and SiO, [9] the plasmon resonance is plotted in Figure 2.8(b).

2.4.4 Birefringent Material

Consider a birefringent thin film material embedded in the film stack in Figure 2.9(a) with a dielectric tensor

$$\epsilon = \begin{bmatrix} 2 & 0 & 0 \\ 0 & 2.2 & 0 \\ 0 & 0 & 2 \end{bmatrix}. \tag{2.140}$$

The reflectance and transmittance of this film stack for light incident through the high index substrate for several different azimuthal angles are shown for both *p*- and *s*-polarizations.

2.4.5 Magneto-Optic Media

Magneto-optical (MO) disks were developed in the 1980s as a rewritable complement to the read-only compact disks. Data were stored on the MO disk in small magnetic domains within a magnetic thin film. The domains were magnetized in either up or down directions perpendicular to the plane of the film and read using the magneto-optic Kerr effect. John Kerr was a Scottish physicist who in 1877 discovered this effect (he also discovered the electrostatic Kerr effect which is different) in which the plane of polarization of linearly polarized light incident upon a magnetic medium is slightly rotated upon reflection. It is the complement of the Faraday effect in which the plane of polarization is rotated in transmission through a magnetic material or a material like glass upon which a strong magnetic field is applied.

Magnetic media are circularly birefringent. The MO effect can be calculated by working in a circular coordinate system in which the refractive index of the magnetic medium is simply a scalar quantity, or by using the Cartesian coordinate system as in this chapter in which case the magnetic medium is represented by a dielectric tensor.

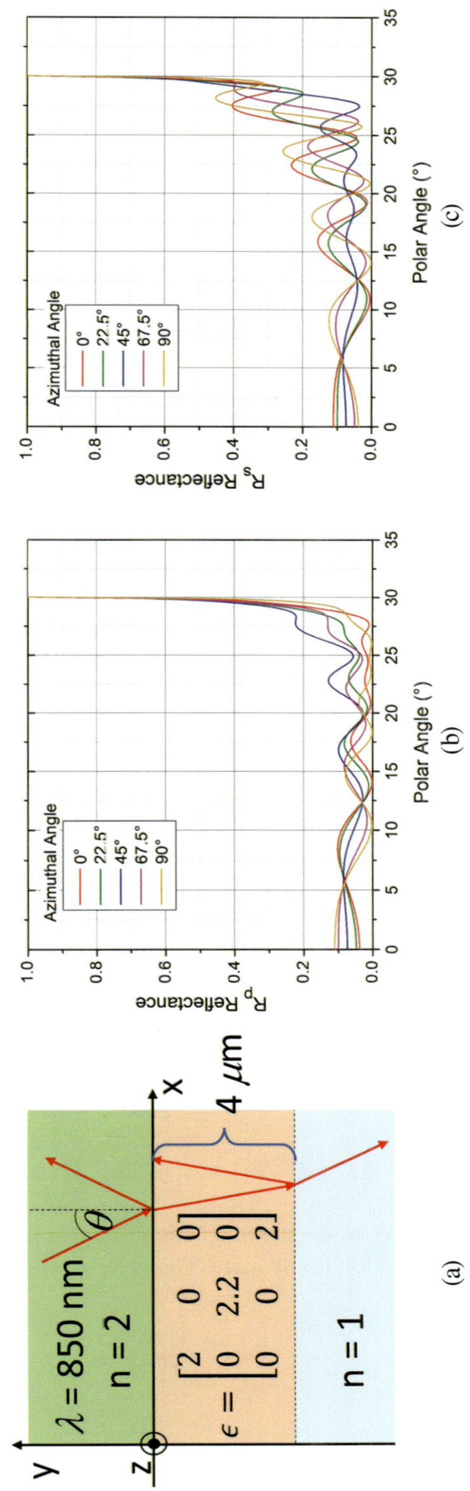

FIGURE 2.9 (a) Birefringent medium. (b) *p*- and (c) *s*-polarized reflectance as a function of angle of incidence.

The signal amplitude from light reflected by the MO media is proportional to the reflectance of the medium multiplied by the square root of the sum of the squares of the MO rotation and MO ellipticity [10]. This value is a common figure of merit for MO media.

$$FOM = R\sqrt{\vartheta_K^2 + \varepsilon_K^2}\,. \tag{2.141}$$

By placing the magnetic thin film within an interference film stack, it is possible to reduce the reflectance of the film which has the effect of increasing the MO rotation and ellipticity. For large reflectances the signal-to-noise ratio (SNR) is determined by the shot noise of the light incident upon the detector. If the reflectance is too small, however, the dark noise of the detector or electronic noise of the amplifier can dominate, thereby reducing the SNR of the media. In practice, it has been found that the SNR is maximized for a thin film reflectance of 20–25%.

The materials used for the magnetic film in MO media are typically rare earth – transition metal amorphous alloys deposited by vacuum sputtering. The most common alloy is composed of Tb, Fe, and Co. For Tb concentrations ~20% the alloy is a ferrimagnet in which the magnetic moment of the Tb atoms is opposed to that of the Fe and Co atoms. The concentration of Tb is typically adjusted so that at room temperature or thereabouts the total magnetic moment of the alloy is almost zero because the moment of the Tb ions is equal to, but in the opposite direction to, that of the transition metal ions. As a result, the applied field required to flip the state of the magnetic domain becomes extremely large, thereby ensuring the stability of the recorded domain and of the stored data. The Fe/Co ratio is adjusted to place the Curie point, the temperature at which the magnetism of the alloy disappears, at slightly above 200°C. It is easy with a focused laser beam at a wavelength of 830 nm to quickly heat the recording medium in a very small spot above the Curie point so that the magnetic state of the domain can be fixed in the direction of the relatively small applied field as the medium cools.

The standard physics definition for right (left) circularly polarized light is that the electric vector rotates clockwise (anti-clockwise) when viewed by an observer looking back in the direction from which the light is coming [11]. The dielectric constants in the circular coordinate system are

$$\epsilon_\pm = \epsilon_{xx} \mp i\epsilon_{xy} = \left[(n \pm \Delta n) + i(K \pm \Delta K)\right]^2 \tag{2.142}$$

for right (+) and left (–) circular polarizations [12]. Solving for the off-diagonal dielectric constant,

$$\pm\epsilon_{xy} = i\left\{\left[(n \pm \Delta n) + i(K \pm \Delta K)\right]^2 - \epsilon_{xx}\right\} \tag{2.143}$$

or to first order in Δn and ΔK,

$$\epsilon_{xy} = -2(n \cdot \Delta K + K \cdot \Delta n) + 2i(n \cdot \Delta n - K \cdot \Delta K). \tag{2.144}$$

The magnitudes of Δn and ΔK relative to the nonmagnetic refractive indices n and K are generally very small, so a first order approximation is adequate. The inverse relations are

$$\Delta n = \frac{-K\,\mathrm{Re}(\epsilon_{xy}) + n\,\mathrm{Im}(\epsilon_{xy})}{2(n^2 + K^2)} \tag{2.145}$$

and

$$\Delta K = -\frac{n\,\mathrm{Re}(\epsilon_{xy}) + K\,\mathrm{Im}(\epsilon_{xy})}{2(n^2 + K^2)}, \tag{2.146}$$

where

$$n^2 = \frac{\mathrm{Re}(\epsilon) + \sqrt{\mathrm{Re}(\epsilon)^2 + \mathrm{Im}(\epsilon)^2}}{2} \tag{2.147}$$

and

$$K^2 = \frac{-\mathrm{Re}(\epsilon) + \sqrt{\mathrm{Re}(\epsilon)^2 + \mathrm{Im}(\epsilon)^2}}{2}. \tag{2.148}$$

The refractive index of a typical TbFeCo alloy used for MO media at a wavelength of 830 nm is ~3.5 + i (3.6) [13]. In circular coordinates, the change in the refractive index at room temperature of the magnetized medium at this wavelength is $n_{\pm} = \pm 0.052 \pm i\,(0.032)$. For a thick alloy film as shown in Figure 2.10(a), the MO rotation is −0.38°, the MO ellipticity is 0.03°, and the reflectivity is 0.58. With such a large reflectivity, less than half of the incident laser light is absorbed, so it becomes difficult to heat the medium above its Curie point.

A quadrilayer film stack can be tuned to obtain the desired reflectance while maximizing the MO signal. One example of an air-incident quadrilayer film stack is shown in Figure 2.10(b) after [5]. The reflectance of the film stack as a function of angle of incidence for both p- and s-incident polarizations is plotted in Figure 2.11(a) for comparison to Figure 2(b) of [5]. The z-component of the Poynting vector is plotted in Figure 2.11(b). About 5.5% of the incident power is reflected at the surface of the film stack. Most of the incident light is absorbed in the thin MO layer. Only about 2% of the incident light reaches, and is absorbed by, the aluminum reflector/heat sink.

The MO rotation and ellipticity for this film stack are plotted in Figure 2.12 as a function of angle of incidence for both incident (linear) polarizations for comparison to Figures 2(a) and 2(c) of [5]. The normal incidence reflectivity of this quadrilayer film stack is 0.0553, the MO rotation is 1.76°, and the MO ellipticity is 1.74°. The figure of merit from (2.141) is 0.137.

However, a real MO disk is coated such that the laser light is incident from the substrate side like a CD to better protect the thin films and to ensure that scratches or dirt on the surface of the substrate are out of focus to the readback system. Polycarbonate is a common disk substrate material with a somewhat larger refractive index of 1.58. A typical MO film stack is shown in Figure 2.13(a). This film stack has a reflectivity of 22.5%, a MO rotation of 0.93°, and a MO ellipticity of −0.18°. Its FOM is 0.21, not much different from the bare MO film in air-incidence or the bare film in substrate-incidence either.

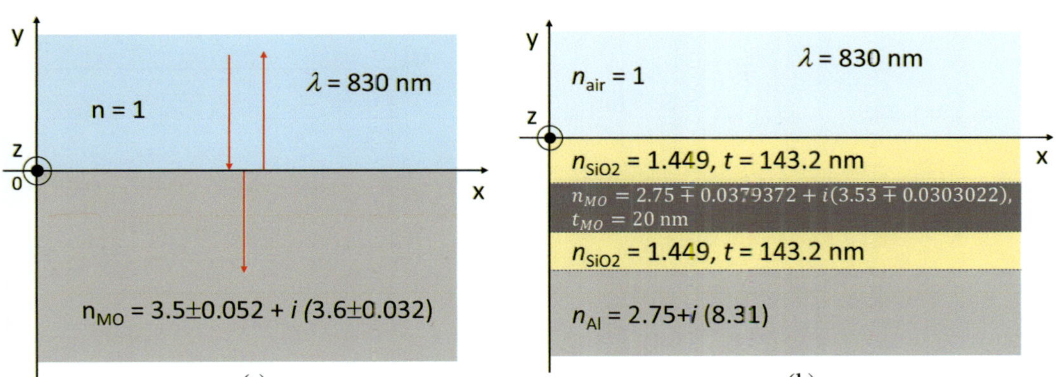

FIGURE 2.10 (a) Light incident normally onto a thick TbFeCo film from vacuum. (b) MO quadrilayer film stack for air-incidence consisting of a quarter-wave anti-reflecting layer of SiO_2, a thin 20 nm MO film of TbFeCo film, a second quarter-wave anti-reflecting SiO_2 layer, and a final opaque aluminum reflector/heat sink. MO media is typically read at a wavelength of 830 nm.

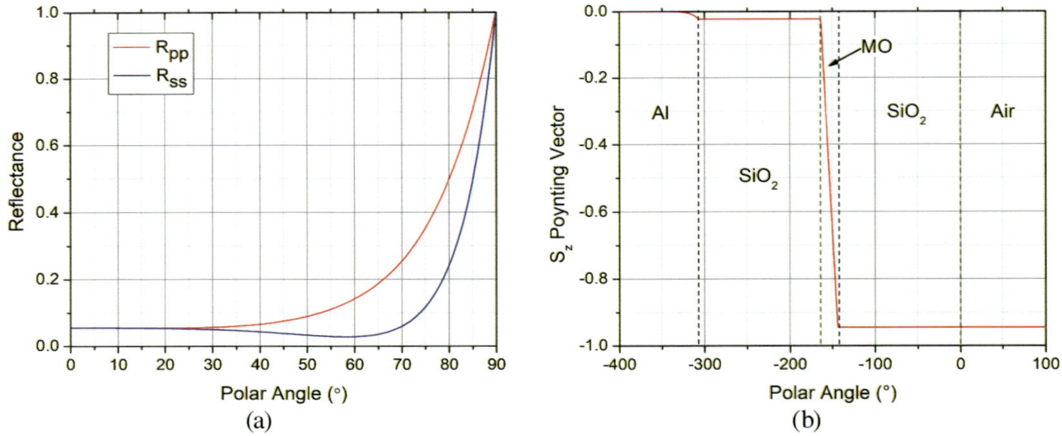

FIGURE 2.11 (a) R_{pp} is the reflected *p*-polarization for incident *p*-polarization, and R_{ss} is the reflected *s*-polarization for incident *s*-polarization. There are also very small reflected components of the opposite polarization. (b) *z*-component of the Poynting vector showing normalized power flow/area in the negative *z* direction.

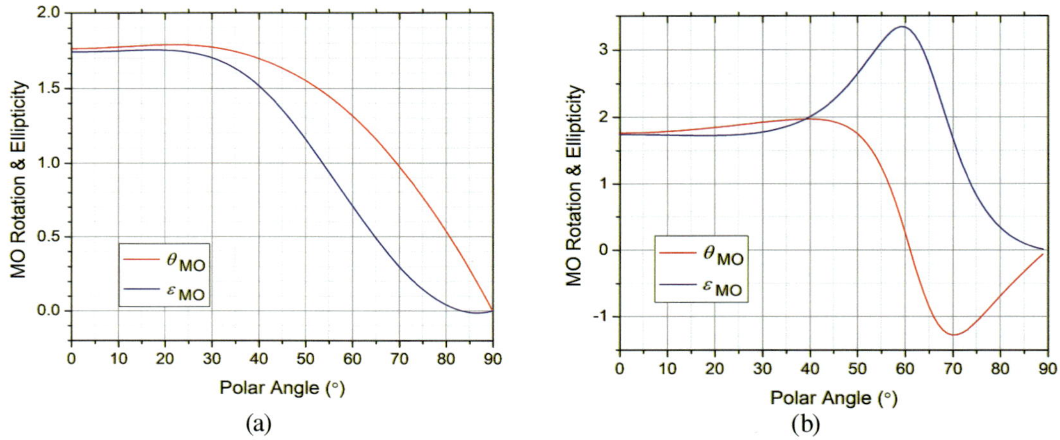

FIGURE 2.12 For the air-incident quadrilayer film stack of Figure 2.10: MO rotation, θ_{MO}, and ellipticity, ε_{MO}, as a function of polar angle of incidence for incident (a) *p*-polarized and (b) *s*-polarized light.

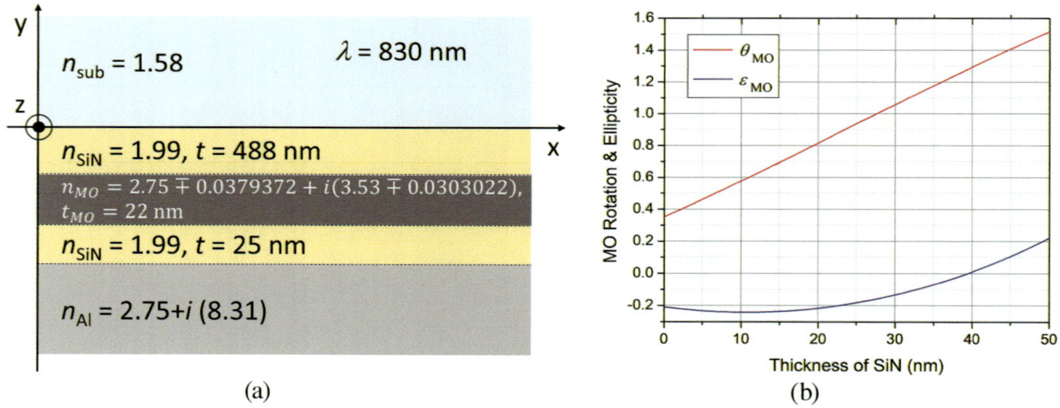

FIGURE 2.13 (a) MO quadrilayer film stack for substrate-incidence consisting of a layer of SiN, a thin 20 nm MO film of TbFeCo film, a second layer of SiN, and a final opaque aluminum reflector/heat sink. (b) MO rotation and ellipticity as a function of the SiN thickness between the MO film and the Al heat sink.

Some of the early MO disk drives were only designed to read Kerr rotation, not Kerr ellipticity. The FOM in (2.141) is only for disk drives that incorporate a quarter waveplate or some other phase compensator to combine the effects of rotation and ellipticity in generating the measured signal. It is also possible to design a film stack with no ellipticity as shown in Figure 2.24. At a SiN thickness of 39.5 nm for the layer next to the aluminum reflector, the reflected MO effect is entirely a rotation of the plane of polarization. The FOM is 0.20, but the reflectivity of the film stack has dropped to ~0.155, which may be too small to operate in the shot noise limit of the detectors.

2.4.6 Quarter-Wave Reflector With and Without Center Defect Layer

Consider the quarter-wave reflector film stack in Figure 2.14(a) designed for a wavelength of 633 nm at normal incidence. The general equation for the thickness of each layer with refractive index n is

$$\frac{\lambda}{4} = \frac{nt}{\cos\theta} = \frac{nt}{\sqrt{n^2 - (n_i \sin\theta_i)^2}} \tag{2.149}$$

or

$$t = \frac{\lambda\sqrt{n^2 - (n_i \sin\theta_i)^2}}{4n} \tag{2.150}$$

where θ_i is the angle of incidence in a medium with refractive index n_i. Assuming there is no dispersion of the refractive indices with wavelength, the calculated reflectance and transmittance of this film stack is shown in Figure 2.14(b). The reflectance remains > 99% from 583 nm to 692 nm.

Now consider the effect of doubling the thickness of a center low index layer as shown in Figure 2.15(a). This extra thick layer has been compared to an impurity state inside the bandgap of a semiconductor. The reflectance now has a very narrow dip that drops down to 0.42 at the design wavelength of 633 nm as shown in Figure 2.15(b).

If it is desired to minimize the reflectance in the notch, an optimization routine can be applied. By varying the thicknesses of the five center layers of film stack in Figure 2.15(a) a better film stack can be found as shown in Figure 2.16.

The wavelength-dependent reflectance is shown in Figure 2.17. The minimum reflectance at 633 nm is less than 5%.

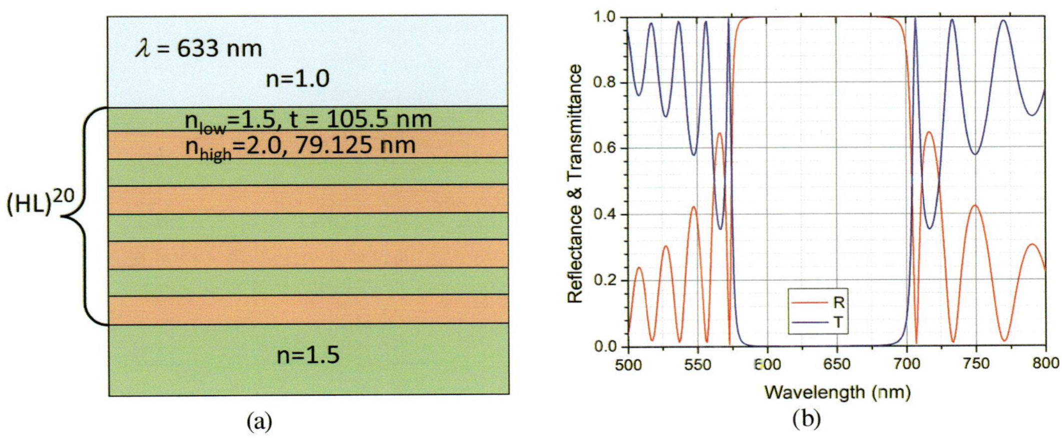

FIGURE 2.14 (a) Quarter-wave dielectric reflector film stack consisting of 20 low and 20 high refractive index thin films alternately deposited onto a low refractive index substrate. (b) Reflectance and transmittance.

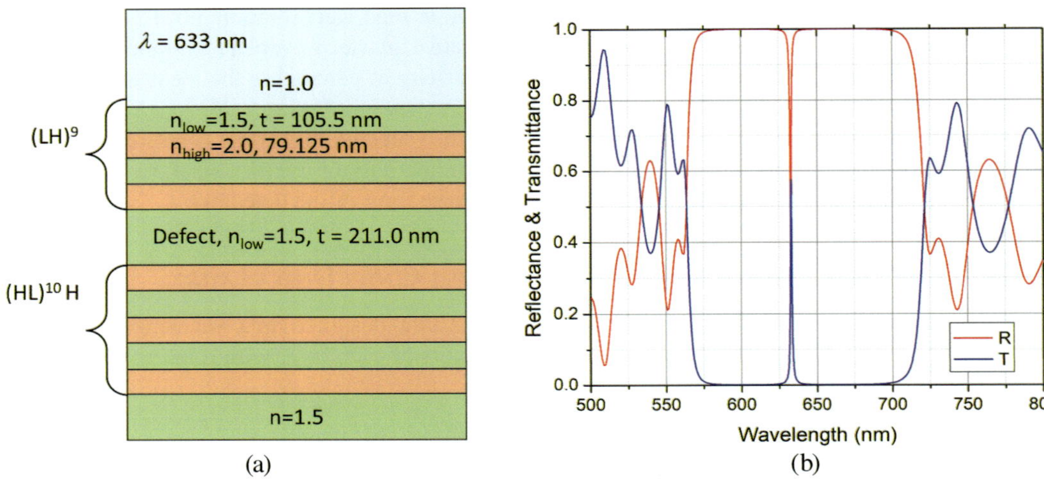

FIGURE 2.15 (a) Quarter-wave dielectric reflector film stack with mid-stack defect consisting of 9 dual layers of low/high refractive index, followed by a half-wave thickness low index layer, followed by 10 dual high/low refractive index films, a final high index layer, and a low refractive index substrate. (b) Reflectance and transmittance of quarter-wave film stack with half-wave "defect" thin film in center.

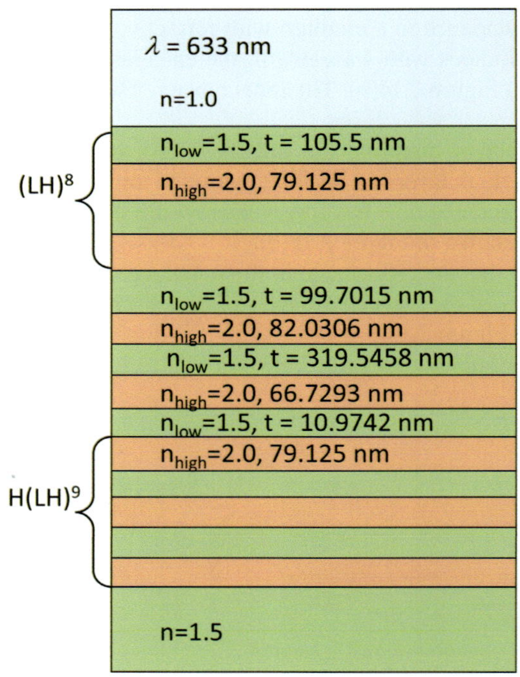

FIGURE 2.16 Defect state film stack optimized from the starting point of Figure 2.15(a) by varying the thicknesses of the center five layers.

2.4.7 Anti-Reflector

Although single layer anti-reflecting films are often suitable to reduce reflectance below 2% at a single wavelength, making use of more layers can provide broader wavelength anti-reflecting properties. The three-layer AR film stack consists of an outer low index material with a $\lambda/4$ optical thickness, a high index middle layer with a $\lambda/2$ optical thickness, and an intermediate index material

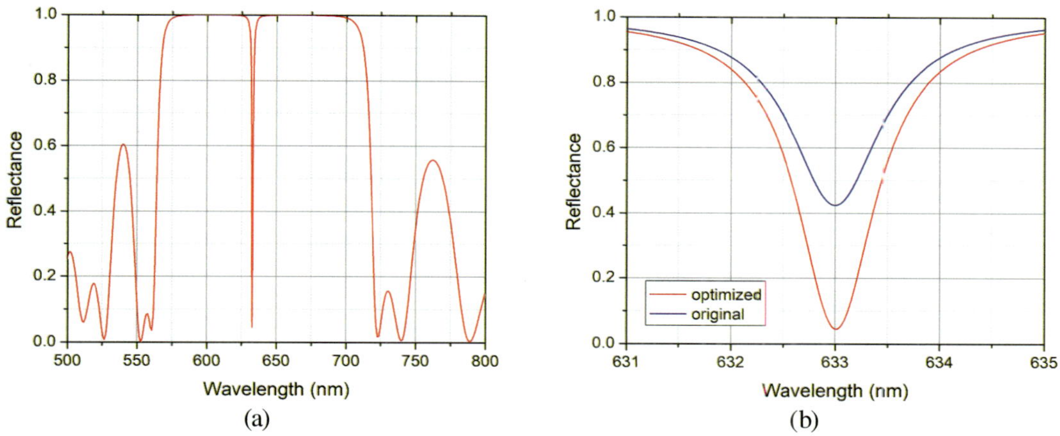

FIGURE 2.17 (a) Defect state in quarter-wave reflector film stack with layer thicknesses optimized for minimum reflectance at 633 nm. (b) Comparison of the reflectances of the original film stack in Figure 2.15(b) to the optimized film stack.

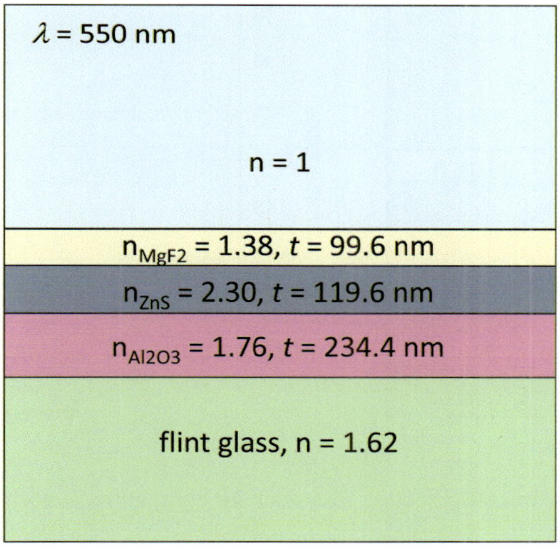

FIGURE 2.18 Three-layer anti-reflecting film stack for flint glass.

bottom layer against the substrate with a $3\lambda/4$ optical thickness [14]. Flint glass with a refractive index of 1.62 has a bare reflectance of 5.6%. Consider the three-layer film stack in Figure 2.18.

The reflectance and transmittance curves for normal incidence are plotted in Figure 2.19(a). At the design wavelength of 550 nm, the reflectance is $< 10^{-5}$. The variation of the reflectance with angle of incidence is plotted in Figure 2.19(b). The reflectance is less than 1% up to an angle of 55° for *p*-polarization and up to 41° for *s*-polarization.

Anti-reflection coatings can also be designed from three-layer film stacks in which the optical thickness of each layer is a quarter wavelength. Consider the anti-reflecting film stack for germanium from [3, p. 119] shown in Figure 2.20(a). The theoretical reflectance for this film stack is shown in Figure 2.20(b).

Anti-reflecting film stacks with more layers can be designed with better properties. Examples of 5- and 7-layer film stacks on glass are shown in Figure 2.21.

The resulting reflectance curves are shown in Figure 2.22.

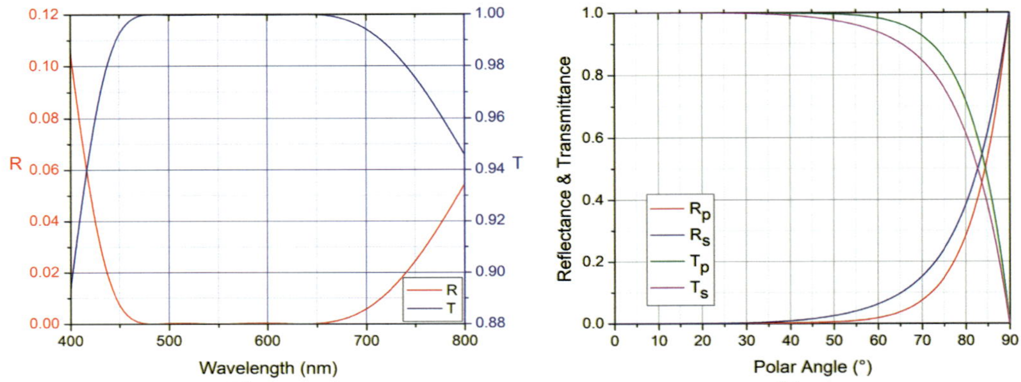

(a) (b)

FIGURE 2.19 (a) Reflectance and transmittance as a function of wavelength for the three-layer anti-reflecting film stack in Figure 2.18. (b) Reflectance and transmittance for the same film stack as a function of angle of incidence.

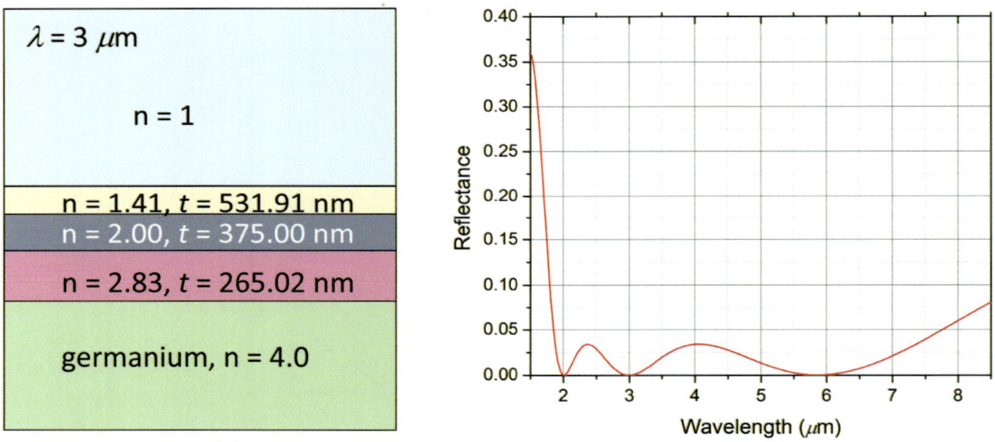

(a) (b)

FIGURE 2.20 (a) Three-layer anti-reflecting film stack for germanium at normal incidence. (b) Reflectance at normal incidence.

Compare to Figure 4.25 of [3].

2.4.8 Bandpass Filter

Bandpass filters are typically designed using many thin film layers. An example of a bandpass filter from Figure 8.26 of [3] is shown in Figure 2.23(a). The reflectance and transmittance around the design wavelength are plotted in Figure 2.23(b). This filter design exhibits square shoulders.

2.4.9 Liquid Crystal

Consider the film stack in Figure 2.24(a) after Figure 9 in [8].

Nematic liquid crystals have the optical properties of uniaxial crystals. An applied voltage can cause the orientation of this axis of the extraordinary index to vary through the thickness of a film. Therefore, to model a nematic liquid crystal thin film under an applied field, the layer must be divided into many thin film regions each characterized by a specific directrix that slowly varies through the layers. The liquid crystal properties for the model developed in [8] are

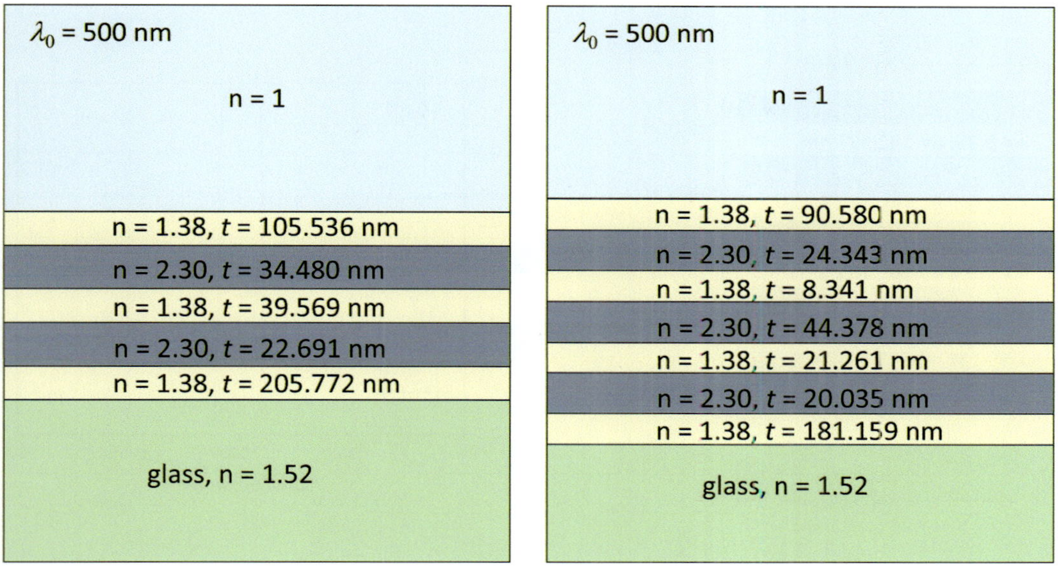

(a) (b)

FIGURE 2.21 (a) 5-layer, and (b) 7-layer anti-reflecting film stacks in the visible for glass substrate after Figure 4.48 in [3].

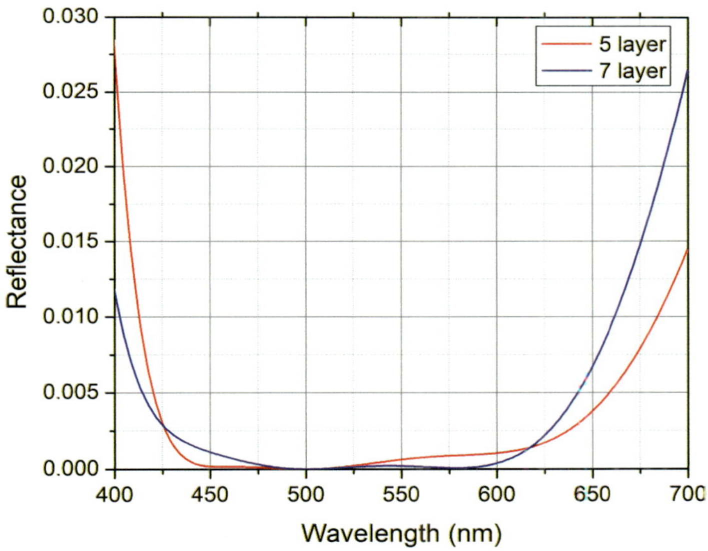

FIGURE 2.22 Anti-reflecting film stacks on glass designed for a center wavelength of 500 nm.

Ordinary permittivity	$\varepsilon_o = 2.3058$	$n_o = 1.518486 + i\,(0.000100)$
Extraordinary index	$\varepsilon_e = 3.0240$	$n_e = 1.738965 + i\,(0.000114)$
Splay elastic constant	$k_{11} = 11.40 \times 10^{-12}$ N/m	
Bend elastic constant	$k_{33} = 18.20 \times 10^{-12}$ N/m	
Dielectric constant (parallel, 0 Hz)		19.100
Dielectric constant (perpendicular, 0 Hz)		5.1200
Layer thickness		8.95 μm

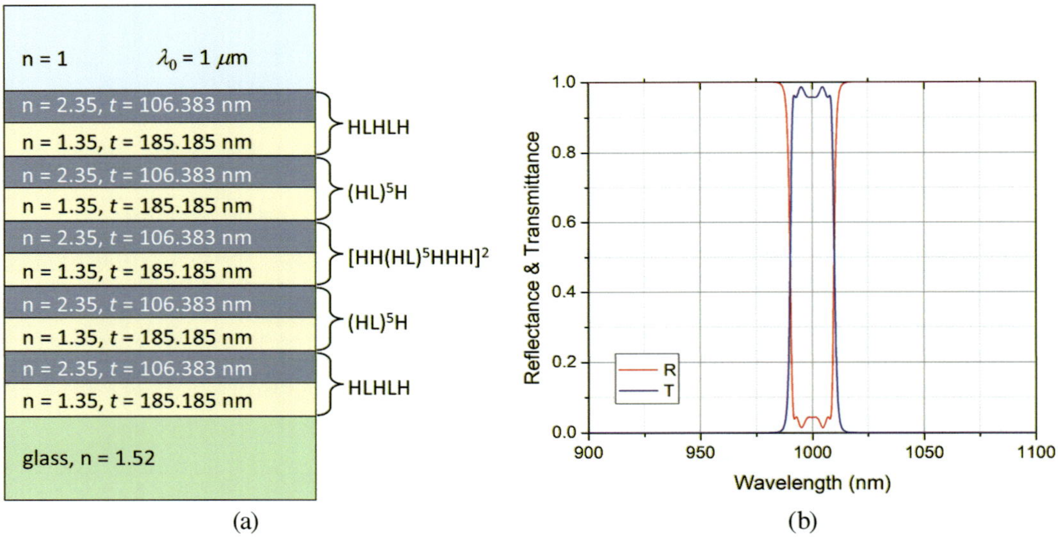

FIGURE 2.23 (a) Film stack for bandpass filter. (b) Reflectance and transmittance for bandpass filter.

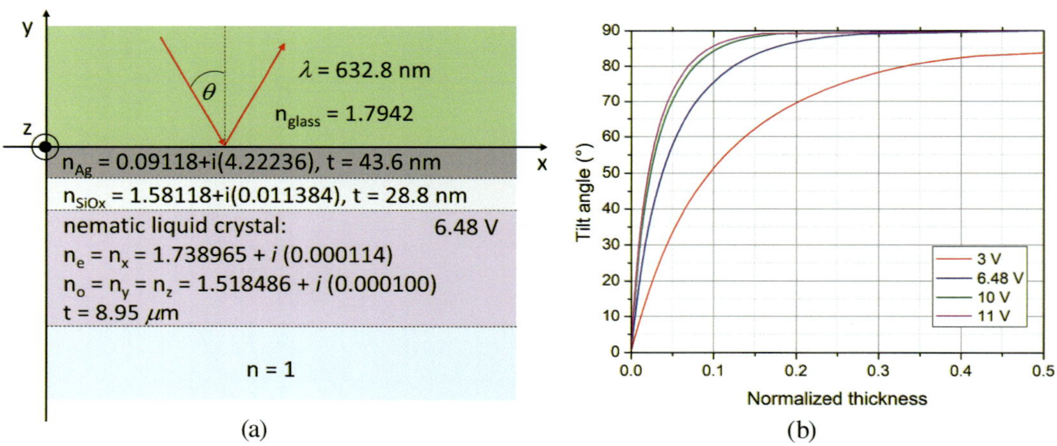

FIGURE 2.24 (a) Film stack for measurement of nematic liquid crystal using surface plasmon resonance after [8]. (b) Tilt angle of nematic director through the thickness of the liquid crystal film for several applied voltages.

If the maximum tilt angle for a given applied voltage is φ_{m} then let

$$Y \equiv \tan^2 \varphi_m . \tag{2.151}$$

Furthermore, the threshold voltage is given by

$$V_0 = \pi \sqrt{\frac{k_{11}}{\varepsilon_0 \left(\varepsilon_{\parallel} - \varepsilon_{\perp}\right)}} . \tag{2.152}$$

Let

$$\kappa = \frac{k_{33} - k_{11}}{k_{11}} \tag{2.153}$$

and

$$\gamma = \frac{\epsilon_{\parallel} - \epsilon_{\perp}}{\epsilon_{\perp}}. \tag{2.154}$$

For a given applied voltage, the value of Y can be calculated by solving the equation

$$\frac{V}{V_0} = \frac{1}{\pi}\sqrt{1+Y(1+\gamma)}\int_0^{\infty}\sqrt{\frac{(1+Y)(1+W)+\kappa YW}{\left[(1+Y)(1+W)+\gamma YW\right](1+Y+W)(1+W)W}}\,dW \tag{2.155}$$

which determines the maximum tilt angle. The value for z that corresponds to a specific tilt angle $\varphi \leq \varphi_m$ can be found by solving for W_l,

$$W_l = \frac{tan^2\varphi(1+Y)}{Y - tan^2\varphi}. \tag{2.156}$$

This value for W_l becomes a limit of integration in the equation

$$z = \frac{L}{2}\left\{\frac{\int_0^{W_L}\sqrt{\frac{\left[(1+Y)(1+W)+\kappa YW\right]\left[(1+Y)(1+W)+\gamma YW\right]}{(1+Y+W)(1+Y)(1+W)W}}\frac{dW}{1+W}}{\int_0^{\infty}\sqrt{\frac{\left[(1+Y)(1+W)+\kappa YW\right]\left[(1+Y)(1+W)+\gamma YW\right]}{(1+Y+W)(1+Y)(1+W)W}}\frac{dW}{1+W}}\right\} \tag{2.157}$$

where L is the total liquid crystal thickness. For this example, the tilt angle of the nematic director is plotted for several different applied voltages in Figure 2.24(b).

Once the tilt angle of the nematic director has been calculated through the thickness of the liquid crystal, the variation of the dielectric tensor can also be found using the rotation matrices,

$$\tilde{\epsilon}(\varphi) = \tilde{\boldsymbol{R}}^{-1}\tilde{\epsilon}(0)\tilde{\boldsymbol{R}} \tag{2.158}$$

where

$$\tilde{\epsilon}(0) = \begin{bmatrix} n_{ex}^2 & 0 & 0 \\ 0 & n_{ord}^2 & 0 \\ 0 & 0 & n_{ord}^2 \end{bmatrix} \tag{2.159}$$

and

$$\tilde{\boldsymbol{R}} = \begin{bmatrix} \cos\varphi & 0 & \sin\varphi \\ 0 & 1 & 0 \\ -\sin\varphi & 0 & \cos\varphi \end{bmatrix}. \tag{2.160}$$

The result is

$$\tilde{\epsilon}(0) = \begin{bmatrix} n_{ex}^2 \cos^2\varphi + n_{ord}^2 \sin^2\varphi & 0 & \left(n_{ex}^2 - n_{ord}^2\right)\sin\varphi\cos\varphi \\ 0 & n_{ord}^2 & 0 \\ \left(n_{ex}^2 - n_{ord}^2\right)\sin\varphi\cos\varphi & 0 & n_{ex}^2 \sin^2\varphi + n_{ord}^2 \cos^2\varphi \end{bmatrix}. \tag{2.161}$$

The liquid crystal layer is subdivided into multiple thinner layers. In [8], 90 sublayers are used, evenly divided in tilt angle of the nematic director. As a result, the thicknesses of each sublayer vary substantially. For the case of an applied voltage of 6.48 V, the tilt angle and thickness for each sublayer is given in Table 2.1. As the nematic director tilts towards 90° in the center of the liquid crystal, rotating from the z-axis towards the x-axis, the refractive index tensor component n_{zz} also increases towards the extraordinary index, while the refractive index component n_{xx} decreases towards the ordinary index. Therefore, the electric field in the center of the liquid crystal for any modes propagating along the interface between the gold film and the liquid crystal interacts with the n_{zz} index. However, the outer edges of the liquid crystal have a smaller refractive index closer to the ordinary index. Essentially, the liquid crystal becomes a waveguide with a high index core able to support a multitude of modes which propagate along and through the liquid crystal waveguide.

The reflectance of this film stack is shown in Figure 2.25. It is quite an interesting reflectance in which the resonant waveguide modes are seen interspersed with the dominant surface plasmon mode observed in Figure 2.7(b). Some of the resonant modes become peaks in the reflectance curve rather than dips. Narrow resonances interacting with a broad resonance, giving rise to characteristic asymmetric line shapes, are called "Fano resonances." They occur in many different areas of physics. One interesting example observed by the author during his doctoral thesis research involved the interaction of narrow phonon resonant lines with a broad charge density wave energy gap in the semimetal compound $NbSe_3$ when cooled to ~ 2 K as shown in Figure 2.26.

TABLE 2.1
Tilt Angle as a Function of Layer Within the Liquid Crystal

Sublayer	Thickness (nm)	Tilt angle (°)
1	8.50762	0.99923
2	8.5524	2.99768
3	8.64192	4.99614
4	8.77616	6.99459
5	8.9551	8.99305
6	9.17879	10.99151
7	9.44735	12.98996
8	9.76105	14.98842
9	10.12034	16.98687
10	10.52591	18.98533
11	10.97871	20.98378
12	11.48	22.98224
13	12.03142	24.98069
14	12.63501	26.97915
15	13.29328	28.97761
16	14.00926	30.97606
17	14.78657	32.97452
18	15.62954	34.97297
19	16.54331	36.97143
20	17.53394	38.96988

TABLE 2.1 (*Continued*)
Tilt Angle as a Function of Layer Within the Liquid Crystal

Sublayer	Thickness (nm)	Tilt angle (°)
21	18.60863	40.96834
22	19.77595	42.96679
23	21.04606	44.96525
24	22.43116	46.96371
25	23.94593	48.96216
26	25.60815	50.96062
27	27.43955	52.95907
28	29.46693	54.95753
29	31.72365	56.95598
30	34.25176	58.95444
31	37.105	60.95289
32	40.35304	62.95135
33	44.08783	64.9498
34	48.43321	66.94826
35	53.55996	68.94672
36	59.71012	70.94517
37	67.23844	72.94363
38	76.68615	74.94208
39	88.92145	76.94054
40	105.42884	78.93899
41	128.96977	80.93745
42	165.31707	82.9359
43	228.82713	84.93436
44	367.30302	86.93282
45	2477.37352	88.93127

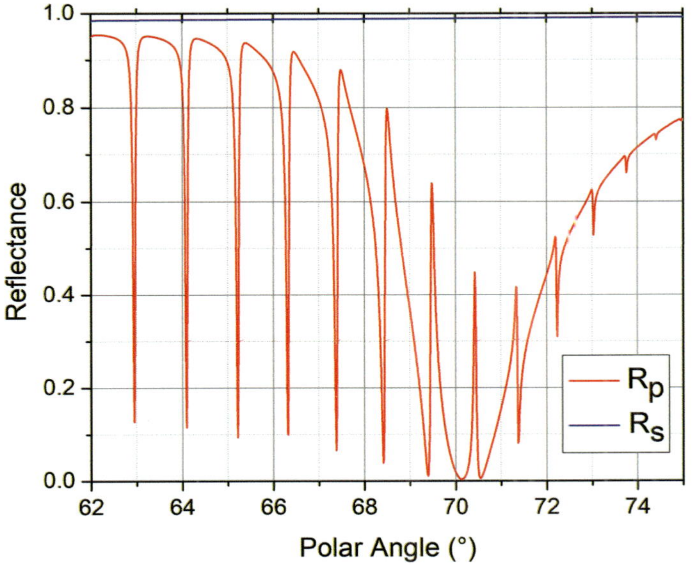

FIGURE 2.25 Reflectance of film stack in Figure 1.38 for applied voltage of 6.48 V across the liquid crystal. Compare to Figure 8b in [8].

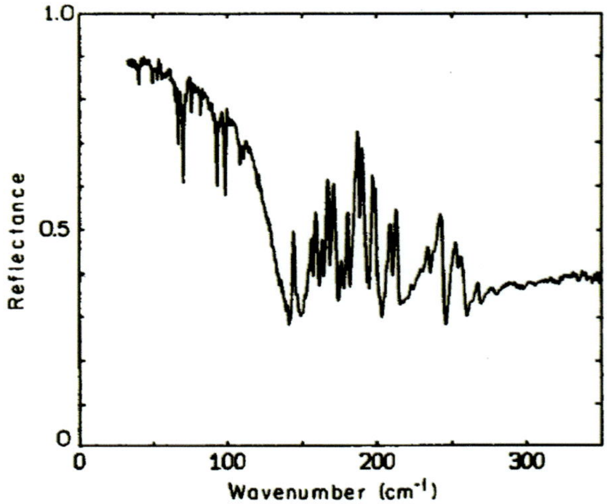

FIGURE 2.26 Reflectance of $NbSe_3$ at 2 K exhibiting many narrow optical phonon absorption lines superimposed on a plasma resonance from a charge density wave energy gap. Note how phonons above the energy gap at ~140 cm^{-1} generate peaks in the reflectance.

Reprinted from *Sol. St. Comm.* **52** W. A. Challener, and P. L. Richards, "Far infrared optical properties of $NbSe_3$," 117–121 (1984) with permission from Elsevier.

REFERENCES

1 D. W. Berreman, "Optics in stratified and anisotropic media: 4 × 4-matrix formulation," *J. Opt. Soc. Am.* **62** (1972) 502–510.
2 P. Yeh, *Optical Waves in Layered Media* (New York: Wiley, 1988).
3 H. A. Macleod, *Thin-Film Optical Filters*, 5th ed. (Boca Raton: CRC Press, 2020).
4 D. Maystre, "A new general integral theory for dielectric coated gratings," *J. Opt. Soc. Am.* **72** (1982) 1385–1392.
5 M. Mansuripur, "Analysis of multilayer thin-film structures containing magneto-optic and anisotropic media at oblique incidence using 2 × 2 matrices," *J. Appl. Phys.* **67** (1990) 6466–6475.
6 E. Kretschmann and H. Raether, "Radiative decay of nonradiative surface plasmons excited by light," *Z. Naturforsch.* **23a** (1968) 2135.
7 D. Sarid and W. Challener, *Modern Introduction to Surface Plasmons* (Cambridge: Cambridge University Press, 2010) 258.
8 K. R. Welford, J. R. Sambles, and M. G. Clark, "Guided modes and surface plasmon-polaritons observed with a nematic liquid crystal using attenuated total reflection," *Liq. Cryst.* **2** (1987) 91–105.
9 E. D. Palik, ed., *Handbook of Optical Constants of Solids* (New York: Academic Press, Inc., 1985) 350–357, 765–769.
10 W. A. Challener and T. A. Rinehart, "Jones matrix analysis of magnetooptical media and read-back systems," *Appl. Opt.* **26** (1987) 3974–3980.
11 M. Born and E. Wolf, *Principles of Optics*, 5th ed. (Oxford: Pergamon Press, 1975) 28.
12 R. F. Wallis and M. Balkanski, *Many-Body Aspects of Solid State Spectroscopy* (New York: North-Holland, 1986) 245.
13 W. A. Challener and S. L. Grove, "Refractive indices of reactive magnetooptical thin films," *Appl. Opt.* **29** (1990) 3040–3045.
14 O. S. Heavens, *Optical Properties of Thin Solid Films* (Mineola, NY: Dover, 1991).

3 Diffraction Gratings

3.1 INTRODUCTION

There are several different techniques for computing the reflected and transmitted diffracted fields from gratings with multilayer conformal thin films, each with its own strengths and weaknesses. The technique described in this chapter was introduced by Chandezon et al. [1] The heart of this technique is a coordinate transformation into a space in which the grating is a planar surface. A Fourier expansion of the fields in this transformed coordinate system is found to converge much more quickly to the correct electric and magnetic fields than the older technique of Fourier expansion in the Cartesian coordinate system. Elston et al. [2] extended the Chandezon technique for reflected diffracted orders in the full conical mount (i.e., any polar and azimuthal angle of incidence), and this chapter primarily follows their derivation. The technique of Elston has also been extended in this chapter to include the calculation of transmitted order intensities. Cotter et al. [3] extended the technique of Chandezon to a scattering matrix approach that avoids losing numerical precision from exceptionally thick films on the grating. Their scattering matrix approach has been further extended in this chapter to the full conical mount. Li [4] provided several additional extensions to the technique and some of his results are also included herein. Various additional references to diffraction theory are found in [5–14].

3.2 TRANSFER MATRIX FORMALISM

3.2.1 Grating Parameters

Consider the multilayer grating in Figure 3.1. The grooves are parallel to the z-axis and the shape of the grooves is defined by the periodic function $y = a(x)$. Each thin film interface j is defined by

$$y_j = a(x) + c_j \tag{3.1}$$

where c_j is a constant specifying the vertical offset of the interface. d_j is the thickness of the jth layer where $j = 0$ is the transmitting medium, and $j = 1$ is the bottom thin film layer. Let there be N total thin film layers so layer $N + 1$ is the incident medium. Assuming a time-dependence of $\exp(-i\omega\, t)$ for the fields, the incident single wavelength field vector is

$$\boldsymbol{F} = \boldsymbol{F}^{(i)} e^{ik\left(x\sin,_i\cos\phi_i - \left(y - y_{N+1}\right)\cos\theta_i + z\sin\theta_i\sin\phi_i\right)} \tag{3.2}$$

where $\boldsymbol{F}$ is either the electric or magnetic field, θ_i is the polar angle of incidence (measured from the vertical y-axis), ϕ_i is the azimuthal angle of incidence (measured from the x-axis), and the incident wavevector is

$$k = n_i k_0. \tag{3.3}$$

The grating period is L, so the grating vector is

$$\boldsymbol{K} \equiv \frac{2\pi}{L}\hat{\boldsymbol{x}}. \tag{3.4}$$

DOI: 10.1201/9781003714231-3

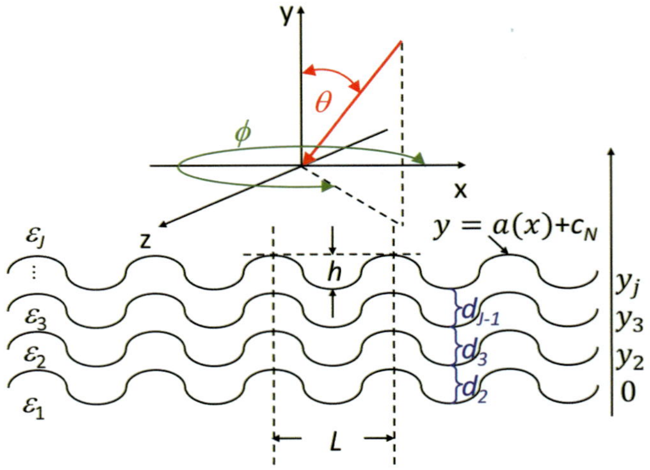

FIGURE 3.1 Multi conformal thin films on a grating surface. Light is incident at polar angle θ_i and azimuthal angle ϕ_i.

$\hat{\boldsymbol{x}}$ is a unit vector along the x-axis. Outside of the grooves, the diffracted field can be expanded in a Fourier series,

$$\boldsymbol{F}_d = \sum_m \boldsymbol{B}_m e^{i(\alpha_m x + \beta_m y + \gamma z)} \tag{3.5}$$

where

$$\alpha_m \equiv n_i k_0 \sin\theta_i \cos\phi_i + mK \tag{3.6}$$

$$\beta_m = \sqrt{(n_i k_0)^2 - \alpha_m^2 - \gamma^2} \tag{3.7}$$

$$\gamma = n_i k_0 \ \sin\theta_i \ \sin\phi_i. \tag{3.8}$$

$\boldsymbol{B}_m$ are the scattered field vectors and n_i is the refractive index of the incident medium. If β_m is real, the mth scattered field is a propagating diffracted order. If β_m is imaginary, then the scattered field is an evanescent order. (3.6)–(3.8) can be solved for the angles of the diffracted modes from the wavevector components.

$$\cos\theta_m = \pm \frac{\beta_m}{n_i k_0} \tag{3.9}$$

$$\tan\phi_m = \frac{\gamma}{\alpha_m}. \tag{3.10}$$

The polar angles of the transmitted fields may differ from those of the incident fields if the refractive indices of the incident and transmitted media are different. Snell's law requires

$$n_i \ \sin\theta_i = n_t \ \sin\theta_t. \tag{3.11}$$

However, the azimuthal angle of the zero[th] order is unchanged throughout the film stack. As a result, the wavevector component along the z-axis parallel to the grooves, γ, is constant throughout the film stack and is determined by the incident wavevector. The wavevector component along the x-axis, α_0, is also constant throughout the film stack by Snell's law, which means that the α_n are the same for each layer and given by (3.6). However, the wavevector components along the y-axis for each transmitted order are

$$\beta_{t,m} = \sqrt{\left(n_t k_0\right)^2 - \alpha_m^2 - \gamma^2}. \tag{3.12}$$

The corresponding angles for the transmitted orders are found from (3.9) and (3.10) in which $\beta_{t,m}$ is substituted for β_m and n_t is substituted for n_i. In the outer media if the modes are evanescent, then β_m and $\cos\theta_m$ are imaginary. In other words, all propagating orders must have a real β value.

3.2.2 Maxwell's Equations in the Transformed Coordinate System

Maxwell's equations (assuming $e^{-i\omega t}$ time-dependence) are

$$\nabla \times \boldsymbol{E} = -\frac{\partial \boldsymbol{B}}{\partial t} = i\omega \boldsymbol{B} = ick_0\mu_0 \boldsymbol{H} = ik_0\eta_0 \boldsymbol{H} \tag{3.13}$$

$$\nabla \times \boldsymbol{H} = \frac{\partial \boldsymbol{D}}{\partial t} = -i\omega\epsilon_j \boldsymbol{E} = -i\frac{k_0}{\sqrt{\epsilon_0\mu_0}} n_j^2\epsilon_0 \boldsymbol{E} = -i\frac{n_j^2 k_0}{\eta_0}\boldsymbol{E} \tag{3.14}$$

where μ_0 is the vacuum permeability, ϵ_0 is the vacuum permittivity, n_j is the refractive index for the jth layer, and

$$\eta_0 = \sqrt{\frac{\mu_0}{\epsilon_0}} \tag{3.15}$$

is the vacuum impedance. For the jth layer, the permittivity ϵ_j is related to the refractive index n_j by

$$\epsilon_j = n_j^2\epsilon_0. \tag{3.16}$$

A nonorthogonal curvilinear coordinate system is defined by

$$v = y - a(x), \tag{3.17}$$

$$u = x, \tag{3.18}$$

and

$$\omega = z. \tag{3.19}$$

In the transformed coordinate frame, the curl of a vector $\boldsymbol{V}$ is given by

$$\nabla \times \boldsymbol{V} = \left(\hat{\varepsilon}^{(1)}\frac{\partial}{\partial u} + \hat{\varepsilon}^{(2)}\frac{\partial}{\partial v} + \hat{\varepsilon}^{(3)}\frac{\partial}{\partial \omega}\right)\left(V_u\hat{\varepsilon}_{(1)} + V_v\hat{\varepsilon}_{(2)} + V_\omega\hat{\varepsilon}_{(3)}\right) \tag{3.20}$$

where

$$\hat{\varepsilon}^{(1)} = \hat{\boldsymbol{x}}, \tag{3.21}$$

$$\hat{\varepsilon}^{(2)} = \hat{\boldsymbol{y}} - a'\hat{\boldsymbol{x}}\,, \tag{3.22}$$

$$\hat{\varepsilon}^{(3)} = \hat{\boldsymbol{z}}\,, \tag{3.23}$$

$$\hat{\varepsilon}_{(1)} = \hat{\boldsymbol{x}} + a'\hat{\boldsymbol{y}}\,, \tag{3.24}$$

$$\hat{\varepsilon}_{(2)} = \hat{\boldsymbol{y}}\,, \tag{3.25}$$

$$\hat{\varepsilon}_{(3)} = \hat{\boldsymbol{z}}\,, \tag{3.26}$$

and

$$a' = \frac{\partial a}{\partial u}\,. \tag{3.27}$$

Therefore,

$$\nabla \times \boldsymbol{V} =$$

$$\begin{aligned}&\left(\frac{\partial V_\omega}{\partial v} - a'\frac{\partial V_u}{\partial \omega} - \frac{\partial V_v}{\partial \omega}\right)\hat{\boldsymbol{x}} + \left(a'\frac{\partial V_\omega}{\partial v} + \frac{\partial V_u}{\partial \omega} - \frac{\partial V_\omega}{\partial u}\right)\hat{\boldsymbol{y}}\\ &+\left(a'\frac{\partial V_u}{\partial u} + a''E_u + \frac{\partial V_v}{\partial u} - \left(1 + a'^2\right)\frac{\partial V_u}{\partial v} - a'\frac{\partial V_v}{\partial v}\right)\hat{\boldsymbol{z}}.\end{aligned} \tag{3.28}$$

Using (3.21)–(3.27) to transform coordinates in Maxwell's equations (3.13) and (3.14), the results are

$$\frac{\partial \boldsymbol{B}}{\partial t} = -ik_0\eta_0\left(H_x\hat{\boldsymbol{x}} + H_y\hat{\boldsymbol{y}} + H_z\hat{\boldsymbol{z}}\right) = -ik_0\eta_0\left[H_u\left(\hat{\boldsymbol{x}} + a'\hat{\boldsymbol{y}}\right) + H_v\hat{\boldsymbol{y}} + H_\omega\hat{\boldsymbol{z}}\right] = -\nabla \times \boldsymbol{E}, \tag{3.29}$$

and

$$\frac{\partial \boldsymbol{D}}{\partial t} = -\frac{in_j^2 k_0}{\eta_0}\left[E_u\left(\hat{\boldsymbol{x}} + a'\hat{\boldsymbol{y}}\right) + E_v\hat{\boldsymbol{y}} + E_\omega\hat{\boldsymbol{z}}\right] = \nabla \times \boldsymbol{H}. \tag{3.30}$$

Applying (3.28) for the curl of the field to the equations in the transformed frame (3.29) and (3.30) and recalling the time-dependence is proportional to $e^{-i\omega t}$ results in six equations.

$$\frac{\partial E_\omega}{\partial v} - a'\frac{\partial E_u}{\partial \omega} - \frac{\partial E_v}{\partial \omega} = ik_0\eta_0 H_u\,, \tag{3.31}$$

$$a'\frac{\partial E_\omega}{\partial v} + \frac{\partial E_u}{\partial \omega} - \frac{\partial E_\omega}{\partial u} = ik_0\eta_0\left(a'H_u + H_v\right), \tag{3.32}$$

$$a'\frac{\partial E_u}{\partial u} + a''E_u + \frac{\partial E_v}{\partial u} - \left(1 + a'^2\right)\frac{\partial E_u}{\partial v} - a'\frac{\partial E_v}{\partial v} = ik_0\eta_0 H_\omega\,, \tag{3.33}$$

$$\frac{\partial H_\omega}{\partial v} - a'\frac{\partial H_u}{\partial \omega} - \frac{\partial H_v}{\partial \omega} = -\frac{in_j^2 k_0}{\eta_0}E_u\,, \tag{3.34}$$

$$a'\frac{\partial H_\omega}{\partial v}+\frac{\partial H_u}{\partial \omega}-\frac{\partial H_\omega}{\partial u}=-\frac{in_j^2 k_0}{\eta_0}\left(a'E_u+E_v\right), \tag{3.35}$$

and

$$a'\frac{\partial H_u}{\partial u}+a''H_u-\left(1+a'^2\right)\frac{\partial H_u}{\partial v}+\left(\frac{\partial}{\partial u}-a'\frac{\partial}{\partial v}\right)H_v=-\frac{in_j^2 k_0}{\eta_0}E_\omega . \tag{3.36}$$

A generalized field vector $\boldsymbol{G}$ that combines electric and magnetic field components is defined:

$$\boldsymbol{G}=\begin{bmatrix}E_u\\E_v\\E_\omega\\H_u\\H_v\\H_\omega\end{bmatrix}. \tag{3.37}$$

Let

$$\tilde{\boldsymbol{C}}\equiv\begin{bmatrix}-a'\dfrac{\partial}{\partial\omega} & -i\gamma & \dfrac{\partial}{\partial v}\\ i\gamma & 0 & a'\dfrac{\partial}{\partial v}-\dfrac{\partial}{\partial u}\\ \left[a'\dfrac{\partial}{\partial u}+a''-\left(1+a'^2\right)\dfrac{\partial}{\partial v}\right] & \left(\dfrac{\partial}{\partial u}-a'\dfrac{\partial}{\partial v}\right) & 0\end{bmatrix} \tag{3.38}$$

and

$$\tilde{\boldsymbol{W}}\equiv\begin{bmatrix}1&0&0\\a'&1&0\\0&0&1\end{bmatrix}. \tag{3.39}$$

Then (3.31)–(3.38) can be written in matrix format,

$$\begin{bmatrix}\tilde{0}&\tilde{\boldsymbol{C}}\\\tilde{\boldsymbol{C}}&\tilde{0}\end{bmatrix}\tilde{\boldsymbol{G}}=\frac{ik_0}{\eta_0}\begin{bmatrix}-n_j^2\tilde{\boldsymbol{W}}&\tilde{0}\\\tilde{0}&\eta_0^2\tilde{\boldsymbol{W}}\end{bmatrix}\tilde{\boldsymbol{G}}. \tag{3.40}$$

This result also recognizes that the fields in the z-direction must have an $e^{-i\omega t}$ or, equivalently, an $e^{+i\gamma z}$ dependence. Therefore, the general field in the transformed coordinate system can be written

$$\boldsymbol{F}\left(u,v,\omega\right)=\boldsymbol{F}\left(u,v\right)e^{i\gamma z}. \tag{3.41}$$

3.2.3 Maxwell's Equations in Terms of Tangential Field Components

It is helpful now to reduce the six field components to four components lying in the plane of the grating. The $H_\omega = H_z$ component lies along the grooves of the grating:

$$E_\omega=E_z, \tag{3.42}$$

and

$$H_\omega = H_z. \tag{3.43}$$

The field components in the surface of the grating and orthogonal to the grooves are

$$E_\parallel = \left(1+a'^2\right)E_u + a'E_v, \tag{3.44}$$

and

$$H_\parallel = \left(1+a'^2\right)H_u + a'H_v. \tag{3.45}$$

The other field components E_u, E_v, H_u, and H_v can be eliminated from (3.31)–(3.36), expressing these equations only in terms of the surface field components E_ω , $E_\parallel$, H_ω , and $H_\parallel$. From (3.44) and (3.45),

$$\frac{\partial E_\parallel}{\partial v} = -ik_0\eta_0 H_\omega + a'\frac{\partial E_u}{\partial u} + a''E_u + \frac{\partial E_v}{\partial u} = -ik_0\eta_0 H_\omega + \frac{\partial}{\partial u}\left(a'E_u + E_v\right), \tag{3.46}$$

and

$$\frac{\partial H_\parallel}{\partial v} = \frac{ik_0 n_j^2}{\eta_0}E_\omega + a'\frac{\partial H_u}{\partial u} + a''H_u + \frac{\partial H_v}{\partial u} = \frac{ik_0 n_j^2}{\eta_0}E_\omega + \frac{\partial}{\partial u}\left(a'H_u + H_v\right). \tag{3.47}$$

Comparing to (3.7) in Elston, it must be true that

$$a'E_u + E_v = \left(\frac{a'}{1+a'^2}\right)E_\parallel - \frac{1}{k_0 n_j^2\left(1+a'^2\right)}\left(\gamma\eta_0 H_\parallel + i\eta_0\frac{\partial H_\omega}{\partial u}\right) \tag{3.48}$$

and comparing to (3.9) in Elston, it must be true that

$$a'H_u + H_v = \left(\frac{a'}{1+a'^2}\right)H_\parallel + \frac{1}{k_0\eta_0\left(1+a'^2\right)}\left(\gamma E_\parallel + i\frac{\partial E_\omega}{\partial u}\right). \tag{3.49}$$

To verify (3.48), $E_\parallel$ and $H_\parallel$ are replaced using (3.44) and (3.45).

$$a'E_u + E_v = a'E_u + \left(\frac{a'^2}{1+a'^2}\right)E_v - \frac{\gamma\eta_0}{k_0 n_j^2}H_u - \frac{\gamma\eta_0 a'}{k_0 n_j^2\left(1+a'^2\right)}H_v - \frac{i\eta_0}{k_0 n_j^2\left(1+a'^2\right)}\frac{\partial H_\omega}{\partial u}, \tag{3.50}$$

which reduces to

$$E_v + \frac{\gamma\eta_0 a'}{k_0 n_j^2}H_v = -\frac{\gamma\eta_0\left(1+a'^2\right)}{k_0 n_j^2}H_u - \frac{i\eta_0}{k_0 n_j^2}\frac{\partial H_\omega}{\partial u}. \tag{3.51}$$

From (3.31),

$$E_v = -a'E_u + \left(\frac{1}{i\gamma}\right)\frac{\partial E_\omega}{\partial v} - \frac{k_0\eta_0}{\gamma}H_u. \tag{3.52}$$

From (3.32),

$$H_v = -a'H_u + \left(\frac{1}{i\gamma}\right)\frac{\partial H_\omega}{\partial v} + \frac{k_0 n_j^2}{\gamma Z_0}E_u. \tag{3.53}$$

Substituting (3.52) and (3.53) into (3.51) and solving for H_u gives

$$H_u = \frac{ik_0 n_j^2}{\eta_0\left(\gamma^2 - k_0^2 n_j^2\right)}\frac{\partial E_\omega}{\partial v} + \frac{i\gamma}{\left(\gamma^2 - k_0^2 n_j^2\right)}\left(a'\frac{\partial H_\omega}{\partial v} - \frac{\partial H_\omega}{\partial u}\right). \tag{3.54}$$

From (3.34) and (3.35),

$$a'E_u + E_v = \frac{i\eta_0}{k_0 n_j^2}\left(a'\frac{\partial H_\omega}{\partial v} - \frac{\partial H_\omega}{\partial u} + i\gamma H_u\right) \tag{3.55}$$

and from (3.52),

$$a'E_u + E_v = \left(\frac{1}{i\gamma}\right)\frac{\partial E_\omega}{\partial v} - \frac{k_0\eta_0}{\gamma}H_u. \tag{3.56}$$

Equating (3.55) and (3.56),

$$\left(-\frac{\gamma\eta_0}{k_0 n_j^2} + \frac{k_0\eta_0}{\gamma}H_u\right) = \left(\frac{1}{i\gamma}\right)\frac{\partial E_\omega}{\partial v} + \frac{i\eta_0}{k_0\epsilon_j}\frac{\partial H_\omega}{\partial u} - \frac{i\eta_0}{k_0\epsilon_j}a'\frac{\partial H_\omega}{\partial v} \tag{3.57}$$

or

$$H_u = \frac{ik_0 n_j^2}{\eta_0\left(\gamma^2 - k_0^2 n_j^2\right)}\frac{\partial E_\omega}{\partial v} + \frac{i\gamma}{\left(\gamma^2 - k_0^2 n_j^2\right)}\left(a'\frac{\partial H_\omega}{\partial v} - \frac{\partial H_\omega}{\partial u}\right). \tag{3.58}$$

(3.54) and (3.58) are identical, thereby verifying the relation (3.48). Hence,

$$\frac{\partial E_\parallel}{\partial v} = -ik_0\eta_0 H_\omega + \frac{\partial}{\partial u}\left(\frac{a'}{1+a'^2}E_\parallel\right) - \frac{\partial}{\partial u}\left(\frac{\gamma\eta_0}{k_0 n_j^2\left(1+a'^2\right)}H_\parallel + \frac{i\eta_0}{k_0 n_j^2\left(1+a'^2\right)}\frac{\partial H_\omega}{\partial u}\right). \tag{3.59}$$

To verify (3.49), we substitute for $E_\parallel$ and $H_\parallel$ from (3.44) and (3.45).

$$k_0\eta_0 H_v = \left(1+a'^2\right)\gamma E_u + \gamma a' E_v + i\frac{\partial E_\omega}{\partial u}. \tag{3.60}$$

Substituting (3.52) and (3.53) into (3.59) and solving for E_u gives

$$E_u = \frac{i\gamma}{\left(\gamma^2 - k_0^2 n_j^2\right)}\left(a'\frac{\partial E_\omega}{\partial v} - \frac{\partial E_\omega}{\partial u}\right) - \frac{ik_0\eta_0}{\left(\gamma^2 - k_0^2 n_j^2\right)}\frac{\partial H_\omega}{\partial v}. \tag{3.61}$$

From (3.32) and (3.33),

$$a'H_u + H_v = \left(\frac{1}{i\gamma}\right)\frac{\partial H_\omega}{\partial v} + \frac{k_0\epsilon_j}{\gamma\eta_0}E_u, \tag{3.62}$$

and

$$a'H_u + H_v = \frac{1}{ik_0\eta_0}\left(a'\frac{\partial E_\omega}{\partial v} - \frac{\partial E_\omega}{\partial u} + i\gamma E_u\right) \tag{3.63}$$

giving

$$E_u = \frac{i\gamma}{\left(\gamma^2 - k_0^2 n_j^2\right)}\left(a'\frac{\partial E_\omega}{\partial v} - \frac{\partial E_\omega}{\partial u}\right) - \frac{ik_0\eta_0}{\left(\gamma^2 - k_0^2 n_j^2\right)}\frac{\partial H_\omega}{\partial v}. \tag{3.64}$$

(3.61) is identical with (3.64), thereby verifying the relation (3.50). Hence,

$$\frac{\partial H_\parallel}{\partial v} = \frac{ik_0 n_j^2}{\eta_0}E_\omega + \frac{\partial}{\partial u}\left(\frac{a'}{1+a'^2}H_\parallel\right) + \frac{\partial}{\partial u}\left(\frac{\gamma}{k_0\eta_0\left(1+a'^2\right)}E_\parallel + \frac{i}{k_0\eta_0\left(1+a'^2\right)}\frac{\partial E_\omega}{\partial u}\right). \tag{3.65}$$

From (3.31),

$$\frac{\partial E_\omega}{\partial v} = i\gamma a'E_u + ik_0\eta_0 H_u + i\gamma E_v. \tag{3.66}$$

Substituting from (3.44) for E_v,

$$a'\frac{\partial E_\omega}{\partial v} = -i\gamma E_u + i\gamma E_\parallel + ik_0\eta_0 a'H_u. \tag{3.67}$$

Substituting for E_u and H_u from (3.57) and (3.63),

$$\frac{\partial H_\omega}{\partial v} = \frac{i\left(\gamma^2 - k_0^2 n_j^2\right)}{k_0\eta_0\left(1+a'^2\right)}E_\parallel - \frac{\gamma}{k_0\eta_0\left(1+a'^2\right)}\frac{\partial E_\omega}{\partial u} + \frac{a'}{\left(1+a'^2\right)}\frac{\partial H_\omega}{\partial u}. \tag{3.68}$$

From (3.34),

$$\frac{\partial H_\omega}{\partial v} = i\gamma a'H_u - \frac{ik_0 n_j^2}{\eta_0}E_u + i\gamma H_v. \tag{3.69}$$

Substituting from (3.45) for H_v and for E_u and H_u from (3.57) and (3.64),

$$\frac{\partial E_\omega}{\partial v} = \frac{a'}{\left(1+a'^2\right)}\frac{\partial E_\omega}{\partial u} - \frac{i\eta_0\left(\gamma^2 - k_0^2 n_j^2\right)}{k_0 n_j^2\left(1+a'^2\right)}H_\parallel + \frac{\gamma\eta_0}{k_0 n_j^2\left(1+a'^2\right)}\frac{\partial H_\omega}{\partial u}. \tag{3.70}$$

(3.59), (3.65), (3.68), and (3.70) are the required field derivatives.

3.2.4 Generalized Field Vector and Eigenvector Equation for Interior Thin Film Layers

There are two periodic functions that now must be expanded in a Fourier series.

$$C(u) \equiv \frac{1}{1+a'^2} = \sum_p C_p e^{ipKu} \tag{3.71}$$

where

$$C_p = \int_0^L C(u) e^{-ipKu} du. \tag{3.72}$$

Also,

$$D_p(u) \equiv \frac{a'}{1+a'^2} = \sum_p D_p e^{ipKu} \tag{3.73}$$

where

$$D_p = \int_0^L D(u) e^{-ipKu} du. \tag{3.74}$$

The groove profile for a sinusoidal grating is

$$a(y) = \frac{h}{2}\left[1-\cos(Kx)\right] = \frac{h}{2}\left[1-\cos(Ku)\right] \tag{3.75}$$

and its derivative is

$$a' = \frac{\partial a}{\partial u} = \frac{Kh}{2}\sin(Ku) = s\sin(Ku) \tag{3.76}$$

where h is the groove depth, and

$$s \equiv \frac{Kh}{2}. \tag{3.77}$$

Then

$$C(u) = \frac{1}{1+s^2\sin^2(Ku)} = \sum_{p=-\infty}^{\infty} C_p e^{ipKu} \tag{3.78}$$

where

$$C_p = \frac{1}{L}\int_0^L \frac{1}{1+s^2\sin^2(Ku)} e^{-ipKu} du = \begin{cases} 0 & p\,\text{odd} \\ \dfrac{1}{\sqrt{1+s^2}}\left(\dfrac{s}{1+\sqrt{1+s^2}}\right)^{|p|} & p\,\text{even} \end{cases}. \tag{3.79}$$

Also,

$$D(u)=\frac{s\sin(Ku)}{1+s^2\sin^2(Ku)}=\sum_{p=-\infty}^{\infty}D_p e^{ipKu}, \tag{3.80}$$

and

$$D_p=\frac{1}{L}\int_0^L D(u)e^{-ipKu}du=\begin{cases}0 & p\,\text{even}\\ sgn(p)\dfrac{-i}{\sqrt{1+s^2}}\left(\dfrac{s}{1+\sqrt{1+s^2}}\right)^{|p|} & p\,\text{odd}\end{cases}. \tag{3.81}$$

The terms C_p and D_p need to be evaluated for twice as many terms as the number of orders used in the calculation. These integral relations are given in Appendix B of [4].

The fields can be expanded in a Fourier series.

$$F(u,\nu)=\sum_m F_m(\nu)e^{i\alpha_m u}. \tag{3.82}$$

The expansions (3.78)–(3.81) are substituted into (3.46).

$$\sum_m \frac{\partial}{\partial v}E_{\|m}(\nu)e^{i\alpha_m u}=-ik_0\eta_0\sum_{m,p}\delta_{mp}H_{\omega p}(\nu)e^{i\alpha_m u}-\frac{\gamma\eta_0}{k_0\epsilon_j}\sum_{m,p}i\alpha_{p+m}C_p e^{i\alpha_{p+m}u}H_{\|m}(\nu)$$

$$+\frac{i\eta_0}{k_0\epsilon_j}\sum_{m,p}\alpha_m\alpha_{p+m}C_p e^{i\alpha_{p+m}u}H_{\omega m}(\nu)+\sum_{m,p}i\alpha_{p+m}D_p e^{i\alpha_{p+m}u}E_{\|m}(\nu) \tag{3.83}$$

where

$$pK+\alpha_m=\alpha_0+pK+mK=\alpha_{p+m}. \tag{3.84}$$

The sums in (3.83) are over all values of m and p. In order to express the partial derivatives of the fields on the left-hand sides of these equations in terms of a sum over all fields on the right-hand side, the dummy variables m and p in the last three terms on the right-hand side are interchanged according to

$$m\to p \tag{3.85}$$

$$p\to m-p \tag{3.86}$$

$$m+p\to m. \tag{3.87}$$

Then the summations over m on both sides of the equations can be dropped and all the $e^{i\alpha_m u}$ terms eliminated. (3.83) becomes

$$-i\frac{\partial}{\partial v}E_{\|m}(\nu)=\sum_P\left[\left(\frac{\eta_0\alpha_p\alpha_m}{k_0n_j^2}\right)C_{m-p}H_{\omega p}(\nu)-(k_0\eta_0)\delta_{mp}H_{\omega p}(\nu)\right.$$
$$\left.-\left(\frac{\gamma\eta_0}{k_0n_j^2}\right)\alpha_m C_{m-p}H_{\|p}(\nu)+\alpha_m D_{m-p}E_{\|p}(\nu)\right]. \tag{3.88}$$

Following the same procedure for (3.47), (3.68), and (3.70), gives [15]

$$-i\frac{\partial}{\partial v}E_{\omega m}(\nu)=$$
$$\sum_p\left[\alpha_p D_{m-p}E_{\omega p}(\nu)+\left(k_0\eta_0-\frac{\gamma^2\eta_0}{k_0 n_j^2}\right)C_{m-p}H_{\parallel p}(\nu)+\left(\frac{\gamma\eta_0}{k_0 n_j^2}\right)\alpha_p C_{m-p}H_{\omega p}(\nu)\right], \tag{3.89}$$

$$-i\frac{\partial}{\partial v}H_{\parallel m}(\nu)=$$
$$\sum_p\left[\delta_{\text{mp}}\left(\frac{k_0 n_j^2}{\eta_0}\right)E_{\omega p}(\nu)-\left(\frac{\alpha_p\alpha_m}{k_0\eta_0}\right)C_{m-p}E_{\omega p}(\nu)+\left(\frac{\gamma}{k_0\eta_0}\right)\alpha_m C_{m-p}E_{\parallel p}(\nu)+\alpha_m D_{m-p}H_{\parallel p}(\nu)\right], \tag{3.90}$$

and

$$-i\frac{\partial}{\partial v}H_{\omega m}(\nu)=$$
$$\sum_p\left[\alpha_p D_{m-p}H_{\omega p}(\nu)+\left(\frac{\gamma^2}{k_0\eta_0}-\frac{k_0 n_j^2}{\eta_0}\right)C_{m-p}E_{\parallel p}(\nu)-\left(\frac{\gamma}{k_0\eta_0}\right)\alpha_p C_{m-p}E_{\omega p}(\nu)\right]. \tag{3.91}$$

In order to evaluate these equations numerically, the infinite sum is truncated at $\pm N$ terms. A general field vector for layer j which includes the parallel and perpendicular components of electric and magnetic fields can now be defined,

$$\boldsymbol{F}^{(j)}(\nu)\equiv\begin{bmatrix}\boldsymbol{E}_{\parallel}\\ \boldsymbol{E}_{\omega}\\ \boldsymbol{H}_{\parallel}\\ \boldsymbol{H}_{\omega}\end{bmatrix} \tag{3.92}$$

where each element of the vector is itself a vector expanded over $\pm N$ diffraction orders. The total number of elements in $\boldsymbol{F}^{(j)}(\nu)$ is $8N+4$. This field vector is formed from field components tangential to the surfaces of each layer and, therefore, the boundary conditions require continuity across the interfaces. Then (3.88)–(3.91) can be written

$$\frac{\partial}{\partial\nu}\boldsymbol{F}^{(j)}(\nu)=i\tilde{\boldsymbol{T}}_j(\nu)\boldsymbol{F}^{(j)}(\nu) \tag{3.93}$$

where for layer j,

$$\tilde{\boldsymbol{T}}^{(j)}(\nu)=\begin{bmatrix}\tilde{\boldsymbol{A}} & \tilde{\boldsymbol{0}} & \tilde{\boldsymbol{C}} & \tilde{\boldsymbol{D}}\\ \tilde{\boldsymbol{0}} & \tilde{\boldsymbol{F}} & \tilde{\boldsymbol{G}} & \tilde{\boldsymbol{H}}\\ \tilde{\boldsymbol{I}} & \tilde{\boldsymbol{J}} & \tilde{\boldsymbol{A}} & \tilde{\boldsymbol{0}}\\ \tilde{\boldsymbol{M}} & \tilde{\boldsymbol{N}} & \tilde{\boldsymbol{0}} & \tilde{\boldsymbol{F}}\end{bmatrix} \tag{3.94}$$

with

$$\tilde{\boldsymbol{A}} \equiv \alpha_m D_{m-p}, \tag{3.95}$$

$$\tilde{\boldsymbol{C}} \equiv -\left(\frac{\gamma\eta_0\alpha_m}{k_0 n_j^2}\right) C_{m-p}, \tag{3.96}$$

$$\tilde{\boldsymbol{D}} \equiv \left(\frac{\eta_0\alpha_p\alpha_m}{k_0 n_j^2}\right) C_{m-p} - \left(k_0\eta_0\right)\delta_{\mathrm{mp}}, \tag{3.97}$$

$$\tilde{\boldsymbol{F}} \equiv \alpha_p D_{m-p}, \tag{3.98}$$

$$\tilde{\boldsymbol{G}} \equiv -\eta_0\left(\frac{\gamma^2 - k_0^2 n_j^2}{k_0 n_j^2}\right) C_{m-p}, \tag{3.99}$$

$$\tilde{\boldsymbol{H}} \equiv \left(\frac{\gamma\eta_0}{k_0 n_j^2}\right)\alpha_p C_{m-p}, \tag{3.100}$$

and

$$\begin{bmatrix} \tilde{\boldsymbol{I}} & \tilde{\boldsymbol{J}} \\ \tilde{\boldsymbol{M}} & \tilde{\boldsymbol{N}} \end{bmatrix} = -\frac{n_j^2}{\eta_0^2}\begin{bmatrix} \tilde{\boldsymbol{A}} & \tilde{\boldsymbol{D}} \\ \tilde{\boldsymbol{G}} & \tilde{\boldsymbol{H}} \end{bmatrix}. \tag{3.101}$$

The matrices $\tilde{\boldsymbol{A}} - \tilde{\boldsymbol{N}}$ are square matrices of rank $2N + 1$. The matrices $\tilde{\boldsymbol{0}}$ are also square matrices of the same rank filled with zeroes. Thus, the matrix $\tilde{\boldsymbol{T}}$ is a square matrix of rank $8N + 4$. The m index indicates row number, while the p index indicates column number. (3.93) is converted into an eigenvector equation for each layer by setting

$$\boldsymbol{F}^{(j)}(\nu) = \sum\nolimits_q b_q^{(j)} \boldsymbol{V}_q^{(j)} e^{ir_q^{(j)}(\nu - y_j)} \tag{3.102}$$

where q is the eigenvector index, $\boldsymbol{V}_q^{(j)}$ are the normalized eigenvectors for layer j, $r_q^{(j)}$ are the associated eigenvalues, and the $b_q^{(j)}$ are the unknown amplitudes of each eigenvector. Inserting (3.102) into (3.93) gives

$$\frac{\partial}{\partial\nu}\boldsymbol{F}^{(j)}(\nu) = i\sum_q r_q^{(j)} b_q^{(j)} \boldsymbol{V}_q^{(j)} e^{ir_q^{(j)}(\nu - y_j)} = i\tilde{\boldsymbol{T}}^{(j)}(\nu)\sum\nolimits_q b_q^{(j)} \boldsymbol{V}_q^{(j)} e^{ir_q^{(j)}(\nu - y_j)} \tag{3.103}$$

which is an eigenvector equation with solution

$$\tilde{\boldsymbol{T}}^{(j)}(\nu)\boldsymbol{V}_q^{(j)} = \boldsymbol{V}_q^{(j)}\,\tilde{\boldsymbol{\rho}}^{(j)}(\nu). \tag{3.104}$$

$\tilde{\boldsymbol{\rho}}^{(j)}(\nu)$ is a phase matrix for layer j given by

$$\tilde{\boldsymbol{\rho}}^{(j)}(\nu) \equiv \begin{bmatrix} e^{ir_1^{(j)}(\nu - y_j)} & 0 & 0 & 0 \\ 0 & e^{ir_2^{(j)}(\nu - y_j)} & 0 & 0 \\ 0 & 0 & \ddots & 0 \\ 0 & 0 & 0 & e^{ir_{8N+4}^{(j)}(\nu - y_j)} \end{bmatrix} \tag{3.105}$$

where the phase is defined relative to the bottom of each layer as shown in Figure 3.1. Therefore, $\tilde{\boldsymbol{\rho}}^{(j)}(\nu)$ is the identity matrix at the bottom of layer j and is the matrix

$$\tilde{\boldsymbol{\rho}}^{(j)}(\nu) = \begin{bmatrix} e^{ir_1^{(j)} d_j} & 0 & 0 & 0 \\ 0 & e^{ir_2^{(j)} d_j} & 0 & 0 \\ 0 & 0 & \ddots & 0 \\ 0 & 0 & 0 & e^{ir_{8N+4}^{(j)} d_j} \end{bmatrix} \tag{3.106}$$

at the top of layer j where d_j is the thickness of that layer.

$\tilde{\boldsymbol{Q}}^{(j)}$ is a square matrix of rank $8N + 4$ for layer j in which each column is an eigenvector $\boldsymbol{V}_q^{(j)}$.

$$\tilde{\boldsymbol{Q}}^{(j)} \equiv \left(\boldsymbol{V}_1^{(j)} \quad \boldsymbol{V}_2^{(j)} \quad \ldots \quad \boldsymbol{V}_{8N+3}^{(j)} \quad \boldsymbol{V}_{8N+4}^{(j)} \right). \tag{3.107}$$

Then for each layer the generalized field vector in layer j is a matrix product,

$$\boldsymbol{F}^{(j)}(\nu) = \sum\nolimits_q \boldsymbol{V}_q^{(j)} e^{ir_q^{(j)}(\nu - y_j)} b_q^{(j)} = \tilde{\boldsymbol{Q}}^{(j)} \tilde{\boldsymbol{\rho}}^{(j)}(\nu) \boldsymbol{b}^{(j)}. \tag{3.108}$$

$\tilde{\boldsymbol{Q}}^{(j)}$ and $\tilde{\boldsymbol{\rho}}^{(j)}(\nu)$ are square matrices of rank $8N + 4$. $\boldsymbol{b}_q^{(j)}$ is a vector of $8N + 4$ elements of the unknown field amplitudes, $b_q^{(j)}$, which are found by matching boundary conditions between layers.

3.2.5 Propagating and Evanescent Fields in the Incident and Transmitted Media

The eigenvector expansion has been determined within each thin film layer as well as the outer two media. However, in the outer two media the fields should be expressed as plane waves, so the plane wave expansion in the transformed coordinate system for each propagating and evanescent mode in the outer media must be determined.

3.2.5.1 Incident Field

The generalized incident field in the transformed coordinate system is

$$\boldsymbol{F}_\omega^{(i)} e^{i\gamma\omega - i\beta_0 v} e^{i\alpha_0 u - i\beta_0 a(u)} \tag{3.109}$$

where the superscript (i) now refers to the incident field. A Fourier series is defined as

$$f(u) = \sum\nolimits_m L_m e^{imKu} \tag{3.110}$$

where

$$L_m = \frac{1}{L} \int_0^L f(u) e^{-imKu} du. \tag{3.111}$$

So

$$e^{-ia(u)\beta_0} = \sum\nolimits_m L_m e^{imKu} \tag{3.112}$$

where

$$L_m(\beta_0) = \frac{1}{L}\int_0^L e^{-ia(u)\beta_0} e^{-imKu} du = \begin{cases} 1 & m = 0, \beta_0 = 0 \\ 0 & m \neq 0, \beta_0 = 0. \\ i^m e^{-\frac{ih\beta_0}{2}} J_m\left(\frac{h\beta_0}{2}\right) & \text{otherwise} \end{cases} \tag{3.113}$$

J_m is the Bessel function of the first kind. This integral relation is provided in Appendix B of [4]. (Note that there is a sign difference in equation (B4) of [4] which may be due to an inverted groove profile.) Therefore,

$$F_\omega = F_\omega^{(i)} e^{i\gamma\omega - i\beta_0 v} e^{i\alpha_0 u} e^{-ia(u)\beta_0} = F_\omega^{(i)} \sum\nolimits_m L_m(\beta_0) e^{i\gamma\omega - i\beta_0 v} e^{i\alpha_m u} \tag{3.114}$$

where

$$\alpha_m \equiv \alpha_0 + mK. \tag{3.115}$$

Differentiating (3.114) with respect to v gives

$$\frac{\partial F_\omega}{\partial \nu} = -i\beta_0 F_\omega^{(i)} e^{i\gamma\omega - i\beta_0 v} e^{i\alpha_0 u - i\beta_0 a(u)} = -i\beta_0 F_\omega = -i\beta_0 F_\omega^{(i)} \sum\nolimits_m L_m(\beta_0) e^{i\gamma\omega - i\beta_0 v} e^{i\alpha_m u}. \tag{3.116}$$

Differentiating the left side of (3.114) with respect to u,

$$\frac{\partial F_\omega}{\partial u} = (i\alpha_0 - i\beta_0 a') F_\omega^{(i)} e^{i\gamma\omega - i\beta_0 v} e^{i\alpha_0 u - i\beta_0 a(u)}$$

$$= i\alpha_0 F_\omega - i\beta_0 a' F_\omega^{(i)} \sum_m L_m(\beta_0) e^{i\gamma\omega - i\beta_0 v} e^{i\alpha_m u}$$

$$= \sum_m (i\alpha_0 - i\beta_0 a') L_m(\beta_0) e^{i\gamma\omega - i\beta_0 v} e^{i\alpha_m u} F_\omega^{(i)}$$

$$= \sum\nolimits_m i\alpha_0 L_m(\beta_0) e^{i\gamma\omega - i\beta_0 v} e^{i\alpha_m u} F_\omega^{(i)} - i\beta_0 a' F_\omega \tag{3.117}$$

and the right side of (3.114),

$$\frac{\partial F_\omega}{\partial u} = iF_\omega^{(i)} \sum_m (\alpha_0 + mK) L_m(\beta_0) e^{i\gamma\omega - i\beta_0 v} e^{i\alpha_m u}$$

$$= i\alpha_0 F_\omega + iF_\omega^{(i)} \sum\nolimits_m mKL_m(\beta_0) e^{i\gamma\omega - i\beta_0 v} e^{i\alpha_m u}. \tag{3.118}$$

Equating (3.117) and (3.118),

$$\beta_0 a' F_\omega = -\sum_m mKL_m(\beta_0) e^{i\gamma\omega - i\beta_0 v} e^{i\alpha_m u} F_\omega^{(i)}. \tag{3.119}$$

Substituting (3.119) back into (3.117),

$$\frac{\partial F_\omega}{\partial u} = \sum_m i\alpha_0 L_m(\beta_0) e^{i\gamma\omega - i\beta_0 v} e^{i\alpha_m u} F_\omega^{(i)} + i\sum_m mKL_m(\beta_0) e^{i\gamma\omega - i\beta_0 v} e^{i\alpha_m u} F_\omega^{(i)} \tag{3.120}$$

or

$$\frac{\partial F_\omega}{\partial u} = i\sum_m (\alpha_0 + mK) L_m(\beta_0) e^{i\gamma\omega - i\beta_0 v} e^{i\alpha_m u} F_\omega^{(i)}. \tag{3.121}$$

Substituting (3.119) into (3.118) also gives

$$\frac{\partial F_\omega}{\partial u} = i(\alpha_0 - \beta_0 a') F_\omega. \tag{3.122}$$

From (3.116) and (3.122),

$$(1 + a'^2)\frac{\partial H_\omega}{\partial v} - a'\frac{\partial H_\omega}{\partial u} = -i(1 + a'^2)\beta_0 H_\omega - ia'(\alpha_0 - \beta_0 a') H_\omega = -i(\beta_0 + a'\alpha_0) H_\omega. \tag{3.123}$$

Solving for $E_\parallel^{(i)}$ from (3.68),

$$E_\parallel^{(i)} = \frac{(1 + a'^2)}{f_1}\frac{\partial H_\omega}{\partial v} - \frac{a'}{f_1}\frac{H_\omega}{\partial u} + \frac{1}{f_1}\left(\frac{\gamma}{k_0\eta_0}\right)\frac{\partial E_\omega}{\partial u} \tag{3.124}$$

where

$$f_1 \equiv \frac{i(\gamma^2 - k_0^2 n_i^2)}{k_0\eta_0}, \tag{3.125}$$

and n_i is the refractive index of the incident medium. Substituting (3.116) and (3.123) into (3.125),

$$E_\parallel^{(i)} = -\frac{i}{f_1}\left(\beta_0 + \beta_0 a'\frac{\alpha_0}{\beta_0}\right) H_\omega + \frac{1}{f_1}\left(\frac{\gamma}{k_0\eta_0}\right)\frac{\partial E_\omega}{\partial u}. \tag{3.126}$$

Substituting (3.114) and (3.122) for the H_ω terms, and (3.123) for the E_ω term,

$$\begin{aligned} E_\parallel^{(i)} &= \frac{i}{f_1}\sum_m \left(\frac{\alpha_0}{\beta_0} mK - \beta_0\right) L_m(\beta_0) e^{i\gamma\omega - i\beta_0 v} e^{i\alpha_m u} H_\omega^{(i)} \\ &+ \frac{i}{f_1}\left(\frac{\gamma}{k_0\eta_0}\right)\sum_m (\alpha_0 + mK) L_m(\beta_0) e^{i\gamma\omega - i\beta_0 v} e^{i\alpha_m u} E_\omega^{(i)} \end{aligned} \tag{3.127}$$

$$= \frac{i}{f_1} \sum_m \left[\left(\frac{\alpha_0}{\beta_0} mK - \beta_0 \right) H_\omega^{(i)} + \left(\frac{\gamma \alpha_m}{k_0 \eta_0} \right) E_\omega^{(i)} \right] L_m(\beta_0) e^{i\gamma\omega - i\beta_0 v} e^{i\alpha_m u}. \tag{3.128}$$

Following the same procedure starting from (3.70) and solving for the incident magnetic field, $H_\parallel^{(i)}$,

$$H_\parallel^{(i)} = \frac{(1+a'^2)}{f_2} \frac{\partial E_\omega}{\partial v} - \left(\frac{a'}{f_2} \right) \frac{E_\omega}{\partial u} - \frac{1}{f_2} \left(\frac{\gamma \eta_0}{k_0 n_i^2} \right) \frac{\partial H_\omega}{\partial u} \tag{3.129}$$

$$= \frac{i}{f_2} \sum_m \left[\left(\frac{\alpha_0}{\beta_0} mK - \beta_0 \right) E_\omega^{(i)} - \left(\frac{\eta_0 \gamma \alpha_m}{k_0 n_i^2} \right) H_\omega^{(i)} \right] L_m(\beta_0) e^{i\gamma\omega - i\beta_0 v} e^{i\alpha_m u} \tag{3.130}$$

where

$$f_2 \equiv i k_0 \eta_0 - \frac{i\gamma^2 \eta_0}{k_0 n_i^2} = \frac{i(k_0^2 n_i^2 - \gamma^2)\eta_0}{k_0 n_i^2} = -\frac{i\eta_0^2}{n_i^2} \left(\frac{\gamma^2 - k_0^2 n_i^2}{k_0 \eta_0} \right) = -\frac{Z_0^2}{n_i^2} f_1. \tag{3.131}$$

Also, from (3.112),

$$E_\omega^{(i)} = \sum_m E_\omega^{(i)} L_m(\beta_0) e^{i\gamma\omega - i\beta_0 v} e^{i\alpha_m u}, \tag{3.132}$$

and

$$H_\omega^{(i)} = \sum_m H_\omega^{(i)} L_m(\beta_0) e^{i\gamma\omega - i\beta_0 v} e^{i\alpha_m u}. \tag{3.133}$$

(3.128), (3.130), (3.132), and (3.133) are the Fourier expansions of the incident field in the transformed coordinate system. The incident field components along the z-axis in terms of s- and p-polarizations can be written (TM, or transverse magnetic, corresponds to p-polarization, and TE, or transverse electric, corresponds to s-polarization),

$$E_\omega^{(i)} = \frac{Z_0}{n_i} (H_{\text{TM}} \sin\phi_i \cos\theta_i + H_{\text{TE}} \cos\phi_i), \tag{3.134}$$

and

$$H_\omega^{(i)} = H_{\text{TM}} \cos\phi_i - H_{\text{TE}} \sin\phi_i \cos\theta_i. \tag{3.135}$$

The refractive index of the incident medium is not included in (11) of [2] but it must be explicitly included in (3.134) when it differs from 1. Inserting (3.134) and (3.135) into (3.128),

$$E_\parallel^{(i)} =$$

$$\frac{i}{f_1} \sum_m \left\{ \left[\left(\frac{\alpha_0}{\beta_0} mK - \beta_0 \right) \cos\phi_i + \left(\frac{\gamma \alpha_m}{n_i k_0} \right) \sin\phi_i \cos\theta_i \right] H_{\text{TM}} \right. \tag{3.136}$$

$$\left. + \left[- \left(\frac{\alpha_0}{\beta_0} mK - \beta_0 \right) \sin\phi_i \cos\theta_i + \left(\frac{\gamma \alpha_m}{n_i k_0} \right) \right] \cos\phi_i H_{\text{TE}} \right\} L_m(\beta_0) e^{i\gamma\omega - i\beta_0 v} e^{i\alpha_m u}$$

Inserting (3.134) into (3.132),

$$E_{\omega}^{(i)} = \sum_m \frac{\eta_0}{n_i}\left(H_{\text{TM}}\sin\phi_i\cos\theta_i + H_{\text{TE}}\cos\phi_i\right)L_m\left(\beta_0\right)e^{i\gamma\omega - i\beta_0 v}e^{i\alpha_m u}. \tag{3.137}$$

Inserting (3.134) and (3.135) into (3.130),

$$\begin{aligned} H_{\parallel}^{(i)} = \frac{i\eta_0}{n_i f_2}\sum_m &\left[\left(\left(\frac{\alpha_0}{\beta_0}mK - \beta_0\right)\sin\phi_i\cos\theta_i - \left(\frac{\gamma\alpha_m}{k_0 n_i}\right)\cos\phi_i\right)H_{\text{TM}} + \left(\left(\frac{\alpha_0}{\beta_0}mK - \beta_0\right)\cos\phi_i\right.\right. \\ &\left.\left. + \left(\frac{\gamma\alpha_m}{k_0 n_i}\right)\sin\phi_i\cos\theta_i\right)H_{\text{TE}}\right]L_m\left(\beta_0\right)e^{i\gamma\omega - i\beta_0 v}e^{i\alpha_m u} \end{aligned}. \tag{3.138}$$

Inserting (3.135) into (3.133),

$$H_{\omega}^{(i)} = \sum_m \left(H_{\text{TM}}\cos\phi_i - H_{\text{TE}}\sin\phi_i\cos\theta_i\right)L_m\left(\beta_0\right)e^{i\alpha_m u}e^{i\gamma\omega - i\beta_0 v}. \tag{3.139}$$

A generalized incident field vector $\tilde{\boldsymbol{F}}_i$ of 4(2N + 1) elements is written,

$$\tilde{\boldsymbol{F}}_i = \begin{bmatrix} \boldsymbol{E}_{\parallel}^i \\ \boldsymbol{E}_{\omega}^i \\ \boldsymbol{H}_{\parallel}^i \\ \boldsymbol{H}_{\omega}^i \end{bmatrix} = \begin{bmatrix} \kappa_{-N}^{(1)} \\ \vdots \\ \kappa_{+N}^{(1)} \\ \kappa_{-N}^{(2)} \\ \vdots \\ \kappa_{+N}^{(2)} \\ \kappa_{-N}^{(3)} \\ \vdots \\ \kappa_{+N}^{(3)} \\ \kappa_{-N}^{(4)} \\ \vdots \\ \kappa_{+N}^{(4)} \end{bmatrix} e^{i\gamma\omega - i\beta_0 v} = \boldsymbol{B}_i e^{i\gamma\omega - i\beta_0 v} \tag{3.140}$$

where

$$\begin{aligned} \kappa_m^{(1)} = \frac{i}{f_1}&\left\{\left[\left(\frac{\alpha_0}{\beta_0}mK - \beta_0\right)\cos\phi_i + \left(\frac{\gamma\alpha_m}{n_i k_0}\right)\sin\phi_i\cos\theta_i\right]H_{\text{TM}}\right. \\ &\left. + \left[-\left(\frac{\alpha_0}{\beta_0}mK - \beta_0\right)\sin\phi_i\cos\theta_i + \left(\frac{\gamma\alpha_m}{n_i k_0}\right)\cos\phi_i\right]H_{\text{TE}}\right\}L_m\left(\beta_0\right)e^{i\alpha_m u}, \end{aligned} \tag{3.141}$$

$$\kappa_m^{(2)} = \frac{\eta_0}{n_i}\left(H_{\text{TM}}^{(i)}\sin\phi_i\cos\theta_i + H_{\text{TE}}^{(i)}\cos\phi_i\right)L_m\left(\beta_0\right)e^{\alpha_m u}, \tag{3.142}$$

$$\kappa_m^{(3)} = \frac{i\eta_0}{n_i f_2}\left\{\left[\left(\frac{\alpha_0}{\beta_0}mK - \beta_0\right)\sin\phi_i\cos\theta_i - \left(\frac{\gamma\alpha_m}{n_i k_0}\right)\cos\phi_i\right]H_{\mathrm{TM}}\right.$$
$$\left. + \left[\left(\frac{\alpha_0}{\beta_0}mK - \beta_0\right)\cos\phi_i + \left(\frac{\gamma\alpha_m}{n_i k_0}\right)\sin\phi_i\cos\theta_i\right]H_{\mathrm{TE}}\right\}L_m(\beta_0)e^{i\alpha_m u}, \tag{3.143}$$

and

$$\kappa_m^{(4)} = \left(H_{\mathrm{TM}}\cos\phi_i - H_{\mathrm{TE}}\sin\phi_i\cos\theta_i\right)L_m(\beta_0)e^{i\alpha_m u}. \tag{3.144}$$

3.2.5.2 Reflected Diffracted Fields

Similar equations can be obtained for the reflected diffracted fields. It should be noted that in [2] only the propagating reflected fields are calculated using the Fourier components technique of the previous section for the incident field. The "reflected" evanescent fields in the incident medium are evaluated in the transformed coordinate system. In the technique described next, the Fourier technique has been used to calculate both the reflected (and transmitted) propagating and evanescent fields which simplifies the algorithm.

It might be asked at this point why the Fourier transform approach used for the incident field is not applied for the calculation of the fields within each interior layer as well, thereby completely avoiding the complexities of the coordinate transformation. The Fourier transform approach is found to converge more slowly and does not accurately provide the field amplitudes within the grooves of the grating so the technique breaks down when the grooves become relatively deep and the evanescent fields become larger in amplitude. In particular, the refractive index within the level of the grooves varies from inside the grooves to outside the grooves, so the wavevector is not constant in this region. The coordinate transform technique, on the other hand, continues to properly calculate the field amplitudes even within the grooves. Deep grooves still require the inclusion of more diffracted orders in the calculation and at some point, the limited computational precision causes the algorithm even in the transformed coordinate system to fail. In [2] all of the evanescent fields in the reflected medium and all of the fields, both evanescent and propagating, in the transmitting medium are calculated in the transformed coordinate system, which should improve the accuracy of the calculation especially for deep grooves. In the second part of this chapter the scattering matrix approach is developed and, in that case, the evanescent field vectors in the transformed coordinate system for the outer media are used.

Let n be the propagating or evanescent reflected field index. No modes that propagate in the downward (transmitted) direction or that grow exponentially as $y \rightarrow +\infty$ are included. Let there be N_r total number of propagating and evanescent modes, for which $\mathrm{Re}(\beta_n) > 0$ or $\mathrm{Im}(\beta_n) > 0$, respectively. There should be $4N + 2$ of these modes. The propagation angles for the nth mode are found from (3.9) and (3.10).

$$\cos\theta_n = \pm\frac{\beta_n}{n_i k_0} \tag{3.145}$$

and

$$\tan\phi_n = \left(\frac{\gamma}{\alpha_n - nK}\right). \tag{3.146}$$

The outgoing reflected diffracted fields are composed of TE and TM modes according to

$$E_{\omega,n}^{(r)} = \frac{\eta_0}{n_i}\left(-\sin\phi_n \cos\theta_n H_{\text{TM}}^{(n)} + \cos\phi_n H_{\text{TE}}^{(n)}\right) \tag{3.147}$$

and

$$H_{\omega,n}^{(r)} = \cos\phi_n H_{\text{TM}}^{(n)} + \sin\phi_n \cos\theta_n H_{\text{TE}}^{(n)}, \tag{3.148}$$

where $H_{\text{TE}}^{(n)}$ and $H_{\text{TM}}^{(n)}$ are the unknown reflected field amplitudes for the nth diffracted order.

Let the propagating and evanescent reflected field amplitudes be defined by a vector with N_r total elements, indexed by n for each reflected mode.

$$\boldsymbol{F}_{\omega}^{(r)} = \begin{bmatrix} F_{\omega,1} \\ F_{\omega,2} \\ \vdots \\ F_{\omega,N_r} \end{bmatrix}. \tag{3.149}$$

The Fourier expansion of these modes is

$$F_{\omega}^{(r)} = \sum_{n=1}^{N_r} F_{\omega,n}^{(r)} \sum_{m=-N}^{N} L_{m-n}\left(-\beta_n\right) e^{i\left(\alpha_m u + \beta_n v - \gamma\omega\right)} \tag{3.150}$$

where $N \to \infty$. Differentiating,

$$\frac{\partial F_{\omega}^{(r)}}{\partial v} = i\sum_n \beta_n F_{\omega,n}^{(r)} \sum_m L_{m-n}\left(-\beta_n\right) e^{i\left(\alpha_m u + \beta_n v + \gamma\omega\right)}, \tag{3.151}$$

$$\frac{\partial F_{\omega}^{(r)}}{\partial u} = i\sum_n F_{\omega,n}^{(r)} \sum_{m,n} \alpha_m L_{m-n}\left(-\beta_n\right) e^{i\left(\alpha_m u + \beta_n v + \gamma\omega\right)} \tag{3.152}$$

where

$$L_{m-n}\left(-\beta_n\right) = \frac{1}{L}\int_0^L e^{ia(u)\beta_n} e^{i(m-n)Ku}\, du = \begin{cases} 1 & m = n,\ \beta_n = 0 \\ 0 & m \neq n,\ \beta_n = 0\,. \\ i^{m-n} e^{\frac{ih\beta_n}{2}} J_{m-n}\left(-\frac{h\beta_n}{2}\right) & \text{otherwise} \end{cases} \tag{3.153}$$

By comparison to (3.117) we also have

$$\beta_n a' \sum_{m=-N}^{N} L_{m-n}\left(-\beta_n\right) e^{i\gamma\omega + i\beta_n v} e^{i\alpha_m u} = \sum_{m=-N}^{N} (m-n) K L_{m-n}\left(-\beta_n\right) e^{i\gamma\omega + i\beta_n v} e^{i\alpha_m u}. \tag{3.154}$$

Inserting (3.151) and (3.152) into (3.68),

$$E_{\parallel} = \frac{\left(1 + a'^2\right)}{f_1} i\sum_n \beta_n H_{\omega,n}^{(r)} \sum_m L_{m-n}\left(-\beta_n\right) e^{i\left(\alpha_m u + \beta_n v + \gamma\omega\right)} -$$

$$\frac{a'}{f_1} i\sum_n H_{\omega,n}^{(r)} \sum_m \alpha_m L_{m-n}\left(-\beta_n\right) e^{i\left(\alpha_m u + \beta_n v + \gamma\omega\right)}$$

$$+\frac{\gamma}{f_1 k_0 Z_0} i \sum_n E_{\omega,n}^{(r)} \sum_m \alpha_m L_{m-n}(-\beta_n) e^{i(\alpha_m u+\beta_n v+\gamma\omega)} \tag{3.155}$$

which simplifies to

$$E_{\|} = \frac{i}{f_1} \sum_{n,m} \left[(1+a'^2)\beta_n H_{\omega,n}^{(r)} - a'\alpha_m H_{\omega,n}^{(r)} + \frac{\gamma}{k_0 Z_0} E_{\omega,n}^{(r)} \alpha_m \right] L_{m-n}(-\beta_n) e^{i(\alpha_m u+\beta_n v+\gamma\omega)}. \tag{3.156}$$

Recalling that from (3.154),

$$\alpha_m = \alpha_n + (m-n)K, \tag{3.157}$$

then

$$E_{\|} = \frac{i}{f_1} \sum_{n,m} \Big\{ \big[(1+a'^2)\beta_n - a'[\alpha_n + (m-n)K] \big] H_{\omega,n}^{(r)} + \frac{\gamma}{k_0 Z_0} [\alpha_n + (m-n)K] E_{\omega,n}^{(r)} \Big\} L_{m-n}(-\beta_n) e^{i(\alpha_m u+\beta_n v+\gamma\omega)} \tag{3.158}$$

or

$$\begin{aligned} E_{\|} &= \frac{i}{f_1} \sum_{n,m} (1+a'^2)\beta_n L_{m-n}(-\beta_n) e^{i(\alpha_m u+\beta_n v+\gamma\omega)} H_{\omega,n}^{(r)} \\ &- \frac{i}{f_1} \sum_{n,m} a'\alpha_n L_{m-n}(-\beta_n) e^{i(\alpha_m u+\beta_n v+\gamma\omega)} H_{\omega,n}^{(r)} \\ &- \frac{i}{f_1} a' \sum_{n,m} (m-n) K L_{m-n}(-\beta_n) e^{i(\alpha_m u+\beta_n v+\gamma\omega)} H_{\omega,n}^{(r)} \\ &+ \frac{i}{f_1} \sum_{n,m} \frac{\gamma}{k_0 \eta_0} [\alpha_n + (m-n)K] L_{m-n}(-\beta_n) e^{i(\alpha_m u+\beta_n v+\gamma\omega)} E_{\omega,n}^{(r)}. \end{aligned} \tag{3.159}$$

Substituting for the *(m-n) K* $H_{\omega,n}^{(r)}$ (third) term on the right-hand side of this equation from (3.154),

$$\begin{aligned} E_{\|} &= \frac{i}{f_1} \sum_{n,m} \beta_n L_{m-n}(-\beta_n) e^{i(\alpha_m u+\beta_n v+\gamma\omega)} H_{\omega,n}^{(r)} \\ &- \frac{i}{f_1} \sum_n \frac{\alpha_n}{\beta_n} \beta_n a' \sum_m L_{m-n}(-\beta_n) e^{i(\alpha_m u+\beta_n v+\gamma\omega)} H_{\omega,n}^{(r)} \\ &+ \frac{i}{f_1} \sum_{n,m} \frac{\gamma}{k_0 \eta_0} [\alpha_n + (m-n)K] L_{m-n}(-\beta_n) e^{i(\alpha_m u+\beta_n v+\gamma\omega)} E_{\omega,n}^{(r)}. \end{aligned} \tag{3.160}$$

Again, substituting from (3.154) into the second term on the right-hand side,

$$E_{\parallel} = \frac{i}{f_1}\sum_{n,m}\beta_n L_{m-n}\left(-\beta_n\right)e^{i\left(\alpha_m u+\beta_n v+\gamma\omega\right)}H^{(r)}_{\omega,n}$$

$$-\frac{i}{f_1}\sum\nolimits_{n,m}\frac{\alpha_n}{\beta_n}\left(m-n\right)KL_{m-n}\left(-\beta_n\right)e^{i\gamma\omega+i\beta_n v}e^{i\alpha_m u}H^{(r)}_{\omega,n} \tag{3.161}$$

$$+\frac{i}{f_1}\sum\nolimits_{n,m}\frac{\gamma}{k_0\eta_0}\left[\alpha_n+\left(m-n\right)K\right]L_{m-n}\left(-\beta_n\right)e^{i\left(\alpha_m u+\beta_n v+\gamma\omega\right)}E^{(r)}_{\omega,n}.$$

Consolidating,

$$E^{(r)}_{\parallel} = \begin{cases} \dfrac{i}{f_1}\sum\nolimits_{n}\dfrac{\gamma}{k_0\eta_0}\alpha_n e^{i\left(\alpha_n u+\gamma\omega\right)}E^{(r)}_{\omega,n} & \beta_n=0, m=n \\ 0 & \beta_n=0, m\neq n. \\ \dfrac{i}{f_1}\sum\nolimits_{n,m}\left\{\left[\beta_n-\dfrac{\alpha_n}{\beta_n}\left(m-n\right)K\right]H^{(r)}_{\omega,n}+\dfrac{\gamma}{k_0\eta_0}\left[\alpha_n+\left(m-n\right)K\right]E^{(r)}_{\omega,n}\right\} & \text{otherwise} \\ \quad\times L_{m-n}\left(-\beta_n\right)e^{i\left(\alpha_m u+\beta_n v+\gamma\omega\right)} & \end{cases} \tag{3.162}$$

Following a similar procedure starting with (3.70) gives the corresponding equation for the magnetic field,

$$H^{(r)}_{\parallel} = \begin{cases} -\dfrac{i}{f_2}\sum\nolimits_{n}\left(\dfrac{\eta_0\gamma}{k_0\epsilon_i}\right)\alpha_n H^{(r)}_{\omega}e^{i\left(\alpha_m u+\gamma\omega\right)} & \beta_n=0, m=n \\ 0 & \beta_n=0, m\neq n \\ \dfrac{i}{f_2}\sum\nolimits_{n,m}\left[\left(\beta_n-\dfrac{\alpha_n}{\beta_n}\left(m-n\right)K\right)E^{(r)}_{\omega}-\left(\dfrac{\eta_0\gamma}{k_0 n_i^2}\right)\left(\alpha_n+\left(m-n\right)K\right)H^{(r)}_{\omega}\right] & \text{otherwise} \\ \quad\times L_{m-n}\left(-\beta_n\right)e^{i\left(\alpha_m u+\beta_n v+\gamma\omega\right)} & \end{cases}. \tag{3.163}$$

The final two equations for the tangential fields come directly from (3.147) and (3.148),

$$E^{(r)}_{\omega} = \sum\nolimits_{n,m}E^{(r)}_{\omega,n}L_{m-n}\left(-\beta_n\right)e^{i\left(\alpha_m u+\beta_n v+\gamma\omega\right)}, \tag{3.164}$$

and

$$H^{(r)}_{\omega} = \sum\nolimits_{n,m}H^{(r)}_{\omega,n}L_{m-n}\left(-\beta_n\right)e^{i\left(\alpha_m u+\beta_n v+\gamma\omega\right)}. \tag{3.165}$$

Therefore, (3.162)–(3.165) are the desired Fourier expansions of the propagating reflected fields in terms of the field amplitudes in the transformed coordinate system. Substituting (3.147) and (3.148) into (3.162) gives

$$E_{\parallel}^{(r)} = \frac{i}{f_1}\sum\nolimits_{n,m}\left\{\left[\left(\beta_n - \frac{\alpha_n}{\beta_n}(m-n)K\right)\left(\cos\phi_n H_{\mathrm{TM}}^{(n)} + \sin\phi_n\cos\theta_n H_{\mathrm{TE}}^{(n)}\right)\right.\right.$$

$$\left.\left. + \frac{\gamma}{k_0\eta_0}\left[\alpha_n + (m-n)K\right]\frac{\eta_0}{n_i}\left(-\sin\phi_n\cos\theta_n H_{\mathrm{TM}}^{(n)} + \cos\phi_n H_{\mathrm{TE}}^{(n)}\right)\right]\right\} L_{m-n}(-\beta_n)e^{i(\alpha_m u+\beta_n v+\gamma\omega)} \tag{3.166}$$

or

$$E_{\parallel,n}^{(r)} = \sum\nolimits_m \left(\eta_{m,n}^{(1)} H_{\mathrm{TM}}^{(n)} + \chi_{m,n}^{(1)} H_{\mathrm{TE}}^{(n)}\right) \tag{3.167}$$

where

$$\chi_{m,n}^{(1)} \equiv \begin{cases} \dfrac{i}{f_1}\left(\dfrac{\gamma\alpha_n}{n_i k_0}\right) cos\phi_n e^{i(\alpha_m u+\gamma\omega)} & \beta_n = 0, m = n \\ 0 & \beta_n = 0, m \neq n. \\ \dfrac{i}{f_1}\left[\left(\beta_n - \dfrac{\alpha_n}{\beta_n}(m-n)K\right)\sin\phi_n\cos\theta_n + \left(\dfrac{\gamma\alpha_m}{n_i k_0}\right)\cos\phi_n\right] L_{m-n}(-\beta_n)e^{i(\alpha_m u+\beta_n v+\gamma\omega)} & \text{otherwise} \end{cases} \tag{3.168}$$

$$\eta_{m,n}^{(1)} \equiv \begin{cases} -\dfrac{i}{f_1}\left(\dfrac{\gamma\alpha_n}{n_i k_0}\right)\sin\phi_n\cos\theta_n e^{i(\alpha_m u+\gamma\omega)} & \beta_n = 0, m = n \\ 0 & \beta_n = 0, m \neq n. \\ \dfrac{i}{f_1}\left[\left(\beta_n - \dfrac{\alpha_n}{\beta_n}(m-n)K\right)\cos\phi_n - \left(\dfrac{\gamma\alpha_m}{n_i k_0}\right)\sin\phi_n\cos\theta_n\right] L_{m-n}(-\beta_n)e^{i(\alpha_m u+\beta_n v+\gamma\omega)} & \text{otherwise} \end{cases} \tag{3.169}$$

This equation can be written in matrix format,

$$\boldsymbol{E}_{\parallel}^{(r)} = \tilde{\boldsymbol{M}}_{E\parallel}^{(r)}\boldsymbol{B}_r \tag{3.170}$$

where $\boldsymbol{E}_{\acute{E}}^{(r)}$ is the vector of reflected diffracted propagating and evanescent fields with $4N + 2$ elements,

$$\tilde{\boldsymbol{M}}_{E\parallel}^{(r)} \equiv \begin{bmatrix} \chi_{-N,n_1}^{(1)} & \chi_{-N,n_2}^{(1)} & \cdots & \chi_{-N,n_{N_r}}^{(1)} & \eta_{-N,n_1}^{(1)} & \eta_{-N,n_2}^{(1)} & \cdots & \eta_{-N,n_{N_r}}^{(1)} \\ \chi_{-N+1,n_1}^{(1)} & \chi_{-N+1,n_2}^{(1)} & \cdots & \chi_{-N+1,n_{N_r}}^{(1)} & \eta_{-N+1,n_1}^{(1)} & \eta_{-N+1,n_2}^{(1)} & \cdots & \eta_{-N+1,n_{N_r}}^{(1)} \\ \vdots & \vdots & \vdots & \vdots & \vdots & \vdots & \vdots & \vdots \\ \chi_{+N,n_1}^{(1)} & \chi_{+N,n_2}^{(1)} & \cdots & \chi_{+N,n_{N_r}}^{(1)} & \eta_{+N,n_1}^{(1)} & \eta_{+N,n_2}^{(1)} & \cdots & \eta_{+N,n_{N_r}}^{(1)} \end{bmatrix} \tag{3.171}$$

is a matrix with $2N + 1$ rows and $4N + 2$ columns, and

$$\boldsymbol{B}_r = \begin{bmatrix} H_{\mathrm{TE}}^{(1)} \\ H_{\mathrm{TE}}^{(2)} \\ \vdots \\ H_{\mathrm{TE}}^{(N_r)} \\ H_{\mathrm{TM}}^{(1)} \\ H_{\mathrm{TM}}^{(2)} \\ \vdots \\ H_{\mathrm{TM}}^{(N_r)} \end{bmatrix} \tag{3.172}$$

is an amplitude vector with $4N + 2$ elements for the unknown reflected field amplitudes, one for each of the propagating and evanescent modes. The elements of $\tilde{\boldsymbol{M}}_{E\omega}^{(r)}$ are taken from (3.168) and (3.169) where the m is the row number index and n is the column number index.

Similarly, substituting (3.147) and (3.148) into (3.164) gives

$$H_{\parallel,n}^{(r)} = \sum\nolimits_m \left(\eta_{m,n}^{(2)} H_{\mathrm{TM}}^{(n)} + \chi_{m,n}^{(2)} H_{\mathrm{TE}}^{(n)} \right) \tag{3.173}$$

where

$$\chi_{m,n}^{(2)} \equiv \begin{cases} -\dfrac{i\eta_0}{n_i f_2}\left(\dfrac{\gamma \alpha_n}{n_i k_0}\right)\sin\phi_n \cos\theta_n e^{i(\alpha_m u + \gamma\omega)} & \beta_n = 0, m = n \\ 0 & \beta_n = 0, m \neq n \\ \dfrac{i\eta_0}{n_i f_2}\left[\left(\beta_n - \dfrac{\alpha_n}{\beta_n}(m-n)K\right)\cos\phi_n - \left(\dfrac{\gamma}{n_i k_0}\right)\left[\alpha_n + (m-n)K\right]\sin\phi_n \cos\theta_n\right] & \text{otherwise} \\ \times L_{m-n}\left(-\beta_n\right) e^{i(\alpha_m u + \beta_n v + \gamma\omega)} & \end{cases} \tag{3.174}$$

$$\eta_{m,n}^{(2)} \equiv \begin{cases} -\dfrac{i\eta_0}{n_i f_2}\left(\dfrac{\gamma \alpha_n}{n_i k_0}\right)\cos\phi_n e^{i(\alpha_m u + \gamma\omega)} & \beta_n = 0, m = n \\ 0 & \beta_n = 0, m \neq n \\ -\dfrac{i\eta_0}{n_i f_2}\left[\left(\beta_n - \dfrac{\alpha_n}{\beta_n}(m-n)K\right)\sin\phi_n \cos\theta_n + \left(\dfrac{\gamma}{n_i k_0}\right)\left[\alpha_n + (m-n)K\right]\cos\phi_n\right] & \text{otherwise} \\ \times L_{m-n}\left(-\beta_n\right) e^{i(\alpha_m u + \beta_n v + \gamma\omega)} & \end{cases} \tag{3.175}$$

or

$$\boldsymbol{H}_{\parallel}^{(r)} = \tilde{\boldsymbol{M}}_{H\parallel}^{(r)} \boldsymbol{B}_r \tag{3.176}$$

where

$$\tilde{\boldsymbol{M}}_{H\|}^{(r)} \equiv \begin{bmatrix} \chi_{-N,n_1}^{(2)} & \chi_{-N,n_2}^{(2)} & \cdots & \chi_{-N,n_{N_r}}^{(2)} & \eta_{-N,n_1}^{(2)} & \eta_{-N,n_2}^{(2)} & \cdots & \eta_{-N,n_{N_r}}^{(2)} \\ \chi_{-N+1,n_1}^{(2)} & \chi_{-N+1,n_2}^{(2)} & \cdots & \chi_{-N+1,n_{N_r}}^{(2)} & \eta_{-N+1,n_1}^{(2)} & \eta_{-N+1,n_2}^{(2)} & \cdots & \eta_{-N+1,n_{N_r}}^{(2)} \\ \vdots & \vdots & \vdots & \vdots & \vdots & \vdots & \vdots & \vdots \\ \chi_{+N,n_1}^{(2)} & \chi_{+N,n_2}^{(2)} & \cdots & \chi_{+N,n_{N_r}}^{(2)} & \eta_{+N,n_1}^{(2)} & \eta_{+N,n_2}^{(2)} & \cdots & \eta_{+N,n_{N_r}}^{(2)} \end{bmatrix} \tag{3.177}$$

Substituting (3.148) into (3.167) gives

$$E_{\omega}^{(n)} = \sum\nolimits_{m} \left(\eta_{m,n}^{(3)} H_{\mathrm{TM}}^{(n)} + \chi_{m,n}^{(3)} H_{\mathrm{TE}}^{(n)} \right) \tag{3.178}$$

and

$$\boldsymbol{E}_{\omega}^{(r)} = \tilde{\boldsymbol{M}}_{E,\omega}^{(r)} \boldsymbol{B}_r \tag{3.179}$$

where

$$\tilde{\boldsymbol{M}}_{E,\omega}^{(r)} \equiv \begin{bmatrix} \chi_{-N,n_1}^{(3)} & \chi_{-N,n_2}^{(3)} & \cdots & \chi_{-N,n_{N_r}}^{(3)} & \eta_{-N,n_1}^{(3)} & \eta_{-N,n_2}^{(3)} & \cdots & \eta_{-N,n_{N_r}}^{(3)} \\ \chi_{-N+1,n_1}^{(3)} & \chi_{-N+1,n_2}^{(3)} & \cdots & \chi_{-N+1,n_{N_r}}^{(3)} & \eta_{-N+1,n_1}^{(3)} & \eta_{-N+1,n_2}^{(3)} & \cdots & \eta_{-N+1,n_{N_r}}^{(3)} \\ \vdots & \vdots & \vdots & \vdots & \vdots & \vdots & \vdots & \vdots \\ \chi_{+N,n_1}^{(3)} & \chi_{+N,n_2}^{(3)} & \cdots & \chi_{+N,n_{N_r}}^{(3)} & \eta_{+N,n_1}^{(3)} & \eta_{+N,n_2}^{(3)} & \cdots & \eta_{+N,n_{N_r}}^{(3)} \end{bmatrix}, \tag{3.180}$$

and the elements of $\tilde{\boldsymbol{M}}_{E,\omega}^{(r)}$ are

$$\chi_{m,n}^{(3)} \equiv \frac{\eta_0}{n_i} \cos\phi_n L_{m-n}\left(-\beta_n\right) e^{i\left(\alpha_m u + \beta_n v + \gamma\omega\right)} \tag{3.181}$$

and

$$\eta_{m,n}^{(3)} \equiv -\frac{\eta_0}{n_i} \sin\phi_n \cos\theta_n L_{m-n}\left(-\beta_n\right) e^{i\left(\alpha_m u + \beta_n v + \gamma\omega\right)}. \tag{3.182}$$

Substituting (3.148) into (3.165) gives

$$H_{\omega}^{(r)} = \sum\nolimits_{n,m} H_{\omega,n}^{(r)} L_{m-n}\left(-\beta_n\right) e^{i\left(\alpha_m u + \beta_n v + \gamma\omega\right)}. \tag{3.183}$$

$$H_{\omega}^{(n)} = \sum\nolimits_{m} \left(\eta_{m,n}^{(4)} H_{\mathrm{TM}}^{(n)} + \chi_{m,n}^{(4)} H_{\mathrm{TE}}^{(n)} \right), \tag{3.184}$$

and

$$\boldsymbol{H}_{\omega}^{(d)} = \tilde{\boldsymbol{M}}_{H,\omega}^{(r)} \boldsymbol{B}_r \tag{3.185}$$

where

$$\tilde{\boldsymbol{M}}_{H,\omega}^{(r)} \equiv \begin{bmatrix} \chi_{-N,n_1}^{(4)} & \chi_{-N,n_2}^{(4)} & \cdots & \chi_{-N,n_{N_r}}^{(4)} & \eta_{-N,n_1}^{(4)} & \eta_{-N,n_2}^{(4)} & \cdots & \eta_{-N,n_{N_r}}^{(4)} \\ \chi_{-N+1,n_1}^{(4)} & \chi_{-N+1,n_2}^{(4)} & \cdots & \chi_{-N+1,n_{N_r}}^{(4)} & \eta_{-N+1,n_1}^{(4)} & \eta_{-N+1,n_2}^{(4)} & \cdots & \eta_{-N+1,n_{N_r}}^{(4)} \\ \vdots & \vdots & \vdots & \vdots & \vdots & \vdots & \vdots & \vdots \\ \chi_{+N,n_1}^{(4)} & \chi_{+N,n_2}^{(4)} & \cdots & \chi_{+N,n_{N_r}}^{(4)} & \eta_{+N,n_1}^{(4)} & \eta_{+N,n_2}^{(4)} & \cdots & \eta_{+N,n_{N_r}}^{(4)} \end{bmatrix} \tag{3.186}$$

and

$$\chi_{m,n}^{(4)} \equiv \sin\phi_n \cos\theta_n L_{m-n}\left(-\beta_n\right) e^{i(\alpha_m u+\beta_n v+\gamma\omega)}, \tag{3.187}$$

$$\eta_{m,n}^{(4)} \equiv \cos\phi_n L_{m-n}\left(-\beta_n\right) e^{i(\alpha_m u+\beta_n v+\gamma\omega)}. \tag{3.188}$$

Combining (3.170), (3.176), (3.179), and (3.185) into a single matrix equation gives finally the generalized reflected diffracted field vector,

$$\boldsymbol{F}^{(r)} = \begin{bmatrix} \boldsymbol{E}_{\|}^{(r)} \\ \boldsymbol{E}_{\omega}^{(r)} \\ \boldsymbol{H}_{\|}^{(r)} \\ \boldsymbol{H}_{\omega}^{(r)} \end{bmatrix} = \begin{bmatrix} \tilde{\boldsymbol{M}}_{E\|}^{(r)} \\ \tilde{\boldsymbol{M}}_{E,\omega}^{(r)} \\ \tilde{\boldsymbol{M}}_{H\|}^{(r)} \\ \tilde{\boldsymbol{M}}_{H,\omega}^{(r)} \end{bmatrix} \boldsymbol{B}_r = \tilde{\boldsymbol{M}}_r \boldsymbol{B}_r \tag{3.189}$$

where $\tilde{\boldsymbol{M}}_r$ is a matrix having $8N + 4$ rows and $4N + 2$ columns and $\boldsymbol{B}_r$ contains $4N + 2$ unknown TE and TM magnetic field amplitudes according to (3.172).

3.2.5.3 Diffracted Transmitted Fields

The generalized field vector with $8N + 4$ unknown amplitudes in each layer has been previously determined. In the transmitting medium, these modes propagate with the factor $e^{-in_t k_0 y \cos\theta_{t,n}}$. If the imaginary part of $\cos\theta_{t,n}$ is positive, then the modes decay exponentially as $y \to -\infty$. If the imaginary part of $\cos\theta_{t,n}$ is negative, then these modes grow exponentially as $y \to -\infty$, and are unphysical in the transmitted medium. There will be $4N + 2$ downward propagating or evanescent modes in the transmitted medium. If $\cos\theta_{t,n}$ (or β_n) is real, then the mode is a propagating mode. Hence, the diffracted transmitted and evanescent modes must all have $\mathrm{Im}(\beta_n) \geq 0$.

By Snell's law, α_n and γ are independent of layer number and are given by (3.6) and (3.8), respectively. Therefore, the transmitted azimuthal angles ϕ_n for each order n are the same as the reflected azimuthal angles. However, the wavevector along the y-axis is different which also changes the polar angles of the transmitted orders.

$$\beta_n^{(t)} = \sqrt{\left(n_t k_0\right)^2 - \alpha_n^2 - \gamma^2} \tag{3.190}$$

and

$$\cos\theta_{t,n} = \pm\frac{\beta_n^{(t)}}{n_t k_0}. \tag{3.191}$$

The general field for a propagating transmitted mode may be expanded as

$$F_{\omega}^{(t)}=\sum_{n}F_{\omega,n}^{(t)}\sum_{m}L_{m-n}\left(+\beta_{n}^{(t)}\right)e^{i\left(\alpha_{m}u-\beta_{n}^{(t)}\nu+\gamma\omega\right)} \tag{3.192}$$

where

$$L_{m-n}^{(t)}\left(+\beta_{n}^{(t)}\right)=\frac{1}{L}\int_{0}^{L}e^{-ia(u)\beta_{n}^{(t)}}e^{-i(m-n)Ku}du$$

$$=\begin{cases}1 & m=n,\beta_{n}^{(t)}=0\\ 0 & m\neq n,\beta_{n}^{(t)}=0\,.\\ i^{m-n}e^{-\frac{ih\beta_{n}^{(t)}}{2}}J_{m-n}\left(+\frac{h\beta_{n}^{(t)}}{2}\right) & \text{otherwise}\end{cases} \tag{3.193}$$

From (3.192),

$$\frac{\partial F_{\omega}^{(t)}}{\partial\nu}=-i\sum_{n}\beta_{n}^{(t)}F_{\omega,n}^{(t)}\sum_{m}L_{m-n}\left(+\beta_{n}^{(t)}\right)e^{i\left(\alpha_{m}u-\beta_{n}^{(t)}\nu+\gamma\omega\right)} \tag{3.194}$$

and

$$\frac{\partial F_{\omega}^{(t)}}{\partial u}=i\sum_{n}F_{\omega,n}^{(t)}\sum_{m,n}\alpha_{m}L_{m-n}\left(+\beta_{n}^{(t)}\right)e^{i\left(\alpha_{m}u-\beta_{n}^{(t)}\nu+\gamma\omega\right)}. \tag{3.195}$$

By comparison to (3.119) and (3.153),

$$\begin{aligned}&\beta_{n}^{(t)}a'\sum_{m=-N}^{N}L_{m-n}\left(+\beta_{n}^{(t)}\right)e^{i\gamma\omega-i\beta_{n}^{(t)}v}e^{i\alpha_{m}u}\\ &=-\sum_{m=-N}^{N}(m-n)KL_{m-n}\left(+\beta_{n}^{(t)}\right)e^{i\gamma\omega-i\beta_{n}^{(t)}v}e^{i\alpha_{m}u}.\end{aligned} \tag{3.196}$$

Substituting (3.194) and (3.195) into (3.68) and solving for $E_{\parallel}^{(t)}$,

$$\begin{aligned}E_{\parallel}^{(t)}=&-\frac{\left(1+a'^{2}\right)}{f_{t1}}i\sum_{n}\beta_{n}^{(t)}H_{\omega,n}^{(t)}\sum_{m}L_{m-n}\left(+\beta_{n}^{(t)}\right)e^{i\left(\alpha_{m}u-\beta_{n}^{(t)}v+\gamma\omega\right)}\\ &-\frac{a'}{f_{t1}}i\sum_{n}H_{\omega,n}^{(t)}\sum_{m,n}\alpha_{m}L_{m-n}\left(+\beta_{n}^{(t)}\right)e^{i\left(\alpha_{m}u-\beta_{n}^{(t)}v+\gamma\omega\right)}\\ &+\frac{1}{f_{t1}}\left(\frac{\gamma}{k_{0}\eta_{0}}\right)i\sum_{n}E_{\omega,n}^{(t)}\sum_{m,n}\alpha_{m}L_{m-n}\left(+\beta_{n}^{(t)}\right)e^{i\left(\alpha_{m}u-\beta_{n}^{(t)}v+\gamma\omega\right)}\end{aligned} \tag{3.197}$$

where

$$f_{t1} = \frac{i\left(\gamma^2 - n_t^2 k_0^2\right)}{k_0 \eta_0}. \tag{3.198}$$

Substituting for α_m from (3.157),

$$\begin{aligned} E_{\parallel}^{(t)} = &-\frac{\left(1+a'^2\right)}{f_{t1}} i \sum_n \beta_n^{(t)} H_{\omega,n}^{(t)} \sum_m L_{m-n}\left(+\beta_n^{(t)}\right) e^{i\left(\alpha_m u - \beta_n^{(t)} v + \gamma\omega\right)} \\ &-\frac{a'}{f_{t1}} i \sum_n H_{\omega,n}^{(t)} \sum_{m,n} \left[\alpha_n + (m-n)K\right] L_{m-n}\left(+\beta_n^{(t)}\right) e^{i\left(\alpha_m u - \beta_n^{(t)} v + \gamma\omega\right)} \\ &+\frac{1}{f_{t1}}\left(\frac{\gamma}{k_0\eta_0}\right) i \sum_n E_{\omega,n}^{(t)} \sum_{m,n} \left[\alpha_n + (m-n)K\right] L_{m-n}\left(+\beta_n^{(t)}\right) e^{i\left(\alpha_m u - \beta_n^{(t)} v + \gamma\omega\right)}. \end{aligned} \tag{3.199}$$

Substituting (3.192) into the second term on the right-hand side of this equation,

$$\begin{aligned} E_{\parallel}^{(t)} = &-\frac{i}{f_{t1}} \sum_n H_{\omega,n}^{(t)} \beta_n^{(t)} \sum_m L_{m-n}\left(+\beta_n^{(t)}\right) e^{i\left(\alpha_m u - \beta_n^{(t)} v + \gamma\omega\right)} \\ &-\frac{i}{f_{t1}} \sum_n \alpha_n H_{\omega,n}^{(t)} a' \sum_{m,n} L_{m-n}\left(+\beta_n^{(t)}\right) e^{i\left(\alpha_m u - \beta_n^{(t)} v + \gamma\omega\right)} \\ &+\frac{i}{f_{t1}} \sum_n \left(\frac{\gamma}{k_0\eta_0}\right) E_{\omega,n}^{(t)} \sum_{m,n} \left[\alpha_n + (m-n)K\right] L_{m-n}\left(+\beta_n^{(t)}\right) e^{i\left(\alpha_m u - \beta_n^{(t)} v + \gamma\omega\right)}, \end{aligned} \tag{3.200}$$

and again substituting (3.192) into the second term on the right-hand side of this equation,

$$\begin{aligned} E_{\parallel}^{(t)} = &-\frac{i}{f_{t1}} \sum_{n,m} H_{\omega,n}^{(t)} \beta_n^{(t)} L_{m-n}\left(+\beta_n^{(t)}\right) e^{i\left(\alpha_m u - \beta_n^{(t)} v + \gamma\omega\right)} \\ &+\frac{i}{f_{t1}} \sum_{n,m} \frac{\alpha_n}{\beta_n^{(t)}} H_{\omega,n}^{(t)} (m-n) K L_{m-n}\left(+\beta_n^{(t)}\right) e^{i\gamma\omega - i\beta_n^{(t)} v} e^{i\alpha_m u} \\ &+\frac{i}{f_{t1}} \sum_{n,m} \left(\frac{\gamma}{k_0\eta_0}\right) E_{\omega,n}^{(t)} \left[\alpha_n + (m-n)K\right] L_{m-n}\left(+\beta_n^{(t)}\right) e^{i\left(\alpha_m u - \beta_n^{(t)} v + \gamma\omega\right)}. \end{aligned} \tag{3.201}$$

Simplifying,

$$E_{\parallel}^{(t)} =$$

$$\begin{cases} \dfrac{i}{f_{t1}}\sum_n \dfrac{\gamma}{k_0\eta_0}\alpha_n e^{i(\alpha_n u+\gamma\omega)}E^{(t)}_{\omega,n} & \beta^{(t)}_n = 0, m=n \\ 0 & \beta^{(t)}_n = 0, m\neq n \\ \dfrac{i}{f_{t1}}\sum_{n,m}\left\{\left[-\beta^{(t)}_n + \dfrac{\alpha_n}{\beta^{(t)}_n}(m-n)K\right]H^{(t)}_{\omega,n} + \dfrac{\gamma}{k_0\eta_0}\left[\alpha_n + (m-n)K\right]E^{(t)}_{\omega,n}\right\} & \text{otherwise} \\ \times L_{m-n}\left(+\beta^{(t)}_n\right)e^{i\left(\alpha_m u-\beta^{(t)}_n v+\gamma\omega\right)} & \end{cases}. \quad (3.202)$$

By the same procedure starting from (3.70) and solving for $H^{(t)}_{\parallel}$,

$$H^{(t)}_{\parallel} = \begin{cases} -\dfrac{i}{f_{t2}}\sum_n \left(\dfrac{\eta_0\gamma}{k_0 n_t^2}\right)\alpha_n H^{(t)}_{\omega} e^{i(\alpha_m u+\gamma\omega)} & \beta_n = 0, m=n \\ 0 & \beta_n = 0, m\neq n \\ \dfrac{i}{f_{t2}}\sum_{n,m}\left\{\left[-\beta^{(t)}_n + \dfrac{\alpha_n}{\beta^{(t)}_n}(m-n)K\right]E^{(t)}_{\omega,n} - \left(\dfrac{\eta_0\gamma}{n_t^2 k_0}\right)\left[\alpha_n + (m-n)K\right]H^{(t)}_{\omega,n}\right\} & \text{otherwise} \\ \times L_{m-n}\left(+\beta^{(t)}_n\right)e^{i\left(\alpha_m u-\beta^{(t)}_n v+\gamma\omega\right)} & \end{cases} \quad (3.203)$$

where

$$f_{t2} \equiv -\frac{\eta_0^2}{n_t^2} f_{t1}. \quad (3.204)$$

The final two equations for the tangential fields are simply taken from (3.147) and (3.148),

$$E^{(t)}_{\omega} = \sum_{n,m} E^{(t)}_{\omega,n} L_{m-n}\left(+\beta^{(t)}_n\right)e^{i\left(\alpha_m u-\beta^{(t)}_n v+\gamma\omega\right)} \quad (3.205)$$

$$H^{(t)}_{\omega} = \sum_{n,m} H^{(t)}_{\omega,n} L_{m-n}\left(+\beta^{(t)}_n\right)e^{i\left(\alpha_m u-\beta^{(t)}_n v+\gamma\omega\right)}. \quad (3.206)$$

The transmitted fields in vector format are

$$\boldsymbol{F}_t = \begin{bmatrix} \boldsymbol{E}^{(t)}_{\parallel} \\ \boldsymbol{E}^{(t)}_{\omega} \\ \boldsymbol{H}^{(t)}_{\parallel} \\ \boldsymbol{H}^{(t)}_{\omega} \end{bmatrix}. \quad (3.207)$$

There are $2N + 1$ terms in each expansion for which the fields do not grow exponentially as $y \to -\infty$, for which $\beta^{(t)} \geq 0$.

The transmitted fields in terms of s- and p-polarization can be written as sums over the diffracted orders,

$$E_{\omega,n}^{(t)} = \frac{\eta_0}{n_t}\left(\sin\phi_{t,n}\cos\theta_{t,n}H_{\mathrm{TM}}^{t} + \cos\phi_{t,n}H_{\mathrm{TE}}^{t}\right) \tag{3.208}$$

$$H_{\omega,n}^{(t)} = \cos\phi_{t,n}H_{\mathrm{TM}}^{t} - \sin\phi_{t,n}\cos\theta_{t,n}H_{\mathrm{TE}}^{t} \tag{3.209}$$

where, as previously found,

$$\alpha_n = n_i k_0 \sin\theta_i \cos\phi_i + nK = n_t k_0 \sin\theta_t \cos\phi_i + nK\ , \tag{3.210}$$

$$\gamma = n_i k_0 \sin\theta_i \sin\phi_i = n_t k_0 \sin\theta_t \sin\phi_i\ , \tag{3.211}$$

$$\beta_n^{(t)} = \sqrt{(n_i k_0)^2 - \alpha_n^2 - \gamma^2}\ , \tag{3.212}$$

$$\cos\theta_{t,n} = \pm\frac{\beta_n^{(t)}}{n_t k_0}\ , \tag{3.213}$$

and

$$\tan\phi_n = \frac{\gamma}{\alpha_n}. \tag{3.214}$$

Let $\boldsymbol{B}_t$ be the unknown amplitude vector for the $4N + 2$ propagating and evanescent transmitted modes of both polarizations.

$$\boldsymbol{B}_t = \begin{bmatrix} H_{\mathrm{TE}}^{(t_1)} \\ H_{\mathrm{TE}}^{(t_2)} \\ \vdots \\ H_{\mathrm{TE}}^{(t_{N_t})} \\ H_{\mathrm{TM}}^{(t_1)} \\ H_{\mathrm{TM}}^{(t_2)} \\ \vdots \\ H_{\mathrm{TM}}^{(t_{N_t})} \end{bmatrix}. \tag{3.215}$$

The complete generalized transmitted field vector with $8N + 4$ elements is then given by

$$\boldsymbol{F}^{(t)} = \tilde{\boldsymbol{M}}_t \boldsymbol{B}_t. \tag{3.216}$$

$\tilde{\boldsymbol{M}}_t$ is a rectangular matrix with $8N + 4$ rows and $4N + 2$ columns. To find $\tilde{\boldsymbol{M}}_t$, (3.208) and (3.209) are inserted into (3.202). Consolidating the unknown parameters $E_{\mathrm{TM}}^{(n)}$ and $H_{\mathrm{TM}}^{(n)}$,

$$E_{\parallel}^{(t)} = \frac{i}{f_{t1}} \sum\nolimits_{n,m} \left\{ \left[\left[-\beta_n^{(t)} + \frac{\alpha_n}{\beta_n^{(t)}} (m-n) K \right] \cos\phi_{t,n} + \left(\frac{\gamma}{n_t k_0} \right) \left[\alpha_n + (m-n) K \right] \sin\phi_{t,n} \cos\theta_{t,n} \right] H_{\mathrm{TM}}^{(t)} \right.$$

$$\left. + \left[\left(\frac{\gamma}{n_t k_0} \right) \left[\alpha_n + (m-n) K \right] \cos\phi_{t,n} + \left[\beta_n^{(t)} - \frac{\alpha_n}{\beta_n^{(t)}} (m-n) K \right] \sin\phi_{t,n} \cos\theta_{t,n} \right] H_{\mathrm{TE}}^{(t)} \right\} \tag{3.217}$$

$$L_{m-n}\left(+\beta_n^{(t)}\right) e^{i\left(\alpha_m u - \beta_n^{(t)} v + \gamma w\right)}$$

Therefore,

$$E_{\parallel,n}^{(t)} = \sum\nolimits_m \left(\tau_{m,n}^{(1)} H_{\mathrm{TM},n}^{(t)} + \upsilon_{m,n}^{(1)} H_{\mathrm{TE},n}^{(t)} \right) \tag{3.218}$$

where

$$\tau_{m,n}^{(1)} = \begin{cases} \dfrac{i}{f_{t1}} \left(\dfrac{\gamma}{n_t k_0} \right) \alpha_n \sin\phi_{t,n} \cos\theta_{t,n} e^{i(\alpha_m u + \gamma w)} & m = n, \beta_n^{(t)} = 0 \\ 0 & m \neq n, \beta_n^{(t)} = 0 \\ \dfrac{i}{f_{t1}} \left[\left(-\beta_n^{(t)} + \dfrac{\alpha_n}{\beta_n^{(t)}} (m-n) K \right) \cos\phi_{t,n} + \left(\dfrac{\gamma}{n_t k_0} \right) \left[\alpha_n + (m-n) K \right] \sin\phi_{t,n} \cos\theta_{t,n} \right] & \text{otherwise} \\ \times L_{m-n}\left(+\beta_n^{(t)}\right) e^{i\left(\alpha_m u - \beta_n^{(t)} v + \gamma w\right)} & \end{cases} \tag{3.219}$$

$$\upsilon_{m,n}^{(1)} = \begin{cases} \dfrac{i}{f_{t1}} \left(\dfrac{\gamma}{n_t k_0} \right) \alpha_n \sin\phi_{t,n} \cos\theta_{t,n} e^{i(\alpha_m u + \gamma w)} & m = n, \beta_n^{(t)} = 0 \\ 0 & m \neq n, \beta_n^{(t)} = 0 \\ \dfrac{i}{f_{t1}} \left[\left(-\beta_n^{(t)} + \dfrac{\alpha_n}{\beta_n^{(t)}} (m-n) K \right) \cos\phi_{t,n} + \left(\dfrac{\gamma}{n_t k_0} \right) \left[\alpha_n + (m-n) K \right] \sin\phi_{t,n} \cos\theta_{t,n} \right] & \text{otherwise} \\ \times L_{m-n}\left(+\beta_n^{(t)}\right) e^{i\left(\alpha_m u - \beta_n^{(t)} v + \gamma w\right)} & \end{cases} \tag{3.220}$$

Therefore,

$$\boldsymbol{E}_{\parallel}^{(t)} = \tilde{\boldsymbol{M}}_{E\parallel}^{(t)} \boldsymbol{B}_t \tag{3.221}$$

where $\boldsymbol{E}_{\parallel}^{(t)}$ is the vector of the transmitted diffracted propagating and evanescent fields with $4N + 2$ elements.

$$\boldsymbol{B}_t \equiv \begin{bmatrix} H_{\mathrm{TE}}^{(-N)} \\ \vdots \\ H_{\mathrm{TE}}^{(+N)} \\ H_{\mathrm{TM}}^{(-N)} \\ \vdots \\ H_{\mathrm{TM}}^{(+N)} \end{bmatrix} \tag{3.222}$$

is also a vector of 4*N* + 2 elements, and

$$\tilde{\boldsymbol{M}}_{E\|}^{(t)} \equiv \begin{bmatrix} \upsilon_{-N,-N}^{(1)} & \cdots & \upsilon_{-N,+N}^{(1)} & \tau_{-N,-N}^{(1)} & \cdots & \tau_{-N,+N}^{(1)} \\ \vdots & \vdots & \vdots & \vdots & \vdots & \vdots \\ \upsilon_{+N,-N}^{(1)} & \cdots & \upsilon_{+N,+N}^{(1)} & \tau_{+N,-N}^{(1)} & \cdots & \tau_{+N,+N}^{(1)} \end{bmatrix} \tag{3.223}$$

is a matrix with 2*N* + 1 rows and 4*N* + 2 columns. The *m* index in (3.219) and (3.220) provides the row number of the elements of $\tilde{\boldsymbol{M}}_{E\|}^{(t)}$ while the *n* index provides the column number.

Similarly, inserting (3.208) and (3.209) into (3.206) and consolidating the unknown parameters $E_{\mathrm{TM}}^{(n)}$ and $H_{\mathrm{TM}}^{(n)}$,

$$H_{\|}^{(t)} =$$

$$\frac{i}{f_2}\sum_{n,m}\Bigg\{\frac{\eta_0}{n_t}\left(-\beta_n^{(t)}+\frac{\alpha_n}{\beta_n^{(t)}}(m-n)K\right)\sin\phi_{t,n}\cos\theta_{t,n}-\left(\frac{\eta_0\gamma}{n_t k_0\epsilon_t}\right)\left[\alpha_n+(m-n)K\right]\cos\phi_{t,n}\,H_{\mathrm{TM}}^{(t)}$$

$$+\left[\frac{\eta_0}{n_t}\left(-\beta_n^{(t)}+\frac{\alpha_n}{\beta_n^{(t)}}(m-n)K\right)\cos\phi_{t,n}+\left(\frac{\eta_0\gamma}{n_t k_0\epsilon_t}\right)\left[\alpha_n+(m-n)K\right]\sin\phi_{t,n}\cos\theta_{t,n}\right]H_{\mathrm{TE}}^{(t)}\Bigg\}$$

$$\cdot L_{m-n}\left(+\beta_n^{(t)}\right)e^{i\left(\alpha_m u-\beta_n^{(t)}v+\gamma w\right)}. \tag{3.224}$$

So

$$H_{\|,n}^{(t)} = \sum_m\left(\tau_{m,n}^{(2)}H_{\mathrm{TM}}^{(t)}+\upsilon_{m,n}^{(2)}H_{\mathrm{TE}}^{(t)}\right) \tag{3.225}$$

where

$$\tau_{m,n}^{(2)} =$$

$$\begin{cases} -\dfrac{i\gamma\eta_0}{n_t^2 k_0 f_{t2}}\alpha_n\cos\phi_{t,n}e^{i(\alpha_m u+\gamma w)} & m\neq n,\beta_n^{(t)}=0 \\ 0 & m\neq n,\beta_n^{(t)}=0 \\ \left\{\dfrac{i\eta_0}{n_t f_{t2}}\left[-\left(\beta_n^{(t)}-\dfrac{a_n}{\beta_n^{(t)}}(m-n)K\right)\sin\phi_{t,n}\cos\theta_{t,n}\right.\right. & \text{otherwise} \\ \left.\left.-\left(\dfrac{\gamma}{n_t k_0}\right)\left[\alpha_n+(m-n)K\right]\cos\phi_{t,n}\right]L_{m-n}\left(+\beta_n^{(t)}\right)e^{i\left(\alpha_m u-\beta_n^{(t)}v+\gamma w\right)}\right\} & \end{cases} \tag{3.226}$$

$$\upsilon_{m,n}^{(2)} =$$

$$\begin{cases} \dfrac{i\eta_0}{n_t f_{t2}}\left(\dfrac{\gamma}{n_t k_0}\right)\alpha_n \sin\phi_{t,n}\cos\theta_{t,n} e^{i(\alpha_m u+\gamma\omega)} & m=n, \beta_n^{(t)}=0 \\ 0 & m=n, \beta_n^{(t)}=0 \\ \left\{\dfrac{i\eta_0}{n_t f_{t2}}\left[-\left(\beta_n^{(t)}-\dfrac{\alpha_n}{\beta_n^{(t)}}(m-n)K\right)\cos\phi_{t,n} \right.\right. & \text{otherwise} \\ \left.\left.+\left(\dfrac{\gamma}{n_t k_0}\right)\left[\alpha_n+(m-n)K\right]\sin\phi_{t,n}\cos\theta_{t,n}\right] L_{m-n}\left(+\beta_n^{(t)}\right)e^{i\left(\alpha_m u-\beta_n^{(t)}v+\gamma\omega\right)}\right\} & \end{cases} \tag{3.227}$$

The matrix equation is

$$\boldsymbol{H}_{\parallel}^{(t)} = \tilde{\boldsymbol{M}}_t \boldsymbol{B}_t \tag{3.228}$$

where $\boldsymbol{H}_{\parallel}^{(t)}$ is a magnetic field amplitude vector with $4N + 2$ elements for the unknown transmitted magnetic field amplitudes, one for each of the propagating and evanescent modes. The elements of $\tilde{\boldsymbol{M}}_{H\parallel}^{(t)}$ are taken from (3.226) and (3.227) where m is the row number index and n is the column number index.

$$\tilde{\boldsymbol{M}}_{H\parallel}^{(t)} \equiv \begin{bmatrix} \upsilon_{-N,-N}^{(2)} & \cdots & \upsilon_{-N,+N}^{(2)} & \tau_{-N,-N}^{(2)} & \cdots & \tau_{-N,+N}^{(2)} \\ \vdots & \vdots & \vdots & \vdots & \vdots & \vdots \\ \upsilon_{+N,-N}^{(2)} & \cdots & \upsilon_{+N,+N}^{(2)} & \tau_{+N,-N}^{(2)} & \cdots & \tau_{+N,+N}^{(2)} \end{bmatrix}. \tag{3.229}$$

Inserting (3.208) into (3.205),

$$\boldsymbol{E}_{\omega}^{(t)} = \sum_{n,m}\frac{\eta_0}{n_t}\left(\sin\phi_{t,n}\cos\theta_{t,n}H_{\text{TM}}^{t} + \cos\phi_{t,n}H_{\text{TE}}^{t}\right)L_{m-n}\left(+\beta_n^{(t)}\right)e^{i\left(\alpha_m u-\beta_n^{(t)}v+\gamma\omega\right)}$$

$$\equiv \sum\nolimits_{n,m}\left(\tau_{m,n}^{(3)}H_{\text{TM}}^{(t)} + \upsilon_{m,n}^{(3)}H_{\text{TE}}^{(t)}\right) \tag{3.230}$$

so

$$\boldsymbol{E}_{\omega}^{(t)} = \tilde{\boldsymbol{M}}_{E,\omega}^{(t)}\boldsymbol{B}_t \tag{3.231}$$

where

$$\tilde{\boldsymbol{M}}_{E,\omega}^{(t)} \equiv \begin{bmatrix} \upsilon_{-N,-N}^{(3)} & \cdots & \upsilon_{-N,+N}^{(3)} & \tau_{-N,-N}^{(3)} & \cdots & \tau_{-N,+N}^{(3)} \\ \vdots & \vdots & \vdots & \vdots & \vdots & \vdots \\ \upsilon_{+N,-N}^{(3)} & \cdots & \upsilon_{+N,+N}^{(3)} & \tau_{+N,-N}^{(3)} & \cdots & \tau_{+N,+N}^{(3)} \end{bmatrix} \tag{3.232}$$

is a matrix of $2N + 1$ rows and $4N + 2$ columns, and

$$\tau_{m,n}^{(3)} = \frac{\eta_0}{n_t}\sin\phi_{t,n}\cos\theta_{t,n}L_{m-n}\left(+\beta_n^{(t)}\right)e^{i\left(\alpha_m u-\beta_n^{(t)}v+\gamma\omega\right)}, \tag{3.233}$$

$$\upsilon_{m,n}^{(3)} = \frac{\eta_0}{n_t}\cos\phi_{t,n} L_{m-n}\left(+\beta_n^{(t)}\right) e^{i\left(\alpha_m u - \beta_n^{(t)} v + \gamma\omega\right)}. \tag{3.234}$$

Inserting (3.209) into (3.206),

$$\begin{aligned} \boldsymbol{H}_\omega^{(t)} &= \sum\nolimits_{n,m}\left(\cos\phi_{t,n} H_{\mathrm{TM}}^t - \sin\phi_{t,n}\cos\theta_{t,n} H_{\mathrm{TE}}^t\right) L_{m-n}\left(+\beta_n^{(t)}\right) e^{i\left(\alpha_m u - \beta_n^{(t)} v + \gamma\omega\right)} \\ &\equiv \sum\nolimits_m \left(\tau_{m,n}^{(4)} H_{\mathrm{TM}}^{(t)} + \upsilon_{m,n}^{(4)} H_{\mathrm{TE}}^{(t)}\right) \end{aligned} \tag{3.235}$$

so

$$\boldsymbol{H}_\omega^{(t)} = \tilde{\boldsymbol{M}}_{H,\omega}^{(t)} \boldsymbol{B}_t \tag{3.236}$$

where

$$\tilde{\boldsymbol{M}}_{H,\omega}^{(t)} \equiv \begin{bmatrix} \upsilon_{-N,-N}^{(4)} & \cdots & \upsilon_{-N,+N}^{(4)} & \tau_{-N,-N}^{(4)} & \cdots & \tau_{-N,+N}^{(4)} \\ \vdots & \vdots & \vdots & \vdots & \vdots & \vdots \\ \upsilon_{+N,-N}^{(4)} & \cdots & \upsilon_{+N,+N}^{(4)} & \tau_{+N,-N}^{(4)} & \cdots & \tau_{+N,+N}^{(4)} \end{bmatrix} \tag{3.237}$$

is a matrix of $2N + 1$ rows and $4N + 2$ columns, and

$$\tau_{m,n}^{(4)} = \cos\phi_{t,n} L_{m-n}\left(+\beta_n^{(t)}\right) e^{i\left(\alpha_m u - \beta_n^{(t)} v + \gamma\omega\right)}, \tag{3.238}$$

and

$$\upsilon_{m,n}^{(4)} = -\sin\phi_{t,n}\cos\theta_{t,n} L_{m-n}\left(+\beta_n^{(t)}\right) e^{i\left(\alpha_m u - \beta_n^{(t)} v + \gamma\omega\right)}. \tag{3.239}$$

Combining (3.221), (3.228), (3.231), and (3.236) gives finally the generalized transmitted diffracted field vector,

$$\boldsymbol{F}^{(t)} = \begin{bmatrix} \boldsymbol{E}_\parallel^{(t)} \\ \boldsymbol{E}_\omega^{(t)} \\ \boldsymbol{H}_\parallel^{(t)} \\ \boldsymbol{H}_\omega^{(t)} \end{bmatrix} = \begin{bmatrix} E_{\parallel,-N}^{(t)} \\ \vdots \\ E_{\parallel,+N}^{(t)} \\ E_{\omega,-N}^{(t)} \\ \vdots \\ E_{\omega,+N}^{(t)} \\ H_{\parallel,-N}^{(t)} \\ \vdots \\ H_{\parallel,+N}^{(t)} \\ H_{\omega,-N}^{(t)} \\ \vdots \\ H_{\omega,+N}^{(t)} \end{bmatrix} = \begin{bmatrix} \tilde{\boldsymbol{M}}_{E\parallel}^{(t)} \\ \tilde{\boldsymbol{M}}_{E,\omega}^{(t)} \\ \tilde{\boldsymbol{M}}_{H\parallel}^{(t)} \\ \tilde{\boldsymbol{M}}_{H,\omega}^{(t)} \end{bmatrix} \begin{bmatrix} H_{\mathrm{TE},-N}^{(t)} \\ \vdots \\ H_{\mathrm{TE},+N}^{(t)} \\ H_{\mathrm{TM},-N}^{(t)} \\ \vdots \\ H_{\mathrm{TM},+N}^{(t)} \end{bmatrix} = \tilde{\boldsymbol{M}}_t \boldsymbol{B}_t \tag{3.240}$$

where $\tilde{M}_t$ is a matrix having 8*N* + 4 rows and 4*N* + 2 columns.

3.2.6 Total Fields in Outer Media

For the upper medium of the incident field, the total generalized field vector including incident, reflected, and evanescent fields is

$$F_N(\nu) = F_i + \tilde{M}_r B_r. \tag{3.241}$$

F_i is the generalized incident field vector in (3.140). It has 8*N* + 4 elements. $\tilde{M}_r$ is the matrix from (3.189) and B_r is the generalized vector of 8*N* + 4 unknown propagating or evanescent reflected field amplitudes in the upper medium.

The total fields in the transmitted medium are

$$F^{(t)}(\nu) = \tilde{M}_t B_t \tag{3.242}$$

where $\tilde{M}_t$ is the matrix in (3.240) with 8*N* + 4 rows and 4*N* + 2 columns, and B_t is the vector of the 4*N* + 2 unknown transmitted propagating and evanescent field amplitudes.

The generalized field vectors for the remaining interior layers are given by (3.108),

$$F^{(j)}(\nu) = \tilde{Q}^{(j)} \tilde{\rho}^{(j)}(\nu) b^{(j)}. \tag{3.243}$$

$\tilde{Q}^{(j)}$ and $\tilde{\rho}^{(j)}(\nu)$ are square matrices of rank 8*N* + 4. $b^{(j)}$ is the vector of unknown field amplitudes in layer *j* with 8*N* + 4 elements.

3.2.7 Interface Boundary Conditions

The layers are numbered from bottom up, that is, the transmitting medium is layer 1, the next layer up is 2, and the top, incident field medium is layer *J* where there are *J* total layers including the outer media. The generalized field vector is composed of the tangential surface fields which must be continuous across every interface. Field continuity between the bottom medium and the lowest thin film layer requires

$$F_2(0) = \tilde{Q}^{(2)} b^{(2)} = \tilde{M}_t B_t = F_1(0). \tag{3.244}$$

Continuity between the lowest film and the next film up requires

$$\begin{aligned} F_3(0) = \tilde{Q}^{(3)} b^{(3)} = F_2(d_2) = \tilde{Q}^{(2)} \tilde{\rho}(d_2) b^{(2)} &= \tilde{Q}^{(2)} \tilde{\rho}^{(2)}(d_2) \left(\tilde{Q}^{(2)}\right)^{-1} \left(\tilde{M}_t B_t\right) \\ &= \tilde{H}^{(2)} \tilde{H}^{(1)} \left(\tilde{M}_t B_t\right) = \tilde{H}^{(2)} \tilde{H}^{(1)} F_1(0) \end{aligned} \tag{3.245}$$

where $\tilde{H}^{(1)}$ is just the identity matrix,

$$\tilde{H}^{(2)} = \tilde{Q}^{(2)} \tilde{\rho}^{(2)}(d_2) \left(\tilde{Q}^{(2)}\right)^{-1}, \tag{3.246}$$

and d_2 is the thickness of the bottom thin film layer. In general, let d_j be the thickness of the jth layer, and define the general transfer matrix,

$$\tilde{\boldsymbol{H}}^{(j)} \equiv \tilde{\boldsymbol{Q}}^{(j)}\tilde{\boldsymbol{\rho}}^{(j)}\left(d_j\right)\left(\tilde{\boldsymbol{Q}}^{(j)}\right)^{-1}. \tag{3.247}$$

The fields at the bottom of layer $j + 1$ are given by

$$\boldsymbol{F}_{j+1}(0) = \boldsymbol{F}_j\left(d_j\right) = \tilde{\boldsymbol{Q}}^{(j)}\tilde{\boldsymbol{\rho}}^{(j)}\left(d_j\right)\boldsymbol{b}^{(j)} = \tilde{\boldsymbol{Q}}^{(j)}\tilde{\boldsymbol{\rho}}^{(j)}\left(d_j\right)\left(\tilde{\boldsymbol{Q}}^{(j)}\right)^{-1}\boldsymbol{F}_j(0) = \tilde{\boldsymbol{H}}^{(j)}\boldsymbol{F}_j(0). \tag{3.248}$$

Transfer matrix $\tilde{\boldsymbol{H}}^{(j)}$ connects the fields at the bottom of layer j to the fields at the bottom of layer $j + 1$. From (3.245) there is also the relation,

$$\tilde{\boldsymbol{Q}}^{(j+1)}\boldsymbol{b}^{(j+1)} = \tilde{\boldsymbol{Q}}^{(j)}\tilde{\boldsymbol{\rho}}^{(j)}\left(d_j\right)\boldsymbol{b}^{(j)} \tag{3.249}$$

or

$$\boldsymbol{b}^{(j)} = \tilde{\boldsymbol{I}}^{(j+1)}\boldsymbol{b}^{(j+1)} \tag{3.250}$$

where

$$\tilde{\boldsymbol{I}}^{(j+1)} \equiv \left(\tilde{\boldsymbol{\rho}}^{(j)}\left(d_j\right)\right)^{-1}\left(\tilde{\boldsymbol{Q}}^{(j)}\right)^{-1}\tilde{\boldsymbol{Q}}^{(j+1)}. \tag{3.251}$$

The $\tilde{\boldsymbol{I}}$ matrix is used in the scattering matrix formalism of Cotter et al. to be discussed in the next section. It is a square matrix of rank $8N + 4$. Repeating the process that generated (3.248) for each layer, the generalized field vector in the incident medium is then

$$\boldsymbol{F}_J(0) = \boldsymbol{F}_i + \tilde{\boldsymbol{M}}_r\boldsymbol{B}_r = \boldsymbol{F}_{J-1}\left(d_{J-1}\right) = \tilde{\boldsymbol{H}}^{(J-1)}\tilde{\boldsymbol{H}}^{(J-2)}\ldots\tilde{\boldsymbol{H}}^{(2)}\tilde{\boldsymbol{H}}^{(1)}\tilde{\boldsymbol{M}}_1\boldsymbol{B}_t = \tilde{\boldsymbol{H}}\left(\tilde{\boldsymbol{M}}_t\boldsymbol{B}_t\right). \tag{3.252}$$

where

$$\tilde{\boldsymbol{H}} \equiv \tilde{\boldsymbol{H}}^{(J-1)}\tilde{\boldsymbol{H}}^{(J-2)}\ldots\tilde{\boldsymbol{H}}^{(2)}\tilde{\boldsymbol{H}}^{(1)}. \tag{3.253}$$

On the right-hand side of (3.252) there are $4N + 2$ unknown amplitudes in $\boldsymbol{B}_t$. On the left-hand side of (3.252) there are also $4N + 2$ unknown amplitudes in $\boldsymbol{B}_r$. There is a total of $8N + 4$ independent equations in (3.252). Therefore, these can be uniquely solved for all the unknown field amplitudes in the reflected and transmitted media. (3.252) is rewritten as

$$\tilde{\boldsymbol{M}}_r\boldsymbol{B}_r - \tilde{\boldsymbol{H}}\tilde{\boldsymbol{M}}_t\boldsymbol{B}_t = -\boldsymbol{F}_i \tag{3.254}$$

and a new super matrix $\tilde{\boldsymbol{S}}$ is defined,

$$\tilde{\boldsymbol{S}} \equiv \left(\tilde{\boldsymbol{M}}_r, -\tilde{\boldsymbol{H}}\tilde{\boldsymbol{M}}_t\right). \tag{3.255}$$

This is a square matrix of rank $8N + 4$. Then

$$\begin{bmatrix}\boldsymbol{B}_r\\ \boldsymbol{B}_t\end{bmatrix} = -\left(\tilde{\boldsymbol{S}}\right)^{-1}\boldsymbol{F}_i. \tag{3.256}$$

All of the unknown field amplitudes in the outer two media are now determined as a function of the incident field.

The total reflectance and transmittance of the diffracted fields must be calculated carefully. The results are not simply ratios of the square of the absolute value of the field amplitudes because this does not take into account the polar angles at which the different diffracted orders are reflected. If the refractive index of the transmitted medium is different from that of the incident medium, then the transmitted modes are also refracted into different polar angles. Both of these effects change the cross section of the beam and, therefore, the energy density. For the reflected orders, the square of the absolute value of the electric field must be multiplied by the cosine of the diffracted polar angle and divided by the cosine of the incident polar angle. For the transmitted orders, the square of the absolute value of the electric field must be multiplied by the product of the cosine of the diffracted polar angle and the refractive index of the incident medium, and divided by the product of the cosine of the incident polar angle and the refractive index of the transmitted medium.

3.3 SCATTERING MATRIX FORMALISM

3.3.1 Transfer Matrices

Although the solution just provided is correct, it can lead to numerical instability when the layers are reasonably thick (i.e., ≥ ~1 μm). As the layer thickness increases, the phase factors involving eigenvalues $\exp\left[ir_q^{(j)}\left(\nu - y_j\right)\right]$ with an imaginary part can increase exponentially, thereby losing sufficient numerical precision in the calculation. To circumvent this problem, Cotter et al. developed a scattering matrix formalism [3]. The transfer matrix $\tilde{\boldsymbol{I}}$ connects the field amplitudes of adjacent layers.

$$\boldsymbol{b}^{(j)} = \begin{bmatrix} \boldsymbol{b}_+^{(j)} \\ \boldsymbol{b}_-^{(j)} \end{bmatrix} = \tilde{\boldsymbol{I}}^{(j+1)} \begin{bmatrix} \boldsymbol{b}_+^{(j+1)} \\ \boldsymbol{b}_-^{(j+1)} \end{bmatrix} = \tilde{\boldsymbol{I}}^{(j+1)} \boldsymbol{b}^{(j+1)}. \tag{3.257}$$

The field amplitude vectors are ordered such that the elements of the top half of the vectors correspond to eigenvalues for the upward propagating waves denoted by the "+" subscript and the bottom half of the vectors correspond to the downward propagating waves denoted by the "–" subscript.

The generalized field vector in layer j is given in (3.108).

$$\boldsymbol{F}^{(j)}\left(\nu\right) = \sum\nolimits_q \boldsymbol{V}_q^{(j)} e^{ir_q^{(j)}\left(\nu - y_j\right)} b_q^{(j)} = \tilde{\boldsymbol{Q}}^{(j)} \tilde{\boldsymbol{\rho}}^{(j)}\left(\nu\right) \boldsymbol{b}^{(j)}. \tag{3.258}$$

$\tilde{\boldsymbol{Q}}_j$ is a square matrix of rank $8N + 4$ for layer j in which each column is an eigenvector $\boldsymbol{V}_q^{(j)}$. From (3.105),

$$\tilde{\boldsymbol{Q}}_j \equiv \left(\boldsymbol{V}_1^{(j)} \quad \boldsymbol{V}_2^{(j)} \quad \ldots \quad \boldsymbol{V}_{8N+3}^{(j)} \quad \boldsymbol{V}_{8N+4}^{(j)}\right). \tag{3.259}$$

$\tilde{\boldsymbol{\rho}}^{(j)}\left(\nu\right)$ is a diagonal matrix of rank $8N + 4$. At the top of layer j, from (3.106),

$$\tilde{\boldsymbol{\rho}}^{(j)}\left(d_j\right) = \begin{bmatrix} e^{ir_1^{(j)} d_j} & 0 & 0 & 0 \\ 0 & e^{ir_2^{(j)} d_j} & 0 & 0 \\ 0 & 0 & \ddots & 0 \\ 0 & 0 & 0 & e^{ir_{8N+4}^{(j)} d_j} \end{bmatrix}. \tag{3.260}$$

In order to use the ordered amplitude sub-vectors in (3.258) with the matrices $\tilde{\boldsymbol{Q}}_j$ and $\tilde{\boldsymbol{\rho}}^{(j)}\left(d_j\right)$, the columns (eigenvectors) of $\tilde{\boldsymbol{Q}}_j$ must be reordered to correspond to the upward "+" and downward

"–" amplitude vectors. The first $2P_j$ columns of $\tilde{\boldsymbol{Q}}_j$ are the eigenvectors for upward propagating waves in which the eigenvalues $r^{(j)}$ are real and positive. The next $4N + 2 - 2P_j$ columns are the eigenvectors corresponding to upward decaying waves or downward exponentially growing waves for which the eigenvalues $r^{(j)}$ are imaginary and positive. These waves decay to zero as $y \to +\infty$. In the bottom transmitted region, these waves grow exponentially and are unphysical. The next $2P_j$ columns of $\tilde{\boldsymbol{Q}}_j$ are the eigenvectors for downward propagating waves in which the eigenvalues $r^{(j)}$ are real and negative. In the top medium the incident field corresponds to such a wave. Finally, the last $4N + 2 - 2P_j$ columns are the eigenvectors corresponding to downward decaying waves or upward exponentially growing waves for which the eigenvalues $r^{(j)}$ are imaginary and negative. In the top medium these waves increase exponentially as $y \to \infty$ and so are unphysical. In the transmitted medium, these waves decay evanescently as $y \to -\infty$. From (3.249),

$$\boldsymbol{b}^{(j)} = \left[\tilde{\boldsymbol{\rho}}^{(j)}(d_2)\right]^{-1}\left(\tilde{\boldsymbol{Q}}_j^{(j)}\right)^{-1}\tilde{\boldsymbol{Q}}_j^{(j+1)}\boldsymbol{b}^{(j+1)}. \tag{3.261}$$

Therefore, the transfer matrix is

$$\tilde{\boldsymbol{I}}^{(j+1)} = \left[\tilde{\boldsymbol{\rho}}^{(j)}(d_2)\right]^{-1}\left(\tilde{\boldsymbol{Q}}^{(j)}\right)^{-1}\tilde{\boldsymbol{Q}}^{(j+1)} = \left[\tilde{\boldsymbol{Q}}^{(j)}\tilde{\boldsymbol{\rho}}^{(j)}(d_2)\right]^{-1}\tilde{\boldsymbol{Q}}^{(j+1)}. \tag{3.262}$$

This equation corresponds to (17) in [3]. The amplitude vectors $\boldsymbol{b}^{(j)}$ and $\boldsymbol{b}^{(j+1)}$ both have $8N + 4$ elements. At the top surface, $j = N + 1$, the first $2P_r$ columns of $\tilde{\boldsymbol{Q}}^{(N+1)}$ correspond to upward propagating reflected waves after reordering. The final $4N + 2$ columns correspond to the downward waves and this part of the matrix will get replaced by the incident field expansion, which is the only downward wave in the problem.

The $\tilde{\boldsymbol{Q}}^{(1)}$ and $\tilde{\boldsymbol{\rho}}_1$ matrices must also be reordered for the thin film layer. In this case, the first $4N + 2$ columns correspond to upward waves having positive imaginary eigenvalues so that the amplitude decreases as $y \to +\infty$. The next $4N + 2$ columns are for the downward waves corresponding to negative imaginary eigenvalues so that the amplitude decreases as $y \to -\infty$. Finally, the $\tilde{\boldsymbol{Q}}^{(0)}$ matrix for the transmitted medium is reordered. In this case, the first $2P_t$ columns correspond to upward propagating modes corresponding to positive real eigenvalues which do not exist in this problem. The next $4N + 2 - 2P_t$ columns correspond to exponentially growing waves as $y \to -\infty$ for which the imaginary part of the eigenvalue is positive. These are unphysical. The next $2P_t$ columns correspond to downward propagating waves which have negative real eigenvalues. These are the transmitted orders for which the amplitudes need to be determined. Finally, the last $4N + 2 - 2P_t$ columns correspond the evanescent waves which decay as $y \to -\infty$ corresponding to eigenvalues for which the imaginary part of the eigenvalue is negative. $\tilde{\boldsymbol{\rho}}^{(0)}$ is the identity matrix, so

$$\tilde{\boldsymbol{I}}^{(1)} = \left(\tilde{\boldsymbol{Q}}^{(0)}\right)^{-1}\tilde{\boldsymbol{Q}}^{(1)}. \tag{3.263}$$

Also,

$$\tilde{\boldsymbol{I}}^{(2)} = \left[\tilde{\boldsymbol{Q}}^{(1)}\tilde{\boldsymbol{\rho}}^{(1)}(d_2)\right]^{-1}\tilde{\boldsymbol{Q}}^{(2)}. \tag{3.264}$$

Each transfer matrix is subdivided into four matrices of rank $4N + 2$.

$$\tilde{\boldsymbol{I}}^{(j)} = \begin{bmatrix} \tilde{\boldsymbol{I}}_{11}^{(j)} & \tilde{\boldsymbol{I}}_{12}^{(j)} \\ \tilde{\boldsymbol{I}}_{21}^{(j)} & \tilde{\boldsymbol{I}}_{22}^{(j)} \end{bmatrix}. \tag{3.265}$$

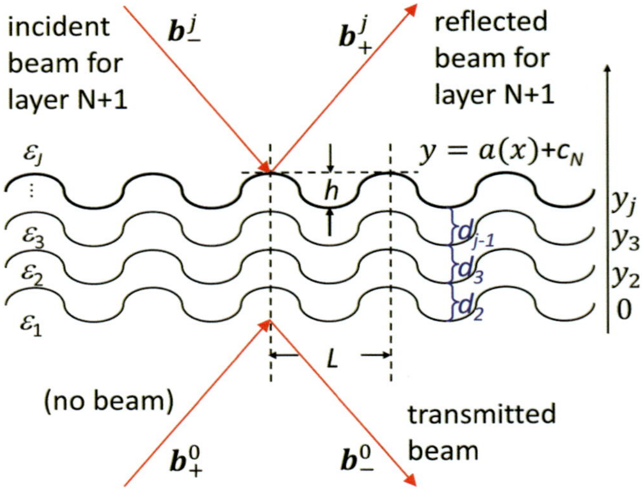

FIGURE 3.2 Scattering matrix geometry.

3.3.2 Scattering Matrices

The scattering matrix is now defined. Consider the multilayer film stack as shown in Figure 3.2.

The scattering matrix connects outgoing waves in layer j and in the bottom layer 0 to incoming waves in those layers according to

$$\begin{bmatrix} \boldsymbol{b}_+^{(j)} \\ \boldsymbol{b}_-^{(0)} \end{bmatrix} = \tilde{\boldsymbol{S}}^{(0,j)} \begin{bmatrix} \boldsymbol{b}_+^{(0)} \\ \boldsymbol{b}_-^{(j)} \end{bmatrix}. \tag{3.266}$$

The amplitude vectors are again ordered with the amplitudes for the upward propagating waves in the top half of each vector, and the amplitudes for the downward propagating waves in the bottom half of each vector. The scattering matrix is also subdivided into four square submatrices of rank $4N + 2$.

$$\tilde{\boldsymbol{S}}^{(0,j)} = \begin{bmatrix} \tilde{\boldsymbol{S}}_{11}^{(0,j)} & \tilde{\boldsymbol{S}}_{12}^{(0,j)} \\ \tilde{\boldsymbol{S}}_{21}^{(0,j)} & \tilde{\boldsymbol{S}}_{22}^{(0,j)} \end{bmatrix}. \tag{3.267}$$

Inserting (3.267) into (3.266) gives two matrix equations for layer j.

$$\boldsymbol{b}_+^{(j)} = \tilde{\boldsymbol{S}}_{11}^{(0,j)} \boldsymbol{b}_+^{(0)} + \tilde{\boldsymbol{S}}_{12}^{(0,j)} \boldsymbol{b}_-^{(j)}. \tag{3.268}$$

$$\boldsymbol{b}_-^{(0)} = \tilde{\boldsymbol{S}}_{21}^{(0,j)} \boldsymbol{b}_+^{(0)} + \tilde{\boldsymbol{S}}_{22}^{(0,j)} \boldsymbol{b}_-^{(j)}. \tag{3.269}$$

For layer $j + 1$,

$$\boldsymbol{b}_+^{(j+1)} = \tilde{\boldsymbol{S}}_{11}^{(0,j+1)} \boldsymbol{b}_+^{(0)} + \tilde{\boldsymbol{S}}_{12}^{(0,j+1)} \boldsymbol{b}_-^{(j+1)}, \tag{3.270}$$

and

$$\boldsymbol{b}_-^{(0)} = \tilde{\boldsymbol{S}}_{21}^{(0,j+1)}\boldsymbol{b}_+^{(0)} + \tilde{\boldsymbol{S}}_{22}^{(0,j+1)}\boldsymbol{b}_-^{(j+1)}. \tag{3.271}$$

Inserting (3.267) into (3.257) also gives two matrix equations.

$$\boldsymbol{b}_+^{(j)} = \tilde{\boldsymbol{I}}_{11}^{(j+1)}\boldsymbol{b}_+^{(j+1)} + \tilde{\boldsymbol{I}}_{12}^{(j+1)}\boldsymbol{b}_-^{(j+1)} \tag{3.272}$$

$$\boldsymbol{b}_-^{(j)} = \tilde{\boldsymbol{I}}_{21}^{(j+1)}\boldsymbol{b}_+^{(j+1)} + \tilde{\boldsymbol{I}}_{22}^{(j+1)}\boldsymbol{b}_-^{(j+1)}. \tag{3.273}$$

$\boldsymbol{b}_-^{(j)}$ can be eliminated from (3.268) and (3.269),

$$\boldsymbol{b}_-^{(j)} = \left(\tilde{\boldsymbol{S}}_{12}^{(0,j)}\right)^{-1}\boldsymbol{b}_+^{(j)} - \left(\tilde{\boldsymbol{S}}_{12}^{(0,j)}\right)^{-1}\tilde{\boldsymbol{S}}_{11}^{(0,j)}\boldsymbol{b}_+^{(0)} = \left(\tilde{\boldsymbol{S}}_{22}^{(0,j)}\right)^{-1}\boldsymbol{b}_-^{(0)} - \left(\tilde{\boldsymbol{S}}_{22}^{(0,j)}\right)^{-1}\tilde{\boldsymbol{S}}_{21}^{(0,j)}\boldsymbol{b}_+^{(0)}. \tag{3.274}$$

Solving for $\boldsymbol{b}_+^{(j)}$,

$$\boldsymbol{b}_+^{(j)} = \tilde{\boldsymbol{S}}_{12}^{(0,j)}\left(\tilde{\boldsymbol{S}}_{22}^{(0,j)}\right)^{-1}\boldsymbol{b}_-^{(0)} + \tilde{\boldsymbol{S}}_{12}^{(0,j)}\left(\tilde{\boldsymbol{S}}_{12}^{(0,j)}\right)^{-1}\tilde{\boldsymbol{S}}_{11}^{(0,j)} - \left(\tilde{\boldsymbol{S}}_{22}^{(0,j)}\right)^{-1}\tilde{\boldsymbol{S}}_{21}^{(0,j)}\boldsymbol{b}_+^{(0)}. \tag{3.275}$$

From (3.272) and (3.274),

$$\boldsymbol{b}_+^{(j)} = \tilde{\boldsymbol{I}}_{11}^{(j+1)}\boldsymbol{b}_+^{(j+1)} + \tilde{\boldsymbol{I}}_{12}^{(j+1)}\boldsymbol{b}_-^{(j+1)} = \tag{3.276}$$

$$\tilde{\boldsymbol{S}}_{12}^{(0,j)}\left(\tilde{\boldsymbol{S}}_{22}^{(0,j)}\right)^{-1}\boldsymbol{b}_-^{(0)} + \tilde{\boldsymbol{S}}_{12}^{(0,j)}\left[\left(\tilde{\boldsymbol{S}}_{12}^{(0,j)}\right)^{-1}\tilde{\boldsymbol{S}}_{11}^{(0,j)} - \left(\tilde{\boldsymbol{S}}_{22}^{(0,j)}\right)^{-1}\tilde{\boldsymbol{S}}_{21}^{(0,j)}\right]\boldsymbol{b}_+^{(0)}.$$

From (3.273) and (3.269),

$$\boldsymbol{b}_-^{(j)} = \tilde{\boldsymbol{I}}_{21}^{(j+1)}\boldsymbol{b}_+^{(j+1)} + \tilde{\boldsymbol{I}}_{22}^{(j+1)}\boldsymbol{b}_-^{(j+1)} = \left(\tilde{\boldsymbol{S}}_{22}^{(0,j)}\right)^{-1}\boldsymbol{b}_-^{(0)} - \left(\tilde{\boldsymbol{S}}_{22}^{(0,j)}\right)^{-1}\tilde{\boldsymbol{S}}_{21}^{(0,j)}\boldsymbol{b}_+^{(0)}. \tag{3.277}$$

Eliminating $\boldsymbol{b}_-^{(0)}$ from (3.276) and (3.277),

$$\boldsymbol{b}_+^{(j+1)} = \left(\tilde{\boldsymbol{I}}_{11}^{(j+1)} - \tilde{\boldsymbol{S}}_{12}^{(0,j)}\tilde{\boldsymbol{I}}_{21}^{(j+1)}\right)^{-1}\left[\tilde{\boldsymbol{S}}_{11}^{(0,j)}\boldsymbol{b}_+^{(0)} + \left(\tilde{\boldsymbol{S}}_{12}^{(0,j)}\tilde{\boldsymbol{I}}_{22}^{(j+1)} - \tilde{\boldsymbol{I}}_{12}^{(j+1)}\right)\boldsymbol{b}_-^{(j+1)}\right]. \tag{3.278}$$

Comparing (3.278) to (3.270), it must be true that

$$\tilde{\boldsymbol{S}}_{11}^{(0,j+1)} = \left(\tilde{\boldsymbol{I}}_{11}^{(j+1)} - \tilde{\boldsymbol{S}}_{12}^{(0,j)}\tilde{\boldsymbol{I}}_{21}^{(j+1)}\right)^{-1}\tilde{\boldsymbol{S}}_{11}^{(0,j)} \tag{3.279}$$

and

$$\tilde{\boldsymbol{S}}_{12}^{(0,j+1)} = \left(\tilde{\boldsymbol{I}}_{11}^{(j+1)} - \tilde{\boldsymbol{S}}_{12}^{(0,j)}\tilde{\boldsymbol{I}}_{21}^{(j+1)}\right)^{-1}\left(\tilde{\boldsymbol{S}}_{12}^{(0,j)}\tilde{\boldsymbol{I}}_{22}^{(j+1)} - \tilde{\boldsymbol{I}}_{12}^{(j+1)}\right) \tag{3.280}$$

These are the results in (22a) and (22b) of [3]. To find $\tilde{\boldsymbol{S}}_{21}^{(0,j+1)}$ and $\tilde{\boldsymbol{S}}_{22}^{(0,j+1)}$, $\boldsymbol{b}_-^{(0)}$ must be expressed in terms of $\boldsymbol{b}_+^{(0)}$ and $\boldsymbol{b}_-^{(j+1)}$. Using (3.268) and (3.269), $\boldsymbol{b}_-^{(j)}$ can be eliminated to solve for $\boldsymbol{b}_-^{(0)}$.

$$\boldsymbol{b}_{-}^{(0)} = \tilde{\boldsymbol{S}}_{22}^{(0,j)} \left(\tilde{\boldsymbol{S}}_{12}^{(0,j)} \right)^{-1} \boldsymbol{b}_{+}^{(j)} + \tilde{\boldsymbol{S}}_{21}^{(0,j)} \boldsymbol{b}_{+}^{(0)} - \tilde{\boldsymbol{S}}_{22}^{(0,j)} \left(\tilde{\boldsymbol{S}}_{12}^{(0,j)} \right)^{-1} \tilde{\boldsymbol{S}}_{11}^{(0,j)} \boldsymbol{b}_{+}^{(0)} . \tag{3.281}$$

Substituting for $\boldsymbol{b}_{+}^{(j)}$ from (3.270),

$$\boldsymbol{b}_{-}^{(0)} =$$

$$\tilde{\boldsymbol{S}}_{22}^{(0,j)} \left(\tilde{\boldsymbol{S}}_{12}^{(0,j)} \right)^{-1} \tilde{\boldsymbol{I}}_{11}^{(j+1)} \boldsymbol{b}_{+}^{(j+1)} + \left[\tilde{\boldsymbol{S}}_{21}^{(0,j)} - \tilde{\boldsymbol{S}}_{22}^{(0,j)} \left(\tilde{\boldsymbol{S}}_{12}^{(0,j)} \right)^{-1} \tilde{\boldsymbol{S}}_{11}^{(0,j)} \right] \boldsymbol{b}_{+}^{(0)} + \tilde{\boldsymbol{S}}_{22}^{(0,j)} \left(\tilde{\boldsymbol{S}}_{12}^{(0,j)} \right)^{-1} \tilde{\boldsymbol{I}}_{12}^{(j+1)} \boldsymbol{b}_{-}^{(j+1)}. \tag{3.282}$$

$\boldsymbol{b}_{+}^{(j+1)}$ is substituted from (3.278) giving

$$\boldsymbol{b}_{-}^{(0)} = \left[\tilde{\boldsymbol{S}}_{22}^{(0,j)} \left(\tilde{\boldsymbol{S}}_{12}^{(0,j)} \right)^{-1} \tilde{\boldsymbol{I}}_{11}^{(j+1)} \tilde{\boldsymbol{S}}_{11}^{(0,j+1)} - \tilde{\boldsymbol{S}}_{22}^{(0,j)} \left(\tilde{\boldsymbol{S}}_{12}^{(0,j)} \right)^{-1} \tilde{\boldsymbol{S}}_{11}^{(0,j)} + \tilde{\boldsymbol{S}}_{21}^{(0,j)} \right] \boldsymbol{b}_{+}^{(0)}$$

$$+ \tilde{\boldsymbol{S}}_{22}^{(0,j)} \left(\tilde{\boldsymbol{S}}_{12}^{(0,j)} \right)^{-1} \left(\tilde{\boldsymbol{I}}_{11}^{(j+1)} \tilde{\boldsymbol{S}}_{12}^{(0,j+1)} + \tilde{\boldsymbol{I}}_{12}^{(j+1)} \right) \boldsymbol{b}_{-}^{(j+1)}. \tag{3.283}$$

Comparing this to (3.271), it must be true that

$$\tilde{\boldsymbol{S}}_{21}^{(0,j+1)} = \tilde{\boldsymbol{S}}_{22}^{(0,j)} \left(\tilde{\boldsymbol{S}}_{12}^{(0,j)} \right)^{-1} \left(\tilde{\boldsymbol{I}}_{11}^{(j+1)} \tilde{\boldsymbol{S}}_{11}^{(0,j+1)} - \tilde{\boldsymbol{S}}_{11}^{(0,j)} \right) + \tilde{\boldsymbol{S}}_{21}^{(0,j)} \tag{3.284}$$

and

$$\tilde{\boldsymbol{S}}_{22}^{(0,j+1)} = \tilde{\boldsymbol{S}}_{22}^{(0,j)} \left(\tilde{\boldsymbol{S}}_{12}^{(0,j)} \right)^{-1} \left(\tilde{\boldsymbol{I}}_{11}^{(j+1)} \tilde{\boldsymbol{S}}_{12}^{(0,j+1)} + \tilde{\boldsymbol{I}}_{12}^{(j+1)} \right). \tag{3.285}$$

(3.284) can be rewritten in a more compact form. From (3.279),

$$\tilde{\boldsymbol{I}}_{11}^{(j+1)} \tilde{\boldsymbol{S}}_{11}^{(0,j+1)} - \tilde{\boldsymbol{S}}_{12}^{(0,j)} \tilde{\boldsymbol{I}}_{21}^{(j+1)} \tilde{\boldsymbol{S}}_{11}^{(0,j+1)} = \tilde{\boldsymbol{S}}_{11}^{(0,j)} . \tag{3.286}$$

Rearranging,

$$\tilde{\boldsymbol{I}}_{11}^{(j+1)} \tilde{\boldsymbol{S}}_{11}^{(0,j+1)} - \tilde{\boldsymbol{S}}_{11}^{(0,j)} = \tilde{\boldsymbol{S}}_{12}^{(0,j)} \tilde{\boldsymbol{I}}_{21}^{(j+1)} \tilde{\boldsymbol{S}}_{11}^{(0,j+1)} , \tag{3.287}$$

or

$$\left(\tilde{\boldsymbol{S}}_{12}^{(0,j)} \right)^{-1} \left(\tilde{\boldsymbol{I}}_{11}^{(j+1)} \tilde{\boldsymbol{S}}_{11}^{(0,j+1)} - \tilde{\boldsymbol{S}}_{11}^{(0,j)} \right) = \tilde{\boldsymbol{I}}_{21}^{(j+1)} \tilde{\boldsymbol{S}}_{11}^{(0,j+1)} . \tag{3.288}$$

From (3.85),

$$\tilde{\boldsymbol{S}}_{22}^{(0,j)} \left(\tilde{\boldsymbol{S}}_{12}^{(0,j)} \right)^{-1} \left(\tilde{\boldsymbol{I}}_{11}^{(j+1)} \tilde{\boldsymbol{S}}_{11}^{(0,j+1)} - \tilde{\boldsymbol{S}}_{11}^{(0,j)} \right) = \tilde{\boldsymbol{S}}_{22}^{(0,j)} \tilde{\boldsymbol{I}}_{21}^{(j+1)} \tilde{\boldsymbol{S}}_{11}^{(0,j+1)} . \tag{3.289}$$

Substituting this into (3.284), gives finally

$$\tilde{\boldsymbol{S}}_{21}^{(0,j+1)} = \tilde{\boldsymbol{S}}_{22}^{(0,j)} \tilde{\boldsymbol{I}}_{21}^{(j+1)} \tilde{\boldsymbol{S}}_{11}^{(0,j+1)} + \tilde{\boldsymbol{S}}_{21}^{(0,j)} \tag{3.290}$$

which is (22c) in [3]. (3.285) can also be simplified. The math is complicated, but the result is

$$\tilde{\boldsymbol{S}}_{22}^{(0,j+1)} = \tilde{\boldsymbol{S}}_{22}^{(0,j)} \tilde{\boldsymbol{I}}_{21}^{(j+1)} \tilde{\boldsymbol{S}}_{12}^{(0,j+1)} + \tilde{\boldsymbol{S}}_{22}^{(0,j)} \tilde{\boldsymbol{I}}_{22}^{(j+1)} \tag{3.291}$$

which is (22d) in [3]. $\tilde{\boldsymbol{S}}^{(0,0)}$ is a unit matrix. The $\tilde{\boldsymbol{I}}$ matrices for each layer have been previously determined, so (3.279), (3.280), (3.290), and (3.304) can be used for each layer in succession to determine all of the $\tilde{\boldsymbol{S}}$ matrices giving finally

$$\begin{bmatrix} \boldsymbol{b}_+^{(N+1)} \\ \boldsymbol{b}_-^{(0)} \end{bmatrix} = \tilde{\boldsymbol{S}}^{(0,N+1)} \begin{bmatrix} \boldsymbol{b}_+^{(0)} \\ \boldsymbol{b}_-^{(N+1)} \end{bmatrix}. \tag{3.292}$$

For the first thin film layer, $j = 1$, it should be noted that the off-diagonal matrices $\tilde{\boldsymbol{S}}_{12}^{(0,0)} = \tilde{\boldsymbol{S}}_{21}^{(0,0)} = \tilde{0}$, which are matrices with all zero elements, and the on-diagonal matrices $\tilde{\boldsymbol{S}}_{11}^{(0,0)} = \tilde{\boldsymbol{S}}_{22}^{(0,0)} = \tilde{\boldsymbol{I}}$ are identity matrices. So

$$\tilde{\boldsymbol{S}}_{11}^{(0,1)} = \left(\tilde{\boldsymbol{I}}_{11}^{(1)}\right)^{-1}, \tag{3.293}$$

$$\tilde{\boldsymbol{S}}_{12}^{(0,1)} = -\left(\tilde{\boldsymbol{I}}_{11}^{(1)}\right)^{-1} \tilde{\boldsymbol{I}}_{12}^{(1)}, \tag{3.294}$$

$$\tilde{\boldsymbol{S}}_{21}^{(0,1)} = \tilde{\boldsymbol{I}}_{21}^{(1)} \tilde{\boldsymbol{S}}_{11}^{(0,1)}, \tag{3.295}$$

and

$$\tilde{\boldsymbol{S}}_{22}^{(0,1)} = \tilde{\boldsymbol{I}}_{21}^{(1)} \tilde{\boldsymbol{S}}_{12}^{(0,1)} + \tilde{\boldsymbol{I}}_{22}^{(1)}. \tag{3.296}$$

3.3.3 Reflected Modes

$\boldsymbol{b}_+^{(0)}(0)$ are upward propagating waves in the transmitting medium which are set to zero. $\boldsymbol{b}_-^{(N+1)}(0)$ is the amplitude vector of the incident field. Therefore, (3.292) gives the reflected and transmitted fields directly in terms of the incident field. Let

$$\tilde{\boldsymbol{S}}^{(0,N+1)} = \begin{bmatrix} \tilde{\boldsymbol{S}}_{11}^{(0,N+1)} & \tilde{\boldsymbol{S}}_{12}^{(0,N+1)} \\ \tilde{\boldsymbol{S}}_{21}^{(0,N+1)} & \tilde{\boldsymbol{S}}_{22}^{(0,N+1)} \end{bmatrix}. \tag{3.297}$$

Then

$$\boldsymbol{b}_+^{(N+1)} = \tilde{\boldsymbol{S}}_{12}^{(0,N+1)} \boldsymbol{b}_-^{(N+1)} \tag{3.298}$$

and

$$\boldsymbol{b}_-^{(0)} = \tilde{\boldsymbol{S}}_{22}^{(0,N+1)} \boldsymbol{b}_-^{(N+1)}. \tag{3.299}$$

Unfortunately, the complete amplitude vector $\boldsymbol{b}_-^{(N+1)}$ for the downward waves in the incident medium in the transformed coordinate system is not yet known. From (3.258), the total fields in the upper medium are determined in the transformed coordinate system by

$$\boldsymbol{F}^{(N+1)} = \tilde{\boldsymbol{Q}}^{(N+1)}\boldsymbol{b}^{(N+1)} = \tilde{\boldsymbol{Q}}^{(N+1)}\begin{bmatrix}\boldsymbol{b}_{+}^{N+1}\\ \boldsymbol{b}_{-}^{N+1}\end{bmatrix}. \tag{3.300}$$

The eigenvector columns of $\tilde{\boldsymbol{Q}}^{(N+1)}$ are arranged using their associated eigenvalues so that the first $2P_r$ columns are for reflected (upward) propagating waves for which $\mathrm{Re}(\lambda) > 0$ and $\mathrm{Im}(\lambda) = 0$, the next $4N + 2 - 2P_r$ columns are for upward evanescent waves for which the associated eigenvalues satisfy $\mathrm{Im}(\lambda) > 0$, and the last $4N + 2$ columns are for the downward propagating and exponentially growing waves. The amplitude vector $\boldsymbol{b}_{+}^{N+1}$ is also composed of two parts. The first $2P_r$ elements represented by the sub-vector $\boldsymbol{b}_r$ are for the reflected propagating wave amplitudes and the remaining $4N + 2 - 2P_r$ elements represented by the sub-vector $\boldsymbol{b}_e$ are for the evanescent field amplitudes. $\tilde{\boldsymbol{Q}}^{(N+1)}$ is now split into four square sub-matrices of rank $4N + 2$,

$$\tilde{\boldsymbol{Q}}^{(N+1)} \equiv \begin{bmatrix}\tilde{\boldsymbol{Q}}_{11}^{(N+1)} & \tilde{\boldsymbol{Q}}_{12}^{(N+1)}\\ \tilde{\boldsymbol{Q}}_{21}^{(N+1)} & \tilde{\boldsymbol{Q}}_{22}^{(N+1)}\end{bmatrix}. \tag{3.301}$$

The total field in the incident medium is, therefore,

$$\boldsymbol{F}^{(N+1)} = \begin{bmatrix}\tilde{\boldsymbol{Q}}_{11}^{(N+1)} & \tilde{\boldsymbol{Q}}_{12}^{(N+1)}\\ \tilde{\boldsymbol{Q}}_{21}^{(N+1)} & \tilde{\boldsymbol{Q}}_{22}^{(N+1)}\end{bmatrix}\begin{bmatrix}\boldsymbol{b}_{+}^{N+1}\\ \boldsymbol{b}_{-}^{N+1}\end{bmatrix}. \tag{3.302}$$

Only $\tilde{\boldsymbol{Q}}_{11}^{(N+1)}$ and $\tilde{\boldsymbol{Q}}_{21}^{(N+1)}$ are required for the remainder of the calculation. The other two matrices, $\tilde{\boldsymbol{Q}}_{12}^{(N+1)}$ and $\tilde{\boldsymbol{Q}}_{22}^{(N+1)}$, represent eigenvectors that are downward propagating or grow exponentially in the upward direction, $y \to +\infty$. The expansion of the incident field will be substituted for these components.

The only downward propagating wave is $\boldsymbol{F}_i$. The upward reflected eigenvectors in the plane wave expansion have the unknown vector amplitudes, $\boldsymbol{b}_r$. Therefore, the total field in the incident medium can also be written as the sum of the incident field plane wave expansion, the reflected propagating field plane wave expansions, and the upward evanescent fields calculated in the transformed coordinate system with unknown vector amplitudes $\boldsymbol{b}_e$. The vector $\boldsymbol{R}$ having $4N + 2$ elements representing the $2P$ reflected propagating amplitudes and the $4N + 2 - 2P$ evanescent field amplitudes is defined as

$$\boldsymbol{R} \equiv \begin{bmatrix}\boldsymbol{b}_r\\ \boldsymbol{b}_e\end{bmatrix}. \tag{3.303}$$

A new matrix $\hat{\boldsymbol{Q}}$ is created for the reflected propagating and evanescent fields by dropping the last $4N + 2$ columns of $\hat{\boldsymbol{Q}}^{(N+1)}$ which correspond to the downward propagating fields. The initial $2P_r$ eigenvector columns in the new matrix are replaced by the expansion components for the reflected plane waves previously calculated. Therefore, $\hat{\boldsymbol{Q}}$ is a rectangular matrix with $8N + 4$ rows and $4N + 2$ columns. $\hat{\boldsymbol{Q}}$ is also split vertically into two square sub-matrices.

$$\hat{\boldsymbol{Q}} = \begin{bmatrix}\hat{\boldsymbol{Q}}_1\\ \hat{\boldsymbol{Q}}_2\end{bmatrix}. \tag{3.304}$$

The total field in the incident medium is

$$\boldsymbol{F}^{(N+1)} = \begin{bmatrix}\hat{\boldsymbol{Q}}_1\\ \hat{\boldsymbol{Q}}_2\end{bmatrix}\boldsymbol{R} + \boldsymbol{F}_i = \begin{bmatrix}\hat{\boldsymbol{Q}}_1\\ \hat{\boldsymbol{Q}}_2\end{bmatrix}\begin{bmatrix}\boldsymbol{b}_r\\ \boldsymbol{b}_e\end{bmatrix} + \begin{bmatrix}\boldsymbol{F}_{i1}\\ \boldsymbol{F}_{i2}\end{bmatrix}. \tag{3.305}$$

Four matrix equations are obtained by equating (3.301) and (3.304),

$$\tilde{\boldsymbol{Q}}_{11}^{(N+1)}\boldsymbol{b}_{+}^{(N+1)}+\tilde{\boldsymbol{Q}}_{12}^{(N+1)}\boldsymbol{b}_{-}^{(N+1)}=\hat{\boldsymbol{Q}}_{1}\boldsymbol{R}+\boldsymbol{F}_{\mathrm{i}1}, \tag{3.306}$$

and

$$\tilde{\boldsymbol{Q}}_{21}^{(N+1)}\boldsymbol{b}_{+}^{(N+1)}+\tilde{\boldsymbol{Q}}_{22}^{(N+1)}\boldsymbol{b}_{-}^{(N+1)}=\hat{\boldsymbol{Q}}_{2}\boldsymbol{R}+\boldsymbol{F}_{\mathrm{i}2} \tag{3.307}$$

where the incident field vector has also been split in half such that there are $4N + 2$ elements in both $\boldsymbol{F}_{\mathrm{i}1}$ and $\boldsymbol{F}_{\mathrm{i}2}$,

$$\boldsymbol{F}_{i}\equiv\begin{bmatrix}\boldsymbol{F}_{\mathrm{i}1}\\ \boldsymbol{F}_{\mathrm{i}2}\end{bmatrix}. \tag{3.308}$$

(3.292) is used to substitute for $\boldsymbol{b}_{+}^{(N+1)}$ in (3.302) and solve for $\boldsymbol{b}_{-}^{(N+1)}$.

$$\boldsymbol{b}_{-}^{(N+1)}=\left(\tilde{\boldsymbol{Q}}_{21}^{(N+1)}\tilde{\boldsymbol{S}}_{12}^{(0,N+1)}+\tilde{\boldsymbol{Q}}_{22}^{(N+1)}\right)^{-1}\left(\hat{\boldsymbol{Q}}_{2}\boldsymbol{R}+\boldsymbol{F}_{\mathrm{i}2}\right). \tag{3.309}$$

This result is used to solve for the transmitted field amplitude vector in (3.292).

$$\boldsymbol{b}_{-}^{(0)}=\tilde{\boldsymbol{S}}_{22}^{(0,N+1)}\left(\tilde{\boldsymbol{Q}}_{21}^{(N+1)}\tilde{\boldsymbol{S}}_{12}^{(0,N+1)}+\tilde{\boldsymbol{Q}}_{22}^{(N+1)}\right)^{-1}\left(\hat{\boldsymbol{Q}}_{2}\boldsymbol{R}+\boldsymbol{F}_{\mathrm{i}2}\right). \tag{3.310}$$

(3.298) is used to substitute for $\boldsymbol{b}_{+}^{(N+1)}$ in (3.307) and solve for $\boldsymbol{R}$.

$$\left(\tilde{\boldsymbol{Q}}_{11}^{(N+1)}\tilde{\boldsymbol{S}}_{12}^{(0,N+1)}+\tilde{\boldsymbol{Q}}_{12}^{(N+1)}\right)\boldsymbol{b}_{-}^{(N+1)}=\hat{\boldsymbol{Q}}_{1}\boldsymbol{R}+\boldsymbol{F}_{\mathrm{i}1}. \tag{3.311}$$

$\boldsymbol{b}_{-}^{(N+1)}$ is replaced using (3.309),

$$\left(\tilde{\boldsymbol{Q}}_{11}^{(N+1)}\tilde{\boldsymbol{S}}_{12}^{(0,N+1)}+\tilde{\boldsymbol{Q}}_{12}^{(N+1)}\right)\left(\tilde{\boldsymbol{Q}}_{21}^{(N+1)}\tilde{\boldsymbol{S}}_{12}^{(0,N+1)}+\tilde{\boldsymbol{Q}}_{22}^{(N+1)}\right)^{-1}\left(\hat{\boldsymbol{Q}}_{2}\boldsymbol{R}+\boldsymbol{F}_{\mathrm{i}2}\right)=\hat{\boldsymbol{Q}}_{1}\boldsymbol{R}+\boldsymbol{F}_{\mathrm{i}1}. \tag{3.312}$$

Solving for $\boldsymbol{R}$,

$$\begin{aligned}\boldsymbol{R}=&\left[\hat{\boldsymbol{Q}}_{1}-\left(\tilde{\boldsymbol{Q}}_{11}^{(N+1)}\tilde{\boldsymbol{S}}_{12}^{(0,N+1)}+\tilde{\boldsymbol{Q}}_{12}^{(N+1)}\right)\left(\tilde{\boldsymbol{Q}}_{21}^{(N+1)}\tilde{\boldsymbol{S}}_{12}^{(0,N+1)}+\tilde{\boldsymbol{Q}}_{22}^{(N+1)}\right)^{-1}\hat{\boldsymbol{Q}}_{2}\right]^{-1}\\ &\cdot\left[\left(\tilde{\boldsymbol{Q}}_{11}^{(N+1)}\tilde{\boldsymbol{S}}_{12}^{(0,N+1)}+\tilde{\boldsymbol{Q}}_{12}^{(N+1)}\right)\left(\tilde{\boldsymbol{Q}}_{21}^{(N+1)}\tilde{\boldsymbol{S}}_{12}^{(0,N+1)}+\tilde{\boldsymbol{Q}}_{22}^{(N+1)}\right)^{-1}\boldsymbol{F}_{\mathrm{i}2}-\boldsymbol{F}_{\mathrm{i}1}\right].\end{aligned} \tag{3.313}$$

$\boldsymbol{R}$ directly provides propagating reflected plane wave amplitudes according to (3.303).

3.3.4 Transmitted Modes

In the transmitted medium, the generalized field vector is

$$\boldsymbol{F}^{(t)}=\tilde{\boldsymbol{Q}}^{(0)}\boldsymbol{b}^{(0)}=\tilde{\boldsymbol{Q}}^{(0)}\begin{bmatrix}0\\ \boldsymbol{b}_{-}^{(0)}\end{bmatrix} \tag{3.314}$$

where $\tilde{\boldsymbol{Q}}_0$ is a square matrix of rank $8N+4$, $\boldsymbol{b}^{(0)}$ is a vector with $8N+4$ elements for which the first $4N+2$ elements are all zero and the remaining $4N+2$ elements compose the previously determined $\boldsymbol{b}_-^{(0)}$ sub-vector.

As previously seen, the transmitted fields can also be expanded in a Fourier series in terms of propagating and evanescent modes. From (3.222) and (3.228),

$$\boldsymbol{F}^{(t)} = \tilde{\boldsymbol{M}}_t \boldsymbol{B}_t = \begin{bmatrix} \tilde{\boldsymbol{M}}_{t1} \\ \tilde{\boldsymbol{M}}_{t2} \end{bmatrix} \begin{bmatrix} \boldsymbol{b}_{\mathrm{TE}}^{(t)} \\ \boldsymbol{b}_{\mathrm{TM}}^{(t)} \end{bmatrix} \tag{3.315}$$

where $\tilde{\boldsymbol{M}}_{t1}$ and $\tilde{\boldsymbol{M}}_{t2}$ are square matrices of rank $4N + 2$, $\boldsymbol{B}_t$ is a vector composed of the $2P_t$ unknown transmitted propagating and $4N + 2 - 2P_t$ evanescent wave amplitudes which are divided into an upper sub-vector for the TE modes and a lower sub-vector for the TM modes. The columns of the matrix $\tilde{\boldsymbol{M}}_t$ and the elements of the sub-vectors are reordered so that the first P_t columns of $\tilde{\boldsymbol{M}}_t$ and the first P_t elements of $\boldsymbol{B}_t$ correspond to the propagating TE modes and the next P_t columns of $\tilde{\boldsymbol{M}}_t$ and the next P_t elements of $\boldsymbol{B}_t$ correspond to the propagating TM modes. A new square matrix $\hat{\boldsymbol{M}}_t$ of rank $8N + 4$ is generated by inserting the first $2P_t$ columns of $\tilde{\boldsymbol{M}}_t$ into columns $4N + 3$ through $4N + 2P_t$ of $\tilde{\boldsymbol{Q}}_0$. There is a new vector $\boldsymbol{b}_t$ of the unknown downward amplitudes for which the first P_t elements correspond to the TE transmitted mode amplitudes, the next P_t elements correspond to the unknown TM transmitted mode amplitudes, and the remaining $4N+2-2P_t$ elements correspond to the downward evanescent mode amplitudes in the transformed coordinate system.

$$\boldsymbol{F}^{(t)} = \hat{\boldsymbol{M}}^{(t)} \begin{bmatrix} 0 \\ \boldsymbol{b}_t \end{bmatrix}. \tag{3.316}$$

$\boldsymbol{b}_t$ is found by equating (3.314) and (3.316) which provides the required TE and TM transmitted mode amplitudes.

$$\boldsymbol{F}^{(t)} = \tilde{\boldsymbol{Q}}^{(0)} \boldsymbol{b}^{(0)} = \begin{bmatrix} \tilde{\boldsymbol{Q}}_{11}^{(0)} & \tilde{\boldsymbol{Q}}_{12}^{(0)} \\ \tilde{\boldsymbol{Q}}_{21}^{(0)} & \tilde{\boldsymbol{Q}}_{22}^{(0)} \end{bmatrix} \begin{bmatrix} \boldsymbol{0} \\ \boldsymbol{b}_-^{(0)} \end{bmatrix} = \hat{\boldsymbol{M}}^{(t)} \begin{bmatrix} \boldsymbol{0} \\ \boldsymbol{b}_t \end{bmatrix} = \begin{bmatrix} \tilde{\boldsymbol{Q}}_{11}^{(0)} & \hat{\boldsymbol{M}}_{12}^{(t)} \\ \tilde{\boldsymbol{Q}}_{21}^{(0)} & \hat{\boldsymbol{M}}_{22}^{(t)} \end{bmatrix} \begin{bmatrix} \boldsymbol{0} \\ \boldsymbol{b}_t \end{bmatrix}. \tag{3.317}$$

All of the transmitted wave amplitudes are contained in $\boldsymbol{b}_t$ which is given by

$$\boldsymbol{b}_t = \left(\hat{\boldsymbol{M}}_{22}^{(t)} \right)^{-1} \tilde{\boldsymbol{Q}}_{22}^{(0)} \boldsymbol{b}_-^{(0)} \tag{3.318}$$

where the first $2P_t$ elements of $\boldsymbol{b}_t$ are the desired TE and TM transmitted mode amplitudes.

3.3.5 Electric and Magnetic Fields in the Film Stack

Sometimes it is of interest to calculate the electric or magnetic field throughout the film stack. For example, when estimating the coupling efficiency of a grating into a waveguide, one must first calculate the electric field through the film stack for coupling into a TE waveguide mode or the magnetic field for coupling into a TM waveguide mode. The fields in layer j are determined from the amplitude vector $\boldsymbol{b}^{(j)}$, the phase matrix $\tilde{\boldsymbol{\rho}}^{(j)}$, and the $\tilde{\boldsymbol{Q}}^{(j)}$ matrix from (3.257) and (3.258).

$$\boldsymbol{F}^{(j)}(\nu) = \begin{bmatrix} E_{\parallel} \\ E_{\omega} \\ H_{\parallel} \\ H_{\omega} \end{bmatrix} = \tilde{\boldsymbol{Q}}^{(j)} \tilde{\boldsymbol{\rho}}^{(j)}(\nu) \boldsymbol{b}^{(j)} = \tilde{\boldsymbol{Q}}^{(j)} \tilde{\boldsymbol{\rho}}^{(j)}(\nu) \begin{bmatrix} \boldsymbol{b}_+^{(j)} \\ \boldsymbol{b}_-^{(j)} \end{bmatrix}. \tag{3.319}$$

To determine the fields throughout the film stack, the vector $\boldsymbol{b}^{(j)}$ must first be calculated for each layer. The simplest approach is to use the transfer matrices in (3.257) and beginning with the previously calculated values for $\boldsymbol{b}^{(0)}$ in the transmitting medium, calculate the amplitude vector for each successive layer up through the film stack.

In the transmitting medium, there are no upward modes so $\boldsymbol{b}_{+}^{(0)}=0$, and $\boldsymbol{b}_{-}^{(0)}$ is defined in (3.310). From (3.257), the inverse of the transfer matrix can be used to calculate the $\boldsymbol{b}^{(j)}$ vector amplitudes for every layer starting from the bottom layer $\boldsymbol{b}^{(0)}$.

$$\boldsymbol{b}^{(j+1)}=\left[\tilde{\boldsymbol{T}}^{(j+1)}\right]^{-1}\boldsymbol{b}^{(j)}. \tag{3.320}$$

Unfortunately, as previously discussed, the transfer matrices tend to lose precision for thicker layers, or for y-axis values that are large. In particular, the calculation may give small, but nonzero, amplitudes for downward evanescent waves in the incident medium. Within the film stack these errors are generally insignificant, but as the y-axis value increases in the incident medium, even small amplitudes for the downward evanescent waves grow exponentially and start to dominate the field values. Therefore, as a work-around for the finite machine precision of the calculation, the program can set all amplitudes of the downward evanescent waves in the incident medium to be identically zero. The phase matrix is given in (3.105) for each layer. In each layer, $0\le \nu_j \le d_j$ and in the transformed coordinate system for a sinusoidal groove profile,

$$y_j=\nu_j+a(x)=\nu_j+\frac{h}{2}\left[1-\cos(Kx)\right]. \tag{3.321}$$

After calculating the generalized field vectors within each layer from (3.319), which provides the values for $E_{\parallel}$, E_{ω}, $H_{\parallel}$, and H_{ω} from (3.92), the TE and TM fields can be calculated within each layer. For downward propagating modes, referring to Figure 3.3, the field conversion equations are

$$E_{\omega}=-E_{\mathrm{TM}}\sin\phi\cos\theta+E_{\mathrm{TE}}\cos\phi=\frac{\eta_0}{n}\left(-H_{\mathrm{TM}}\sin\phi\cos\theta+H_{\mathrm{TE}}\cos\phi\right), \tag{3.322}$$

$$H_{\omega}=\frac{n}{\eta_0}\left(E_{\mathrm{TM}}\cos\phi+E_{\mathrm{TE}}\sin\phi\cos\theta\right)=H_{\mathrm{TM}}\cos\phi+H_{\mathrm{TE}}\sin\phi\cos\theta, \tag{3.323}$$

$$E_{\parallel}=E_{\mathrm{TM}}\cos\phi\cos\theta+E_{\mathrm{TE}}\sin\phi=\frac{\eta_0}{n}\left(H_{\mathrm{TM}}\cos\phi\cos\theta-H_{\mathrm{TE}}\sin\phi\right), \tag{3.324}$$

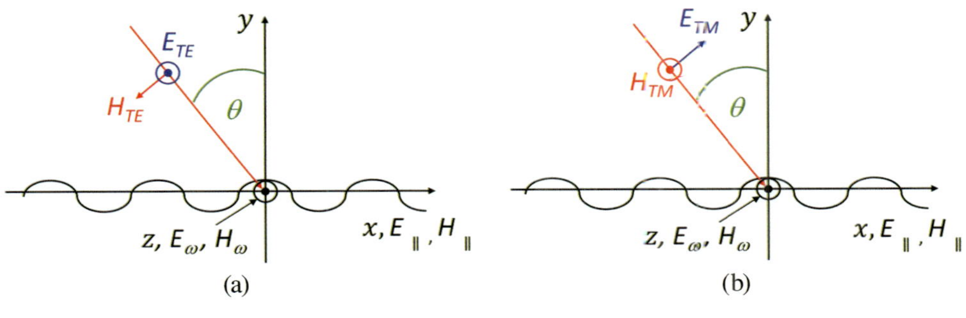

FIGURE 3.3 (a) Downward propagating TE or s-polarized wave. (b) Downward propagating TM or p-polarized wave.

and

$$H_{\parallel} = \frac{n}{\eta_0}\left(E_{\mathrm{TM}} \sin\phi - E_{\mathrm{TE}} \cos\phi \,\cos\theta\right) = H_{\mathrm{TM}} \sin\phi - H_{\mathrm{TE}} \cos\phi \,\cos\theta \,. \tag{3.325}$$

For upward propagating modes, referring to Figure 3.4, the field conversion equations are

$$E_{\omega} = E_{\mathrm{TM}}\sin\phi\cos\theta + E_{\mathrm{TE}}\cos\phi = \frac{\eta_0}{n}\left(H_{\mathrm{TM}}\sin\phi\cos\theta + H_{\mathrm{TE}}\cos\phi\right), \tag{3.326}$$

$$H_{\omega} = \frac{n}{\eta_0}\left(E_{\mathrm{TM}}\cos\phi - E_{\mathrm{TE}}\sin\phi\cos\theta\right) = H_{\mathrm{TM}}\cos\phi - H_{\mathrm{TE}}\sin\phi\cos\theta \,, \tag{3.327}$$

$$E_{\parallel} = -E_{\mathrm{TM}} \cos\phi\cos\theta + E_{\mathrm{TE}} \sin\phi = \frac{\eta_0}{n}\left(-H_{\mathrm{TM}} \cos\phi\cos\theta + H_{\mathrm{TE}} \sin\phi\right), \tag{3.328}$$

and

$$H_{\parallel} = \frac{n}{\eta_0}\left(E_{\mathrm{TM}} \sin\phi + E_{\mathrm{TE}} \cos\phi \,\cos\theta\right) = H_{\mathrm{TM}} \sin\phi + H_{\mathrm{TE}} \cos\phi \,\cos\theta \,. \tag{3.329}$$

For determining the overlap integrals with waveguide modes for coupling efficiency calculations, it is simplest to use the electric field for TE polarization and the magnetic field for TM polarization. For downward propagating modes,

$$E_{\mathrm{TE}} = \left(\frac{\cos\phi}{\cos^2\phi + \sin^2\phi\cos^2\theta}\right)E_{\omega} + \frac{\eta_0}{n}\left(\frac{\sin\phi\cos\theta}{\cos^2\phi + \sin^2\phi\cos^2\theta}\right)H_{\omega}\,, \tag{3.330}$$

and

$$H_{\mathrm{TM}} = \frac{n}{\eta_0}\left(\frac{\sin\phi\cos\theta}{\cos^2\phi + \sin^2\phi\cos^2\theta}\right)E_{\omega} + \left(\frac{\cos\phi}{\cos^2\phi + \sin^2\phi\cos^2\theta}\right)H_{\omega}\,, \tag{3.331}$$

and for upward propagating modes,

$$E_{\mathrm{TE}} = \left(\frac{\cos\phi}{\cos^2\phi + \sin^2\phi\cos^2\theta}\right)E_{\omega} - \frac{\eta_0}{n}\left(\frac{\sin\phi\cos\theta}{\cos^2\phi + \sin^2\phi\cos^2\theta}\right)H_{\omega}\,, \tag{3.332}$$

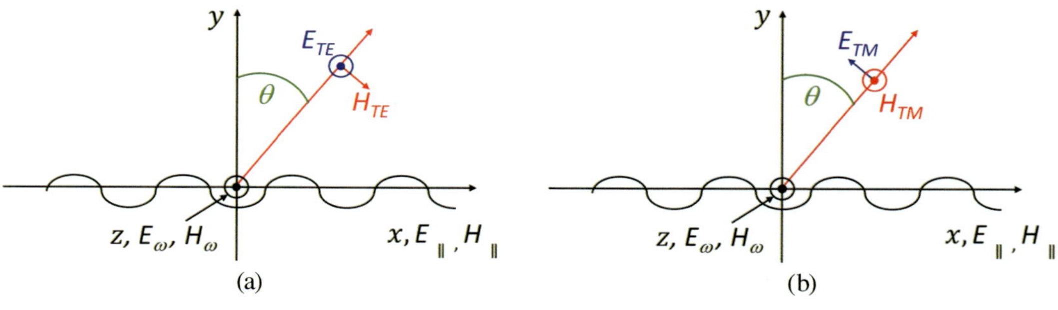

FIGURE 3.4 (a) Upward propagating TE or *s*-polarized wave. (b) Upward propagating TM or *p*-polarized wave.

and

$$H_{\text{TM}} = \frac{n}{\eta_0}\left(\frac{\sin\phi\cos\theta}{\cos^2\phi+\sin^2\phi\cos^2\theta}\right)E_\omega + \left(\frac{\cos\phi}{\cos^2\phi+\sin^2\phi\cos^2\theta}\right)H_\omega. \tag{3.333}$$

3.4 NON-SINUSOIDAL GRATINGS

3.4.1 Skewed Gratings

If the grating is not sinusoidal, then the values for the Fourier components C_n and D_n can be calculated numerically. For example, one equation for a skewed grating is [3]

$$a(x) = h\left[\left(\frac{1-\cos(Kx)}{2}\right)^{\zeta} - \frac{1}{2}\right]. \tag{3.334}$$

where h is the grating depth, K is the grating wavevector, and ζ is the skew factor. If $\zeta = 1$, then $a(x)$ reduces to a sinusoidal grating with peak-to-peak amplitude h. Figure 3.5 shows the effect of skew on the groove cross section.

The derivative of $a(x)$ is

$$a' = \frac{da(x)}{dx} = \frac{K\zeta h}{2^{\zeta}}\left[1-\cos(Kx)\right]^{\zeta-1}\sin(Kx). \tag{3.335}$$

This function is substituted into (3.72) and (3.74) with the help of (3.71) and (3.73) and numerically integrated to obtain values for C_n and D_n. It is also substituted into (3.113), (3.153), and (3.193) to determine L_m, L_{m-n}, and $L^{(t)}_{m-n}$, respectively.

3.4.2 Offset Gratings

An equation for a grating with the groove offset sideways is

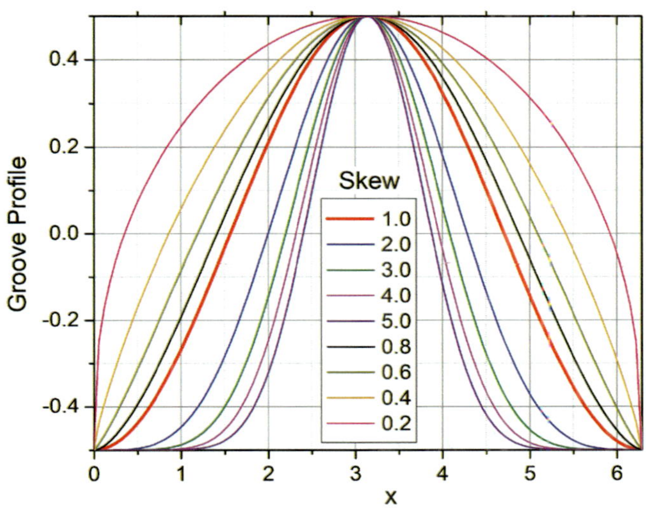

FIGURE 3.5 Effect of the skew parameter ζ.

$$a(x)=h\left[\frac{1-\cos\left(Kx+\xi\sin\left(\frac{Kx}{2}\right)\right)}{2}\right]. \tag{3.336}$$

where h is the grating depth, K is the grating wavevector, and ξ is the offset factor. If $\xi = 0$, then $a(x)$ reduces to a sinusoidal grating with peak-to-peak amplitude h. Figure 3.6 shows the effect of offset on the groove profile. The horizontal symmetry around the peak of the groove is broken by the offset factor. The derivative of $a(x)$ is

$$a'=\frac{da(x)}{dx}=\frac{h}{2}\left[K+\frac{K\xi}{2}\cos\left(\frac{Kx}{2}\right)\right]\sin\left[Kx+\xi\sin\left(\frac{Kx}{2}\right)\right]. \tag{3.337}$$

This function is substituted into (3.72) and (3.74) with the help of (3.71) and (3.73) and numerically integrated to obtain values for C_n and D_n. It is also substituted into (3.113), (3.153), and (3.193) to determine L_m, L_{m-n}, and $L^{(t)}_{m-n}$, respectively.

3.4.3 Tilted Gratings

Tilted gratings are discussed in detail in [16]. Consider a cosine wave that is tilted by angle α as shown in Figure 3.7. Let $x = u(\xi)$ and $a(x) = a(\xi)$ be defined parametrically by the variable ξ,

$$a(\xi)=\frac{h}{2}(1-\cos K\xi). \tag{3.338}$$

$$u(\xi)=\xi+\tan(\alpha)a(\xi)=\xi+\frac{h}{2}(1-\cos K\xi)\tan(\alpha) \tag{3.339}$$

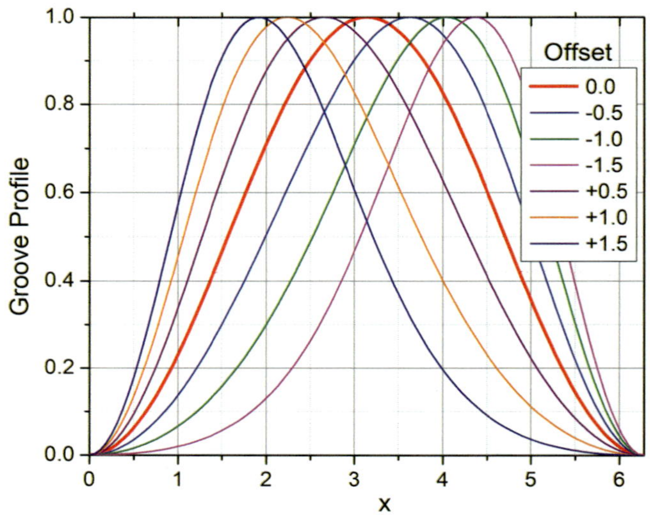

FIGURE 3.6 Effect of the offset parameter ξ.

After a little algebra, it is found that

$$du = \left[1 + \left(\frac{hK}{2}\right)\sin K\xi \tan\alpha\right] d\xi \tag{3.340}$$

and

$$a' = \frac{da}{du} = \frac{\left(\frac{hK}{2}\right)\sin K\xi}{1 + \left(\frac{hK}{2}\right)\sin K\xi \tan\alpha}. \tag{3.341}$$

(3.339)–(3.341) are then substituted into (3.71)–(3.74) to compute C_p and D_p numerically, and into (3.113), (3.153), and (3.193) to compute L_m, L_{m-n}, and $L^{(t)}_{m-n}$, respectively. It should be noted that for a slant angle $\alpha \geq \sim 60°$, the grating in Cartesian coordinates is no longer a function, having multiple values of y for certain values of x.

3.4.4 Linearly Segmented Gratings

If the grating is not sinusoidal, then the values for the Fourier components C_n and D_n can be calculated numerically. For a grating formed from n_s linear segments as shown in Figure 3.8, the slope of the nth segment is

$$a'_n = \frac{y_n - y_{n-1}}{x_n - x_{n-1}}. \tag{3.342}$$

This function is substituted into (3.72) and (3.74) with the help of (3.71) and (3.73) and integrated to obtain values for C_m and D_m. The integral is converted into a summation over the N segments in the groove profile.

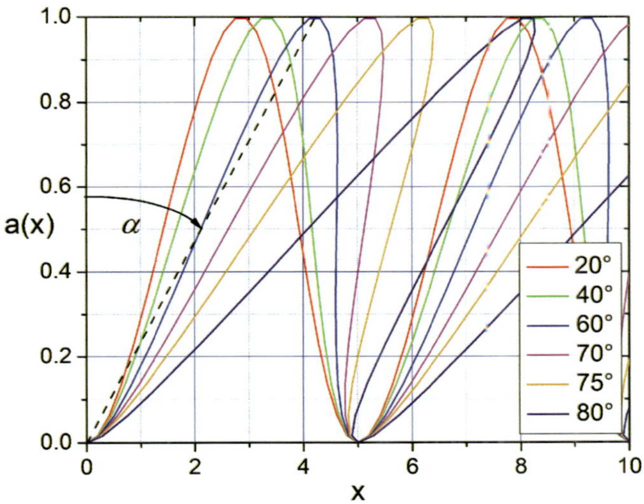

FIGURE 3.7 Form of slanted grating using parametric equations.

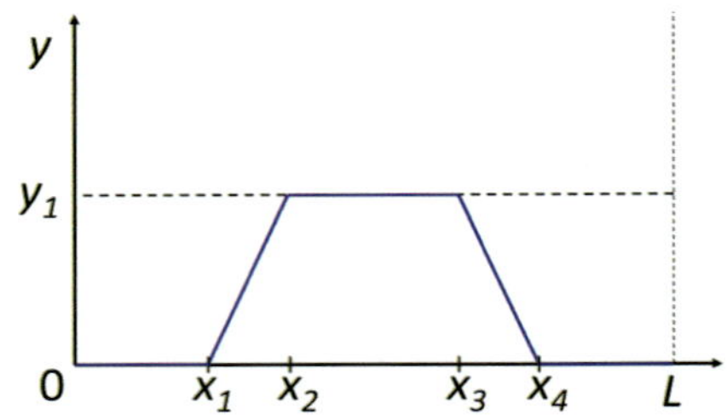

FIGURE 3.8 Trapezoidal groove profile.

$$C_m = \frac{1}{p}\int_0^p \frac{1}{1+(a')^2} e^{-imKu} du = \begin{cases} \sum_{n=1}^{N} \frac{x_n - x_{n-1}}{1+(a'_n)^2} & m = 0 \\ \sum_{n=1}^{N} \left(\frac{i}{pK}\right) \frac{1}{1+(a'_n)^2} \left(e^{-ipKx_n} - e^{-ipKx_{n-1}}\right) & m \neq 0 \end{cases}. \quad (3.343)$$

$$D_m = \frac{1}{p}\int_0^p \frac{a'}{1+(a')^2} e^{-imKu} du = \begin{cases} \sum_{n=1}^{N} \frac{a'_n (x_n - x_{n-1})}{1+(a'_n)^2} & m = 0 \\ \sum_{n=1}^{N} \left(\frac{i}{pK}\right) \frac{a'_n}{1+(a'_n)^2} \left(e^{-ipKx_n} - e^{-ipKx_{n-1}}\right) & m \neq 0 \end{cases}. \quad (3.344)$$

The L_m functions for the incident field expansion must also be calculated from (3.113) for non-sinusoidal groove profiles.

$$L_m(\beta_0) = \frac{1}{L}\int_0^L e^{-ia(u)\beta_0} e^{-imKu} du. \quad (3.345)$$

For the linear segmented groove profile, the integral becomes a summation. For a linear segment with endpoints (x_{n-1}, y_{n-1}) and (x_n, y_n),

$$a_n(u) = \left(\frac{y_n - y_{n-1}}{x_n - x_{n-1}}\right)(u - x_{n-1}) + y_{n-1} \equiv s_n u + t_n \quad (3.346)$$

where

$$s_n \equiv \frac{y_n - y_{n-1}}{x_n - x_{n-1}} \quad (3.347)$$

and

$$t_n \equiv y_{n-1} - s_n x_{n-1}. \quad (3.348)$$

Then,

$$L_m(\beta_0) = \frac{1}{L}\sum_{n=1}^{N} X_n \quad (3.349)$$

where

$$X_n = \begin{cases} e^{-it_n\beta_0}\left(x_n - x_{n-1}\right) & \beta_0 s_n + mK = 0 \\ \dfrac{ie^{-i\beta_0 t_n}}{\left(\beta_0 s_n + mK\right)}\left(e^{-i(\beta_0 s_n + mK)x_n} - e^{-i(\beta_0 s_n + mK)x_{n-1}}\right) & \text{otherwise} \end{cases}. \tag{3.350}$$

The L_{m-n} functions for the reflected field expansion must be calculated from (3.153) for non-sinusoidal groove profiles.

$$L_{m-n}\left(-\beta_n\right) = \frac{1}{L}\int_0^L e^{ia(u)\beta_n} e^{i(m-n)Ku}\,du. \tag{3.351}$$

Therefore,

$$L_{m-n}\left(-\beta_n\right) = \frac{1}{L}\sum\nolimits_{n=1}^{N} X_{m-n} \tag{3.352}$$

where

$$X_n = \begin{cases} e^{+it_n\beta_n}\left(x_n - x_{n-1}\right) & \beta_n s_n + (m-n)K = 0 \\ -\dfrac{ie^{+i\beta_n t_n}}{\beta_n s_n + (m-n)K}\left(e^{+i(\beta_n s_n + (m-n)K)x_n} - e^{+i(\beta_n s_n + (m-n)K)x_{n-1}}\right) & \text{otherwise} \end{cases}. \tag{3.353}$$

Finally, the $L_{m-n}^{(t)}$ functions for the transmitted field expansion must be calculated from (3.196) for non-sinusoidal groove profiles.

$$L_{m-n}^{(t)}\left(+\beta_n^{(t)}\right) = \frac{1}{L}\int_0^L e^{-ia(u)\beta_n^{(t)}} e^{-i(m-n)Ku}\,du. \tag{3.354}$$

Comparison of (3.352) and (3.354) gives immediately

$$L_{m-n}^{(t)}\left(+\beta_n^{(t)}\right) = \frac{1}{L}\sum\nolimits_{n=1}^{N} X_{m-n}^{(t)} \tag{3.355}$$

where

$$X_{m-n}^{(t)} = \begin{cases} e^{-it_n\beta_n^{(t)}}\left(x_n - x_{n-1}\right) & s_n = m - n = 0 \\ \dfrac{ie^{-i\beta_n^{(t)} t_n}}{\beta_n^{(t)} s_n + (m-n)K}\left(e^{-i\left(\beta_n^{(t)} s_n + (m-n)K\right)x_n} - e^{-i\left(\beta_n^{(t)} s_n + (m-n)K\right)x_{n-1}}\right) & \text{otherwise} \end{cases}. \tag{3.356}$$

3.5 EXAMPLES

3.5.1 SURFACE PLASMONS

In Chapter 1 the Kretschmann configuration was considered for launching surface plasmons via a prism coupler. Not surprisingly, diffraction gratings can be used for the same purpose. Consider the grating in Figure 3.9. A glass substrate with a sinusoidal grating that is 55 nm deep with a 0.8 μm period is coated with a 50 nm layer of gold. Light at a wavelength of 1 μm is incident from air onto the grating at a variable angle of incidence in the plane perpendicular to the grooves. The reflectance is plotted in Figure 3.10 by including ±5 diffracted orders.

Just as there is an approximate equation for calculating the surface plasmon resonance angle in the Kretchmann configuration, there is also one for gratings which is again based on the energy and momentum conservation laws. In this case, however, the grating can impart momentum to the photon in ±integral increments of the grating vector $2\pi / p$ where p is the grating period. The general approximate equation, which neglects the imaginary part of the dielectric constant of the metal film, is [17]

$$(n_i \sin\theta)^2 + \frac{2mn_i\lambda}{p}\sin\theta\cos\phi + \left(\frac{m\lambda}{p}\right)^2 = \frac{\epsilon_i \operatorname{Re}(\epsilon_m)}{\epsilon_i + \operatorname{Re}(\epsilon_m)} \tag{3.357}$$

where n_i is the refractive index of the incident medium, and m is a positive or negative integer (corresponding to the number of grating vectors added to or subtracted from the incident photon wavevector). In this example, the incident beam is in the plane perpendicular to the grooves, so $\phi = 0°$ and (3.345) reduces to

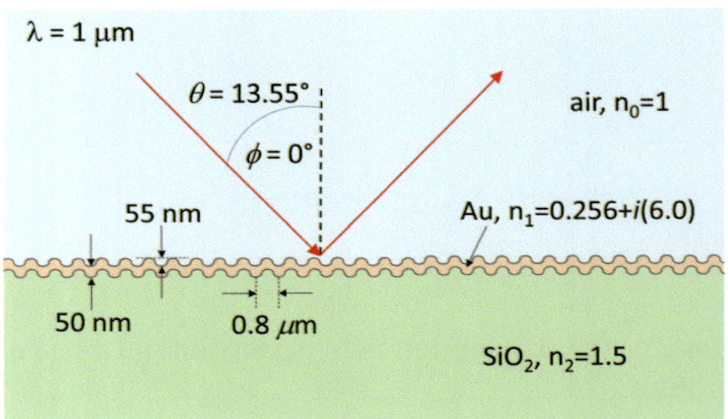

FIGURE 3.9 Diffraction grating in glass substrate with 50 nm gold film deposited conformally.

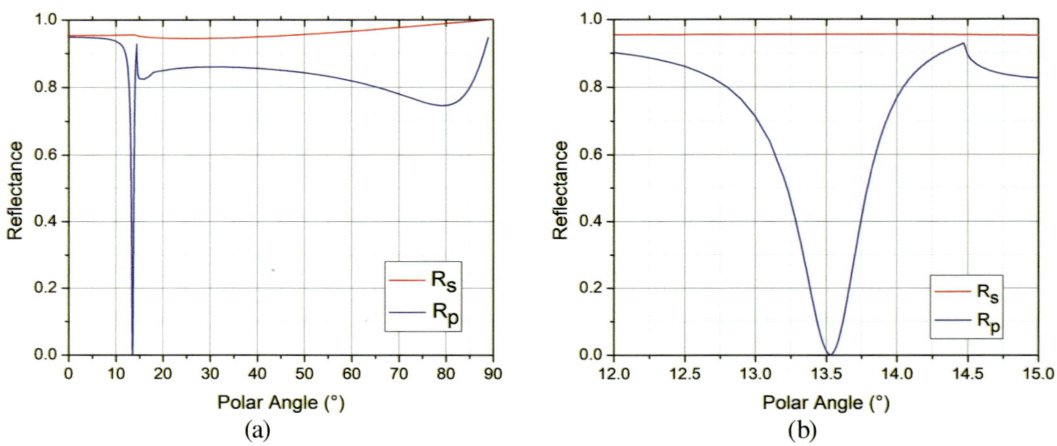

FIGURE 3.10 Incident beam in the plane perpendicular to the grooves. (a) Reflectance for incident s- and p-polarizations as a function of polar angle of incidence. (b) Expanded region around the polar angle for excitation of the surface plasmon.

$$\sin\theta = \frac{m\lambda}{n_i\, p} \pm \sqrt{\frac{\mathrm{Re}(\epsilon_m)}{\epsilon_i + \mathrm{Re}(\epsilon_m)}}. \tag{3.358}$$

The $m = 1$ resonance is predicted from (3.358) at $\theta \sim 13.6°$ which is very close to the more accurate vector resonance calculation in Figure 3.10.

If the plane of incidence is parallel to the grooves ($\phi = 90°$) the resonance equation is

$$\sin\theta = \sqrt{\frac{\mathrm{Re}(\epsilon_m)}{\epsilon_i + \mathrm{Re}(\epsilon_m)} \pm \left(\frac{m\lambda}{p}\right)^2}. \tag{3.359}$$

If the grating period is increased to 2 μm and the groove depth to 140 nm with all other parameters remaining the same, then (3.346) predicts a resonance at ~61.9° which again is very close to the vector calculation result in Figure 3.11(b).

There are some interesting features of these plots. The surface plasmon resonance always occurs for the incident polarization that is perpendicular to the grooves of the grating. In Figure 3.10, the resonance occurs in the blue R_p plot, while in Figure 3.11 it occurs in the red R_s plot. There is another reflectance anomaly in Figure 3.11 at ~10° which is a reflectance peak rather than a dip. This is due to the $m = 2$ result from (3.346) for which the resonance is predicted at ~9.7°. Two grating vectors are coupled to the incident photon to provide the necessary momentum to excite this second order surface plasmon.

As described in [2], it is possible to make use of the surface plasmon effect and the symmetry breaking effect of a diffraction grating to cause conversion of light polarization from S to P and vice versa. Figure 2 in [16] demonstrates polarization conversion for a grating shown in Figure 3.12(a). The permittivity value of the gold substrate is not explicitly given in the reference. The refractive index value in Figure 3.12 is from the literature. When the incident light is polarized perpendicular to the grating with the azimuthal angle $\phi = 0°$, there is no polarization conversion as shown in Figure 3.12(b). When the same grating is oriented at an azimuthal angle of 45° to the incident light, polarization conversion occurs. As shown in Figure 3.13(a) (Figure 2(b) of [2]), a small fraction of the incident TM polarization is reflected as TE polarization at the resonance angle of 18.1°. Polarization conversion also occurs, unsurprisingly, for incident TE polarization as shown in Figure 3.13(b).

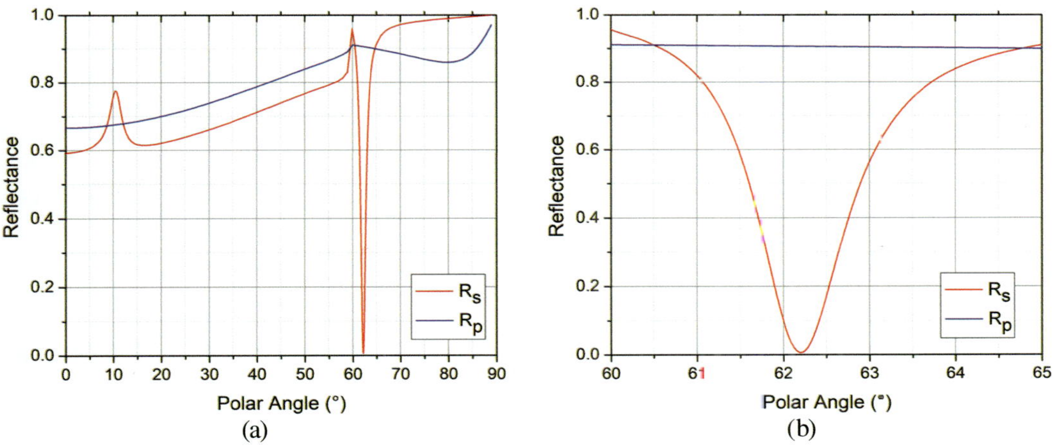

FIGURE 3.11 Incident beam in the plane perpendicular to the grooves. (a) Reflectance for incident *s*- and *p*-polarizations as a function of polar angle of incidence. (b) Expanded region around the polar angle for excitation of the surface plasmon.

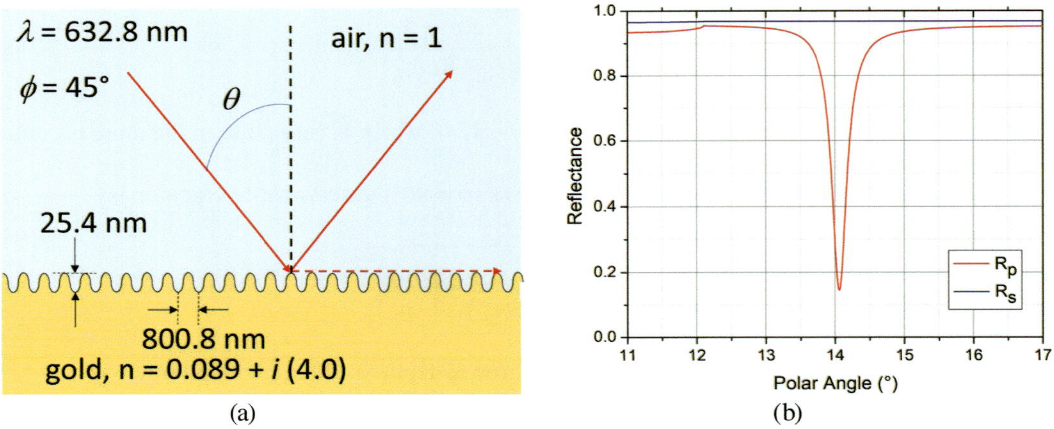

FIGURE 3.12 (a) Gold diffraction grating. (b) Reflectance for an azimuthal angle of 0° (incident plane perpendicular to the grating).

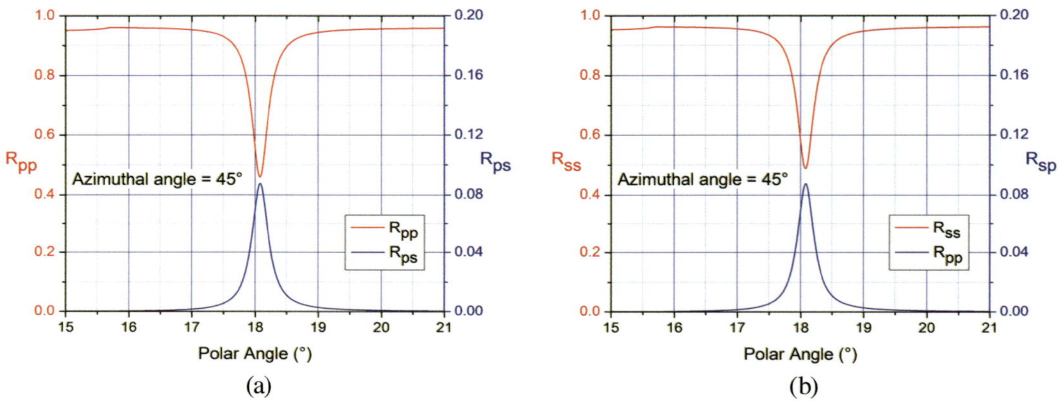

FIGURE 3.13 (a) For incident TM (p) polarization, the zero-order TM (p) and TE (s) reflectances are plotted as a function of polar angle of incidence for 45° azimuthal angle of incidence. (b) Same as (a) but for incident TE (s) polarization.

Polarization conversion as a function of the azimuthal angle of incidence is discussed in [18]. The silver grating structure is shown in Figure 3.14(a). The surface plasmon resonance occurs for a polar angle of 14.1° for incident light perpendicular to the grooves and continuously increases to 41.7° for light incident parallel to the grooves. The reflected light for *s*-polarization as a function of azimuthal angle when the incident light is *p*-polarized is shown in Figure 3.14(b) using ±5 diffracted orders along with a $\sin^2\phi$ curve fit.

3.5.2 Silver Grating With Dielectric Overcoat

[16] provides a second example for which the incident beam is TM-polarized at an azimuthal angle of 45° but the silver grating has a dielectric overcoat. The dielectric material and thickness are not specified in the paper. The parameters in Figure 3.15 are partly taken from those in Figure 4 of [3]. In this case, with the grating oriented at 45° to the plane of incidence, the symmetry for the two orthogonal incident polarizations is broken, thereby enabling conversion of the incident light into the opposite polarization upon reflection as shown in Figure 3.16. With a relatively thick dielectric layer on top of the silver substrate, it is possible to support TE and TM waveguide modes in addition

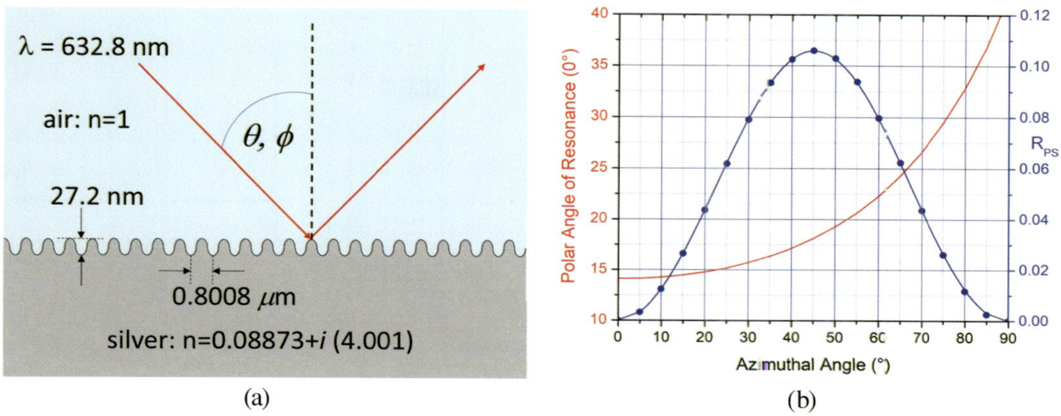

FIGURE 3.14 (a) Silver grating to demonstrate polarization conversion. (b) Polarization conversion from incident TM to reflected TE. The peak conversion occurs at $\phi = 45°$. The solid blue line is a curve fit $\propto \sin^2 \phi$. The red curve is the value of the polar angle at the maximum polarization conversion. As ϕ increases, θ_{max} also increases monotonically.

After Figure 3 of [16].

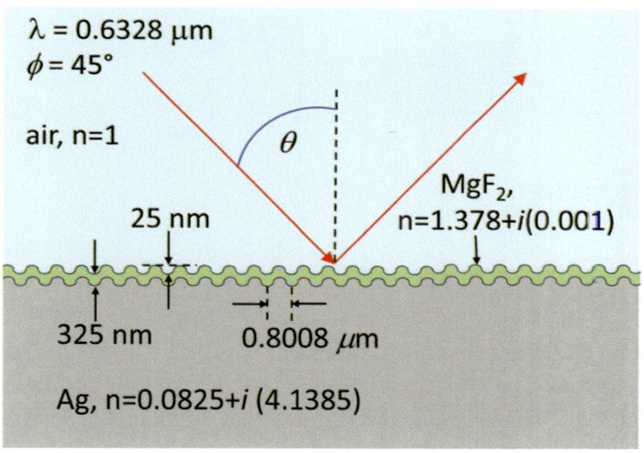

FIGURE 3.15 Example of polarization conversion for silver substrate with dielectric overcoat.

to the surface plasmon mode. Note that the linewidths of the waveguide modes are considerably narrower than that of the surface plasmon. This is due to the fact that the electric fields for the waveguide modes are much smaller inside the silver substrate and are, therefore, much less heavily damped by absorption in the silver.

3.5.3 Sensing With Surface Plasmons and Gratings

The evanescent field from a propagating surface plasmon extends into the neighboring dielectric region. If this region can be modified by an environmental effect, then the surface plasmon system can become an extremely sensitive sensor. As an example of a plasmonic sensor, consider the sensor system in Figure 3.17(a) [19]. In this case, the light source is an LED with a center wavelength of 635 nm. It is focused onto a plastic disk that has an embossed spiral grating with a 0.7 μm period (similar to a CD or DVD). The groove can be approximated as trapezoidal with a width ~0.36 μm and land ~0.30 μm as shown in Figure 3.17(b). The disk substrate is coated with an optically thick layer of silver and protected by a 100 nm overcoat of the transparent dielectric YO_x. A 100 nm layer

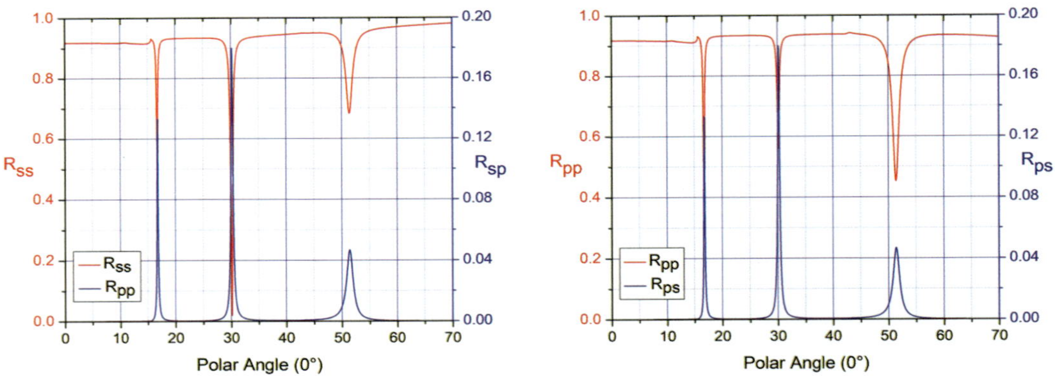

(a) (b)

FIGURE 3.16 For the geometry in Figure 3.15, the polarized mode reflectances are calculated for (a) incident TE (s-)polarization, and (b) incident TM (p-)polarization. After Figure 3 of [16].

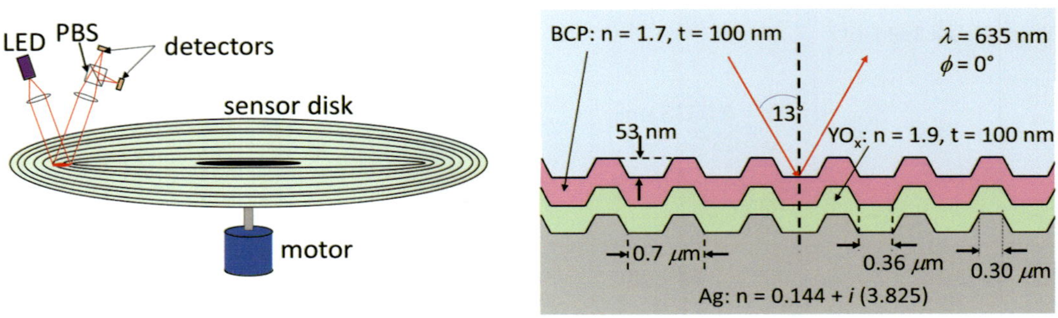

(a) (b)

FIGURE 3.17 (a) Apparatus for sensor disk. A motor spins the disk. The beam from an LED is collimated and reflected from the surface of the disk with a wavelength of 635 nm at a polar angle of 13°. A polarizing beam-splitter (PBS) splits the *p*- and *s*-polarized reflected light, which are detected separately. (b) Grating-based film stack for ammonia sensing. The polycarbonate substrate has an embossed spiral grating structure with a groove depth of 53 nm and period of 0.7 μm that is coated conformally with an optically thick layer of silver (100 nm), a protective 150 nm layer of YO_x dielectric, and a 100 nm layer of the pH indicator bromocresol purple.

of the pH indicator, bromocresol purple (BCP), is evaporated on top of the YO_x. As deposited, the BCP is clear to slightly yellow.

Ammonia in the air reacts with BCP and turns it pink. The refractive index of the BCP in the wavelength range of the LED slightly changes. This affects the resonance condition of the surface plasmon. One way to sense the presence of ammonia is to measure changes in the resonance angle as shown in Figure 3.18(a). This is indeed how many surface plasmon sensors operate. However, an alternate technique was developed for this sensor. In this case, the YO_x film was not a planar film but rather deposited with a thickness wedge so that at one edge of the disk substrate its thickness was only 100 nm, while at the opposite edge of the substrate the thickness was 200 nm. For a fixed angle of incidence of the beam from the LED, there are two spots around the circumference of the disk where the YO_x film is the correct thickness to support a surface plasmon resonance. As the ammonia reacts with the BCP and changes the resonance condition, the two resonance spots on the disk move closer together around the circumference of the disk. The motor spins at a constant rate and the difference in time between when the two resonances are detected is measured. As the resonance spots move closer together on the disk, the time difference between resonance measurements gets shorter. The time for a complete revolution of the disk is also measured and used to normalize the resonance timing so that absolute time measurements are not required. The entire sensor electronics can be constructed very inexpensively with a microcontroller chip. In a test of this system as shown in Figure 3.18(b), as ammonia slowly diffused across the glove box, the background noise level was equivalent to an ammonia concentration of ~5 ppbv after three minutes.

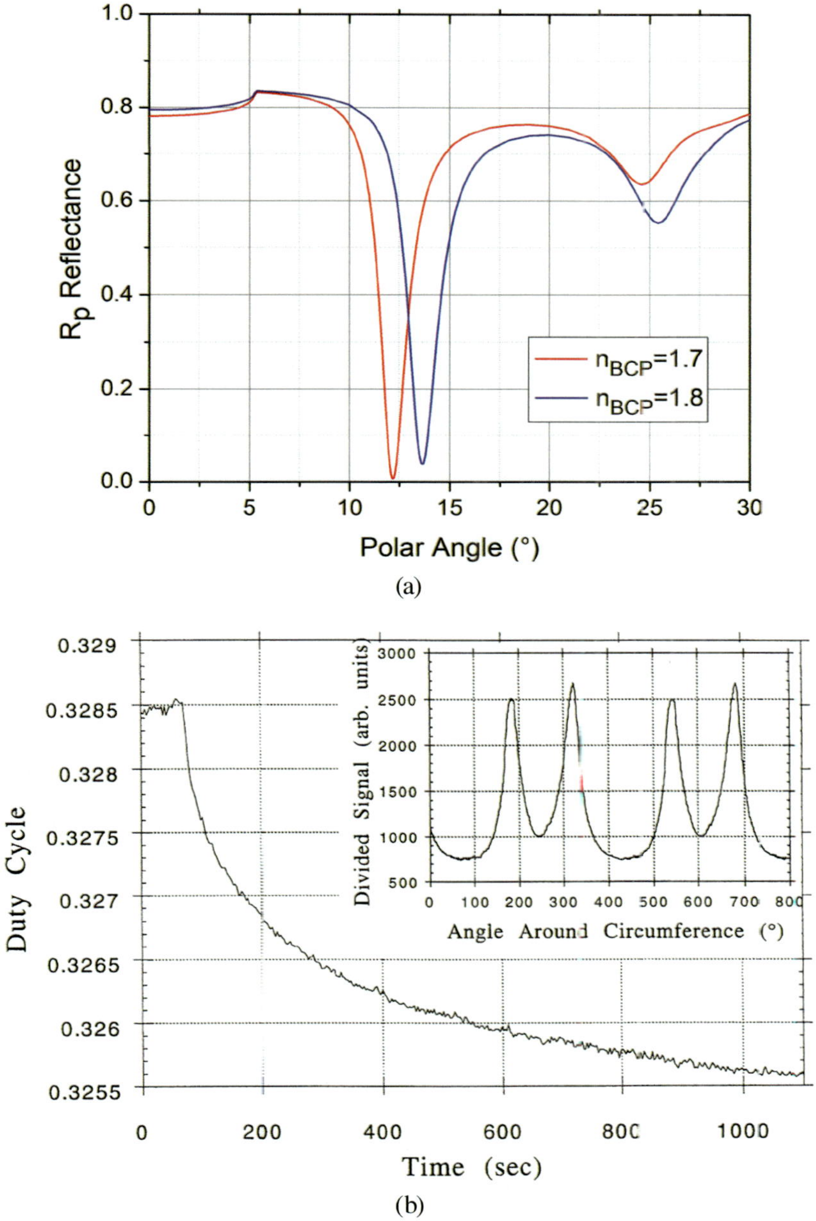

FIGURE 3.18 (a) Surface plasmon resonance angle is strongly affected by a change in refractive index of the dielectric layer above the metal film. In this case, a refractive index change of 0.1 causes the reflectance at 12.2° to change from nearly zero to 0.68. (b) Response of sensor upon exposure to 100 ppbv of ammonia.

Reprinted from *Sens. Act.* B56, W. A. Challener, R. R. Ollmann and K. K. Kam, "A surface plasmon resonance gas sensor in a 'compact disc' format," 254–258 (1999) with permission from Elsevier.

3.5.4 Biosensing With Gratings and Waveguides

Diffraction gratings can also couple light into dielectric waveguide modes. These modes typically give rise to much narrower resonances than surface plasmons because of their much lower optical losses. The evanescent tail of a waveguide mode can also extend into the surrounding medium, making it sensitive to refractive index changes in the medium. Consider the waveguide film stack shown in Figure 3.19(a).

Light incident from air at ~34° into the water solution through a transparent window gets refracted to ~25° from which it can excite a TE planar waveguide mode in the dielectric films as shown in Figure 3.20. The grating also efficiently couples light back out into the water from the mode propagating in the films, generating a narrow peak in the reflectance. Waveguides support modes of both polarizations and thicker waveguides can support multiple modes giving rise to multiple peaks in the reflectance.

The dielectric surface of the waveguide can be functionalized with various biological molecules like antibodies, antigens, biotin, or avidin. Antibodies bind to their corresponding antigen, and avidin

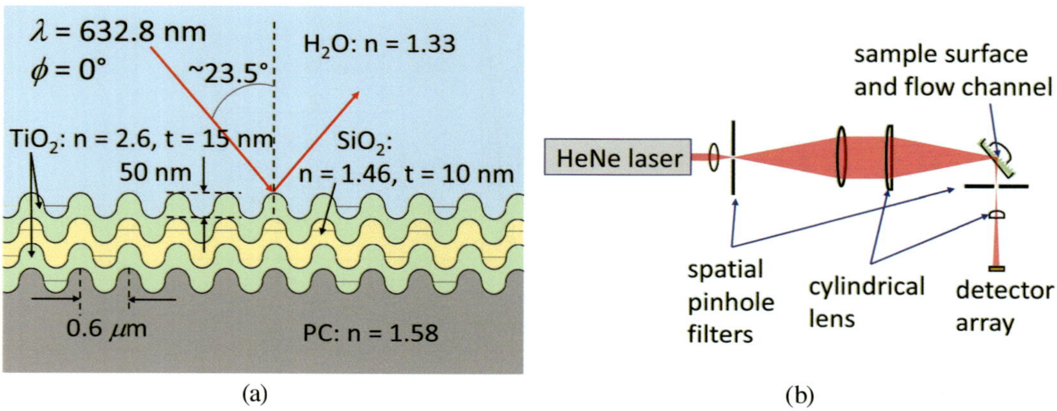

FIGURE 3.19 (a) Three-layer waveguide deposited onto a grating for biosensing. (b) Design of biosensor optics from [18].

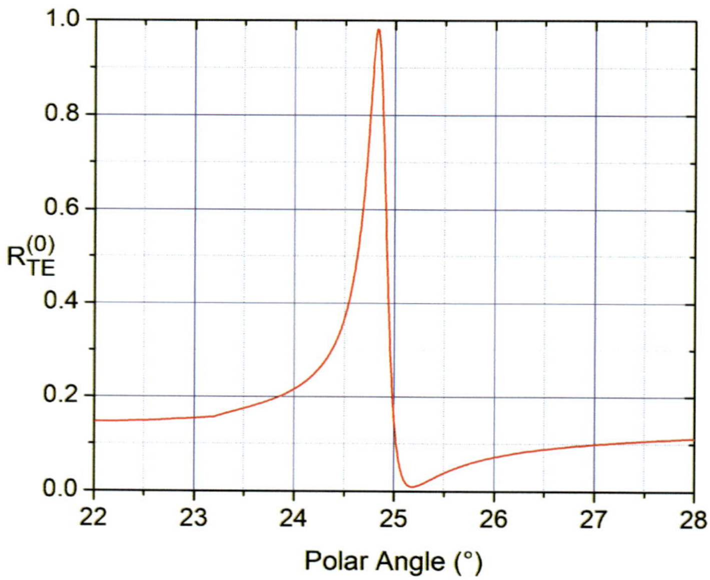

FIGURE 3.20 Reflectance of zero TE order for film stack in Figure 3.19(a) with a sinusoidal grating.

binds to biotin. If a liquid sample flows across a functionalized surface and molecules in the liquid are present that can bind to the surface, these molecules displace water (or buffer solution), thereby creating a larger index of refraction at the surface of the waveguide. The mode index of the waveguide increases slightly, thereby shifting the resonance angle. Biotin is a small molecule with a molecular weight of 244 Da while avidin is a large protein with a molecular weight of 67–68 kDa. When the surface of this sensor is functionalized with biotin and a solution containing avidin is passed across it, there is an easily detectable resonance shift, as shown in Figure 3.21(a) [20]. When the surface is functionalized with avidin, and a solution containing biotin is passed across it, there is a much smaller but still easily detectable shift in resonance angle of ~0.5 mdeg as shown in Figure 3.21(b). The surface sensitivity of this biosensor was estimated as ~0.3 pg/mm^2 with a 0.4 s integration time.

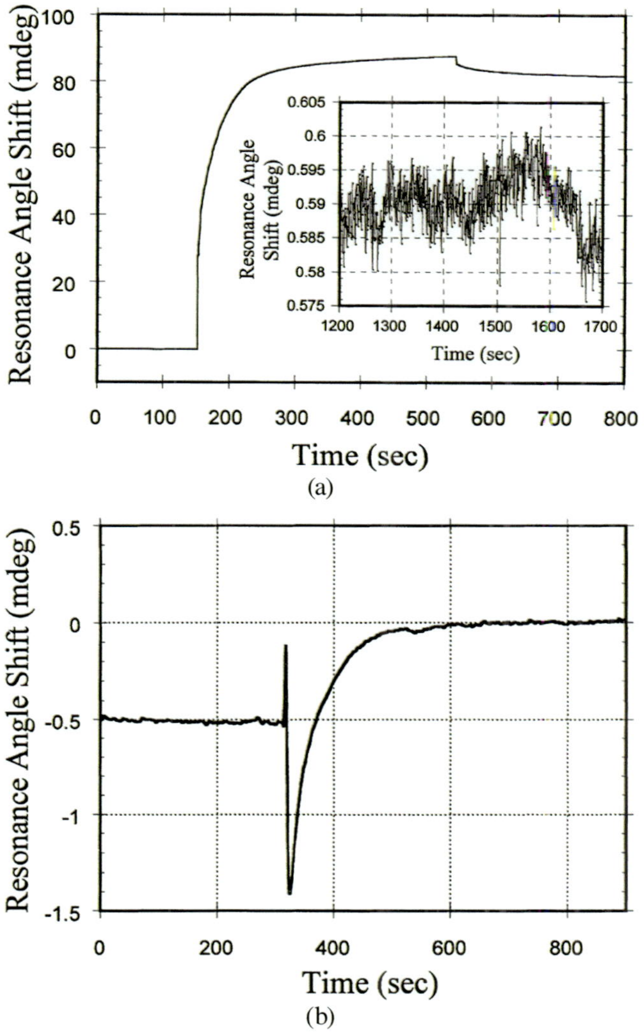

FIGURE 3.21 (a) Shift in resonance angle with time as avidin binds to a biotin-functionalized sensor surface. (b) Shift in resonance angle with time as biotin binds to an avidin-functionalized sensor surface. The initial negative spike in the figure is due to a small change in temperature as the flowing liquid is switched from pure buffer solution to a solution with biotin.

Reprinted from *Sens. Act. B* **71**, W. A. Challener, J. D. Edwards, R. W. McGowan, J. Skorjanec and Z. Yang, "A multilayer grating-based evanescent wave sensing technique," 42–46 (2000) with permission from Elsevier.

3.5.5 Thick Dielectric Overcoat

When a thin film becomes relatively thick (on the order of 1 μm) the exponentially growing terms of the evanescent fields in the expansion cannot be handled with precision by the computer in the transfer matrix method. The scattering matrix method, discussed previously, was developed in [3] to re-order the way in which the mathematical terms are handled so that this is not a problem. To demonstrate this capability, the example in Figure 2 of [3] is shown in Figure 3.22(a) and the reflectance calculated by both the transfer matrix and the scattering matrix techniques is plotted in Figure 3.22(b).

3.5.6 Effect of Skew

The effect of grating skew was also considered in [3]. The film stack used for Figure 3 in that reference is shown in Figure 3.23(a) and the results in (b).

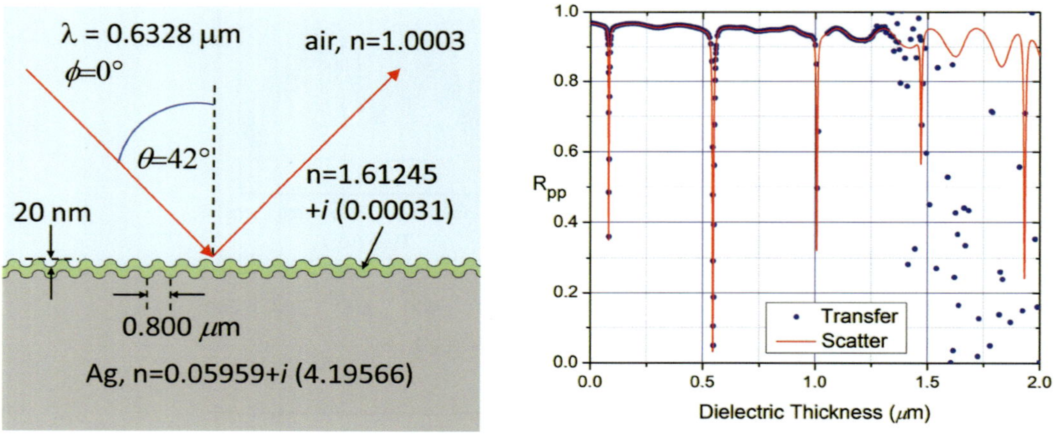

(a) (b)

FIGURE 3.22 (a) Film stack for comparison of transfer matrix method to scattering matrix method. (b) Plot of p-polarized reflectance as a function of dielectric layer thickness. At ~1.25 μm, the transfer matrix method loses accuracy.

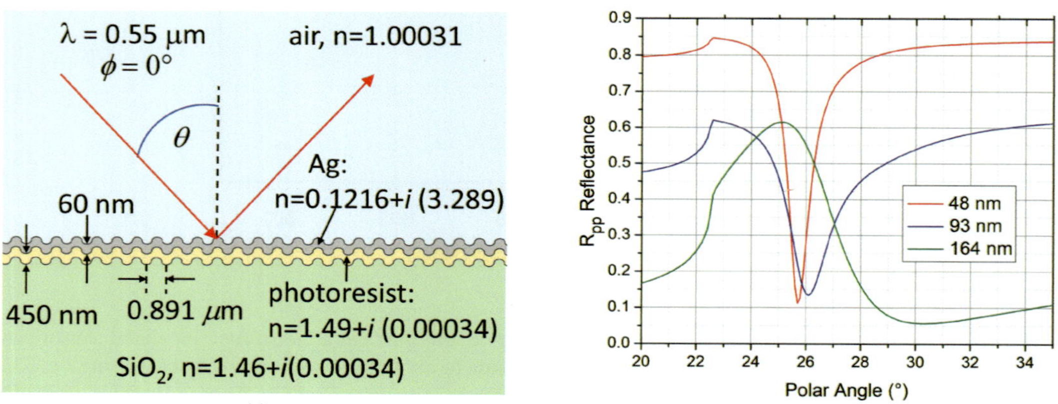

(a) (b)

FIGURE 3.23 (a) Film stack to investigate the effect of skew from Figure 3 in [3]. (b) p-polarized reflectance for groove depths of 48, 93, and 164 nm, respectively, and skews of $\gamma = 0.868$, 1.038, and 0.674, respectively.

3.5.7 Effect of Groove Depth

Consider the gold grating in Figure 3.24(a). The minimum in R_{PP} at a groove depth of 0.09 µm in Figure 3.24(b) is due to the surface plasmon excitation. The other minima are apparently simple interference effects. For an incident wavelength of 700 nm, the interference minima might be expected to occur when the light reflected from the bottom of the groove returns out-of-phase with that at the top of the groove, that is, at groove depths of 0.35 µm + $m\lambda$ where m is an integer. This relation seems to hold well for the TE polarization but is clearly modified for the TM polarization due to the interaction with the surface plasmon mode. At 1.7 µm groove depth, the R_{ss} reflectance is larger than 1, demonstrating that even the scattering matrix algorithm fails for grooves that are too deep. The range of accurate groove depths is not clear from this plot but perhaps extends to ~1 µm. This plot was calculated using only ±5 diffracted orders. By increasing the number of diffracted orders, the calculation can often be improved, especially in the case of deep grooves.

Adding additional orders to the calculation in Figure 3.25 has improved the accuracy of the *s*-polarized reflectance calculation to perhaps a groove depth of ~1.7 µm for the ±10 order calculation. The ±15 order calculation exhibits anomalies between 0.6 and 0.75 µm groove depth indicating that the precision in the calculation is beginning to breakdown. Interestingly, the R_{pp}

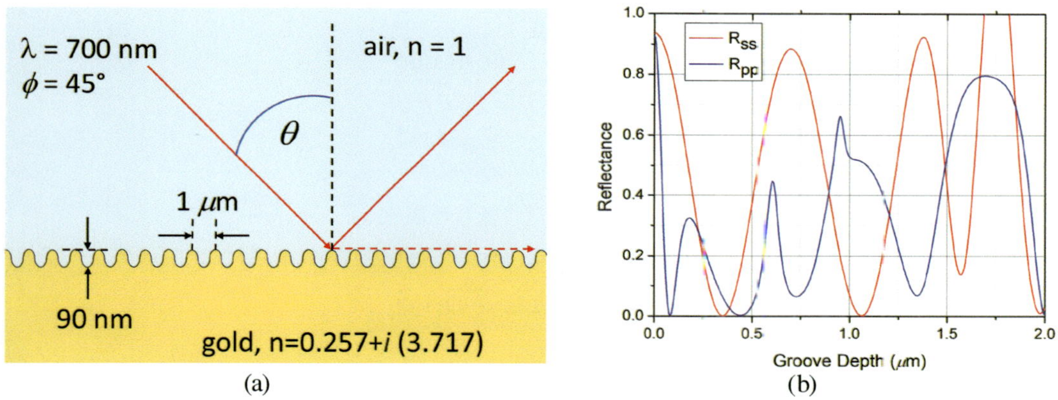

FIGURE 3.24 (a) Simple gold grating. (b) Zero order reflectance for both incident polarizations as a function of groove depth using ±5 diffraction orders in the calculation. $\theta = 20°$. $\phi = 0°$.

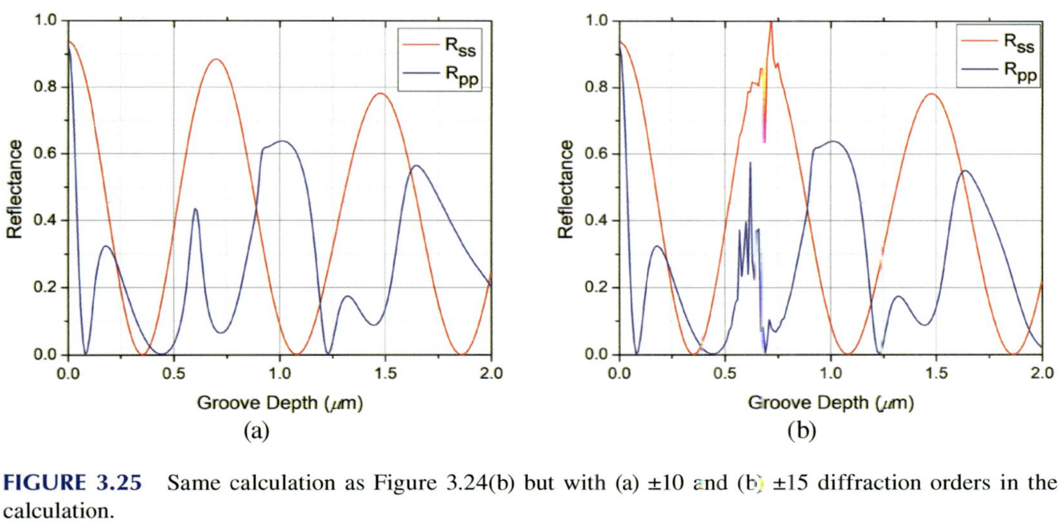

FIGURE 3.25 Same calculation as Figure 3.24(b) but with (a) ±10 and (b) ±15 diffraction orders in the calculation.

results for the ±10 order and ±15 order calculations are also very similar out to ~1.75 μm groove depth except for the region between 0.5 and 0.7 μm groove depth where precision problems again are evident.

Another example of the effect of groove depth is considered for an aluminized grating in Figures 5 and 6 of [1]. The grating profile is shown in Figure 3.26. The zero order and sum of zero and first order reflectance is shown for each incident polarization in Figure 3.27.

3.5.8 Multilayer Littrow Mount Diffraction Efficiency

Multilayer dielectric gratings have been designed for high efficiency in the Littrow mount configuration [21]. For this configuration, the negative first order is reflected back along the path of the incident beam. In the plane of incidence perpendicular to the grating, the equation for the angle of incidence in the Littrow mount is

$$\sin\theta = \frac{\lambda}{2pn_i}. \tag{3.360}$$

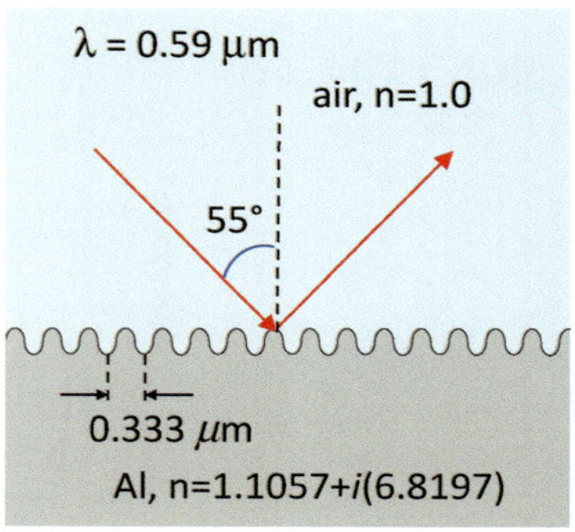

FIGURE 3.26 Aluminized diffraction grating.

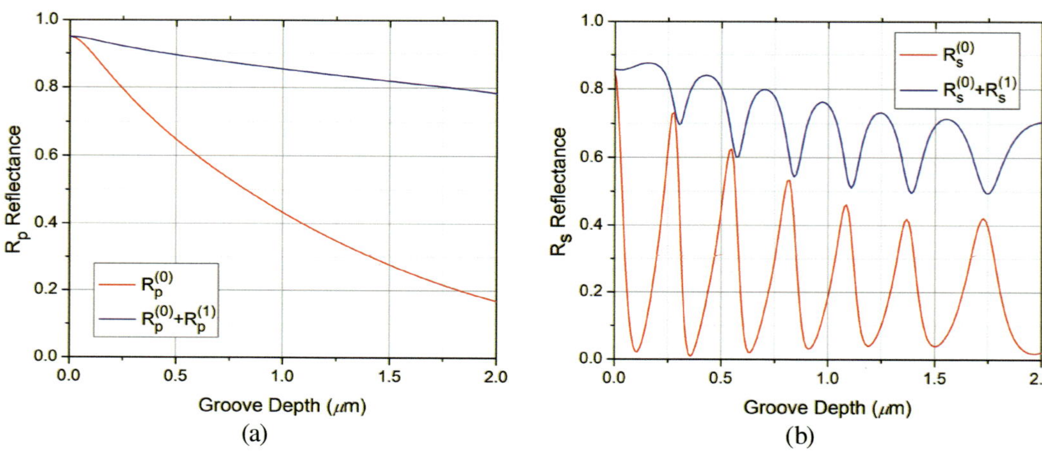

FIGURE 3.27 Efficiency of aluminized diffraction grating as a function of groove depth to grating period ratio for (a) incident TE polarization, and (b) incident TM polarization with the plane of incidence perpendicular to the grooves.

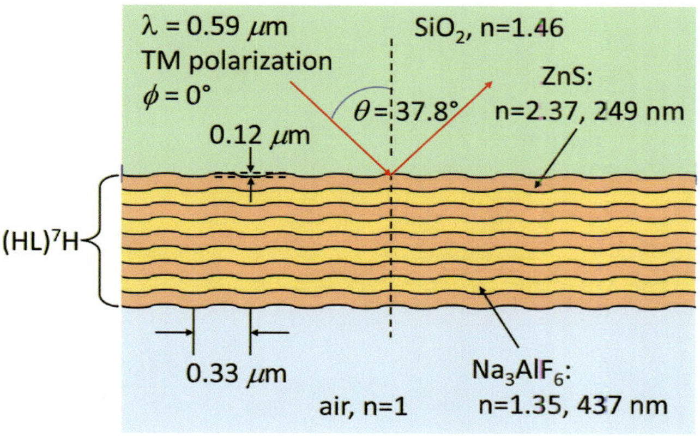

FIGURE 3.28 Multilayer dielectric grating in Littrow mount configuration with light incident from the substrate-side.

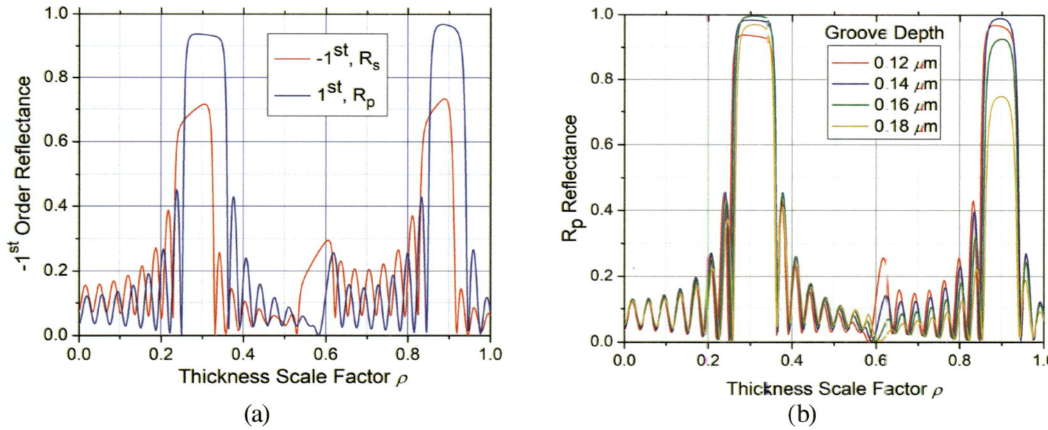

FIGURE 3.29 (a) Calculation of -1st order reflectance for both *s*- and *p*-polarizations for a groove depth $d =$ 0.12 µm. The *p*-polarized efficiency curve is after Figure 1 in Ref. [21]. (b) Same calculation for the *p*-polarization for three groove depths.

Consider the example in Figure 3.28 from Figure 1 in [21]. This film stack is a multilayer quarter-wave stack at the Littrow angle and is especially efficient for TM polarization as shown in Figure 3.29(a). The calculation is repeated for two other groove depths in Figure 3.29(b) demonstrating that a groove depth of ~0.16 µm provides the highest reflectance for the scale factor ~0.33.

A second example of a multilayer film on a diffraction grating, shown in Figure 3.30(a) after Figure 3 in [4], is a reflection coating on a silica grating substrate consisting of alternating layers of high and low index materials at quarter-wave thicknesses. The grating is designed for the Littrow mount in which the -1st diffraction order is reflected back into the incident beam. The multilayer film in this example consists of 15 layers in which the odd layers are high refractive index of ZnS with $n = 2.37$ and the even layers are low index of Na_3AlF_6 (cryolite) with $n = 1.35$, so the topmost and bottom most layers are formed from the high index material. This quarter-wave reflector is designed for incident *p*-polarization. The *s*-polarized reflectance at the Littrow angle is very poor for all scale factors as seen in Figure 3.30(b).

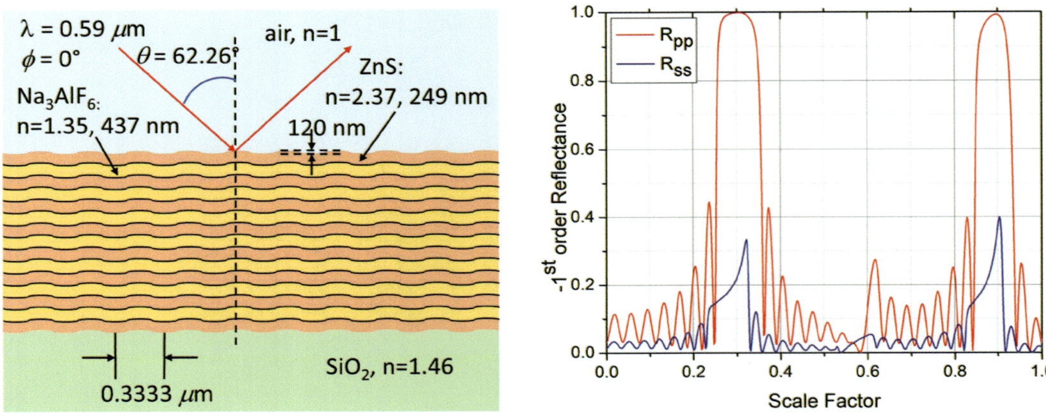

(a) (b)

FIGURE 3.30 (a) Multilayer dielectric grating in Littrow mount configuration with light incident from the air-side after Figure 3 in [4]. (b) *s*- and *p*-polarized reflectance.

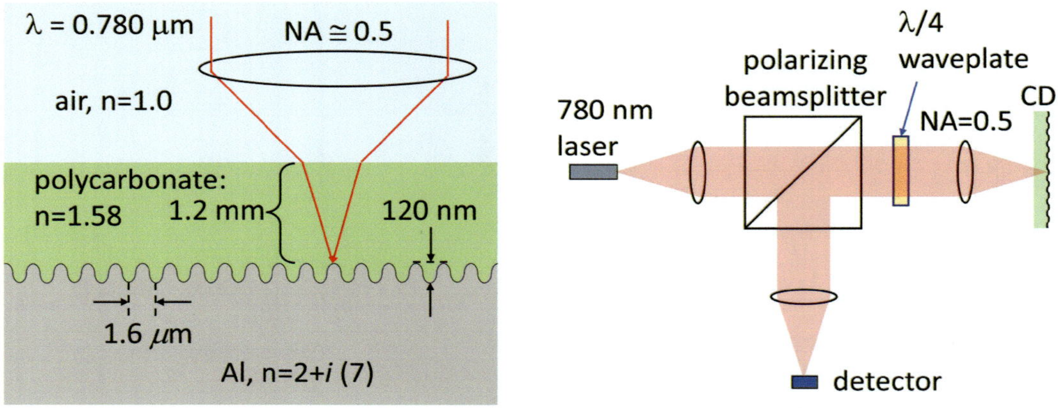

(a) (b)

FIGURE 3.31 (a) A polycarbonate substrate with molded pits and grooves is coated on its back side with an aluminum alloy to make a CD. (b) A simple CD readback system using a quarter-wave plate as an optical isolator.

3.5.9 Tracking Signals From an Optical Disk

Compact disks are typically polycarbonate substrates that are aluminized on the surface away from the optical system as shown in Figure 3.31. The pits form a spiral track on the disk and are typically ~120 nm deep. CD's are read at an optical wavelength of 780 nm. DVDs are read typically at 635 nm, and Blu-ray DVDs at 405 nm. Shorter wavelengths enable a smaller diffraction-limited focused spot corresponding to a higher bit density and storage capacity.

Consider the optical system in Figure 3.32(a) as described in [22]. The laser light is focused onto the disk through a lens with a numerical aperture of 0.4 at a wavelength of 632.8 nm. The disk is mounted backwards so that the light is incident directly through air onto the aluminized surface. The grating is shown in Figure 3.32(b) and consists of an opaque aluminum film deposited onto a thick plastic substrate with a 1.5 μm grating period, 0.5 μm groove width, and 70 nm groove depth.

The electric and magnetic fields of an incident light ray are refracted by a lens towards the focus as shown in Figure 3.33. In Figure 3.33(a), the incident electric field has field components in the $\hat{x}$

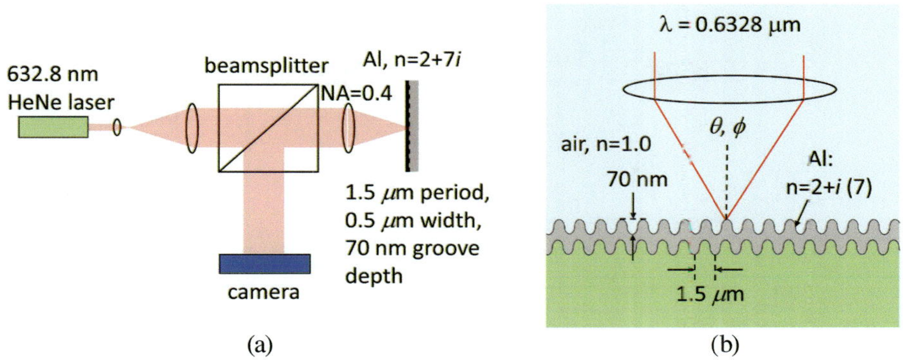

FIGURE 3.32 (a) Optical bench test setup described in [22]. (b) The polycarbonate substrate is coated with an optically thick aluminum layer and mounted "backwards" so that the aluminum layer is facing the optical system.

and $\hat{z}$ directions. Three light rays are diagrammed, for incident azimuthal angles of 0°, 45°, and 90°. The azimuthal angles of the refracted rays remain the same. The corresponding TM and TE components are shown both before and after refraction. The magnetic field components are shown in Figure 3.33(b). The E_ω and H_ω components of the incident light rays have already been determined in terms of the H_{TE} and H_{TM} fields in (3.134) and (3.135). From Figure 3.33(a) the electric fields are

$$E_{TM} = \sqrt{\cos\theta}\left(+\cos\phi E_x - \sin\phi E_y\right) \tag{3.361}$$

and

$$E_{TE} = \sqrt{\cos\theta}\left(+\sin\phi E_x + \cos\phi E_y\right). \tag{3.362}$$

The factor $\sqrt{\cos\theta}$ corrects for energy conservation upon refraction.

The magnetic fields are

$$H_{TM} = -\sin\phi E_x - \cos\phi E_y, \tag{3.363}$$

and

$$H_{TE} = +\cos\phi E_x + \sin\phi E_y. \tag{3.364}$$

For incident light polarized along the x-axis (perpendicular to the grooves),

$$E_\omega^{(i)} = \frac{\eta_0}{n_i}\left(-\sin^2\phi_i\cos\theta_i + \cos^2\phi_i\right), \tag{3.365}$$

and

$$H_\omega^{(i)} = -\sin\phi_i\cos\phi_i\left(1 + \cos\theta_i\right). \tag{3.366}$$

For incident light polarized along the y-axis (parallel to the grooves),

$$E_\omega^{(i)} = \frac{\eta_0}{n_i}\sin\phi_i\cos\phi_i\left(1 - \cos\theta_i\right), \tag{3.367}$$

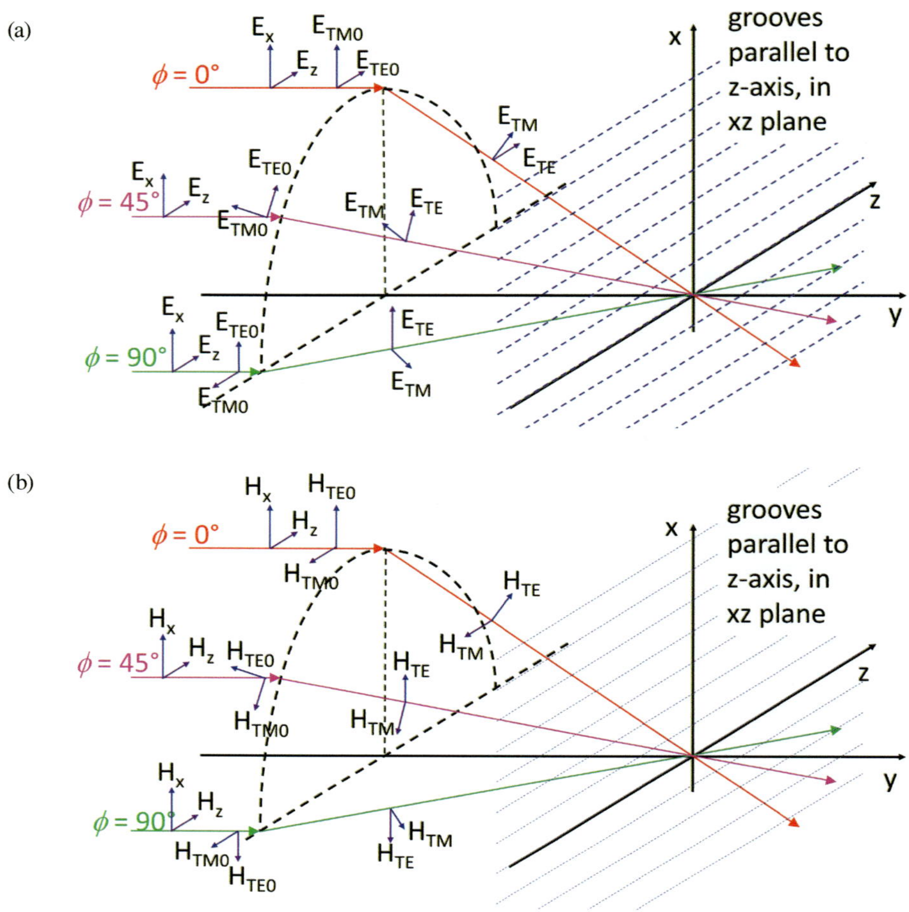

FIGURE 3.33 (a) The incident collimated light beam is refracted by a stigmatic lens to a focal point. The TE- and TM-polarized electric fields are illustrated for the light rays. (b) The TE- and TM-polarized magnetic fields are illustrated for the light rays.

and

$$H_\omega^{(i)} = -\sin^2\phi_i\cos\theta_i - \cos^2\phi_i. \tag{3.368}$$

Incident circularly polarized light has amplitude $(\hat{x} \pm i\,\hat{y})/\sqrt{2}$. From (3.362)–(3.365),

$$E_\omega^{(i)} = \frac{\eta_0}{\sqrt{2}n_i}\left[\left(-\sin^2\phi_i\cos\theta_i + \cos^2\phi_i\right) \pm i\,\sin\phi_i\cos\phi_i\left(1-\cos\theta_i\right)\right], \tag{3.369}$$

and

$$H_\omega^{(i)} = \frac{1}{\sqrt{2}}\left[-\sin\phi_i\cos\phi_i\left(1+\cos\theta_i\right) \mp i\left(\sin^2\phi_i\cos\theta_i + \cos^2\phi_i\right)\right]. \tag{3.370}$$

Two cases are considered to calculate the image on the camera in Figure 3.34, (a) for the incident electric field perpendicular to the grooves in the disk, and (b) for the incident electric field parallel to the grooves in the disk. Each point in the image is formed from light rays reflected and diffracted from the grooved surface of the disk. In particular, the lens recaptures all of the reflected zero order light rays, but only captures a portion of the ± first order light rays which have been shifted in polar angle from the normal by $\sin^{-1}\left(\frac{\lambda}{p}\right)=25°$. The calculation takes some time to complete as it is summing light rays at over 40,000 points in the image plane.

The bright large regions on left and right are caused by the overlap on the detector between the ± first reflected orders and the zero reflected order. Because the beam is centered on the groove, the ± first orders have the same phase and happen to interfere constructively with the zero order, thereby increasing the brightness in these regions of the image plane. At ~9.5°, corresponding to ~(±0.17,0) in the Figure 3.34, there is a bright line in (a) and a dimmer line in (b) that correspond to a surface plasmon resonance in the zero diffracted order. The resonance actually disappears in (b) for $y = 0$ ($\phi = 0°$, 90°) because the incident light is completely polarized parallel to the grooves, but for other azimuthal angles the line appears for incident light polarized parallel to the grooves, because after refraction by the lens there is also a small field component polarized perpendicular to the grooves. This surface plasmon resonance is actually a second order resonance. The approximate resonance angles are determined by conservation of momentum from the dielectric constants of the dielectric medium above the grating and the metallic medium of the grating. The approximate resonance equations were previously given in (3.354)–(3.356). In this example, $\epsilon_d = 1$, $\mathrm{Re}(\epsilon_m) = -45$, $\lambda = 0.6328\,\mu\mathrm{m}$, and $p = 1.5$ μm. The $m = 1$ resonance occurs at $\theta \cong 36.9°$ and is outside of the cone of light illuminated/captured by the 0.4 NA objective lens. The $m = 2$ resonance, however, is predicted to occur at $\theta = \pm 10.3°$ ($\sin\theta = \pm 0.18$) for $\theta = 0°$ on the vertical axis. If the Fourier transform of the grating profile has no second order sinusoidal terms, then this resonance should be very small as is the case here for which the reflectivity only changes by ~1% at the resonance.

The second-order resonance can be greatly increased in amplitude by changing the land/groove duty cycle of the grating from a symmetric 50%. For example, two trapezoidal grating profiles are shown in Figure 3.35. In each case, the land is 0.9 μm wide and the groove is 0.4 μm wide, so the land/groove symmetry has been removed. Also, though the groove profiles are identical, they are laterally shifted from each other. The profile in (a) is still symmetrically placed with respect to the focused spot, but the groove profile in (b) is asymmetric. Therefore, the calculated image should exhibit this asymmetry.

The calculated image on the camera is shown in Figure 3.36 and can be compared to the experimental result in [22]. The Chandezon algorithm as described in section for trapezoidal groove profiles is only suitable for shallow gratings and should be used with care, but works well in this case. There

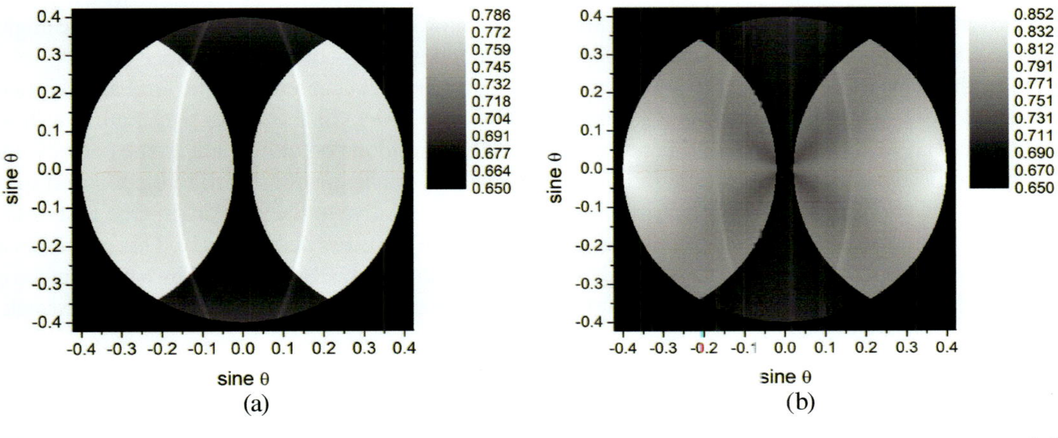

FIGURE 3.34 Calculated camera image for light polarized (a) perpendicular to the grooves and (b) parallel to the grooves.

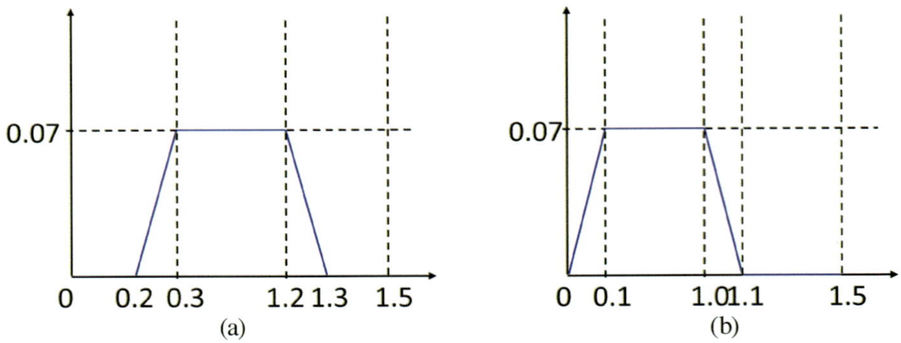

FIGURE 3.35 Trapezoidal grating profiles. (a) Symmetric profile leading to equal phases shifts between + and − orders. (b) Asymmetric profile for which the phase shifts of the ± orders have opposite signs and interfere with the zero order differently.

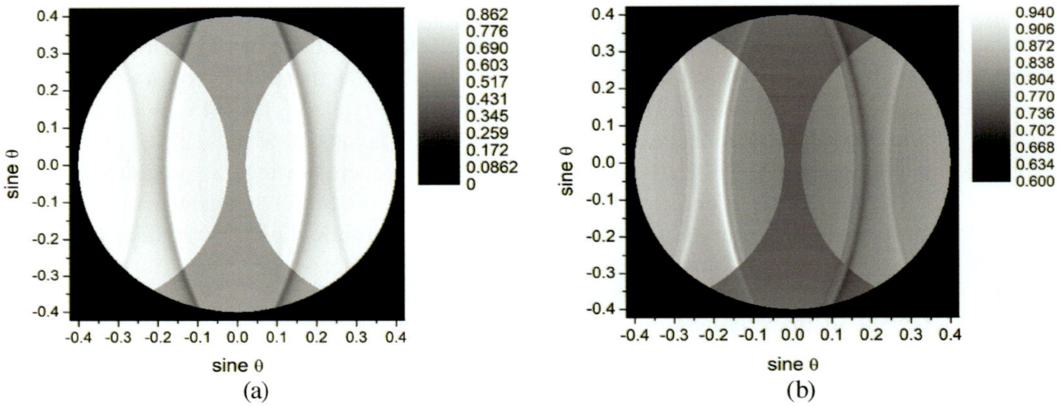

FIGURE 3.36 Calculated camera images for (a) a symmetric, trapezoidal groove profile, and (b) an asymmetric trapezoidal groove profile.

are two clearly visible lines due to surface plasmon resonances on either side of the center line. The $m = 1$ plasmon line is still outside of the capture range of the lens. The $m = 2$ plasmon lines are now visible running through the vertical $\sin\theta = 0$ points at the horizontal values $\sin\theta = \pm 0.17$. From (3.259) the $m = 3$ resonance should appear at $\theta = \pm 14°$ ($\sin\theta = \pm 0.24$) on the horizontal axis, as indeed it is now readily apparent as well. According to (3.256) $m = 2$ and $m = 3$ resonances actually have opposite angle signs, which explains why the resonance lines curve in opposite directions in the image.

For the symmetric groove profile in Figure 3.36(a), the beam is centered on the land and the two first orders have identical phase shifts relative to the zero order. In this case, they both interfere constructively with the zero-order, generating two brighter regions on either side of the center line. However, because the groove profile in Figure 3.36(b) is asymmetric with regard to the focused spot position, the two first orders now have opposite phase shifts relative to the zero order. One first order generally interferes constructively with the zero order, while the other destructively. This generates a left-right brightness imbalance. In a disk drive, this imbalance is easily measured by a split detector and used as feedback to the tracking actuator to maintain the optical stylus on the center of the track.

3.5.10 Field Amplitudes and Grating Coupling Into Waveguides

A final example demonstrates the design of a grating for coupling light into a waveguide. Generally, gratings are designed so that the light couples into the -1st TE diffraction order which creates an electric field within the grating that closely approximates that of the TE_0 mode of the waveguide. The grating period is selected so that the $+1^{st}$ order and all higher orders are evanescent so that there is no light lost into these modes. Light that is reflected into the zero order and any light transmitted through the grating or absorbed in the grating are not coupled into the waveguide and should also be minimized. A reflector layer is often placed below the waveguide cladding of the grating in order to reflect light that is transmitted through the grating on the first pass back through the grating for a second pass. If the reflected light on the second pass through the grating generates a waveguide mode in-phase with the mode generated by the incident light in the first pass through the grating, then the two waves sum constructively to generate a waveguide mode with larger amplitude. To ensure this phase matching criterion, the thickness of the lower cladding between the core and the reflector is adjusted for maximum amplitude of the -1st order.

The angle of incidence is typically chosen as a small negative angle. If the light is incident at 0° (normal incidence), then symmetry requires half of the incident light to propagate in the wrong direction inside the waveguide (unless the grating is blazed or is otherwise asymmetric), automatically reducing the coupling efficiency by half. Because the light is gently focused onto the grating, some of the incident light rays will be at slightly shorter polar angles and some at slightly longer angles, and the grating period and focusing lens should be chosen so that none of these rays generate reflected propagating orders.

If λ is the vacuum wavelength, n_m is the mode index of the desired waveguide mode, n_i is the refractive index of the incident medium, m is the integer mode number (= −1 in this case), and θ is the polar angle of incidence, then the period of the grating is chosen according to

$$p = -\frac{m\lambda}{n_m + n_i \sin\theta}. \tag{3.371}$$

Consider the buried SiN grating/waveguide with SiO_2 cladding shown in Figure 3.37. The planar waveguide program can be used to compute the (symmetric) TE_0 waveguide mode by neglecting the reflector layer beneath the cladding which is generally not present under the waveguide. For a wavelength of 1.5 μm the mode index is 1.5086 for the TE_0 planar waveguide mode. For an angle of incidence of −15°, a grating period of 0.797 μm matches the in-plane component of the incident wavevector to the TE_0 waveguide mode index and makes the $+1^{st}$ order evanescent.

The electric field for TE incident polarization and for each order is next calculated through the film stack using ±5 diffracted orders. The electric field at each point in the grating film stack is, of course, a complex number. The -1st TE diffracted order is the order which couples to the waveguide. As a simplification, the magnitude of the field for this order is plotted in Figure 3.38(a) along the

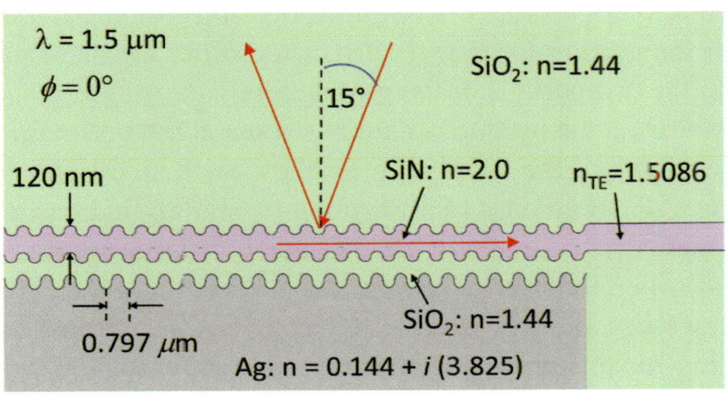

FIGURE 3.37 SiN waveguide with SiO_2 cladding and a reflector underlayer designed to couple light in the -1st order at a wavelength of 1.5 μm into the TE mode of a waveguide. The TE_0 planar waveguide mode index is 1.5086.

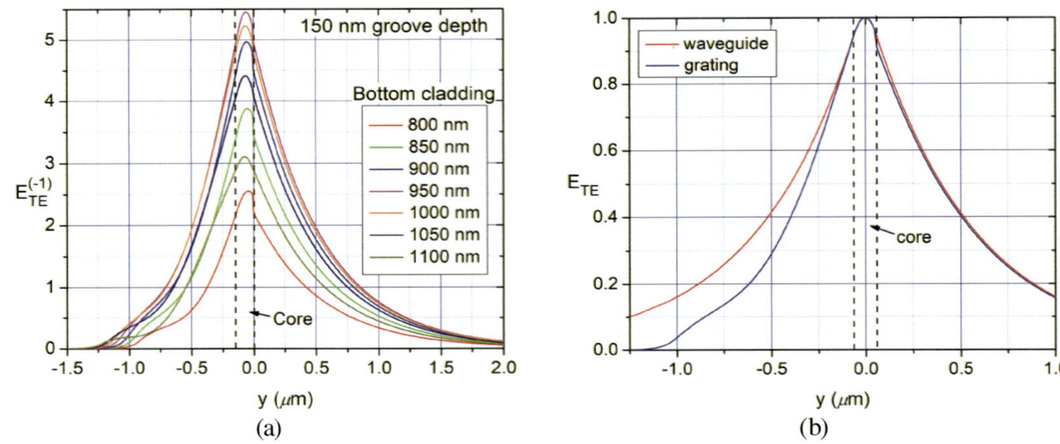

FIGURE 3.38 (a) Magnitude of the electric field through the film stack for the –1st TE order at a 150 nm groove depth for several different lower cladding thicknesses. (b) Comparison of the normalized fields in the grating film stack for 950 nm cladding thickness to that of the TE_0 mode of the planar waveguide (calculated from the planar waveguide program).

y-axis at $x = 0$ for a 150 nm groove depth and several different lower cladding thicknesses. A lower cladding thickness of ~950 nm is found to maximize the amplitude of the –1st order inside the core. A comparison of the normalized TE_0 waveguide mode field, calculated with the planar waveguide program, with the normalized field magnitude for the grating at this thickness and groove depth is plotted in Figure 3.38(b). The overlap is quite good in the region of the core and the upper cladding but begins to diverge close to the lower reflector at –1 μm because the metallic reflector forces the electric field to decay more quickly in the grating film stack than for the waveguide without the reflector. It might be possible to improve the overlap, for example, by increasing the thickness of the lower cladding to the next thickness at which the reflected light from the reflector again generates a –1st order mode that interferes constructively with the mode generated by the incident light, which should be approximately a half wavelength in optical thickness. Some light is also reflected at the outer surface of the cladding. An antireflection coating on top surface of the cladding can eliminate this optical loss.

The peak electric field value at the center of the core can be calculated as a function of both groove depth and lower cladding thickness. This is a quick way to determine the nearly optimum values. The contour plot in Figure 3.39(a) shows that a lower cladding thickness of ~970 nm and a groove depth of ~140 nm should be close to optimum coupling efficiency. It is interesting to compute the electric field within the grating for the –1st diffracted order for these parameters. The result is shown in Figure 3.39(b) for two periods of the grating along the *x* axis.

Although this procedure can be used to obtain a first cut at the optimum grating coupler design, it does not provide an estimate of the coupling efficiency. Such an estimate can be easily obtained, however, by introducing a small amount of absorption by slightly increasing the imaginary part of the refractive index of the core. As the –1st TE diffracted order propagates along the grating, the mode is continuously absorbed in the core layer. If the absorption in the core dominates the absorption in the lower reflector layer, then the drop in reflectance is due almost entirely to the absorption of the power in the TE_{-1} mode. Increasing the refractive index of the core eventually begins to directly affect the reflected and transmitted light and the reflectance begins to again increase. As shown in Figure 3.40, the minimum reflectance for the grating film stack in Figure 3.39(b) is ~0.32 for an imaginary part of the core refractive index of ~0.022 indicating an upper limit on the fraction of power transferred to the waveguide of ~0.68.

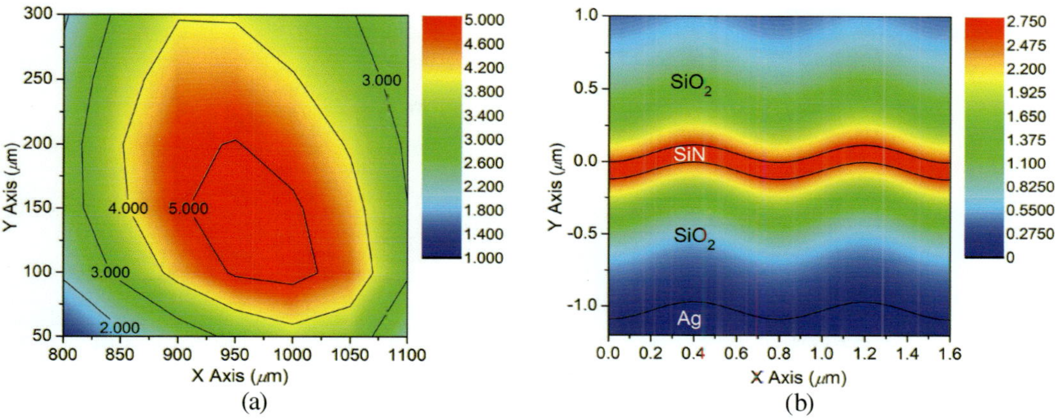

FIGURE 3.39 (a) Contour plot of peak field amplitude for the -1st TE diffraction order at the center of the core. (b) Plot of real part of the TE electric field for -1st diffracted order for 970 nm lower cladding thickness and 140 nm groove depth.

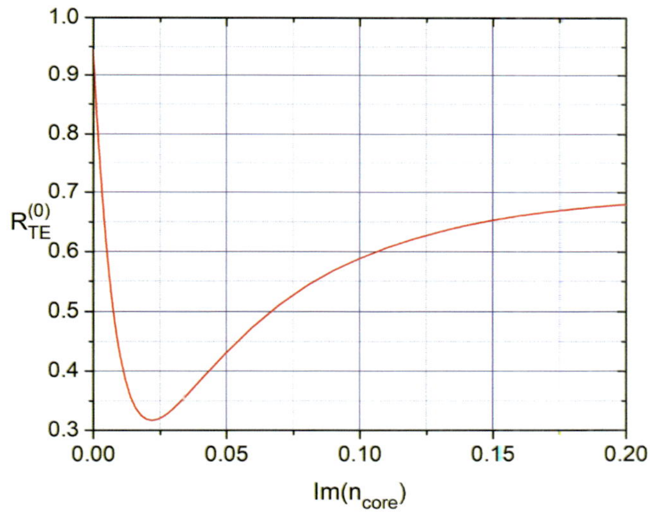

FIGURE 3.40 Variation in reflectance of grating as a function of the imaginary part of the refractive index of the core.

As a second example, consider the grating in Figure 3.41(a) from [23]. This grating is designed for a nominal angle of incidence of −6.5° for which there is only a reflected zero order and for which the -1st order matches the TE_0 waveguide, propagating along the core with a mode index of 1.637. The peak electric field in the core occurs for a groove depth of ~20 nm as shown in Figure 3.41(b).

The electric field calculated for the -1st order is compared to the electric field of the TE_0 waveguide mode calculated from the planar waveguide theory in Figure 3.42(a). The overlap is nearly perfect. In [22] a grating coupler is considered in which a 50:50 rectangular groove is etched only partially into the core from either top or bottom. The thickness of the core, therefore, is not constant, and so the effective index of the core can be considered an average of core and cladding, that is, smaller than that of the core itself. Consequently, as the groove depth increases and the average core index decreases, the angle of incidence for excitation of the $TE^{(-1)}$ diffracted order must increase according

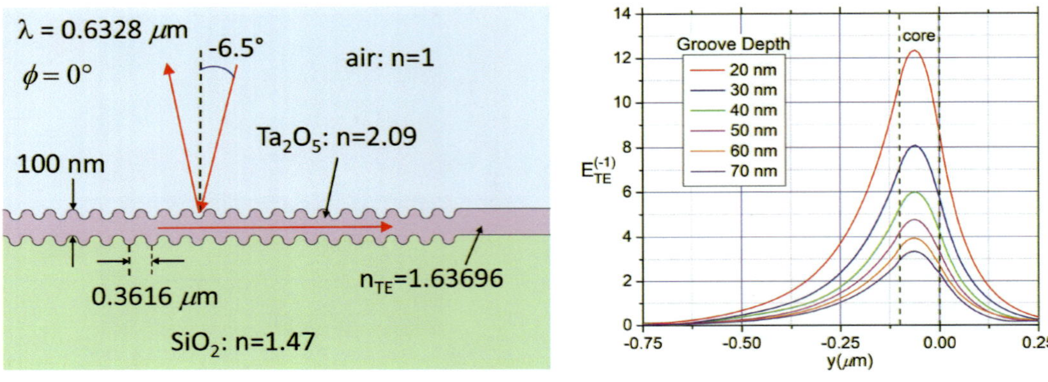

(a) (b)

FIGURE 3.41 (a) Ta_2O_5 waveguide with air and SiO_2 cladding designed to couple light in the -1st order at a wavelength of 0.6328 μm into the TE mode of a waveguide. The TE_0 planar waveguide mode index is 1.63696. (b) Electric field profile along y-axis for several different sinusoidal groove depths.

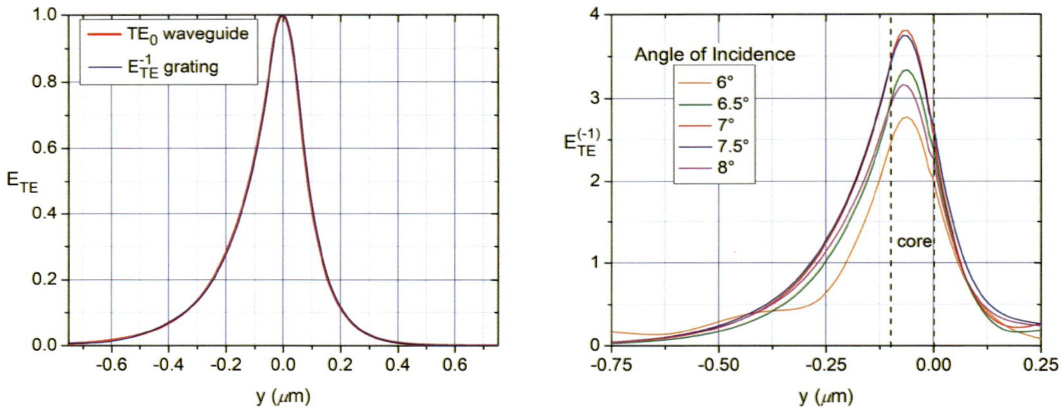

(a) (b)

FIGURE 3.42 (a) Normalized overlap of the electric field calculated for the -1st diffracted order using the diffraction grating program for the grating in Figure 3.41(a) with a 70 nm groove depth, and the TE_0 planar waveguide mode calculated using the planar waveguide program. (b) Field profile of the negative first order inside the grating for several different angles of incidence.

to (3.371) for a fixed grating period. When the grating has a sinusoidal grooved surface on top *and* bottom as shown in Figure 3.41(a), the effective core index does not change with groove depth, and therefore the optimal angle of incidence for exciting the -1st order is not expected to change substantially. This is indeed the case, as shown in Figure 3.42(b) for a fixed groove depth of 70 nm and several different angles of incidence. The optimal angle of incidence as indicated by the peak electric field value is between 7° and 7.5°, not much different than the design angle of 6.5°. On the other hand, in [22] with the etched groove, the optimal angle of incidence is found to increase to ~15.6°.

Again, to estimate the amount of optical power transferred to the waveguide mode, the total absorption (= $1 - R^{(0)} - T^{(0)}$) is plotted as a function of the imaginary part of the refractive index in Figure 3.43(a). The absorption peaks at ~0.4 for a very small imaginary index of ~0.002. This estimate of the coupling efficiency is smaller than that of the prior example due to the lack of a reflector beneath the core. However, it is about twice as large as the coupling efficiency calculated for the one-sided groove profile in Figure 2(b) of [23]. [23] demonstrates how the finite difference time domain (FDTD) method can be used to directly calculate the coupling efficiencies for nearly arbitrary grating

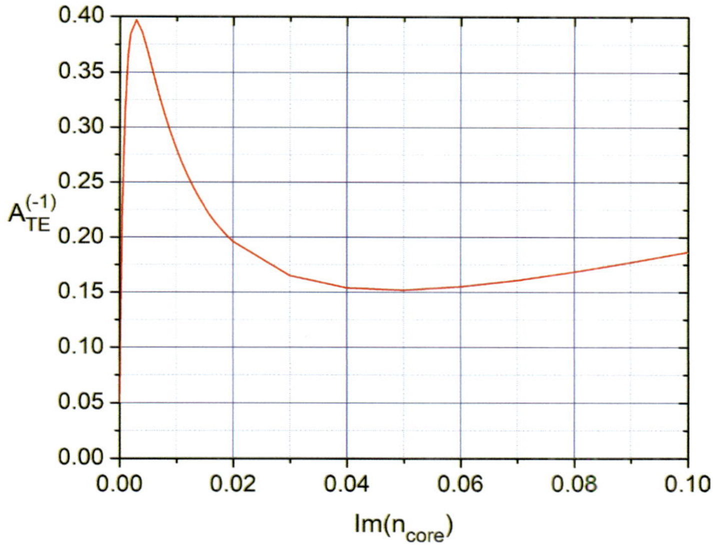

FIGURE 3.43 Plot of absorption (1-*R*-*T*) as a function of the imaginary part of the core refractive index.

profiles as well as finite gratings and focused light beams, though it is a much more tedious technique and will be discussed in more detail in Chapter 9.

REFERENCES

1 J. Chandezon, M. T. Dupuis, G. Cornet, and D. Maystre, "Multicoated gratings: A differential formalism applicable to the entire optical region," *J. Opt. Soc. Am.* **72** (1982) 839–846.
2 S. J. Elston, G. P. Bryan-Brown, and J. R. Sambles, "Polarization conversion from diffraction gratings," *Phys. Rev. B.* **44** (1991) 6393–6400.
3 N. P. K. Cotter, T. W. Priest, and J. R. Sambles, "Scattering-matrix approach to multilayer diffraction," *J. Opt. Soc. Am. A.* **12** (1995) 1097–1103.
4 L. Li, "Multilayer-coated diffraction gratings: Differential method of Chandezon *et. al.* revisited," *J. Opt. Soc. Am. A.* **11** (1994) 2816–2828.
5 D. Maystre, "A new general integral theory for dielectric coated gratings," *J. Opt. Soc. Am.* **72** (1982) 1385–1392.
6 D. Maystre, "A new theory for multiprofile, buried gratings," *Opt. Commun.* **26** (1978) 127–132.
7 J. Chandezon, D. Maystre, and G. Raoult, "A new theoretical method for diffraction gratings and its numerical application," *J. Opt. (Paris).* **11** (1980) 235–241.
8 M. G. Moharam and T. K. Gaylord, "Diffraction analysis of dielectric surface-relief gratings," *J. Opt. Soc. Am.* **72** (1982) 1385–1392.
9 D. Maystre, "Rigorous Vector Theories of Diffraction Gratings," in *Progress in Optics*, E. Wolf, ed. (Amsterdam: Elsevier, 1984) 21, 1–67.
10 E. Popov and L. Mashev, "Conical diffraction mounting generalization of a rigorous differential method," *J. Opt. (Paris).* **17** (1986) 175–180.
11 E. Popov and L. Mashev, "Convergence of Rayleigh-Fourier method and rigorous differential method for relief diffraction gratings," *Opt. Acta.* **33** (1986) 593–605.
12 D. Y. K. Ko and J. R. Sambles, "Scattering matrix method for propagation of radiation in stratified media: attenuated total reflection studies of liquid crystals," *J. Opt. Soc. Am. A.* **5** (1988) 1863–1866.
13 L. F. DeSandre and J. M. Elson, "Extinction-theorem analysis of diffraction anomalies in overcoated gratings," *J. Opt. Soc. Am. A.* **8** (1992) 763–777.
14 W. A. Challener, "Vector diffraction of a grating with conformal thin films," *J. Opt. Soc. Am. A.* **13** (1996) 1859–1869.
15 The γ^2 factors in (3.89) and (3.91) are incorrectly shown without the square in (22) and (24) of [2].

16 N. P. Wanstall, T. W. Priest, W. C. Tan, M. B. Sobnack, and J. R. Sambles, "Standing-wave surface-plasmon resonances with overhanging zero-order metal gratings," *J. Opt. Soc. Am. A.* **15** (1998) 2869–2876.
17 D. Sarid and W. A. Challener, *Modern Introduction to Surface Plasmons* (New York: Cambridge University Press, 2010) chapter 10.
18 S. J. Elston, G. P. Bryan-Brown, T. W. Priest, and J. R. Sambles, "Surface-resonance polarization conversion mediated by broken surface symmetry," *Phys. Rev. B.* **44** (1991) 3483–3485.
19 W. A. Challener, R. R. Ollmann, and K. K. Kam, "A surface plasmon resonance gas sensor in a 'compact disc' format," *Sens. Act. B.* **56** (1999) 254–258.
20 W. A. Challener, J. D. Edwards, R. W. McGowan, J. Skorjanec, and Z. Yang, "A multilayer grating-based evanescent wave sensing technique," *Sens. Act. B.* **71** (2000) 42–46.
21 L. Li and J. Hirsh, "All-dielectric high-efficiency reflection gratings made with multilayer thin-film coatings," *Opt. Lett.* **20** (1995) 1349–1351.
22 R. E. Gerber, L. Li, and M. Mansuripur, "Effect of surface plasmon excitations on the irradiance pattern of the return beam in optical disk data storage," *Appl. Opt.* **34** (1995) 4929–4936.
23 C. Peng and W. A. Challener, "Input-grating couplers for narrow Gaussian beam: Influence of groove depth," *Opt. Exp.* **12** (2004) 6481–6490.

4 Planar Waveguides

4.1 INTRODUCTION

Planar waveguides are perhaps the simplest optical means to support guided wave propagation. Planar waveguides support both transverse electric (TE) and transverse magnetic (TM) modes. In this chapter the equations for calculating mode indices and fields are derived for multilayer planar waveguides characterized by refractive indices or dielectric constants at each wavelength.

4.2 TRANSVERSE MAGNETIC (TM) MODES

4.2.1 Wavevectors

Consider the multilayer planar waveguide film stack in Figure 4.1. The waveguide mode propagates in the +z-direction. The magnetic field only has an x-axis component. The electric field has components along both y- and z-axes. Within each layer the fields are represented by an upward and downward propagating plane wave, except in the outer two layers where there is only a single "propagating" plane wave (though for guided modes the wavevectors in these layers must be imaginary so that the fields actually decay exponentially away from the surface). Each layer is characterized by a refractive index which may be a complex number to represent lossy dielectrics or metals.

An $e^{-i\omega t}$ time dependence is assumed for the fields. The fields for the top medium are

$$\boldsymbol{E}_0 = \left(-E_{0a}\sin\theta_0\,\hat{\boldsymbol{y}} - E_{0a}\cos\theta_0\,\hat{\boldsymbol{z}}\right)e^{-ik_{y0}y+ik_z z}, \tag{4.1}$$

$$\boldsymbol{H}_0 = \left(\frac{n_0}{Z_0}\right)E_{0a}\,\hat{\boldsymbol{x}}\,e^{-ik_{y0}y+ik_z z}. \tag{4.2}$$

where n_0 is the refractive index of the background medium, and $Z_0 \equiv \sqrt{\frac{\mu_0}{\varepsilon_0}} \cong 376.73\Omega$ is the impedance of free space.

The fields for thin film layer m of N total thin film layers are

$$\begin{aligned}\boldsymbol{E}_m &= \left(-E_{ma}\sin\theta_m\,\hat{\boldsymbol{y}} - E_{ma}\cos\theta_m\,\hat{\boldsymbol{z}}\right)e^{-ik_{ym}(y-y_{m-1})+ik_z z}\\ &\quad+\left(-E_{mb}\sin\theta_m\,\hat{\boldsymbol{y}} + E_{mb}\cos\theta_m\,\hat{\boldsymbol{z}}\right)e^{ik_{ym}(y-y_{m-1})-ik_z z}\end{aligned} \tag{4.3}$$

and

$$\boldsymbol{H}_m = \frac{n_m}{Z_0}E_{ma}\,\hat{\boldsymbol{x}}\,e^{-ik_{ym}(y-y_{m-1})+ik_z z} + \frac{n_m}{Z_0}E_{mb}\,\hat{\boldsymbol{x}}\,e^{ik_{ym}(y-y_{m-1})+ik_z z}, \tag{4.4}$$

and the fields for the bottom medium are

$$\boldsymbol{E}_{N+1} = \left(-E_{N+1,b}\sin\theta_{N+1}\,\hat{\boldsymbol{y}} + E_{N+1,b}\cos\theta_{N+1}\,\hat{\boldsymbol{z}}\right)e^{ik_{y,N+1}(y-y_N)+ik_z z} \tag{4.5}$$

DOI: 10.1201/9781003714231-4

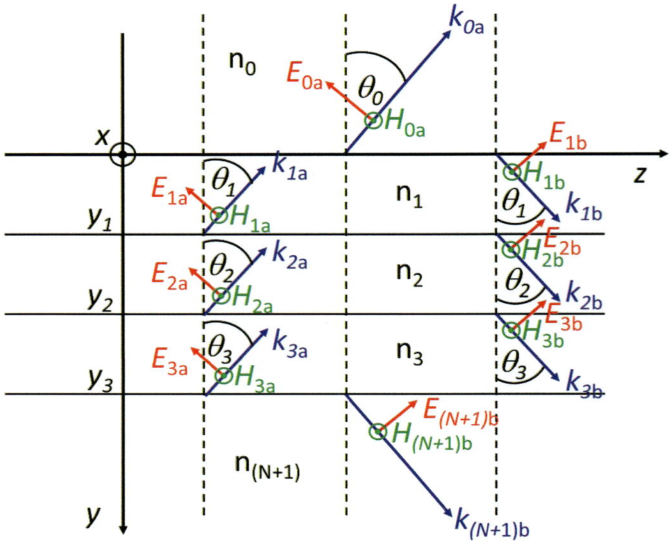

FIGURE 4.1 Multilayer thin film stack in which a TM mode is propagating. The magnetic field has only an x-component orthogonal to the direction of propagation.

and

$$\boldsymbol{H}_{N+1} = \frac{n_{N+1}}{Z_0} E_{N+1,b}\, \hat{\boldsymbol{x}}\, e^{ik_{y,N+1}(y-y_N)-ik_z z}. \tag{4.6}$$

In order to match the component of the wavevector along the z-axis at the interface between layers, Snell's law requires

$$n_0 \sin\theta_0 = n_1 \sin\theta_1 = \cdots = n_m \sin\theta_m. \tag{4.7}$$

The effective index or "mode index" in the direction of propagation along the z-axis, is

$$n_{\text{mode}} = n_m \sin\theta_m \tag{4.8}$$

and the wavevector component along the y-axis is

$$\left|k_{\text{ym}}\right| = n_m k_0 \cos\theta_m = \sqrt{n_m^2 k_0^2 - k_z^2}. \tag{4.9}$$

Because of the square root, there is ambiguity in the sign of k_{ym}. The sign is chosen such that when the imaginary part is positive, k_{ym} corresponds to propagation in the positive y-direction. In the top medium, the exponential factor $e^{-ik_{y0}y}$ goes to zero as $y \to -\infty$. Similarly, in the bottom medium the factor $e^{ik_{y,N+1}(y-y_N)}$ goes to zero as $y \to +\infty$.

4.2.2 Boundary Conditions

Boundary conditions require the tangential component of the fields to be continuous across each interface, so at the top surface

$$-E_0\cos\theta_0 = -E_{1a}\cos\theta_1 + E_{1b}\cos\theta_1 \tag{4.10}$$

and

$$n_0 E_0 = n_1 E_{1a} + n_1 E_{1b}\,. \tag{4.11}$$

Let d_m be the thickness of layer n. So

$$y_{m+1} - y_m = d_m. \tag{4.12}$$

At the bottom of the first layer,

$$-E_{1a}\cos\theta_1 e^{-ik_{y1}d_1} + E_{1b}\cos\theta_1 e^{+ik_{z1}d_1} = -E_{2a}\cos\theta_2 + E_{2b}\cos\theta_2 \tag{4.13}$$

and

$$n_1 E_{1a} e^{-ik_{z1}d_1} + n_1 E_{1b} e^{+ik_{z1}d_1} = n_2 E_{2a} + n_2 E_{2b} \tag{4.14}$$

and so on until at the bottom of the last layer,

$$-E_{\mathrm{Na}}\cos\theta_N e^{-ik_{zN}d_N} + E_{\mathrm{Nb}}\cos\theta_N e^{+ik_{zN}d_N} = \cos\theta_{N+1} E_{N+1,b} \tag{4.15}$$

and

$$n_N E_{\mathrm{Na}} e^{-ik_{zN}d_N} + n_N E_{\mathrm{Nb}} e^{+ik_{zN}d_N} = n_{N+1} E_{N+1,b}\,. \tag{4.16}$$

Recasting these as 2 × 2 matrix equations,

$$\begin{bmatrix} -\cos\theta_0 \\ n_0 \end{bmatrix} E_0 = \begin{bmatrix} -\cos\theta_1 & \cos\theta_1 \\ n_1 & n_1 \end{bmatrix} \begin{bmatrix} E_{1a} \\ E_{1b} \end{bmatrix} \tag{4.17}$$

and

$$\begin{bmatrix} -\cos\theta_1 e^{-ik_{z1}d_1} & \cos\theta_1 e^{+ik_{z1}d_1} \\ n_1 e^{-ik_{z1}d_1} & n_1 e^{+ik_{z1}d_1} \end{bmatrix} \begin{bmatrix} E_{1a} \\ E_{1b} \end{bmatrix} = \begin{bmatrix} -\cos\theta_2 & \cos\theta_2 \\ n_2 & n_2 \end{bmatrix} \begin{bmatrix} E_{2a} \\ E_{2b} \end{bmatrix} \tag{4.18}$$

until finally at the bottom surface after N layers,

$$\begin{bmatrix} -\cos\theta_N e^{-ik_{zN}a_N} & \cos\theta_N e^{+ik_{zN}d_N} \\ n_N e^{-ik_{zN}d_N} & n_N e^{+ik_{zN}d_N} \end{bmatrix} \begin{bmatrix} E_{\mathrm{Na}} \\ E_{\mathrm{Nb}} \end{bmatrix} = \begin{bmatrix} \cos\theta_{N+1} \\ n_{N+1} \end{bmatrix} E_{N+1,b}. \tag{4.19}$$

4.2.3 Mode Index

Define the general matrix $\tilde{\boldsymbol{R}}_m$ for layer m,

$$\tilde{\boldsymbol{R}}_m = \begin{bmatrix} -\cos\theta_m e^{-ik_{zm}d_m} & \cos\theta_m e^{+ik_{zm}d_m} \\ n_m e^{-ik_{zm}d_m} & n_m e^{+ik_{zm}d_m} \end{bmatrix} \tag{4.20}$$

and $\tilde{\boldsymbol{S}}_m$,

$$\tilde{\boldsymbol{S}}_m = \begin{bmatrix} -\cos\theta_m & \cos\theta_m \\ n_m & n_m \end{bmatrix}. \tag{4.21}$$

(4.17)–(4.19) are then

$$\begin{bmatrix} -\cos\theta_0 \\ n_0 \end{bmatrix} E_0 = \tilde{\boldsymbol{S}}_1 \begin{bmatrix} E_{1a} \\ E_{1b} \end{bmatrix} \tag{4.22}$$

$$\tilde{\boldsymbol{R}}_m \begin{bmatrix} E_{\text{ma}} \\ E_{\text{mb}} \end{bmatrix} = \tilde{\boldsymbol{S}}_{m+1} \begin{bmatrix} E_{m+1,a} \\ E_{m+1,b} \end{bmatrix} \text{ or } \begin{bmatrix} E_{\text{ma}} \\ E_{\text{mb}} \end{bmatrix} = \left(\tilde{\boldsymbol{R}}_m\right)^{-1} \tilde{\boldsymbol{S}}_{m+1} \begin{bmatrix} E_{m+1,a} \\ E_{m+1,b} \end{bmatrix} \tag{4.23}$$

$$\tilde{\boldsymbol{R}}_N \begin{bmatrix} E_{\text{Na}} \\ E_{\text{Nb}} \end{bmatrix} = \begin{bmatrix} \cos\theta_{N+1} \\ n_{N+1} \end{bmatrix} E_{N+1,b} \text{ or } \begin{bmatrix} E_{\text{Na}} \\ E_{\text{Nb}} \end{bmatrix} = \left(\tilde{\boldsymbol{R}}_N\right)^{-1} \begin{bmatrix} \cos\theta_{N+1} \\ n_{N+1} \end{bmatrix} E_{N+1}. \tag{4.24}$$

Combining (4.22)–(4.24),

$$\begin{bmatrix} -\cos\theta_0 \\ n_0 \end{bmatrix} E_0 = \tilde{\boldsymbol{S}}_1 \left(\tilde{\boldsymbol{R}}_1\right)^{-1} \tilde{\boldsymbol{S}}_2 \begin{bmatrix} E_{2,a} \\ E_{2,b} \end{bmatrix} \tag{4.25}$$

$$\begin{bmatrix} -\cos\theta_0 \\ n_0 \end{bmatrix} E_0 = \left\{\left[\tilde{\boldsymbol{S}}_1 \left(\tilde{\boldsymbol{R}}_1\right)^{-1}\right]\left[\tilde{\boldsymbol{S}}_2 \left(\tilde{\boldsymbol{R}}_2\right)^{-1}\right]\cdots\left[\tilde{\boldsymbol{S}}_m \left(\tilde{\boldsymbol{R}}_m\right)^{-1}\right]\cdots\left[\tilde{\boldsymbol{S}}_N \left(\tilde{\boldsymbol{R}}_N\right)^{-1}\right]\right\} \begin{bmatrix} \cos\theta_{N+1} \\ n_{N+1} \end{bmatrix} E_{N+1}. \tag{4.26}$$

Let

$$\tilde{\boldsymbol{M}} \equiv \left\{\left[\tilde{\boldsymbol{S}}_1 \left(\tilde{\boldsymbol{R}}_1\right)^{-1}\right]\left[\tilde{\boldsymbol{S}}_2 \left(\tilde{\boldsymbol{R}}_2\right)^{-1}\right]\cdots\left[\tilde{\boldsymbol{S}}_m \left(\tilde{\boldsymbol{R}}_m\right)^{-1}\right]\cdots\left[\tilde{\boldsymbol{S}}_N \left(\tilde{\boldsymbol{R}}_N\right)^{-1}\right]\right\}. \tag{4.27}$$

Then

$$\begin{aligned} \begin{bmatrix} -\cos\theta_0 \\ n_0 \end{bmatrix} E_0 = \tilde{\boldsymbol{M}} \begin{bmatrix} \cos\theta_{N+1} \\ n_{N+1} \end{bmatrix} E_{N+1} &= \begin{bmatrix} M_{11} & M_{12} \\ M_{21} & M_{22} \end{bmatrix} \begin{bmatrix} \cos\theta_{N+1} \\ n_{N+1} \end{bmatrix} E_{N+1} \\ &= \begin{bmatrix} M_{11}\cos\theta_{N+1} + M_{12} n_{N+1} \\ M_{21}\cos\theta_{N+1} + M_{22} n_{N+1} \end{bmatrix} E_{N+1} \end{aligned} \tag{4.28}$$

or

$$\tilde{\boldsymbol{Q}} \begin{bmatrix} E_0 \\ E_{N+1} \end{bmatrix} = \boldsymbol{0} \tag{4.29}$$

where

$$\tilde{\boldsymbol{Q}} \equiv \begin{bmatrix} \cos\theta_0 & \left(M_{11}\cos\theta_{N+1} + M_{12} n_{N+1}\right) \\ -n_0 & \left(M_{21}\cos\theta_{N+1} + M_{22} n_{N+1}\right) \end{bmatrix}. \tag{4.30}$$

$\cos(\theta_{N+1})$ can be expressed in terms of θ_0 by (4.9).

$$k_z^2 = n_0^2 k_0^2 - (n_0 k_0)^2 \cos^2\theta_0 = n_{N+1}^2 k_0^2 - (n_{N+1} k_0)^2 \cos^2\theta_{N+1} \tag{4.31}$$

or

$$\cos\theta_{N+1} = \frac{\sqrt{n_{N+1}^2 - n_0^2 \sin^2\theta_0}}{n_{N+1}}. \tag{4.32}$$

(4.29) has a solution only if the determinant of $\tilde{\boldsymbol{Q}}$ is zero. This is a transcendental equation that must be solved numerically. For dielectric film stacks, the mode index can never be larger than the largest refractive index in the film stack. It can never be smaller than the smallest value of refractive index of the outer claddings. If the mode index were smaller than this, the light would leak out of the waveguide. A search algorithm can be implemented to scan the mode index between these two limits to find all values that make the determinant equal to zero.

Consider the three-layer planar waveguide in Figure 4.2(a) after the example on p. 307 of [1]. The determinant of Q is plotted between 1.5 and 2.5 in Figure 4.2(b) for the TM mode. There are four zeroes of Det(Q) for mode indices between 1.5 and 2.5 at 1.7374, 2.0759, 2.3167, and 2.4550 which are mode indices for propagating TM waveguide modes.

4.2.4 Field Amplitudes and Poynting Vector

For a specific mode index, the electric field amplitudes E_a and E_b within each layer are calculated from (4.22)–(4.24). The electric field amplitude E_0 in the upper cladding is set to 1 at the interface. Then from (4.22),

$$\begin{bmatrix} E_{1a} \\ E_{1b} \end{bmatrix} = \left(\tilde{\boldsymbol{S}}_1\right)^{-1} \begin{bmatrix} -\cos\theta_0 \\ n_0 \end{bmatrix} E_{0a}. \tag{4.33}$$

The electric field amplitudes in subsequent layers are calculated by repeated application of (4.23),

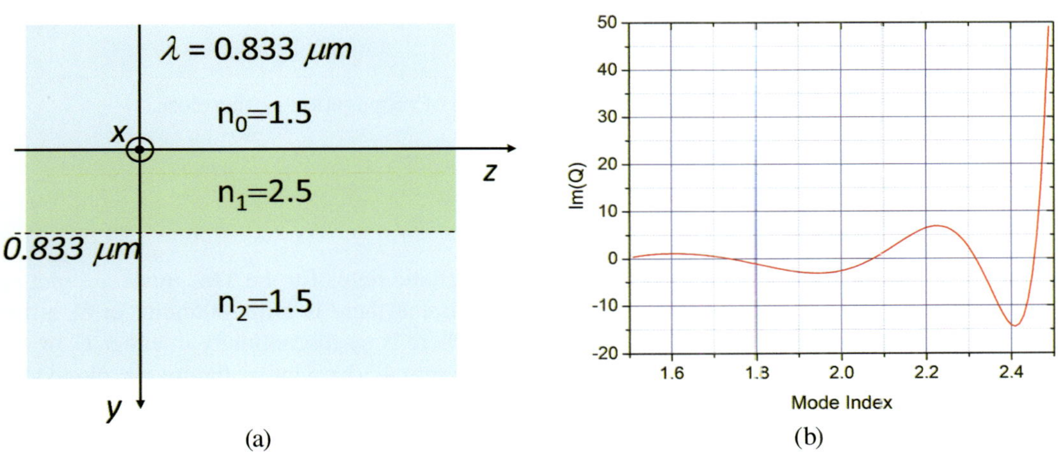

FIGURE 4.2 (a) Symmetric planar waveguide example. (b) Im(Q) as a function of TM mode index.

$$\begin{bmatrix} E_{m+1,a} \\ E_{m+1,b} \end{bmatrix} = \left(\tilde{\boldsymbol{S}}_{m+1}\right)^{-1} \tilde{\boldsymbol{R}}_m \begin{bmatrix} E_{\text{ma}} \\ E_{\text{mb}} \end{bmatrix}. \tag{4.34}$$

Finally, the electric field amplitude in the bottom cladding is obtained from either half of (4.24),

$$E_{N+1,b} = \frac{\tilde{\boldsymbol{R}}_{N,11} E_{\text{Na}} + \tilde{\boldsymbol{R}}_{N,12} E_{\text{Nb}}}{\cos\theta_{N+1}} = \frac{\tilde{\boldsymbol{R}}_{N,21} E_{\text{Na}} + \tilde{\boldsymbol{R}}_{N,22} E_{\text{Nb}}}{\text{n}_{N+1}}. \tag{4.35}$$

The electric and magnetic fields can now be calculated as a function of y from (4.1)–(4.6).

$$\boldsymbol{E}_0 = \left(-E_{0a} \sin\theta_0 \hat{\boldsymbol{y}} - E_{0a} \cos\theta_0 \hat{\boldsymbol{z}}\right) e^{-ik_{y0}y} \tag{4.36}$$

$$\boldsymbol{E}_m = \left(-E_{\text{ma}} \sin\theta_m \hat{\boldsymbol{y}} - E_{\text{ma}} \cos\theta_m \hat{\boldsymbol{z}}\right) e^{-ik_{\text{ym}}(y-y_{m-1})} + \left(-E_{\text{mb}} \sin\theta_m \hat{\boldsymbol{y}} + E_{\text{mb}} \cos\theta_m \hat{\boldsymbol{z}}\right) e^{ik_{\text{ym}}(y-y_{m-1})} \tag{4.37}$$

$$\boldsymbol{E}_{N+1} = \left(-E_{N+1,b} \sin\theta_{N+1} \hat{\boldsymbol{y}} + E_{N+1,b} \cos\theta_{N+1} \hat{\boldsymbol{z}}\right) e^{ik_{y,N+1}(y-y_{N-1})} \tag{4.38}$$

$$\boldsymbol{H}_0 = n_0 E_{0a} \hat{\boldsymbol{x}} e^{-ik_{y0}y} \tag{4.39}$$

$$\boldsymbol{H}_m = \frac{n_m}{Z_0} E_{\text{ma}} \hat{\boldsymbol{x}} e^{-ik_{\text{ym}}(y-y_{m-1})} + \frac{n_m}{Z_0} E_{\text{mb}} \hat{\boldsymbol{x}} e^{ik_{\text{ym}}(y-y_{m-1})} \tag{4.40}$$

$$\boldsymbol{H}_{N+1} = \frac{n_{N+1}}{Z_0} E_{N+1,b} \hat{\boldsymbol{x}} e^{ik_{y,N+1}(y-y_N)}. \tag{4.41}$$

The Poynting vector is given by

$$\boldsymbol{S} = \frac{1}{2}\text{Re}\left(\boldsymbol{E} \times \boldsymbol{H}^*\right) = \frac{1}{2}\text{Re}\left[\hat{\boldsymbol{x}}\left(E_y H_z^* - E_z H_y^*\right) + \hat{\boldsymbol{y}}\left(E_z H_x^* - E_x H_z^*\right) + \hat{\boldsymbol{z}}\left(E_x H_y^* - E_y H_x^*\right)\right]. \tag{4.42}$$

The z-component of the Poynting vector in the direction of propagation is, therefore,

$$S_z = -\frac{\text{Re}\left(E_y H_x^*\right)}{2}, \tag{4.43}$$

because $E_x = 0$ for the TM mode. The electric and magnetic fields for the TM_0 mode are plotted in Figure 4.3(a) for the example in Figure 4.2(a). Note that there is a discontinuity in E_y at the interfaces because it is normal to the interfaces, while there is no discontinuity in either E_z or H_x. The z-component of the Poynting vector is plotted in Figure 4.3(b). Similar figures are plotted for the TM_1 mode and TM_2 modes in Figures 3.4 and 3.5, respectively. Again, to convert to SI units, the magnetic field and the Poynting vector must each be multiplied by $1/\sqrt{\mu_0}$. It should also be noted that the fields are in general complex quantities with both a real and imaginary part. For this example, E_y and H_x are real, and E_z is pure imaginary (90° out-of-phase with E_y).

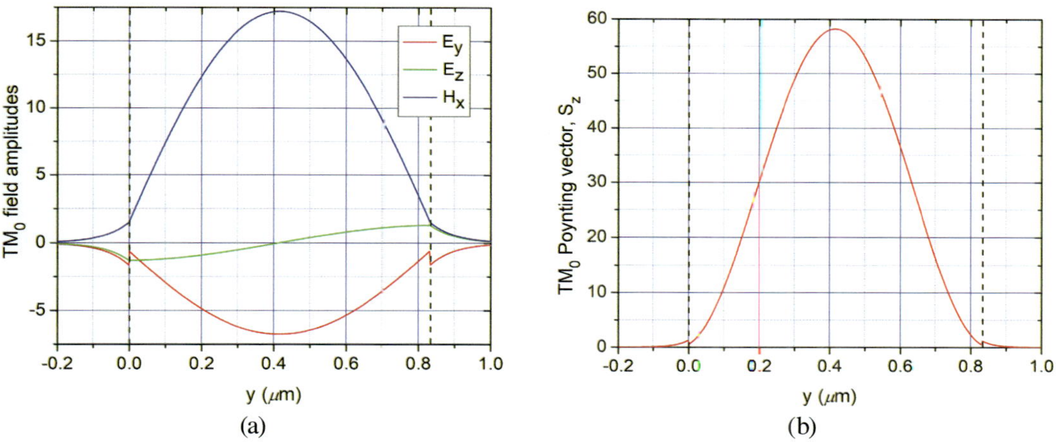

FIGURE 4.3 (a) Field amplitudes through thickness of waveguide for TM_0 mode of Figure 4.2(a) waveguide. (b) z-component of the Poynting vector.

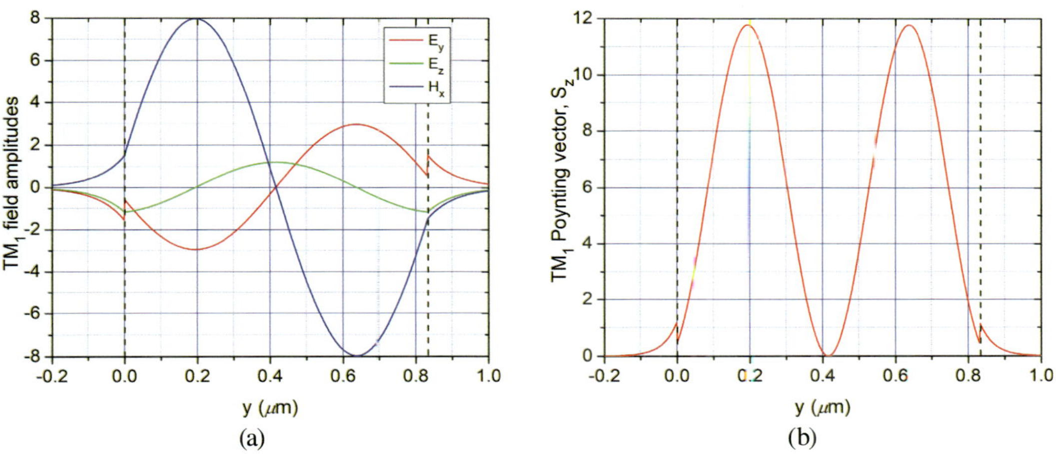

FIGURE 4.4 (a) Field amplitudes through thickness of waveguide for TM_1 mode of Figure 4.2(a) waveguide. (b) z-component of the Poynting vector.

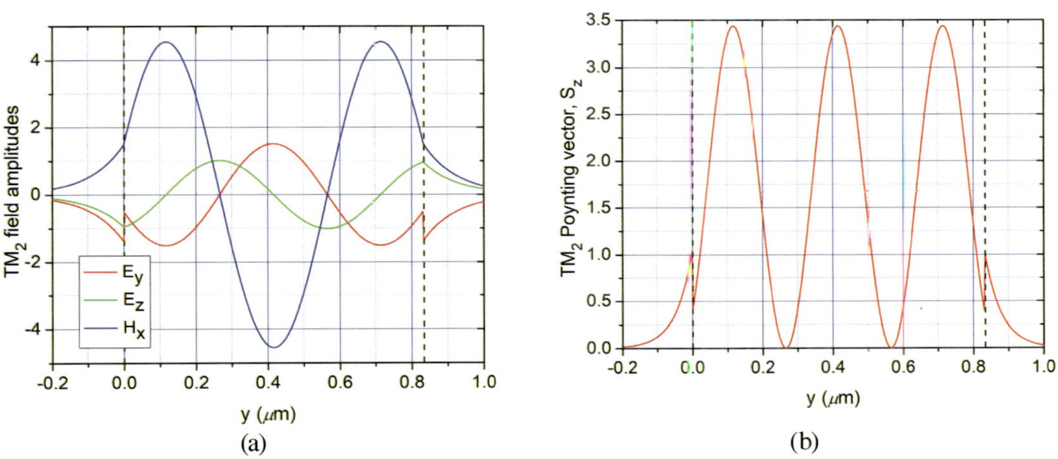

FIGURE 4.5 (a) Field amplitudes through thickness of waveguide for TM_2 mode of Figure 4.2(a) waveguide. (b) z-component of the Poynting vector.

4.3 TRANSVERSE ELECTRIC (TE) MODES

4.3.1 Wavevectors

Consider the multilayer thin film stack in Figure 4.6. The waveguide mode propagates in the $+z$-direction. The electric field only has an x-axis component. The magnetic field has components along both y- and z-axes. Within each layer the fields are represented by an upward and downward propagating plane wave, except in the outer two layers where there is only a single "propagating" plane wave (though for guided modes the wavevectors in these layers must be imaginary so that the fields actually decay exponentially away from the surface). Each layer is characterized by a refractive index which may be a complex number to represent lossy dielectrics or metals.

The TE mode field equations in each layer (assuming an $e^{-i\omega t}$ time dependence) are:

$$\vec{E}_0 = E_{0a}\,\hat{x}\,e^{-ik_{y0}y+ik_z z} \tag{4.44}$$

$$\vec{H}_0 = \frac{n_0}{Z_0}E_{0a}\left(\sin\theta_0\,\hat{y}+\cos\theta_0\,\hat{z}\right)e^{-ik_{y0}x+ik_z z} \tag{4.45}$$

$$\vec{E}_m = E_{\mathrm{ma}}\,\hat{x}\,e^{-ik_{\mathrm{ym}}(y-y_{m-1})+ik_z z} + E_{\mathrm{mb}}\,\hat{x}\,e^{+ik_{\mathrm{ym}}(y-y_{m-1})+ik_z z} \tag{4.46}$$

$$\begin{aligned}\vec{H}_m = {} & \frac{n_1}{Z_0}E_{1a}\left(\sin\theta_m\,\hat{y}+\cos\theta_m\,\hat{z}\right)e^{-ik_{\mathrm{ym}}(y-y_{m-1})+ik_z z} \\ & + \frac{n_m}{Z_0}E_{\mathrm{mb}}(\sin\theta_m\,\hat{y}-\cos\theta_m\,\hat{z})e^{ik_{\mathrm{ym}}(y-y_{m-1})+ik_z z}\end{aligned} \tag{4.47}$$

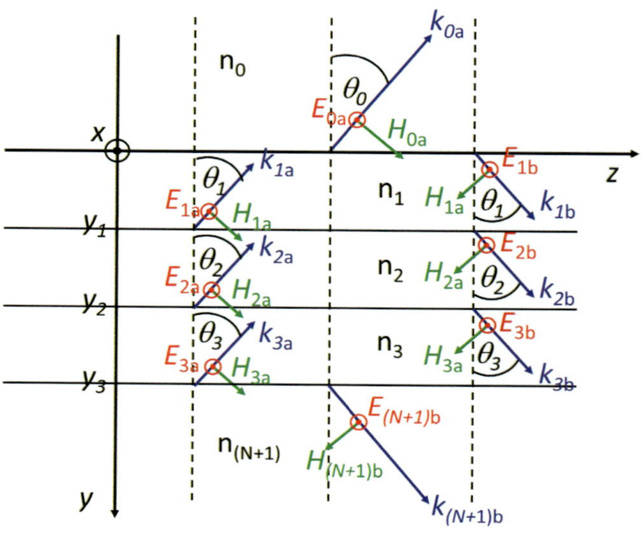

FIGURE 4.6 Multilayer thin film stack in which a TE mode is propagating. The magnetic field has only an x-component orthogonal to the direction of propagation.

$$\vec{E}_{N+1} = E_{N+1,b}\,\hat{x}\,e^{+ik_{y,N+1}(y-y_N)+ik_z z} \tag{4.48}$$

$$\vec{H}_{N+1} = \frac{n_{N+1}}{Z_0} E_{N+1,b}\left(\sin\theta_{N+1}\,\hat{y} - \cos\theta_{N+1}\,\hat{z}\right) e^{ik_{y,N+1}(y-y_N)+ik_z z}. \tag{4.49}$$

In order to match the component of the wavevector along the z-axis at the interface between layers, Snell's law requires

$$n_0 \sin\theta_0 = n_1 \sin\theta_1 = \cdots = n_m \sin\theta_m. \tag{4.50}$$

Therefore, the mode index is

$$n_{\text{mode}} = n_m \sin\theta_m \tag{4.51}$$

and

$$\left|k_{\text{ym}}\right| = n_m k_0 \cos\theta_m = \sqrt{n_m^2 k_0^2 - k_z^2}. \tag{4.52}$$

Because of the square root, there is ambiguity in the sign of k_{ym}. The sign is chosen such that the imaginary part is positive. In that case, k_{ym} corresponds to propagation in the positive y-direction. In the top medium, the exponential factor $e^{-ik_{y0}y}$ goes to zero as $y \to -\infty$. Similarly, in the bottom medium the factor $e^{ik_{y,N+1}(y-y_N)}$ goes to zero as $y \to +\infty$.

4.3.2 Boundary Conditions

Matching boundary conditions at each layer gives,

$$E_{0a} = E_{1a} + E_{1b} \tag{4.53}$$

$$n_0 \cos\theta_0 E_{0a} = n_1 \cos\theta_1 E_{1a} - n_1 \cos\theta_1 E_{1b} \tag{4.54}$$

$$E_{1a} e^{-ik_{y1}d_1} + E_{1b} e^{+ik_{z1}d_1} = E_{2a} + E_{2b} \tag{4.55}$$

$$n_1 \cos\theta_1 e^{-ik_{y1}d_1} E_{1a} - n_1 \cos\theta_1 e^{+ik_{y1}d_1} E_{1b} = n_2 \cos\theta_2 E_{2a} - n_2 \cos\theta_2 E_{2b} \tag{4.56}$$

$$E_{\text{Na}} e^{-ik_{zN}d_N} + E_{\text{Nb}} e^{+ik_{zN}d_N} = E_{N+1,b} \tag{4.57}$$

$$n_N \cos\theta_N e^{-ik_{zN}d_N} E_{\text{Na}} - n_N \cos\theta_N e^{+ik_{zN}d_N} E_{\text{Nb}} = -n_{N+1} \cos\vartheta_{N+1} E_{N+1,b}. \tag{4.58}$$

In matrix format,

$$\begin{bmatrix} 1 \\ n_0 \cos\theta_0 \end{bmatrix} E_{0a} = \begin{bmatrix} 1 & 1 \\ n_1 \cos\theta_1 & -n_1 \cos\theta_1 \end{bmatrix} \begin{bmatrix} E_{1a} \\ E_{1b} \end{bmatrix} \tag{4.59}$$

$$\begin{bmatrix} e^{-ik_{zm}d_1} & e^{+ik_{zm}d_1} \\ n_m \cos\theta_m e^{-ik_{zm}d_1} & -n_m \cos\theta_m e^{+ik_{zm}d_1} \end{bmatrix} \begin{bmatrix} E_{\text{ma}} \\ E_{\text{mb}} \end{bmatrix} = \begin{bmatrix} 1 & 1 \\ n_{m+1} \cos\theta_{m+1} & -n_{m+1} \cos\theta_{m+1} \end{bmatrix} \begin{bmatrix} E_{m+1,a} \\ E_{m+1,b} \end{bmatrix} \tag{4.60}$$

$$\begin{bmatrix} e^{-ik_{zN}d_N} & e^{+ik_{zN}d_N} \\ n_N\cos\theta_N e^{-ik_{zN}d_N} & -n_N\cos\theta_N e^{+ik_{zN}d_N} \end{bmatrix}\begin{bmatrix} E_{\mathrm{Na}} \\ E_{\mathrm{Nb}} \end{bmatrix} = \begin{bmatrix} 1 & 1 \\ n_{N+1}\cos\theta_{N+1} & -n_{N+1}\cos\theta_{N+1} \end{bmatrix}\begin{bmatrix} 0 \\ E_{N+1,b} \end{bmatrix} \tag{4.61}$$

Let

$$\tilde{\boldsymbol{R}}_m \equiv \begin{bmatrix} e^{-ik_{zm}d_m} & e^{+ik_{zm}d_m} \\ n_m\cos\theta_m e^{-ik_{zm}d_m} & -n_m\cos\theta_m e^{+ik_{zm}d_m} \end{bmatrix} \tag{4.62}$$

$$\tilde{\boldsymbol{S}}_m \equiv \begin{bmatrix} 1 & 1 \\ n_m\cos\theta_m & -n_m\cos\theta_m \end{bmatrix} \tag{4.63}$$

then

$$\begin{bmatrix} 1 \\ n_0\cos\theta_0 \end{bmatrix} E_{0a} = \tilde{\boldsymbol{S}}_1 \begin{bmatrix} E_{1a} \\ E_{1b} \end{bmatrix} \tag{4.64}$$

$$\tilde{\boldsymbol{R}}_m \begin{bmatrix} E_{\mathrm{ma}} \\ E_{\mathrm{mb}} \end{bmatrix} = \tilde{\boldsymbol{S}}_{m+1} \begin{bmatrix} E_{m+1,a} \\ E_{m+1,b} \end{bmatrix} \tag{4.65}$$

$$\tilde{\boldsymbol{R}}_N \begin{bmatrix} E_{\mathrm{Na}} \\ E_{\mathrm{Nb}} \end{bmatrix} = \tilde{\boldsymbol{S}}_{N+1} \begin{bmatrix} 0 \\ E_{N+1,b} \end{bmatrix} = \begin{bmatrix} 1 \\ -n_{N+1}\cos\theta_{N+1} \end{bmatrix} E_{N+1,b} \tag{4.66}$$

giving the sequence

$$\begin{bmatrix} 1 \\ n_0\cos\theta_0 \end{bmatrix} E_{0a} = \left[\tilde{\boldsymbol{S}}_1 \left(\tilde{\boldsymbol{R}}_1\right)^{-1} \cdots \tilde{\boldsymbol{S}}_{m+1} \left(\tilde{\boldsymbol{R}}_{m+1}\right)^{-1} \cdots \tilde{\boldsymbol{S}}_N \left(\tilde{\boldsymbol{R}}_N\right)^{-1}\right] \tilde{\boldsymbol{S}}_{N+1} \begin{bmatrix} 0 \\ E_{N+1,b} \end{bmatrix} = \tilde{\boldsymbol{M}}\, \tilde{\boldsymbol{S}}_{N+1} \begin{bmatrix} 0 \\ E_{N+1,b} \end{bmatrix} \tag{4.67}$$

where

$$\tilde{\boldsymbol{M}} \equiv \tilde{\boldsymbol{S}}_1 \left(\tilde{\boldsymbol{R}}_1\right)^{-1} \cdots \tilde{\boldsymbol{S}}_{m+1} \left(\tilde{\boldsymbol{R}}_{m+1}\right)^{-1} \cdots \tilde{\boldsymbol{S}}_N \left(\tilde{\boldsymbol{R}}_N\right)^{-1}. \tag{4.68}$$

4.3.3 Mode Index

(4.68) is the same form of equation as was found for the TM modes only with different definitions for the $\tilde{\boldsymbol{R}}_m$ and $\tilde{\boldsymbol{S}}_m$ matrices. The solution follows similarly.

$$\begin{bmatrix} 1 \\ n_0\cos\theta_0 \end{bmatrix} E_0 = \tilde{\boldsymbol{M}} \begin{bmatrix} \cos\theta_{N+1} \\ n_{N+1} \end{bmatrix} E_{N+1} = \begin{bmatrix} M_{11} & M_{12} \\ M_{21} & M_{22} \end{bmatrix}\begin{bmatrix} 1 \\ -n_{N+1}\cos\theta_{N+1} \end{bmatrix} E_{N+1} =$$
$$\begin{bmatrix} M_{11} - n_{N+1}M_{12}\cos\theta_{N+1} \\ M_{21} - n_{N+1}M_{22}\cos\theta_{N+1} \end{bmatrix} E_{N+1}. \tag{4.69}$$

or

$$\tilde{\boldsymbol{Q}} \begin{bmatrix} E_0 \\ E_{N+1} \end{bmatrix} = 0 \tag{4.70}$$

where

$$\tilde{Q} \equiv \begin{bmatrix} 1 & \left(-M_{11} + n_{N+1} M_{12} \cos\theta_{N+1}\right) \\ n_0 \cos\theta_0 & \left(-M_{21} + n_{N+1} M_{22} \cos\theta_{N+1}\right) \end{bmatrix}. \tag{4.71}$$

$\cos\theta_{N+1}$ is expressed in terms of θ_0 using (4.9).

$$k_z^2 = n_0^2 k_0^2 - (n_0 k_0)^2 \cos^2\theta_0 = n_{N+1}^2 k_0^2 - (n_{N+1} k_0)^2 \cos^2\theta_{N+1} \tag{4.72}$$

or

$$\cos\theta_{N+1} = \frac{\sqrt{n_{N+1}^2 - n_0^2 \sin^2\theta_0}}{n_{N+1}}. \tag{4.73}$$

(4.70) has a solution only if the determinant of $\tilde{Q}$ is zero. This is a transcendental equation that must be solved numerically. The mode index can never be larger than the largest refractive index in the film stack. If can never be smaller than the smallest value of refractive index of the outer claddings. If the mode index were smaller than this, the light would leak out of the waveguide. A search algorithm can be implemented to scan the mode index between these two limits to find all values that make the determinant equal to zero.

Again, consider the symmetric waveguide in Figure 4.2. The determinant of $\tilde{Q}$ in (4.71) is plotted between 1.5 and 2.5 in Figure 4.7. The four roots provide the TE mode indices which are 2.4626, 2.3479, 2.1480, and 1.8496.

4.3.4 Field Amplitudes and Poynting Vector

For a specific mode index, the electric field amplitudes E_a and E_b within each layer are calculated from (4.59)–(4.61). The electric field amplitude E_0 in the upper cladding is set to 1. Then from (4.59),

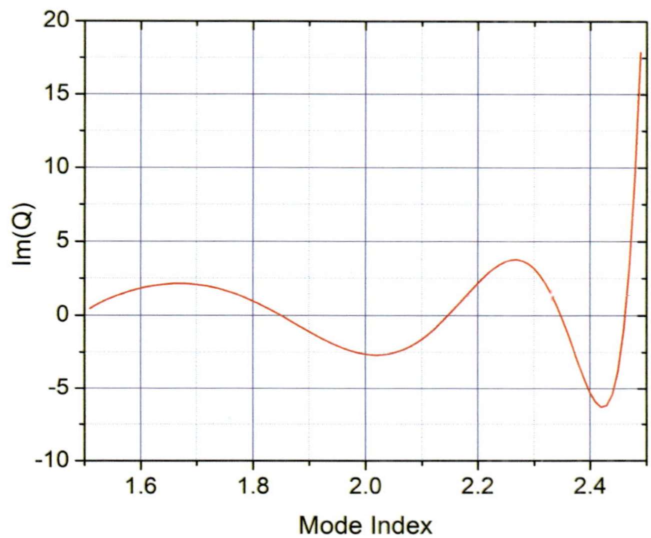

FIGURE 4.7 Im(Q) as a function of mode index for TE modes.

$$\begin{bmatrix} E_{1a} \\ E_{1b} \end{bmatrix} = \left(\tilde{\boldsymbol{S}}_1\right)^{-1} \begin{bmatrix} 1 \\ n_0 \cos\theta_0 \end{bmatrix} E_{0a}. \tag{4.74}$$

The electric field amplitudes in subsequent layers are calculated by repeated application of (4.60). Let

$$\tilde{\boldsymbol{P}}_m = \begin{bmatrix} P_{m,11} & P_{m,12} \\ P_{m,21} & P_{m,22} \end{bmatrix} \equiv \left(\tilde{\boldsymbol{S}}_{m+1}\right)^{-1} \tilde{\boldsymbol{R}}_m \tag{4.75}$$

so

$$\begin{bmatrix} E_{m+1,a} \\ E_{m+1,b} \end{bmatrix} = \left(\tilde{\boldsymbol{S}}_{m+1}\right)^{-1} \tilde{\boldsymbol{R}}_m \begin{pmatrix} E_{\text{ma}} \\ E_{\text{mb}} \end{pmatrix} = \tilde{\boldsymbol{P}}_m \begin{bmatrix} E_{\text{ma}} \\ E_{\text{mb}} \end{bmatrix}. \tag{4.76}$$

Finally, the electric field amplitude in the bottom cladding is obtained from (4.61),

$$\begin{bmatrix} 0 \\ E_{N+1,b} \end{bmatrix} = \left(\tilde{\boldsymbol{S}}_{N+1}\right)^{-1} \tilde{\boldsymbol{R}}_N \begin{bmatrix} E_{\text{Na}} \\ E_{\text{Nb}} \end{bmatrix} = \tilde{\boldsymbol{P}}_N \begin{bmatrix} E_{\text{Na}} \\ E_{\text{Nb}} \end{bmatrix} \tag{4.77}$$

or

$$E_{N+1,b} = \tilde{\boldsymbol{P}}_{N,21} E_{\text{Na}} + \tilde{\boldsymbol{P}}_{N,22} E_{\text{Nb}}. \tag{4.78}$$

The electric and magnetic fields can now be calculated as a function of y from (4.44)–(4.49).

$$\boldsymbol{E}_0 = E_{0a} \hat{\boldsymbol{x}} e^{-ik_{y0}y + ik_z z} \tag{4.79}$$

$$\boldsymbol{E}_m = E_{\text{ma}} \hat{\boldsymbol{x}} e^{-ik_{\text{ym}}(y - y_{m-1}) + ik_z z} + E_{\text{mb}} \hat{\boldsymbol{x}} e^{+ik_{\text{ym}}(y - y_{m-1}) + ik_z z} \tag{4.80}$$

$$\boldsymbol{E}_{N+1} = E_{N+1,b} \hat{\boldsymbol{x}} e^{+ik_{y,N+1}(y - y_N) + ik_z z} \tag{4.81}$$

$$\boldsymbol{H}_0 = \frac{n_0}{Z_0} E_{0a} \left(\sin\theta_0 \hat{\boldsymbol{y}} + \cos\theta_0 \hat{\boldsymbol{z}}\right) e^{-ik_{y0}x + ik_z z} \tag{4.82}$$

$$\begin{aligned} \boldsymbol{H}_m = {} & \frac{n_1}{Z_0} E_{1a} \left(\sin\theta_m \hat{\boldsymbol{y}} + \cos\theta_m \hat{\boldsymbol{z}}\right) e^{-ik_{\text{ym}}(y - y_{m-1}) + ik_z z} \\ & + \frac{n_m}{Z_0} E_{\text{mb}} (\sin\theta_m \hat{\boldsymbol{y}} - \cos\theta_m \hat{\boldsymbol{z}}) e^{ik_{\text{ym}}(y - y_{m-1}) + ik_z z} \end{aligned} \tag{4.83}$$

$$\boldsymbol{H}_{N+1} = \frac{n_{N+1}}{Z_0} E_{N+1,b} \left(\sin\theta_{N+1} \hat{\boldsymbol{y}} - \cos\theta_{N+1} \hat{\boldsymbol{z}}\right) e^{ik_{y,N+1}(y - y_N) + ik_z z}. \tag{4.84}$$

The Poynting vector is given by (4.42) so the z-component of the Poynting vector is just

$$S_z = +\frac{\text{Re}\left(E_x H_y^*\right)}{2} \tag{4.85}$$

as the y-component of the electric field and the x-component of the magnetic field are both zero for the TE mode.

Returning to the waveguide example in Figure 4.2(a), the fields and Poynting vectors for the first three TE modes are plotted in Figures 3.8–3.10.

4.4 EXAMPLES

4.4.1 PIC Chip Symmetric Planar Waveguide

A symmetric planar waveguide in Figure 4.11 is considered in Figures 3.6 and 3.7 of [2] as a typical waveguide in photonic integrated circuit (PIC) chips. There is only one TM mode with mode index 2.0511 and one TE mode with mode index 2.8448. The electric field amplitudes are shown in Figure 4.12(a) and the Poynting vectors in 3.12(b).

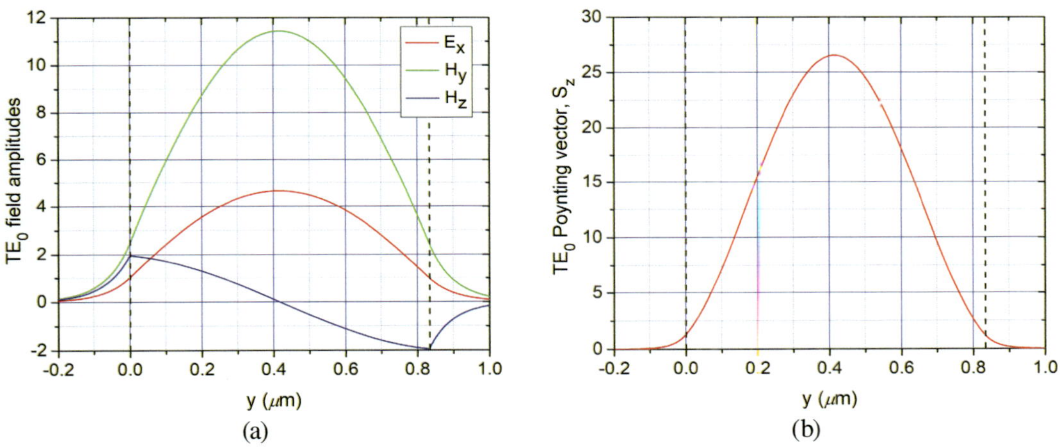

FIGURE 4.8 (a) Field amplitudes through thickness of waveguide for TE_0 mode of Figure 4.2(a) waveguide. (b) z-component of the Poynting vector.

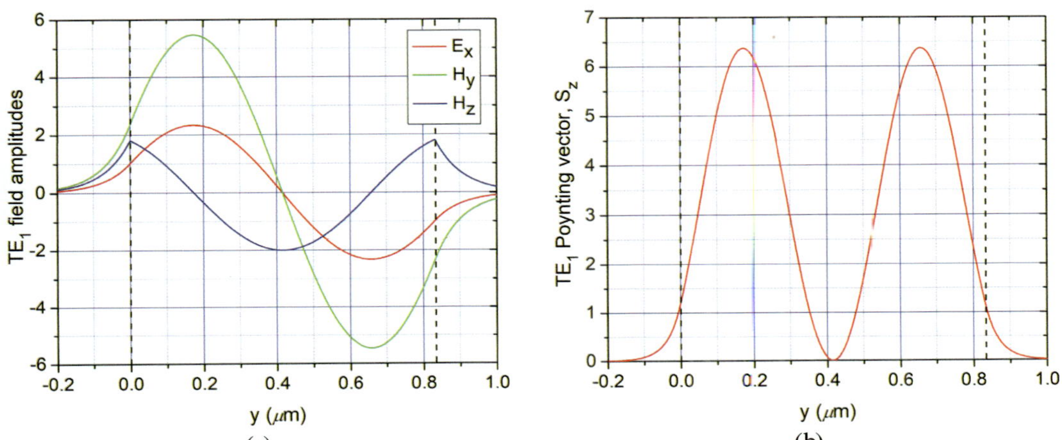

FIGURE 4.9 (a) Field amplitudes through thickness of waveguide for TE_1 mode of Figure 4.2(a) waveguide. (b) z-component of the Poynting vector.

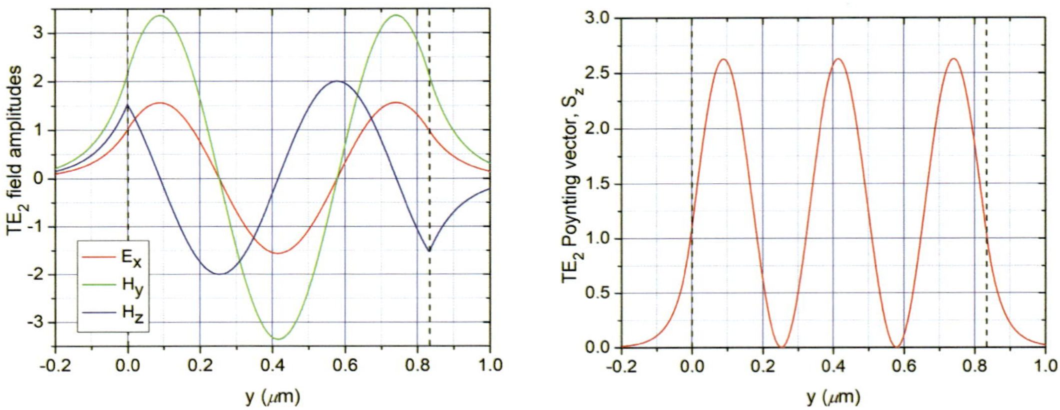

(a) (b)

FIGURE 4.10 (a) Field amplitudes through thickness of waveguide for TE_3 mode of Figure 4.2(a) waveguide. (b) z-component of the Poynting vector.

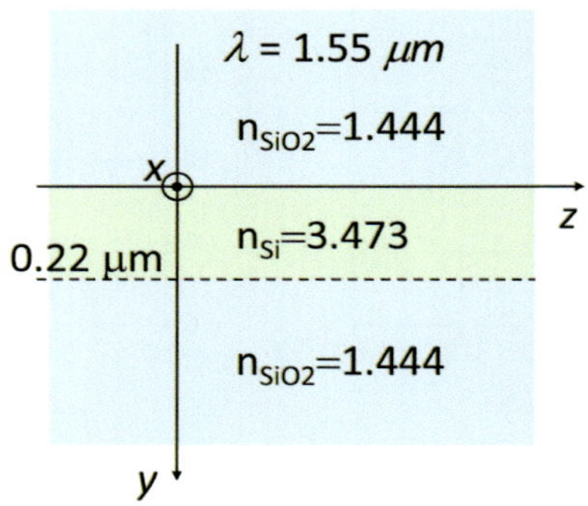

FIGURE 4.11 Symmetric Si waveguide for operation at $\lambda = 1.5\ \mu$m.

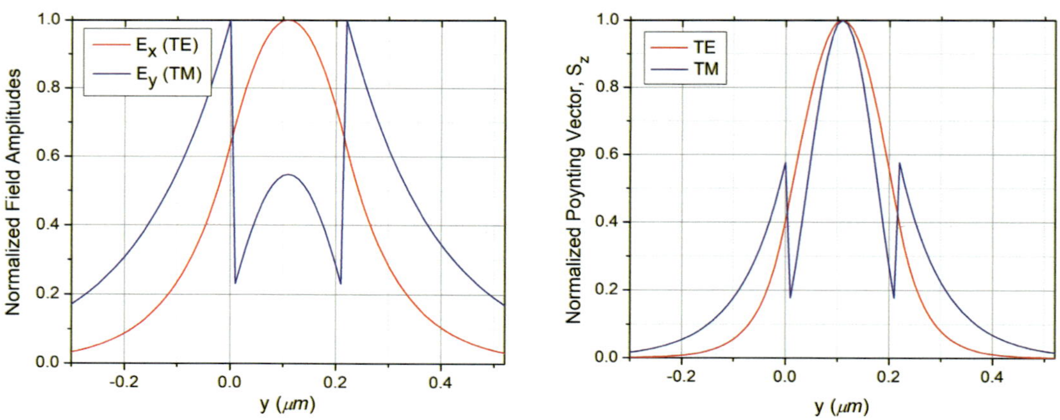

(a) (b)

FIGURE 4.12 (a) Normalized electric field amplitudes and (b) Poynting vector amplitudes for the symmetric planar waveguide in Figure 4.11.

The electric field for the TM_0 mode is considerably broader than that of the TE_0 mode. As a result, it interacts more strongly with the Si substrate underneath the lower cladding and is rather heavily damped. Therefore, most PIC chips only use TE waveguide modes. The TM mode index is slightly different than that given in [2] and the field profile in Figure 4.7 for the TM mode in [1] is incorrect, exhibiting a dip at the center of the waveguide rather than a peak.

4.4.2 Asymmetric Planar Waveguide Dispersion

Consider the asymmetric planar waveguide in Figure 4.13(a). There are three TE and three TM modes when $t = \lambda$. The TE mode indices are 3.4704, 3.3822, and 3.2433, and the TM mode indices are 3.4734, 3.3938, and 3.2640. The dispersion of the mode indices as a function of core thickness is shown in Figure 4.13(b) for comparison to Figure 11.7 in [1]. Note that the TE mode index is always larger than the corresponding TM mode index for the same thickness because the electric field for the TM mode extends further into the lower index claddings than does the TE mode.

4.4.3 Surface Plasmons

Surface plasmons are guided modes that propagate along an interface between a dielectric with a positive dielectric constant and a metal or other material with a negative real part of its dielectric constant [2]. Consider the thin gold film in Figure 4.14. There are two surface plasmon modes that can propagate along the gold layer, a "symmetric mode" for which the longitudinal electric field component, E_z, is symmetric about the center of the gold layer, and an "antisymmetric mode" for which E_z is antisymmetric. Both of these surface plasmon modes are TM modes; there is no E_x field component. The index range of the search algorithm of the mode index must be adjusted to scan between 1 and slightly more than 1 for the real part of the TM mode, and between 0 and slightly more than 0 for the imaginary part of the TM mode.

The two solutions are $n_a = 1.0187582 + i\,(0.00124)$ and $n_s = 1.0743856 + i\,(0.01678)$. The second solution has a larger imaginary part of its mode index indicating that it is absorbed more quickly. The e^{-2} propagation distance d is given by

$$e^{-\mathrm{Im}(n)k_0 d} = e^{-2} \qquad \Rightarrow \qquad d = \frac{\lambda_0}{\pi\,\mathrm{Im}(n)} \tag{4.86}$$

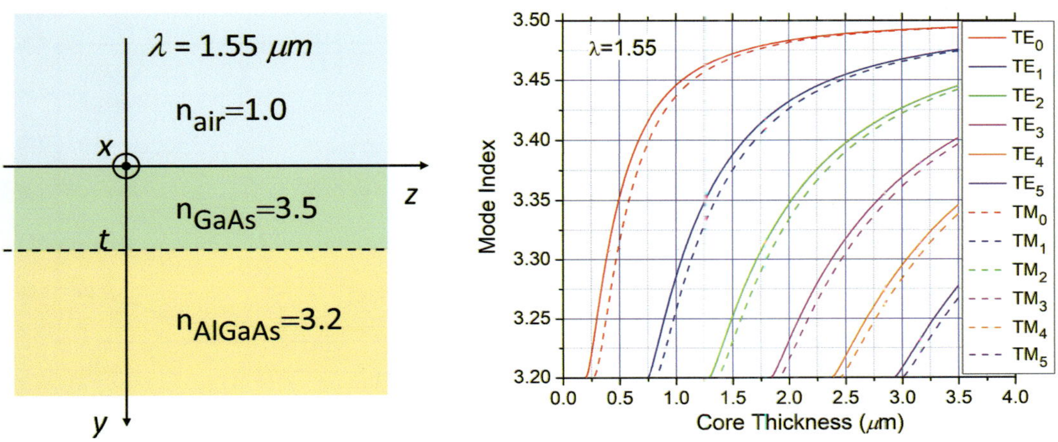

(a) (b)

FIGURE 4.13 (a) Asymmetric planar waveguide after Figure 11.7 in [1]. (b) Mode index dispersion as a function of core thickness.

which is 180 μm for the long-range antisymmetric mode and only 13 μm for the short-range symmetric mode. The electric field profiles are graphed in Figure 4.15. The evanescent electric field in the surrounding air decays more slowly for the antisymmetric mode – more of the optical power in the mode is located in air where there is no absorption – so its propagation distance is correspondingly longer.

As the gold film gets thinner, the absorption of the antisymmetric mode decreases and that of the symmetric mode increases. Calculated propagation distances of the symmetric and antisymmetric modes for several gold film thicknesses are given in Table 4.1.

These surface plasmon modes can be excited via evanescent wave coupling from a prism as shown in Figure 4.16 for a silver film sandwiched between two SiO_2 layers. This excitation technique was invented by Otto [3]. For a 20 nm silver film, the surface plasmon mode indices are 1.5193 + i (0.000219) and 2.1277 + i (0.03360), respectively. Because these modes are so narrow, the search range should be small and/or the search increment small. This latter requirement can mean that the search algorithm may take a longer time to discover the mode indices. The larger imaginary part of the refractive index for the second mode indicates that it is the short-range surface plasmon. The first mode index corresponds to the long-range surface plasmon.

The reflectance of a light wave incident at angle θ from within the higher index prism medium can be computed using the algorithm of Chapter 1. As shown in Figure 4.17, there is a very narrow

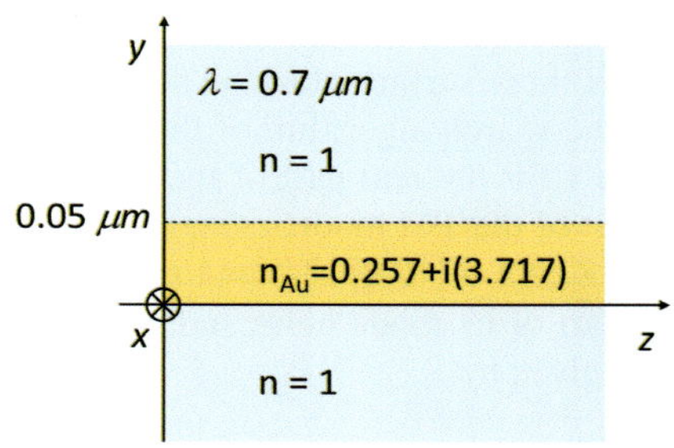

FIGURE 4.14 Surface plasmon modes guided by dielectric/metal surface.

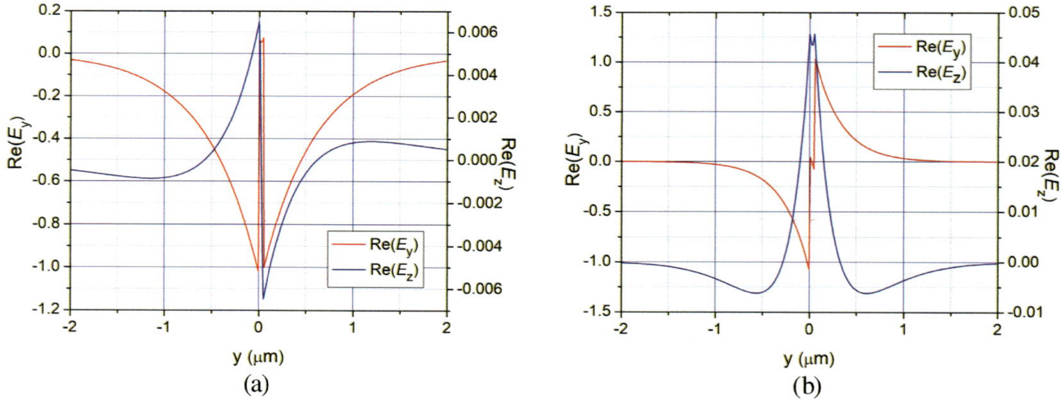

FIGURE 4.15 (a) Real part of E_y and E_z field components of the antisymmetric mode with n_{SP} = 1.019. (b) Real part of E_y and E_z field components of the symmetric mode with n_{SP} = 1.074.

TABLE 4.1
Surface Plasmon Propagation

t (μm)	Re(n_a)	Im(n_a)	d_a(μm)	Re(n_s)	Im(n_s)	d_s(μm)
0.05	1.01876	0.001240	180	1.07439	0.01678	13
0.04	1.01382	0.000723	308	1.09971	0.02360	9.4
0.03	1.00877	0.000344	648	1.15330	0.03739	6.0
0.02	1.00427	0.000121	1840	1.29551	0.07118	3.1
0.01	1.00113	0.000024	9330	1.89030	0.18905	1.2

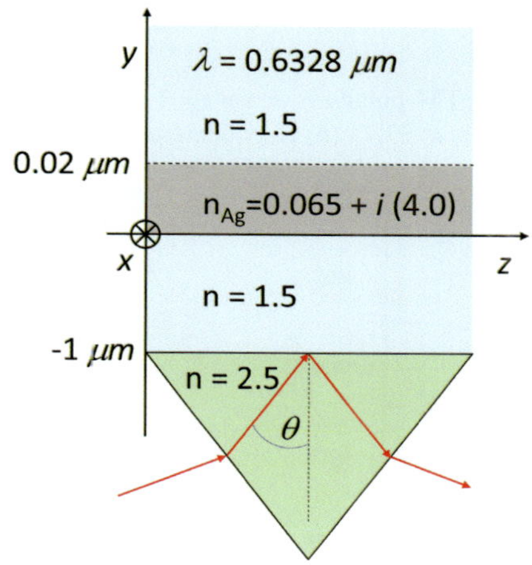

FIGURE 4.16 Prism coupling technique to excite the surface plasmon modes from [3], p. 143 ff.

minimum in the reflectance at 37.4°. An approximate relation for the angle of incidence (inside the prism) for which the TM-polarized beam will excite the SP is

$$\sin\theta_{SP} = \frac{\mathrm{Re}\left(n_{SP}\right)}{n_p} \tag{4.87}$$

where n_p is the refractive index of the prism, and n_{SP} is the mode index of the surface plasmon. The predicted mode index for a resonant mode at 37.4° from (4.87) is 1.518 which corresponds to the long range, antisymmetric surface plasmon mode. The very small extinction coefficient for this mode also explains why the absorption line is so narrow.

The real part of the electric field components for the two surface plasmon modes are plotted in Figure 4.18. The longitudinal component of the electric field (i.e., the E_z component in the direction of propagation) determines the symmetry of the mode even though in this case the E_y components, which exhibit the opposite symmetry, have larger amplitudes. Using the algorithm of Chapter 1, the electric field inside the prism through the film stack for TM-polarized light that is incident at 37.4° is plotted in Figure 4.19. The antisymmetric field profile again indicates that this is the long-range surface plasmon mode.

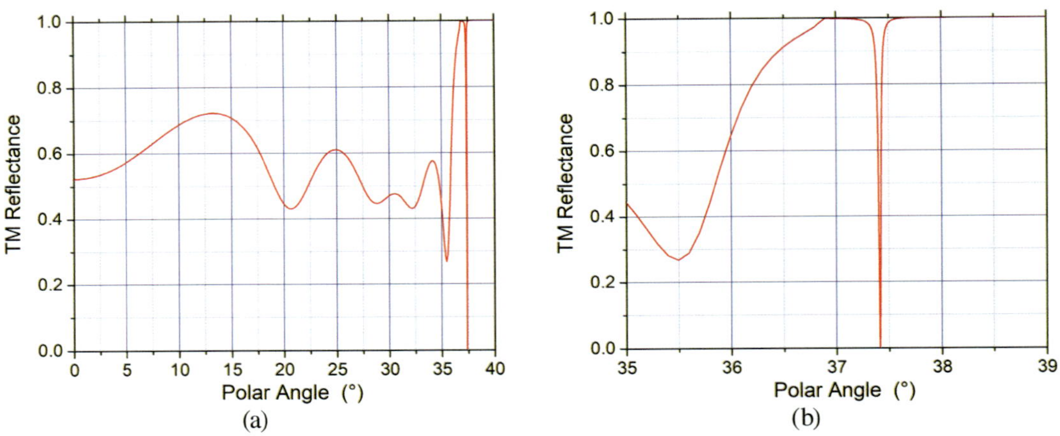

FIGURE 4.17 (a) Reflectance of the TM-polarization for $t = 1$ μm calculated by the algorithm of Chapter 1. The short range plasmon modes occurs at 37.4°. (b) Expanded angle scale.

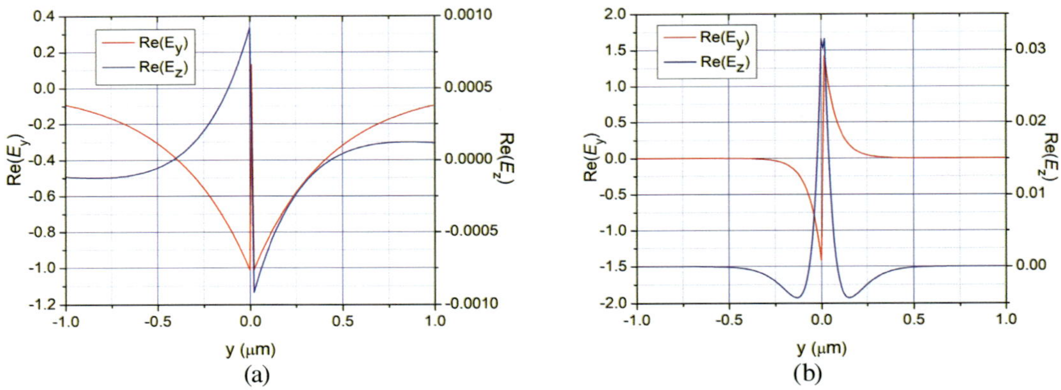

FIGURE 4.18 (a) Real part of the E_y and E_z field components for the long-range, antisymmetric SP mode with $n_{SP} = 1.519$. (b) Real part of the E_y and E_z field components for the short-range, symmetric SP mode with $n_{SP} = 2.128$.

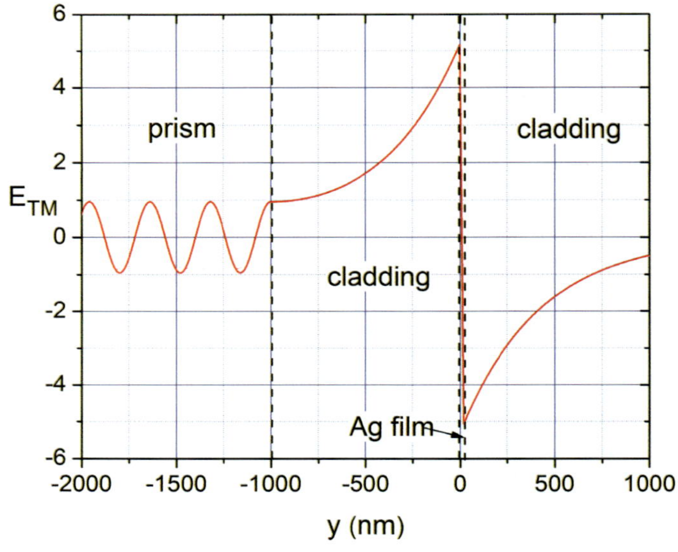

FIGURE 4.19 Electric field through film stack for TM-polarized incident light at the resonant angle of 37.4°.

4.4.4 Multichannel Waveguides

Consider an asymmetric multilayer waveguide consisting of five high index layers spaced by four lower index layers with different outer claddings as shown in Figure 4.20. The mode indices as a function of thickness of the lower index core layers are calculated while the thicknesses of the high index core layers are kept fixed. For more in-depth discussion of multichannel waveguides, see chapter 11 of [1]. The variation in mode index with thickness of the lower index core layers is plotted in Figure 4.21.

As discussed in detail in [1], it is seen that all of the TE and TM modes except for two converge to one of two mode indices for large thicknesses a. The two TE modes and two TM modes which do not converge are surface modes. For example, for $a = 0.55\ \mu m$, there is a TE mode with index

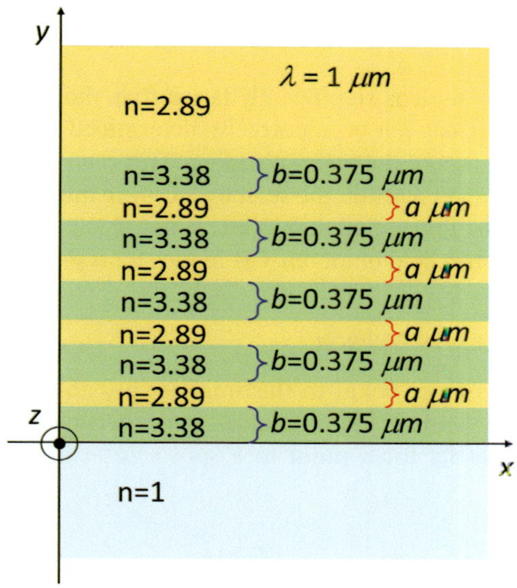

FIGURE 4.20 Asymmetric multichannel waveguide with $b = 3\ \lambda/8$ after Figure 11.12(b) in [1].

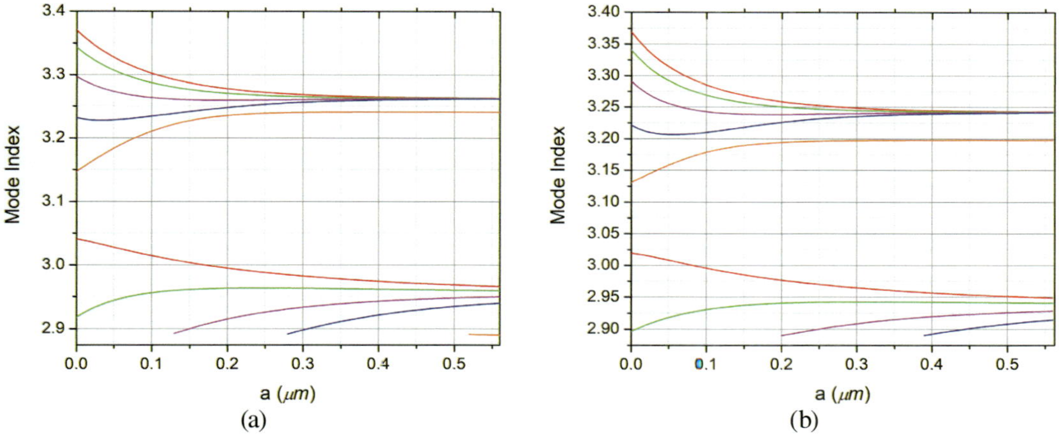

FIGURE 4.21 Variation in mode index for (a) TE, and (b) TM modes for the multichannel waveguide in Figure 4.20.

3.2409 and one with index 2.8905. The calculated mode indices and error function for this core thickness is shown below.

TE Modes

1 n = 3.262798003955056	err = 0.000000000000000	−0.000000001060417
2 n = 3.262458457303604	err = 0.000000000000000	0.000000000548745
3 n = 3.262033394044249	err = 0.000000000000000	−0.000000000831057
4 n = 3.261685016454969	err = 0.000000000000000	0.000000002353934
5 n = 3.240957361273765	err = 0.000000000000000	−0.000010093674064
6 n = 2.966188605677234	err = 0.000000000000000	0.000000000064976
7 n = 2.959464614216019	err = 0.000000000000000	−0.000000000044653
8 n = 2.949508321778011	err = 0.000000000000000	0.000000000043354
9 n = 2.939189071937824	err = 0.000000000000000	−0.000000000035081
10 n = 2.890469430731099	err = 0.000000000000000	−150.092688799706800

The error function for the last mode is significantly larger than those of the other modes. This is an indication that the mode index has not been correctly determined. In fact, the mode index for this mode is slightly smaller than that of the upper cladding. This unusual effect is completely different from all-dielectric waveguides. As a result, the search algorithm must be started with an initial guess that is slightly smaller than the upper cladding index. When this is done, the correct mode index is found to be 2.86652. The electric field of all ten TM modes for $a = 0.55\ \mu$m is shown in Figure 4.22.

4.4.5 Planar Solid Immersion Mirror

One often desires to focus a laser beam to the smallest possible spot or to resolve the smallest possible image. With far field optics, resolution is limited by diffraction. A uniformly illuminated lens focuses light at its focal point to a spot size at its full width at half maximum (FWHM) given by

$$d \cong \frac{0.51\lambda}{NA} \tag{4.88}$$

where λ is the wavelength and *NA* is the numerical aperture of the lens. For higher resolution, the gap between a microscope objective and the substrate can be filled with a high-index oil to boost the numerical aperture of the lens in the focal region by the refractive index of the oil. Even higher resolution can be obtained with a so-called solid immersion lens, [4] which is an objective designed to have no gap between the lens and the substrate so that the numerical aperture is determined by the refractive index of the lens itself.

A similar approach can be taken to focus light waves within planar waveguides [5]. Planar waveguides can be shaped as either solid immersion lenses or solid immersion mirrors. The mode index of the propagating waveguide mode is controlled by the thickness and refractive indices of the core and cladding layers. Snell's law also applies to transitions between waveguide regions with different mode indices to ensure matching of the wavefronts at the interface between the regions. By proper shaping of the interface, a waveguide mode can then be brought to a focus. Unfortunately, it is difficult to precisely control the refractive indices and thickness of the waveguide layers to fabricate waveguide lenses. Waveguide mirrors, however, are much simpler to fabricate. In particular, a parabolic mirror within the planar medium can make use of either total internal reflection or a metallic reflective coating on the edges to focus a planar waveguide mode. The light bending is then determined by the fidelity of the edge shape – how closely it approximates a parabolic curve – rather than by refraction caused by changes in the mode index.

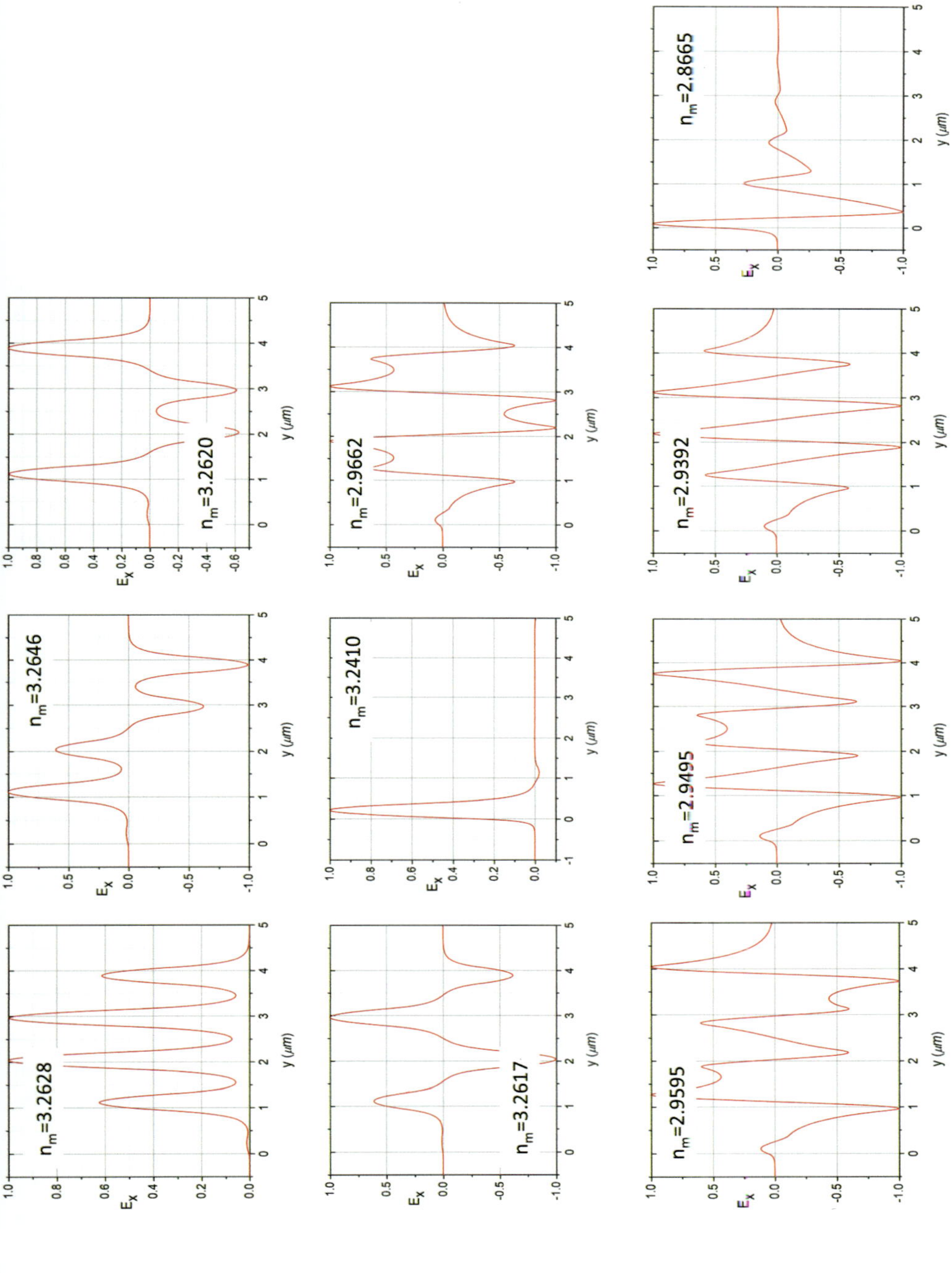

FIGURE 4.22 E_x for the first ten modes with core thickness a = 0.55 μm. (j) is a surface mode after Figure 11.17 of [1].

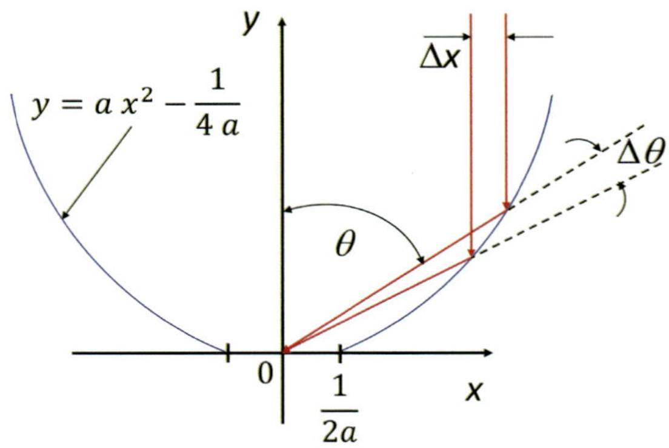

FIGURE 4.23 A planar waveguide mode is incident from the top inside a truncated parabolic waveguide, propagating in the negative direction along the y-axis. If the edge around this region is etched so that the effective index outside of it is close to 1.0 then the waveguide mode undergoes total internal reflection and is brought to a focus at the origin.

A truncated parabolic waveguide mirror is shown in Figure 4.23. The parabola is described by the equation

$$y = a\,x^2 - \frac{1}{4a}. \tag{4.89}$$

Light that reflects from point (x,y) of the parabola is incident upon the focal point at angle θ where

$$\tan\theta = \frac{x}{y}. \tag{4.90}$$

Therefore,

$$\left|\frac{d\theta}{dx}\right| = \frac{1}{2a}\left(\frac{1}{1-\cos\theta}\right). \tag{4.91}$$

The electric field amplitude scales as the square root of the optical power, so the energy conservation scale factor for the electric field amplitude is,

$$\frac{E_f}{E_i} = \frac{1}{\sqrt{1-\cos\theta}} \tag{4.92}$$

where E_f is the field amplitude at the focal point and E_i is the incident field amplitude. If there is an intensity distribution across the opening of the parabolic mirror, an additional scale factor to the focused electric field amplitude can be applied.

Consider a symmetric planar waveguide with a 70 nm core thickness with refractive index 2.1 and with cladding refractive index 1.6. The TE waveguide mode has a mode index of 1.774. The top opening width of the parabola is 50 μm and the length to its truncation is 100 μm resulting in a bottom opening of 6.25 μm. The core and cladding are replaced by air outside of the parabolic region. The wavelength of the incident mode in vacuum is 413 nm.

The minimum angle of incidence θ_{min} is 14° for the upper left and right points of the parabola, and the maximum angle θ_{max} is 90° at the point of intersection of the parabola with the x-axis. The equation for total internal reflection results in

$$\Theta_{tot} = \sin^{-1}\left(\frac{n_{clad}}{n_{mode}}\right) \cong 34.3°. \tag{4.93}$$

The angle of incidence θ in Figure 4.23 is related to Θ_{tot} by

$$\theta = 180° - 2\,\Theta_{tot} \cong 111°. \tag{4.94}$$

So to a first approximation (neglecting scattering at the edge) total internal reflection should occur for all incident angles.

The electric field in the focal region is calculated by summing the planar waveguide TE mode for angles of incidence between 14° and 90° in 1° increments. For a given point (x,y) within the parabola at the center of the core, the electric field E_{TE} of the TE mode that is incident upon the focal point at angle θ has field components along both x- and y-axes.

$$E_x(\theta) = E_{TE}(\theta)\cos\theta\left(e^{ik_z d_1} + e^{ik_z d_2}\right) \tag{4.95}$$

and

$$E_y(\theta) = E_{TE}(\theta)\sin\theta\left(e^{ik_z d_1} - e^{ik_z d_2}\right) \tag{4.96}$$

where

$$d_1 = -x\sin\theta - y\cos\theta \tag{4.97}$$

and

$$d_2 = +x\sin\theta - y\cos\theta. \tag{4.98}$$

The total field at point (x,y) is then found by summing the electric field components in (4.95) and (4.96) at the center of the core over the range of angles incident upon the focal point, 14°–90°, by 1°,

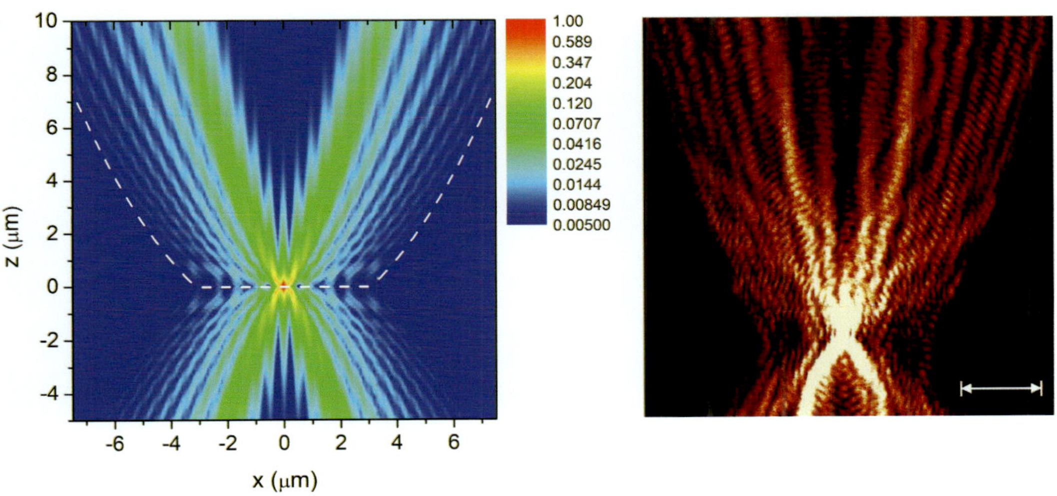

(a) (b)

FIGURE 4.24 (a) Calculated focused TE mode in planar waveguide. (b) Result measured by scanning near field microscopy. The scale bar is 3 μm.

on both left and right halves of the parabola and scaling the field amplitude at each incident angle by (4.92). This calculation does not take into account phase shifts or amplitude changes that may occur at each reflection. The total $|E|^2$ intensity is shown in Figure 4.24(a) along with the intensity pattern measured for this PSIM using a scanning near field microscopy in Figure 4.24(b). The FWHM of the focused spot along this axis is calculated to be ~80 nm, while the SNOM measurement yielded a value of ~94 nm [6].

REFERENCES

1 P. Yeh, *Optical Waves in Layered Media* (New York: John Wiley & Sons, 1988).

2 D. Sarid and W. Challener, *Modern Introduction to Surface Plasmons* (Cambridge: Cambridge University Press, 2010).

3 A. Otto, "Excitation of nonradiative surface plasma waves in silver by the method of frustrated total reflection," *Z. Phys.* **216** (1968) 398.

4 S. M. Mansfield and G. S. Kino, "Solid Immersion Microscope," *Appl. Phys. Lett.* **57** (1990) 2615–2616.

5 R. Ulrich and R. J. Martin, "Geometrical optics in thin film light guides," *Appl. Opt.* **10** (1971) 2077–2085.

6 W. A. Challener, C. Mihalcea, C. Peng, and K. Pelhos, "Miniature solid immersion mirror with focused spot less than a quarter wavelength," *Opt. Exp.* **13** (2005) 7189–7197.

5 General Waveguides

5.1 INTRODUCTION

This chapter is an introduction to one of the powerful techniques for calculating fields within waveguides known as the method of lines. It is not meant to be a complete discussion of general waveguides. For guided wave modeling, Collin's classic book, *Field Theory of Guided Waves*, [1] is recommended. For extensive development of the technique developed in this chapter one should consult Pregla's book, *Analysis of Electromagnetic Fields and Waves*, [2] along with an extensive list of publications on that subject by Pregla and his colleagues. In their publications, this technique is generally developed in terms of electric and magnetic potentials, but in this chapter the equations are derived in terms of electric and magnetic fields which may or may not be easier to understand but is consistent with the approach used in the other chapters of this book.

Unlike the planar waveguides considered in Chapter 3, there are no general analytic solutions for the fields in general dielectric waveguides like rib waveguides or step waveguides. A variety of techniques have been developed for the analysis of these waveguides including the effective index method, finite element and finite difference methods, the mode matching technique, and the method of lines. Some of these techniques are simply approximations that set fields exactly to zero in some regions of space in which the fields are expected to be small, and boundary conditions are ignored. The resulting solutions may be adequate for some applications or to attain a better understanding of the general field profiles. However, they are usually inadequate for applications of photonic integrated chip design. Numerical methods like the finite element method (FEM) and the finite difference time domain method (FDTD) can accurately model nearly any waveguide design, but they are often quite computationally slow. The method of lines combines the analytical and numerical approaches and is, therefore, generally more computationally efficient for waveguide analysis than FEM or FDTD.

As shown in Chapter 3, the modes of *planar* waveguides are either transverse electric (TE), in which there is no component of the electric field in the direction of propagation, or transverse magnetic (TM), in which there is no component of the magnetic field in the direction of propagation. However, dielectric waveguide modes in general cannot be so described as usually all modes have both electric and magnetic field amplitudes in the direction of propagation. Nevertheless, it is often the case that one component is much smaller than the others in which case one can speak of "quasi-TE" or "quasi-TM" modes. Alternatively, these are sometimes referred to as "longitudinal section magnetic" (LSM) or "longitudinal section electric" (LSE) modes. For the LSM mode, the magnetic field in one of the transverse directions is zero. For the LSE mode, the electric field in one of the transverse directions is zero. One can also speak of TM_{pq} or EH_{pq} modes, and TE_{pq} or HE_{pq} modes [3]. The two subscripts refer to the number of nodes of the dominant electric field component in the horizontal (x-axis) and vertical (y-axis) directions, respectively. If E_x is the dominant electric field component, they are denoted HE or quasi-TE modes. If E_y is the dominant electric field component, they are denoted EH or quasi-TM modes.

Consider the simple rectangular waveguide in Figure 5.1 with a core index of 2.0, surrounded by a cladding with index 1.0. There are nine propagating modes for this waveguide. The mode index values are

1. 1.802126628718: HE_{00}
2. 1.732771932078: EH_{00}
3. 1.609638433610: HE_{10}
4. 1.575598315668: EH_{10}

DOI: 10.1201/9781003714231-5

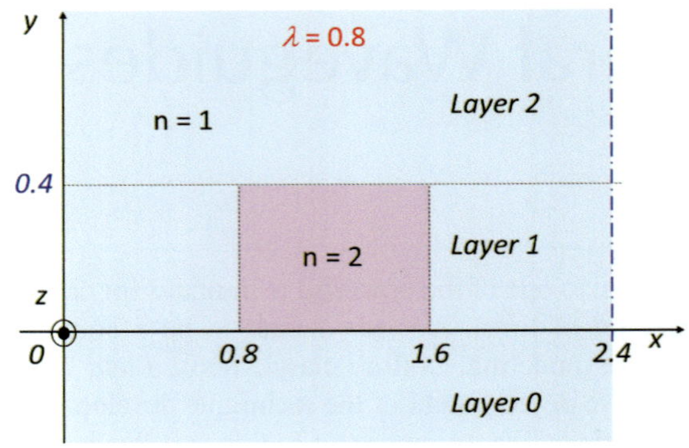

FIGURE 5.1 Example of a simple rectangular strip waveguide for which the core index is 2.0 and the surrounding cladding index is 1.0. The wavelength is 0.8 μm, width of the core is 0.8 μm, and the height of the core is 0.4 μm.

5. 1.345187492236: HE_{01}
6. 1.309261469112: EH_{20}
7. 1.276650699235: HE_{20}
8. 1.112688570321: EH_{01}
9. 1.077827774317: HE_{11}

Of course, the precision in these mode index numbers far exceeds the precision in the input parameters, but it is found numerically that this level of precision is often necessary to reduce the error function to an acceptably small value when searching for the mode index. The E_x and E_y field profiles for all nine modes are plotted in Figure 5.2 along with their designations as HE or EH modes.

5.2 WAVE EQUATIONS

Electromagnetic waves are described by Maxwell's equations. In the absence of free charges and currents,

$$\nabla\times\boldsymbol{E}=i\omega\mu_0\boldsymbol{H} \tag{5.1}$$

$$\nabla\times\boldsymbol{H}=-i\omega\epsilon_0\epsilon\,\boldsymbol{E} \tag{5.2}$$

$$\nabla\cdot\boldsymbol{E}=\nabla\cdot\boldsymbol{B}=0. \tag{5.3}$$

The dielectric constant of the medium is ϵ, the free space permittivity ϵ_0, and free space permeability μ_0. Some useful constants are the speed of light in free space,

$$c=\frac{1}{\sqrt{\mu_0\epsilon_0}}=\frac{\omega}{k_0}, \tag{5.4}$$

where ω is the angular frequency of the radiation, k_0 is the free space wavevector, and λ is the free space wavelength, and the impedance of free space,

$$\eta_0=\sqrt{\frac{\mu_0}{\epsilon_0}}. \tag{5.5}$$

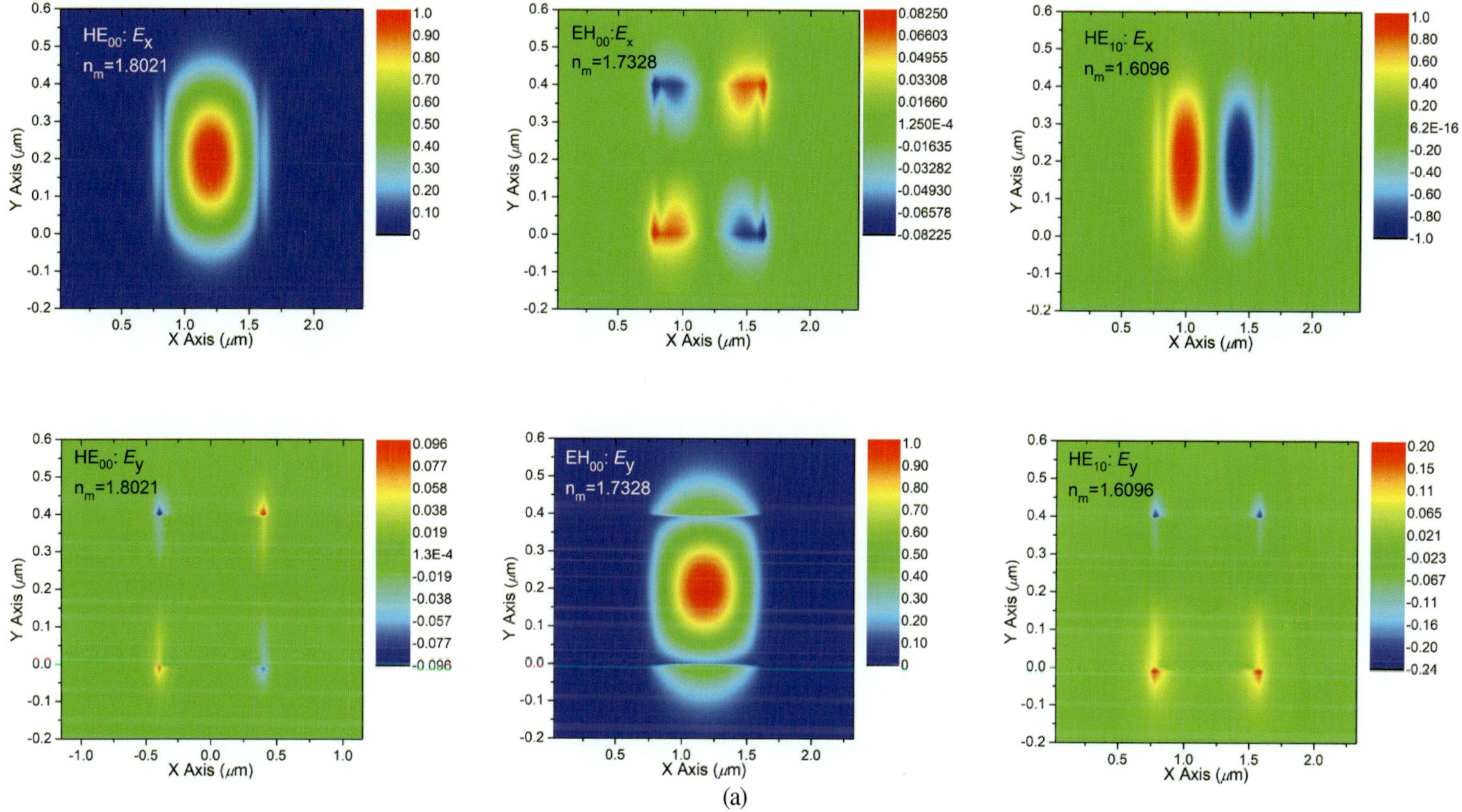

FIGURE 5.2 E_x and E_y field profiles for the nine propagating modes of the strip waveguide in Figure 5.1. (a) Field profiles for HE_{00}, EH_{00}, and HE_{10} modes. (b) Field profiles for EH_{10}, HE_{01}, and EH_{20} modes. (c) Field profiles for HE_{20}, EH_{01}, and HE_{11} modes.

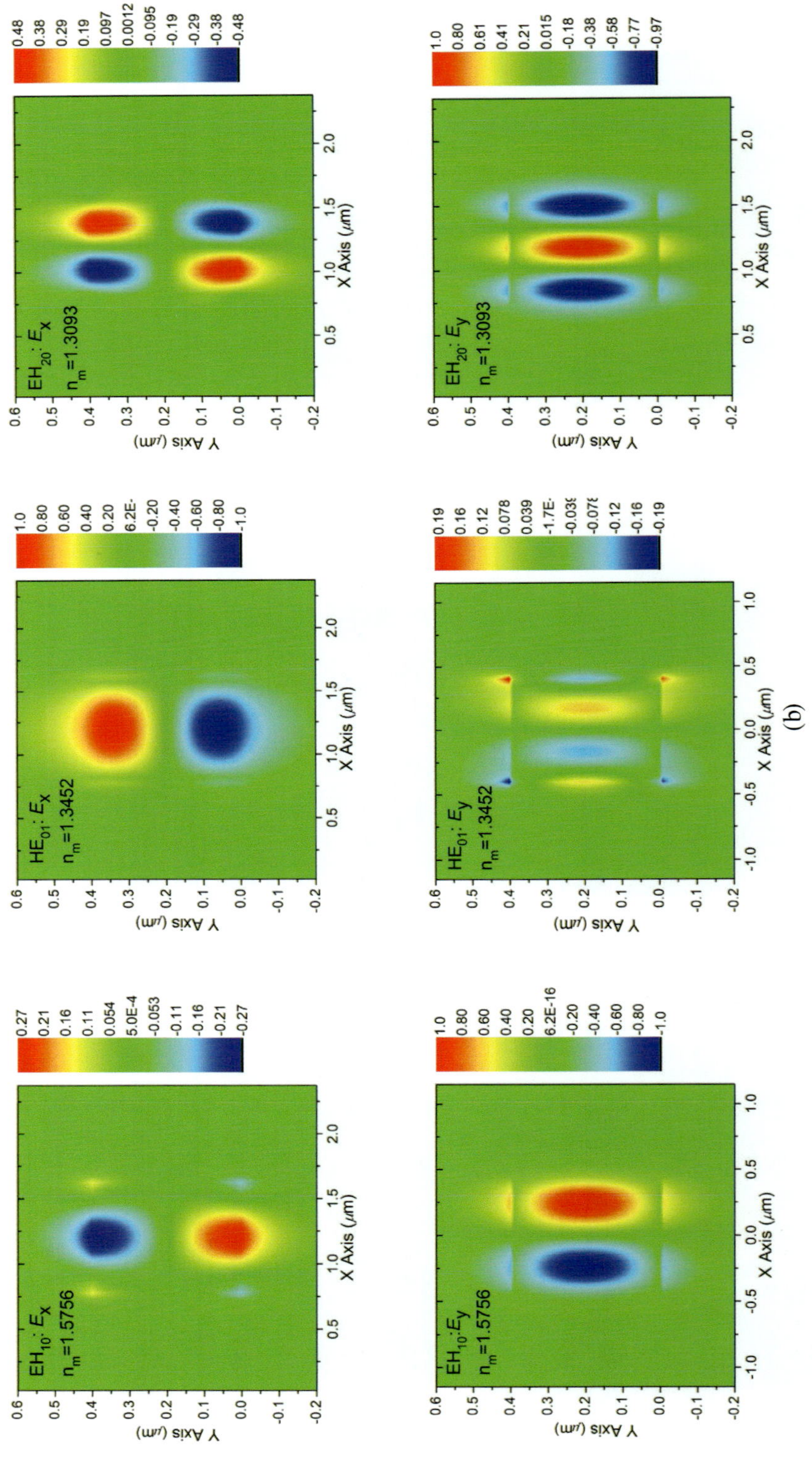

FIGURE 5.2 *(Continued)*

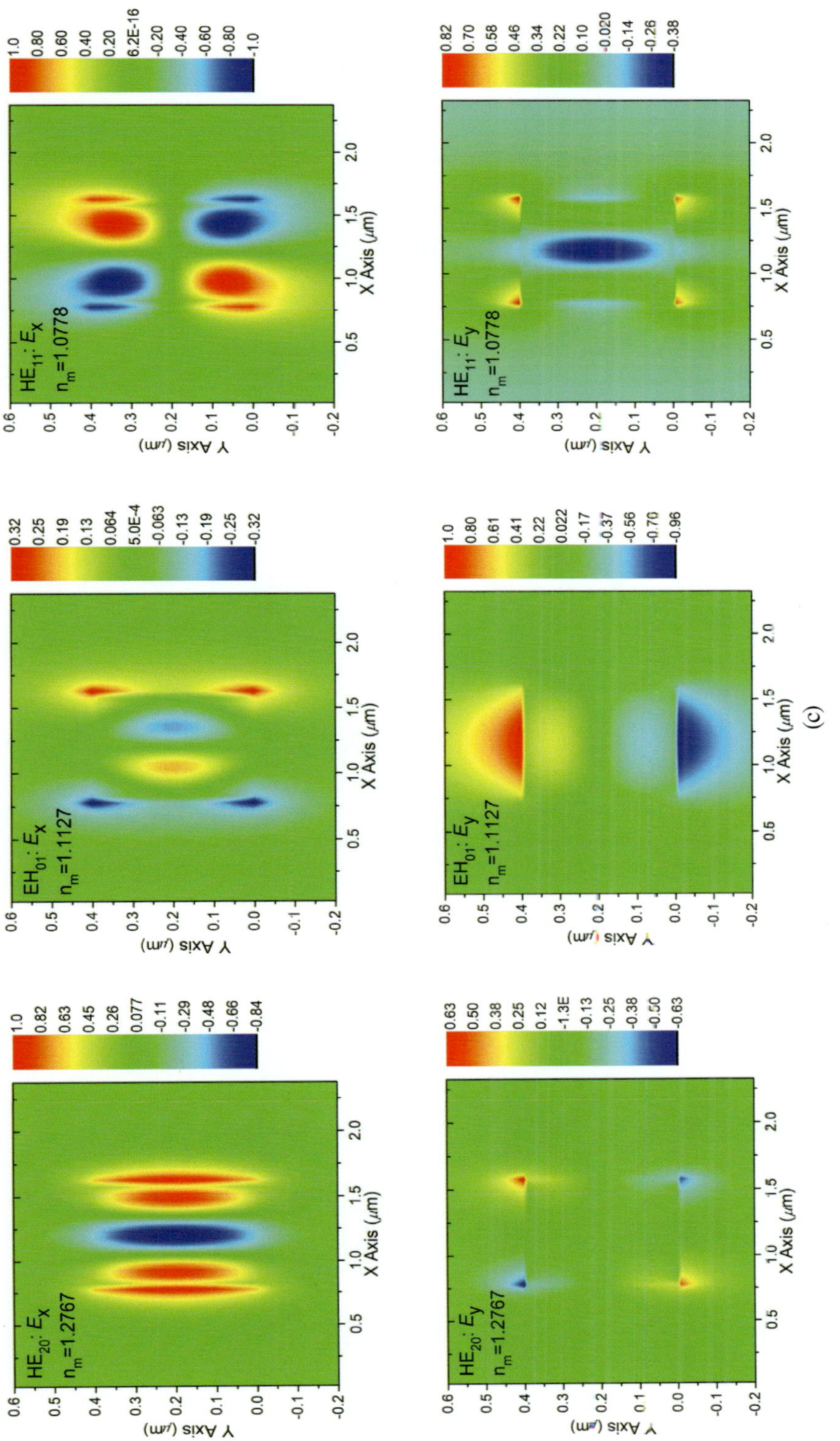

FIGURE 5.2 *(Continued)*

The wave equations are derived from Maxwell's equations for an electromagnetic wave propagating along the z-axis with wavevector k_z. Complex number notation with an $e^{-i\omega t}$ time dependence is used, so the complex fields propagate as $e^{i(k_z z-\omega t)}$.

The separate Cartesian components of (5.1) and (5.2) are

$$-i\omega\epsilon\epsilon_0 E_x = \frac{\partial H_z}{\partial y} - \frac{\partial H_y}{\partial z} = \frac{\partial H_z}{\partial y} - ik_z H_y, \tag{5.6}$$

$$-i\omega\epsilon\epsilon_0 E_y = \frac{\partial H_x}{\partial z} - \frac{\partial H_z}{\partial x} = ik_z H_x - \frac{\partial H_z}{\partial x}, \tag{5.7}$$

$$-i\omega\epsilon\epsilon_0 E_z = \frac{\partial H_y}{\partial x} - \frac{\partial H_x}{\partial y}, \tag{5.8}$$

$$i\omega\mu_0 H_x = \frac{\partial E_z}{\partial y} - \frac{\partial E_y}{\partial z} = \frac{\partial E_z}{\partial y} - ik_z E_y, \tag{5.9}$$

$$i\omega\mu_0 H_y = \frac{\partial E_x}{\partial z} - \frac{\partial E_z}{\partial x} = ik_z E_x - \frac{\partial E_z}{\partial x}, \tag{5.10}$$

and

$$i\omega\mu_0 H_z = \frac{\partial E_y}{\partial x} - \frac{\partial E_x}{\partial y}. \tag{5.11}$$

H_z is substituted from (5.11) into (5.6) and (5.7) to solve for H_x and H_y. The dielectric constant is assumed to be a function of x but not y when applying the partial derivatives.

$$H_x = \frac{1}{k_z k_0 \eta_0}\frac{\partial^2 E_x}{\partial y \partial x} - \frac{1}{k_z k_0 \eta_0}\left(\frac{\partial^2}{\partial x^2} + k_0^2\epsilon\right)E_y \tag{5.12}$$

$$H_y = \frac{1}{k_0 k_z \eta_0}\left(k_0^2\epsilon + \frac{\partial^2}{\partial y^2}\right)E_x - \frac{1}{k_0 k_z \eta_0}\frac{\partial^2 E_y}{\partial y \partial x}. \tag{5.13}$$

E_z from (5.8) is substituted into (5.9) and (5.10) to solve for E_x and E_y.

$$E_z = -\frac{i\eta_0}{k_0}\epsilon^{-1}\frac{\partial H_x}{\partial y} + \frac{i\eta_0}{k_0}\epsilon^{-1}\frac{\partial H_y}{\partial x}. \tag{5.14}$$

$$E_x = -\frac{\eta_0}{k_z k_0}\frac{\partial}{\partial x}\epsilon^{-1}\frac{\partial H_x}{\partial y} + \left(\frac{k_0\eta_0}{k_z} + \frac{\eta_0}{k_z k_0}\frac{\partial}{\partial x}\epsilon^{-1}\frac{\partial}{\partial x}\right)H_y. \tag{5.15}$$

$$E_y = \left(-\frac{\eta_0}{k_z k_0}\epsilon^{-1}\frac{\partial^2}{\partial y^2} - \frac{k_0\eta_0}{k_z}\right)H_x + \frac{\eta_0}{k_z k_0}\epsilon^{-1}\frac{\partial^2 H_y}{\partial y \partial x}. \tag{5.16}$$

Finally, H_x and H_y from (5.12) and (5.13) are substituted into (5.15) and (5.16). This gives one wave equation in E_x alone and a second wave equation in both E_x and E_y.

$$\frac{\partial^2}{\partial y^2}E_x+\frac{\partial}{\partial x}\epsilon^{-1}\frac{\partial}{\partial x}\epsilon E_x-\left(k_z^2-k_0^2\epsilon\right)E_x=0\,. \tag{5.17}$$

$$\frac{\partial^2}{\partial y^2}E_y+\frac{\partial^2}{\partial x^2}E_y-\left(k_z^2-k_0^2\epsilon\right)E_y=\left(\frac{\partial^2}{\partial y\partial x}-\epsilon^{-1}\frac{\partial^2}{\partial y\partial x}\epsilon\right)E_x. \tag{5.18}$$

If the dielectric constant is independent of x, then the term on the right-hand side of (5.18) drops out. The dielectric constant factors in (5.17) also drop out, and (5.17) and (5.18) become identical.

Alternatively, E_x and E_y from (5.15) and (5.16) can be substituted into (5.12) and (5.13). After simplification a wave equation for H_x and a second wave equation that includes both H_x and H_y are obtained.

$$\frac{\partial^2}{\partial y^2}H_x+\frac{\partial^2}{\partial x^2}H_x-\left(k_z^2-k_0^2\epsilon\right)H_x=0\,. \tag{5.19}$$

$$\frac{\partial^2}{\partial y^2}H_y+\epsilon\frac{\partial}{\partial x}\epsilon^{-1}\frac{\partial}{\partial x}H_y-\left(k_z^2-k_0^2\epsilon\right)H_y=\epsilon\frac{\partial}{\partial x}\epsilon^{-1}\frac{\partial H_x}{\partial y}-\frac{\partial^2}{\partial y\partial x}H_x. \tag{5.20}$$

Because there are now two wave equations that can be solved immediately for E_x and H_x, it is useful to derive the additional equations for the other field components in terms of these two field components. H_z can be eliminated from (5.7) and (5.11) and then the resulting equations used to find E_y in terms of E_x and H_x.

$$\left(\frac{\partial^2}{\partial x^2}+k_0^2\epsilon\right)E_y=\frac{\partial^2 E_x}{\partial x\partial y}-k_0\eta_0k_zH_x. \tag{5.21}$$

E_y can also be eliminated from (5.7) and (5.11) to solve for H_z in terms of E_x and H_x.

$$\left(\frac{\partial}{\partial x}\epsilon^{-1}\frac{\partial}{\partial x}+k_0^2\right)H_z=\frac{ik_0}{\eta_0}\frac{\partial E_x}{\partial y}+ik_z\frac{\partial}{\partial x}\epsilon^{-1}H_x. \tag{5.22}$$

Similarly, E_z can be eliminated from (5.8) and (5.10) to solve for H_y in terms of E_x and H_x.

$$\left(\frac{\partial}{\partial x}\epsilon^{-1}\frac{\partial}{\partial x}+k_0^2\right)H_y=\frac{k_0k_z}{\eta_0}E_x+\frac{\partial}{\partial x}\epsilon^{-1}\frac{\partial H_x}{\partial y}. \tag{5.23}$$

Finally, H_y can be eliminated from (5.8) and (5.10) to solve for E_z.

$$\left(\frac{\partial^2}{\partial x^2}+k_0^2\epsilon\right)E_z=ik_z\frac{\partial}{\partial x}E_x-ik_0\eta_0\frac{\partial H_x}{\partial y}. \tag{5.24}$$

5.3 DISCRETIZATION OF THE WAVE EQUATIONS

Consider the waveguide geometry in Figure 5.3, which has been divided into three layers. The middle layer is further divided into three regions. The dielectric constant within each region is

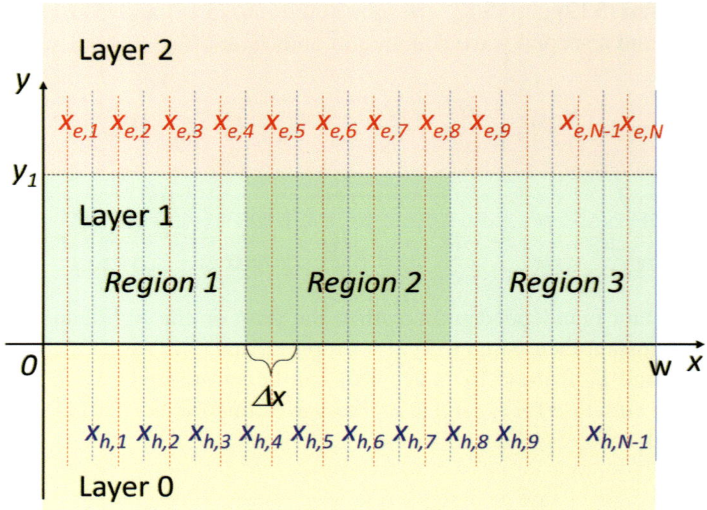

FIGURE 5.3 Example of a rectangular slab waveguide.

a constant. In this example, the center region is the waveguide core. The computation space lies between $x = 0$ and w along the x-axis and from $-\infty$ to $+\infty$ along the y-axis. By using appropriate layers and regions, a variety of different waveguide cross sections can be modeled. In the method of lines, the fields are only computed along the dashed vertical lines. E_x, H_y, and H_z are calculated on the red "electric" lines x_e as shown, while E_y, E_z, and H_x are calculated on the blue "magnetic" lines x_h. The x_e lines are offset from the x_h lines by $\Delta x/2$, so the electric lines and magnetic lines are separately equally spaced by Δx. The magnetic field lines are placed so that they lie along the vertical edges of the regions in each layer. In this example, magnetic lines lie along the x-axis boundaries and so there is one more electric field line than magnetic field line *within* the computation space.

In the method of lines, the fields are only computed along vertical lines as shown in Figure 5.3. The partial derivatives of the fields along the x-axis are thereby discretized as well. The boundaries on the x-axis of the computation space occur at $x = 0$ and w along $x_{h=0}$ and $x_{h=\mathrm{N}}$ lines. E_x, H_y, and H_z are computed along the x_e lines, while E_y, E_z, and H_x are computed along the x_h lines. Thus, the x and y components of the fields are interleaved with offset $\Delta x/2$. In this example there are N x_e lines and N-1 x_h lines. The fields calculated on the x_h lines are forced to zero on the sides of the computation space for the boundary conditions in this calculation, so the space must be wide enough for this to be an accurate approximation. The partial derivatives along the x-axis for the other field components calculated on the x_e lines are also zero on the sides of the computation space.

The dielectric constants within the regions of the computation space are specified in terms of a matrix for each layer in which the diagonal elements are the dielectric constants for each line within that layer. There are two dielectric constant matrices for each layer m, one for the x_e lines and one for the x_h lines. If $n_{h,n}$ is the refractive index in layer m along line n, then the dielectric constant matrix for the x_h lines is

$$\tilde{\epsilon}_h^{(m)} \equiv \begin{bmatrix} n_{h1}^2 & 0 & 0 & \cdots & 0 \\ 0 & n_{h2}^2 & 0 & \cdots & 0 \\ 0 & 0 & n_{h3}^2 & \cdots & 0 \\ \vdots & \vdots & \vdots & \ddots & 0 \\ 0 & 0 & 0 & 0 & n_{h,N-1}^2 \end{bmatrix} \qquad (5.25)$$

which is a square matrix of rank *N-1*. The subscript for each element along the diagonal is numbered according to line number. When an x_h field line falls on a border between two regions of different dielectric constants, the average dielectric constant value is used.

For the electric lines, x_e, the dielectric constant matrix is

$$\tilde{\boldsymbol{\epsilon}}_e^{(n)} \equiv \begin{bmatrix} n_{e1}^2 & 0 & 0 & \cdots & 0 \\ 0 & n_{e2}^2 & 0 & \cdots & 0 \\ 0 & 0 & n_{e3}^2 & \cdots & 0 \\ \vdots & \vdots & \vdots & \ddots & 0 \\ 0 & 0 & 0 & 0 & n_{e,N}^2 \end{bmatrix} \tag{5.26}$$

which is a square matrix of rank *N*.

Let F_h represent one of the field components E_y, E_z, or H_x, calculated on the x_h lines in layer m. The *N-1* F_h field amplitudes on the x_h lines within the computation space for location y within layer m can be combined into a vector,

$$\boldsymbol{F}_h^{(m)}(y) = \begin{bmatrix} F_{h,1}^{(m)}(y) \\ F_{h,2}^{(m)}(y) \\ F_{h,3}^{(m)}(y) \\ \vdots \\ F_{h,(N-1)}^{(m)}(y) \end{bmatrix}. \tag{5.27}$$

The partial derivative of F_h on the electric line $x_{e,n}$ is determined by a linear fit to the field values on the neighboring magnetic lines $x_{h,n}$ and $x_{h,n-1}$.

$$\left.\frac{\partial F_{h,n}(y)}{\partial x}\right|_{x_{e,n}} \approx \frac{F_{h,n}(y) - F_{h,n-1}(y)}{\Delta \mathrm{x}}. \tag{5.28}$$

At $x_{h,0}$ and $x_{h,N}$ the field amplitude is zero, so

$$\left.\frac{\partial F_{h,1}(y)}{\partial x}\right|_{x_{e,1}} = \frac{F_{h,1}(y) - F_{h,0}(y)}{\Delta \mathrm{x}} = \frac{F_{h,1}(y)}{\Delta \mathrm{x}} \tag{5.29}$$

and

$$\left.\frac{\partial F_{h,N}(y)}{\partial x}\right|_{x_{e,N}} = \frac{F_{h,N}(y) - F_{h,N-1}(y)}{\Delta \mathrm{x}} = -\frac{F_{h,N-1}(y)}{\Delta \mathrm{x}}. \tag{5.30}$$

A matrix equation can be used to represent the partial derivatives.

$$\frac{\partial \boldsymbol{F}_h(y)}{\partial x} = \left(\frac{1}{\Delta \mathrm{x}}\right) \begin{bmatrix} F_{\mathrm{h}1}(y) \\ F_{\mathrm{h}2}(y) - F_{\mathrm{h}1}(y) \\ F_{\mathrm{h}3}(y) - F_{\mathrm{h}2}(y) \\ \vdots \\ -F_{h,N-1}(y) \end{bmatrix} = \left(\frac{1}{\Delta \mathrm{x}}\right) \begin{bmatrix} 1 & 0 & \cdots & 0 \\ -1 & 1 & \cdots & 0 \\ 0 & -1 & \ddots & 0 \\ \vdots & \vdots & \ddots & 1 \\ 0 & 0 & 0 & -1 \end{bmatrix} \begin{bmatrix} F_{h,1}(y) \\ F_{h,2}(y) \\ F_{h,3}(y) \\ \vdots \\ F_{h,N-1}(y) \end{bmatrix}$$

$$= \left(\frac{1}{\Delta x}\right) \tilde{\boldsymbol{D}} \begin{bmatrix} F_{h,1}(y) \\ F_{h,2}(y) \\ F_{h,3}(y) \\ \vdots \\ F_{h,N-1}(y) \end{bmatrix} = \left(\frac{1}{\Delta x}\right) \tilde{\boldsymbol{D}} \boldsymbol{F}_h(y) \tag{5.31}$$

where $\tilde{\boldsymbol{D}}$ is a rectangular matrix with N rows and $N - 1$ columns,

$$\tilde{\boldsymbol{D}} \equiv \begin{bmatrix} 1 & 0 & \cdots & 0 \\ -1 & 1 & \cdots & 0 \\ 0 & -1 & \ddots & 0 \\ \vdots & \vdots & \ddots & 1 \\ 0 & 0 & 0 & -1 \end{bmatrix}. \tag{5.32}$$

Acting on the F_h field, it generates a derivative with N elements on the x_e lines. Similarly, the partial derivatives of the F_e fields are discretized according to

$$\left.\frac{\partial F_{e,n}(y)}{\partial x}\right|_{x_{h,n}} \approx \frac{F_{e,n+1}(y) - F_{e,n}(y)}{\Delta x}. \tag{5.33}$$

The N *non-zero* F_e fields on the x_e lines within the computation space in a given layer m can be combined into a vector,

$$\boldsymbol{F}_e^{(m)}(y) = \begin{bmatrix} F_{e1}^{(m)}(y) \\ F_{e2}^{(m)}(y) \\ F_{e3}^{(m)}(y) \\ \vdots \\ F_{e,N}^{(m)}(y) \end{bmatrix}. \tag{5.34}$$

The partial derivative of the F_e fields can be represented as a matrix equation,

$$\frac{\partial \boldsymbol{F}_e(y)}{\partial x} = \left(\frac{1}{\Delta x}\right) \begin{bmatrix} F_{e2}(y) - F_{e1}(y) \\ F_{e3}(y) - F_{e2}(y) \\ F_{e4}(y) - F_{e3}(y) \\ \vdots \\ F_{e,N}(y) - F_{e,N-1}(y) \end{bmatrix} = -\left(\frac{1}{\Delta x}\right) \begin{bmatrix} 1 & -1 & 0 & 0 & \cdots & 0 \\ 0 & 1 & -1 & 0 & \cdots & 0 \\ 0 & 0 & 1 & -1 & \cdots & 0 \\ \vdots & \vdots & \vdots & \ddots & \ddots & 0 \\ 0 & 0 & 0 & 0 & 1 & -1 \end{bmatrix} \begin{bmatrix} F_{e1}(y) \\ F_{e2}(y) \\ F_{e3}(y) \\ \vdots \\ F_{e,N}(y) \end{bmatrix}$$

$$= -\left(\frac{1}{\Delta x}\right) \tilde{\boldsymbol{D}}^t \boldsymbol{F}_e(y) \tag{5.35}$$

where $\tilde{\boldsymbol{D}}^t$ is a matrix of $N - 1$ rows and N columns and is the transpose of $\tilde{\boldsymbol{D}}$,

$$\tilde{\boldsymbol{D}}^t \equiv \begin{bmatrix} 1 & -1 & 0 & 0 & \cdots & 0 \\ 0 & 1 & -1 & 0 & \cdots & 0 \\ 0 & 0 & 1 & -1 & \cdots & 0 \\ \vdots & \vdots & \vdots & \ddots & \ddots & 0 \\ 0 & 0 & 0 & 0 & 1 & -1 \end{bmatrix}. \tag{5.36}$$

Particularly note the minus sign in front of $\tilde{\boldsymbol{D}}^t$ in (5.35)! Let

$$\tilde{\boldsymbol{P}}_h \equiv \tilde{\boldsymbol{D}}^t \tilde{\boldsymbol{D}} = \begin{bmatrix} 1 & -1 & 0 & 0 & \cdots & 0 \\ 0 & 1 & -1 & 0 & \cdots & 0 \\ 0 & 0 & 1 & -1 & \cdots & 0 \\ \vdots & \vdots & \vdots & \ddots & \ddots & 0 \\ 0 & 0 & 0 & 0 & 1 & -1 \end{bmatrix} \begin{bmatrix} 1 & 0 & 0 & \cdots & 0 \\ -1 & 1 & 0 & \cdots & 0 \\ 0 & -1 & 1 & \ddots & 0 \\ 0 & 0 & -1 & \ddots & 0 \\ \vdots & \vdots & \ddots & \ddots & 1 \\ 0 & 0 & 0 & 0 & -1 \end{bmatrix}$$

$$= \begin{bmatrix} 2 & -1 & 0 & \cdots & 0 \\ -1 & 2 & -1 & \cdots & 0 \\ 0 & -1 & 2 & \cdots & 0 \\ \vdots & \vdots & \vdots & \ddots & -1 \\ 0 & 0 & 0 & -1 & 2 \end{bmatrix} \tag{5.37}$$

and

$$\tilde{\boldsymbol{P}}_{h2} \equiv \tilde{\boldsymbol{D}} \tilde{\boldsymbol{D}}^t = \begin{bmatrix} 1 & 0 & 0 & \cdots & 0 \\ -1 & 1 & 0 & \cdots & 0 \\ 0 & -1 & 1 & \ddots & 0 \\ 0 & 0 & -1 & \ddots & 0 \\ \vdots & \vdots & \ddots & \ddots & 1 \\ 0 & 0 & 0 & 0 & -1 \end{bmatrix} \begin{bmatrix} 1 & -1 & 0 & 0 & \cdots & 0 \\ 0 & 1 & -1 & 0 & \cdots & 0 \\ 0 & 0 & 1 & -1 & \cdots & 0 \\ \vdots & \vdots & \vdots & \ddots & \ddots & 0 \\ 0 & 0 & 0 & 0 & 1 & -1 \end{bmatrix}$$

$$= \begin{bmatrix} 1 & -1 & 0 & \cdots & 0 \\ -1 & 2 & -1 & \cdots & 0 \\ 0 & -1 & 2 & \cdots & 0 \\ \vdots & \vdots & \vdots & \ddots & -1 \\ 0 & 0 & 0 & -1 & 1 \end{bmatrix}. \tag{5.38}$$

Furthermore, let

$$\tilde{\boldsymbol{P}}_{e1} \equiv \tilde{\boldsymbol{D}} \left(\tilde{\epsilon}_h\right)^{-1} \tilde{\boldsymbol{D}}^t \boldsymbol{\epsilon}_e \tag{5.39}$$

and

$$\tilde{\boldsymbol{P}}_{e2} \equiv \tilde{\epsilon}_h \tilde{\boldsymbol{D}}^t \left(\tilde{\epsilon}_e\right)^{-1} \tilde{\boldsymbol{D}}. \tag{5.40}$$

The wave equations (5.17)–(5.20) within each layer are now discretized as follows.

$$\frac{\partial^2}{\partial y^2}\boldsymbol{E}_x - \frac{\tilde{\boldsymbol{P}}_{e1}}{\Delta x^2}\boldsymbol{E}_x - \left(k_z^2\tilde{\boldsymbol{I}}_e - k_0^2\tilde{\epsilon}_e\right)\boldsymbol{E}_x = \mathbf{0}, \tag{5.41}$$

$$\frac{\partial^2}{\partial y^2}\boldsymbol{E}_y - \frac{\tilde{\boldsymbol{P}}_h}{\Delta x^2}\boldsymbol{E}_y - \left(k_z^2\tilde{\boldsymbol{I}}_h - k_0^2\tilde{\epsilon}_h\right)\boldsymbol{E}_y = \left(-\frac{\tilde{\boldsymbol{D}}^t}{\Delta x} + \left(\tilde{\epsilon}_h\right)^{-1}\frac{\tilde{\boldsymbol{D}}^t}{\Delta x}\tilde{\epsilon}_e\right)\frac{\partial}{\partial y}\boldsymbol{E}_x \tag{5.42}$$

$$\frac{\partial^2}{\partial y^2}\boldsymbol{H}_x - \frac{\tilde{\boldsymbol{P}}_h}{\Delta x^2}\boldsymbol{H}_x - \left(k_z^2\tilde{\boldsymbol{I}}_h - k_0^2\tilde{\epsilon}_h\right)\boldsymbol{H}_x = 0, \tag{5.43}$$

and

$$\frac{\partial^2}{\partial y^2}\boldsymbol{H}_y - \frac{\tilde{\boldsymbol{P}}_{e2}}{\Delta x^2}\boldsymbol{H}_y - \left(k_z^2\tilde{\boldsymbol{I}}_h - k_0^2\tilde{\epsilon}_h\right)\boldsymbol{H}_y = \left(\tilde{\epsilon}_e\frac{\tilde{\boldsymbol{D}}}{\Delta x}\tilde{\epsilon}_h^{-1} - \frac{\tilde{\boldsymbol{D}}}{\Delta x}\right)\frac{\partial}{\partial y}\boldsymbol{H} \tag{5.44}$$

where $\tilde{\boldsymbol{I}}_e$ is the $N \times N$ identity matrix and $\tilde{\boldsymbol{I}}_h$ is the $(N-1) \times (N-1)$ identity matrix. (5.41) and (5.43) will be solved for $\boldsymbol{E}_x$ and $\boldsymbol{H}_x$, respectively. The remaining fields in each layer are found from the discretized versions of (5.21)–(5.24).

$$\boldsymbol{E}_y(y) = \left(\frac{\tilde{\boldsymbol{P}}_h}{\Delta x^2} - k_0^2\tilde{\epsilon}_h\right)^{-1}\left(\frac{\tilde{\boldsymbol{D}}^t}{\Delta x}\frac{\partial}{\partial y}\boldsymbol{E}_x(y) + k_0\eta_0 k_z\boldsymbol{H}_x(y)\right). \tag{5.45}$$

$$\boldsymbol{E}_z(y) = i\left(\frac{\tilde{\boldsymbol{P}}_h}{\Delta x^2} - k_0^2\tilde{\epsilon}_h\right)^{-1}\left(k_z\frac{\tilde{\boldsymbol{D}}^t}{\Delta x}\boldsymbol{E}_x(y) + k_0\eta_0\frac{\partial}{\partial y}\boldsymbol{H}_x(y)\right) \tag{5.46}$$

$$\boldsymbol{H}_y(y) = \left(-\frac{\tilde{\boldsymbol{D}}}{\Delta x}\tilde{\epsilon}_h^{-1}\frac{\tilde{\boldsymbol{D}}^t}{\Delta x} + k_0^2\tilde{\boldsymbol{I}}_e\right)^{-1}\left(\frac{k_0 k_z}{\eta_0}\boldsymbol{E}_x(y) + \frac{\tilde{\boldsymbol{D}}}{\Delta x}\left(\tilde{\epsilon}_h\right)^{-1}\frac{\partial}{\partial y}\boldsymbol{H}_x(y)\right) \tag{5.47}$$

$$\boldsymbol{H}_z(y) = i\left(-\frac{\tilde{\boldsymbol{D}}}{\Delta x}\tilde{\epsilon}_h^{-1}\frac{\tilde{\boldsymbol{D}}^t}{\Delta x} + k_0^2\tilde{\boldsymbol{I}}_e\right)^{-1}\left(\frac{k_0}{\eta_0}\frac{\partial}{\partial y}\boldsymbol{E}_x(y) + k_z\frac{\tilde{\boldsymbol{D}}}{\Delta x}\left(\tilde{\epsilon}_h\right)^{-1}\boldsymbol{H}_x(y)\right) \tag{5.48}$$

5.4 SOLUTION OF THE WAVE EQUATIONS

The wave equations (5.41) and (5.43) have the same form. Let

$$\tilde{\boldsymbol{X}} \equiv k_0^2\tilde{\epsilon} - \frac{1}{\Delta x^2}\tilde{\boldsymbol{P}}. \tag{5.49}$$

Then (5.41) and (5.43) can be written

$$\frac{\partial^2}{\partial y^2}\boldsymbol{Y}_x - \left(k_z^2\,\tilde{\boldsymbol{I}} - \tilde{\boldsymbol{X}}\right)\boldsymbol{Y}_x = 0 \tag{5.50}$$

where $\boldsymbol{Y}_x$ represents either $\boldsymbol{E}_x$ or $\boldsymbol{H}_x$. Let the eigenvector matrix $\tilde{\boldsymbol{S}}$ and eigenvalue matrix $\tilde{\boldsymbol{d}}$ of $\tilde{\boldsymbol{X}}$ be

$$\tilde{\boldsymbol{X}}\tilde{\boldsymbol{S}} = \tilde{\boldsymbol{S}}\tilde{\boldsymbol{d}}\left(\lambda_m\right) \tag{5.51}$$

where the λ_m are the eigenvalues for each line. $\tilde{\boldsymbol{d}}(\lambda_m)$ is the diagonal matrix,

$$\tilde{\boldsymbol{d}}(\lambda_m) = \begin{bmatrix} \lambda_1 & 0 & 0 & 0 & 0 \\ 0 & \lambda_2 & 0 & 0 & 0 \\ 0 & 0 & \lambda_3 & 0 & 0 \\ 0 & 0 & 0 & \ddots & 0 \\ 0 & 0 & 0 & 0 & \lambda_M \end{bmatrix} \tag{5.52}$$

of rank $N - 1$ for the magnetic lines and N for the electric lines. The solution for the field vector $\boldsymbol{Y}_x$ in the principal axes coordinate system is

$$\boldsymbol{Y}_x(y) = \tilde{\boldsymbol{S}}\boldsymbol{\Phi}(y). \tag{5.53}$$

$\tilde{\boldsymbol{S}}$ is a square matrix of rank N for $\boldsymbol{Y}_x = \boldsymbol{E}_x$ and of rank $N - 1$ for $\boldsymbol{Y}_x = \boldsymbol{H}_x$. Substituting (5.51) and (5.53) into (5.50) and simplifying gives the partial differential equation

$$\frac{\partial^2}{\partial y^2}\boldsymbol{\Phi}(y) - \left(\tilde{\boldsymbol{k}}_y\right)^2 \boldsymbol{\Phi}(y) = 0 \tag{5.54}$$

where

$$\left(\tilde{\boldsymbol{k}}_y\right)^2 \equiv k_z^2\, \tilde{\boldsymbol{I}}_N - \tilde{\boldsymbol{d}}(\lambda). \tag{5.55}$$

(5.54) is a simple harmonic equation with exponential or hyperbolic function solutions. Decaying exponential solutions are chosen for the top and bottom claddings. In particular, for the bottom cladding, layer 0, the solution to (5.54) is

$$\tilde{\boldsymbol{F}}^{(0)}(y) = \begin{bmatrix} e^{+k_{y,1}^{(0)}y} & 0 & 0 & \cdots & 0 \\ 0 & e^{+k_{y,2}^{(0)}y} & 0 & \cdots & 0 \\ 0 & 0 & e^{+k_{y,3}^{(0)}y} & \cdots & 0 \\ \vdots & \vdots & \vdots & \ddots & 0 \\ 0 & 0 & 0 & 0 & e^{+k_{y,M}^{(0)}y} \end{bmatrix} \tag{5.56}$$

for which each term along the diagonal goes to 0 as $y \to -\infty$. This is a square matrix of rank $N - 1$ for H_x and rank N for E_x. The superscript in parentheses indicates the layer number. The lower cladding is defined as layer 0, the upper cladding is layer $L + 1$, and there are L intermediate layers. The total field vector in the lower cladding is

$$\boldsymbol{Y}_x^{(0)}(y) = \tilde{\boldsymbol{S}}^{(0)}\,\tilde{\boldsymbol{F}}^{(0)}(y)\,\boldsymbol{A}^{(0)}. \tag{5.57}$$

$\boldsymbol{A}^{(0)}$ is the unknown field amplitude vector for the lower cladding. Each element of $\boldsymbol{A}^{(0)}$ is the field amplitude of a line in the bottom cladding. For $\boldsymbol{E}_x$ and $\boldsymbol{H}_x$,

$$\boldsymbol{E}_x^{(0)}(y) = \tilde{\boldsymbol{S}}_e^{(0)}\,\tilde{\boldsymbol{F}}_e^{(0)}(y)\,\boldsymbol{A}_e^{(0)} \tag{5.58}$$

and

$$\boldsymbol{H}_x^{(0)}(y) = \tilde{\boldsymbol{S}}_h^{(0)}\,\tilde{\boldsymbol{F}}_h^{(0)}(y)\,\boldsymbol{A}_h^{(0)}. \tag{5.59}$$

The other field components are found from (5.45)–(5.48). In particular,

$$\boldsymbol{E}_z^{(0)}(y)=i\left(\frac{\tilde{\boldsymbol{P}}_h}{\Delta x^2}-k_0^2\tilde{\boldsymbol{\epsilon}}_h^{(0)}\right)^{-1}\left(k_z\frac{\tilde{\boldsymbol{D}}^t}{\Delta x}\tilde{\boldsymbol{S}}_e^{(0)}\tilde{\boldsymbol{F}}_e^{(0)}(y)\boldsymbol{A}_e^{(0)}+k_0\eta_0\tilde{\boldsymbol{S}}_h^{(0)}\tilde{\boldsymbol{k}}_{yh}^{(0)}\tilde{\boldsymbol{F}}_h^{(0)}(y)\boldsymbol{A}_h^{(0)}\right), \tag{5.60}$$

$$\boldsymbol{H}_z^{(0)}(y)=$$

$$i\left(-\frac{\tilde{\boldsymbol{D}}}{\Delta x}\left(\tilde{\boldsymbol{\epsilon}}_h^{(0)}\right)^{-1}\frac{\tilde{\boldsymbol{D}}^t}{\Delta x}+k_0^2\tilde{\boldsymbol{I}}_e\right)^{-1}\left[\frac{k_0}{\eta_0}\tilde{\boldsymbol{S}}_e^{(0)}\tilde{\boldsymbol{k}}_{ye}^{(0)}\tilde{\boldsymbol{F}}_e^{(0)}(y)\boldsymbol{A}_e^{(0)}+k_z\frac{\tilde{\boldsymbol{D}}}{\Delta x}\left(\tilde{\boldsymbol{\epsilon}}_h\right)^{-1}\tilde{\boldsymbol{S}}_h^{(0)}\tilde{\boldsymbol{F}}_h^{(0)}(y)\boldsymbol{A}_h^{(0)}\right]. \tag{5.61}$$

For the upper cladding layer, the solution to (5.54) again uses exponentials.

$$\tilde{\boldsymbol{G}}^{(L+1)}(y)=\begin{bmatrix} e^{-k_{y,1}^{(L+1)}(y-y_L)} & 0 & 0 & \cdots & 0 \\ 0 & e^{-k_{y,2}^{(L+1)}(y-y_L)} & 0 & \cdots & 0 \\ 0 & 0 & e^{-k_{y,3}^{(L+1)}(y-y_L)} & \cdots & 0 \\ \vdots & \vdots & \vdots & \ddots & 0 \\ 0 & 0 & 0 & 0 & e^{-k_{y,M}^{(L+1)}(y-y_L)} \end{bmatrix} \tag{5.62}$$

for which each element along the diagonal goes to 0 as $y \to +\infty$. Again, this matrix is square with rank N for E_x and rank $N-1$ for H_x. $\tilde{\boldsymbol{F}}_e^{(0)}$ is the identity matrix at the bottom interface $y = 0$, and $\tilde{\boldsymbol{G}}^{(L+1)}$ is the identity matrix at the upper surface, $y = y_L$. The total field vector in the upper cladding is

$$\boldsymbol{Y}_x^{(L+1)}(y)=\tilde{\boldsymbol{S}}^{(L+1)}\tilde{\boldsymbol{G}}^{(L+1)}(y)\boldsymbol{B}^{(L+1)} \tag{5.63}$$

where $\boldsymbol{B}^{(L+1)}$ is the unknown field amplitude vector. $\boldsymbol{E}_x$ and $\boldsymbol{H}_x$ in the top layer are

$$\boldsymbol{E}_x^{(L+1)}(y)=\tilde{\boldsymbol{S}}_e^{(L+1)}\tilde{\boldsymbol{G}}_e^{(L+1)}(y)\boldsymbol{B}_e^{(L+1)} \tag{5.64}$$

and

$$\boldsymbol{H}_x^{(L+1)}(y)=\tilde{\boldsymbol{S}}_h^{(L+1)}\tilde{\boldsymbol{G}}_h^{(L+1)}(y)\boldsymbol{B}_h^{(L+1)}. \tag{5.65}$$

The other field components are found from (5.45)–(5.48). In particular,

$$\boldsymbol{E}_z^{(L+1)}(y)=$$

$$i\left(\frac{\tilde{\boldsymbol{P}}_h}{\Delta x^2}-k_0^2\tilde{\epsilon}_h^{(L+1)}\right)^{-1}\left(\frac{k_z}{\Delta x}\tilde{\boldsymbol{D}}^t\,\tilde{\boldsymbol{S}}\tilde{\boldsymbol{G}}_e^{(L+1)}(y)\boldsymbol{B}_e^{(L+1)}-k_0\eta_0\tilde{\boldsymbol{S}}_h^{(L+1)}\tilde{\boldsymbol{k}}_{yh}^{(L+1)}\tilde{\boldsymbol{G}}_h^{(L+1)}(y)\boldsymbol{B}_h^{(L+1)}\right) \tag{5.66}$$

$$\boldsymbol{H}_z^{(L+1)}(y) = i\left(-\frac{\tilde{\boldsymbol{D}}}{\Delta x}\left(\tilde{\boldsymbol{\epsilon}}_h^{(L+1)}\right)^{-1}\frac{\tilde{\boldsymbol{D}}^t}{\Delta x} + k_0^2\tilde{\boldsymbol{I}}_e\right)^{-1}\left[-\frac{k_0}{\eta_0}\tilde{\boldsymbol{S}}_e^{(L+1)}\tilde{\boldsymbol{k}}_{ye}^{(L+1)}\tilde{\boldsymbol{G}}_e^{(L+1)}(y)\boldsymbol{B}_e^{(L+1)}\right.$$
$$\left. + \frac{k_z}{\Delta x}\tilde{\boldsymbol{D}}(\tilde{\boldsymbol{\epsilon}}_h)^{-1}\tilde{\boldsymbol{S}}_h^{(L+1)}\tilde{\boldsymbol{G}}_h^{(L+1)}(y)\boldsymbol{B}_h^{(L+1)}\right] \tag{5.67}$$

The intermediate layers are conveniently expressed in terms of hyperbolic functions. Let

$$\tilde{\boldsymbol{F}}^{(m)}(y) = \begin{bmatrix} \cosh\left[k_{y,1}^{(m)}(y-y_{m-1})\right] & 0 & 0 & \cdots & 0 \\ 0 & \cosh\left[k_{y,2}^{(m)}(y-y_{m-1})\right] & 0 & \cdots & 0 \\ 0 & 0 & \cosh\left[k_{y,3}^{(m)}(y-y_{m-1})\right] & \cdots & 0 \\ \vdots & \vdots & \vdots & \ddots & 0 \\ 0 & 0 & 0 & 0 & \cosh\left[k_{y,M}^{(m)}(y-y_{m-1})\right] \end{bmatrix} \tag{5.68}$$

and

$$\tilde{\boldsymbol{G}}^{(m)}(y) = \begin{bmatrix} \sinh\left[k_{y,1}^{(m)}(y-y_{m-1})\right] & 0 & 0 & \cdots & 0 \\ 0 & \sinh\left[k_{y,2}^{(m)}(y-y_{m-1})\right] & 0 & \cdots & 0 \\ 0 & 0 & \sinh\left[k_{y,3}^{(m)}(y-y_{m-1})\right] & \cdots & 0 \\ \vdots & \vdots & \vdots & \ddots & 0 \\ 0 & 0 & 0 & 0 & \sinh\left[k_{y,M}^{(m)}(y-y_{m-1})\right] \end{bmatrix}. \tag{5.69}$$

The lower surface of layer m is at y_{m-1} and the upper surface at y_m. The y values are offset by the starting y-value for each layer so that at the lower surface of each layer m, $\tilde{\boldsymbol{F}}^{(m)}$ is the identity matrix and $\tilde{\boldsymbol{G}}^{(m)}$ is the zero matrix. As before, $\tilde{\boldsymbol{F}}^{(m)}$ and $\tilde{\boldsymbol{G}}^{(m)}$ are square matrices of rank $N - 1$ for H_x and rank N for E_x. Within each intermediate layer m, the field is

$$\boldsymbol{Y}_x^{(m)}(y) = \tilde{\boldsymbol{S}}^{(m)}\left(\tilde{\boldsymbol{F}}^{(m)}(y)\boldsymbol{A}^{(m)} + \tilde{\boldsymbol{G}}^{(m)}(y)\boldsymbol{B}^{(m)}\right). \tag{5.70}$$

$\boldsymbol{A}^{(m)}$ and $\boldsymbol{B}^{(m)}$ are the unknown field amplitude vectors for layer m for each field E_x and H_x. The total number of unknown $\boldsymbol{A}$ and $\boldsymbol{B}$ amplitude vectors for L intermediate layers plus the top and bottom claddings for E_x and H_x is, therefore, $2(1 + 2L + 1) = 4L + 4$.

The fields for intermediate layers can now be written explicitly from (5.45)–(5.48). In particular,

$$\boldsymbol{E}_x^{(m)}(y) = \tilde{\boldsymbol{S}}_e^{(m)}\tilde{\boldsymbol{F}}_e^{(m)}(y)\boldsymbol{A}_e^{(m)} + \tilde{\boldsymbol{S}}_e^{(m)}\tilde{\boldsymbol{G}}_e^{(m)}(y)\boldsymbol{B}_e^{(m)}, \tag{5.71}$$

$$\boldsymbol{H}_x^{(m)}(y) = \tilde{\boldsymbol{S}}_h^{(m)}\tilde{\boldsymbol{F}}_h^{(m)}(y)\boldsymbol{A}_h^{(m)} + \tilde{\boldsymbol{S}}_h^{(m)}\tilde{\boldsymbol{G}}_h^{(m)}(y)\boldsymbol{B}_h^{(m)}, \tag{5.72}$$

$$E_z^{(m)}(y) = i\left(\frac{\tilde{P}_h}{\Delta x^2} - k_0^2\tilde{\epsilon}_h\right)^{-1}\left[\frac{k_z}{\Delta x}\tilde{D}^t\,\tilde{S}_e^{(m)}\,\tilde{F}_e^{(m)}(y)\,A_e^{(m)} + \frac{k_z}{\Delta x}\tilde{D}^t\,\tilde{S}_e^{(m)}\,\tilde{G}_e^{(m)}(y)\,B_e^{(m)}\right.$$

$$\left. + k_0\eta_0\tilde{S}_h^{(m)}\,\tilde{k}_{\mathrm{yh}}^{(m)}\,\tilde{G}_h^{(m)}(y)\,A_h^{(m)} + k_0\eta_0\tilde{S}_h^{(m)}\,\tilde{k}_{\mathrm{yh}}^{(m)}\,\tilde{F}_h^{(m)}(y)\,B_h^{(m)}\right], \tag{5.73}$$

and

$$H_z^{(m)}(y) = i\left(-\frac{\tilde{D}}{\Delta x}(\tilde{\epsilon}_h)^{-1}\frac{\tilde{D}^t}{\Delta x} + k_0^2\,\tilde{I}_e\right)^{-1}\left[\frac{k_0}{\eta_0}\tilde{S}_e^{(m)}\,\tilde{k}_{\mathrm{ye}}^{(m)}\,\tilde{G}_e^{(m)}(y)\,A_e^{(m)}\right.$$

$$+ \frac{k_0}{\eta_0}\tilde{S}_e^{(m)}\,\tilde{k}_{\mathrm{ye}}^{(m)}\,\tilde{F}_e^{(m)}(y)\,B_e^{(m)} + \frac{k_z}{\Delta x}\tilde{D}(\tilde{\epsilon}_h)^{-1}\,\tilde{S}_h^{(m)}\,\tilde{F}_h^{(m)}(y)\,A_h^{(m)} \tag{5.74}$$

$$\left. + \frac{k_z}{\Delta x}\tilde{D}^t(\tilde{\epsilon}_h)^{-1}\,\tilde{S}_h^{(m)}\,\tilde{G}_h^{(m)}(y)\,B_h^{(m)}\right].$$

The wavevector(s) that satisfy boundary conditions must be determined. For each wavevector solution the unknown vector amplitudes $A_e^{(m)}$, $A_h^{(m)}$, $B_e^{(m)}$, and $B_h^{(m)}$ are then determined. From these amplitudes, E_x and H_x are calculated from (5.71) and (5.72). The remaining field components are then calculated from (5.45)–(5.48). The partial derivatives of E_x and H_x are required for these equations. From (5.58), (5.59), (5.64), (5.65), (5.71), and (5.72),

$$\frac{\partial E_x^{(0)}(y)}{\partial y} = \tilde{S}_e^{(0)}\,\tilde{k}_{\mathrm{ye}}^{(0)}\,\tilde{F}_e^{(0)}(y)\,A_e^{(0)}, \tag{5.75}$$

$$\frac{\partial H_x^{(0)}(y)}{\partial y} = \tilde{S}_h^{(0)}\,\tilde{k}_{\mathrm{yh}}^{(0)}\,\tilde{F}_h^{(0)}(y)\,A_h^{(0)}, \tag{5.76}$$

$$\frac{\partial E_x^{(L+1)}(y)}{\partial y} = -\tilde{S}_e^{(L+1)}\,\tilde{k}_{\mathrm{ye}}^{(L+1)}\,\tilde{G}_e^{(L+1)}(y)\,B_e^{(L+1)}, \tag{5.77}$$

$$\frac{\partial H_x^{(L+1)}(y)}{\partial y} = -\tilde{S}_h^{(L+1)}\,\tilde{k}_{\mathrm{yh}}^{(L+1)}\,\tilde{G}_h^{(L+1)}(y)\,B_h^{(L+1)}, \tag{5.78}$$

$$\frac{\partial E_x^{(m)}(y)}{\partial y} = \tilde{S}_e^{(m)}\,\tilde{k}_{\mathrm{ye}}^{(m)}\,\tilde{G}_e^{(m)}(y)\,A_e^{(m)} + \tilde{S}_e^{(m)}\,\tilde{k}_{\mathrm{ye}}^{(m)}\,\tilde{F}_e^{(m)}(y)\,B_e^{(m)}, \tag{5.79}$$

and

$$\frac{\partial H_x^{(m)}(y)}{\partial y} = \tilde{S}_h^{(m)}\,\tilde{k}_{\mathrm{yh}}^{(m)}\,\tilde{G}_h^{(m)}(y)\,A_h^{(m)} + \tilde{S}_h^{(m)}\,\tilde{k}_{\mathrm{yh}}^{(m)}\,\tilde{F}_h^{(m)}(y)\,B_h^{(m)}. \tag{5.80}$$

Again, note the minus signs in (5.77) and (5.78), as well as the interchange of $\tilde{F}^{(m)}(y)$ and $\tilde{G}^{(m)}(y)$ in (5.79) and (5.80).

5.5 BOUNDARY CONDITIONS

The mode index or mode wavevector k_z is determined through equating the tangential field components at the interfaces between layers. Collecting the previous results, $\boldsymbol{E}_x$ and $\boldsymbol{H}_x$ in each layer are obtained from (5.58), (5.59), (5.64), (5.65), (5.71), and (5.72).

$$\boldsymbol{E}_x^{(0)}(y) = \tilde{\boldsymbol{S}}_e^{(0)}\tilde{\boldsymbol{F}}_e^{(0)}(y)\boldsymbol{A}_e^{(0)}. \tag{5.81}$$

$$\boldsymbol{E}_x^{(L+1)}(y) = \tilde{\boldsymbol{S}}_e^{(L+1)}\tilde{\boldsymbol{G}}_e^{(L+1)}(y)\boldsymbol{B}_e^{(L+1)}. \tag{5.82}$$

$$\boldsymbol{E}_x^{(m)}(y) = \tilde{\boldsymbol{S}}_e^{(m)}\left(\tilde{\boldsymbol{F}}_e^{(m)}(y)\boldsymbol{A}_e^{(m)} + \tilde{\boldsymbol{G}}_e^{(m)}(y)\boldsymbol{B}_e^{(m)}\right). \tag{5.83}$$

$$\boldsymbol{H}_x^{(0)}(y) = \tilde{\boldsymbol{S}}_h^{(0)}\tilde{\boldsymbol{F}}_h^{(0)}(y)\boldsymbol{A}_h^{(0)}. \tag{5.84}$$

$$\boldsymbol{H}_x^{(L+1)}(y) = \tilde{\boldsymbol{S}}_h^{(L+1)}\tilde{\boldsymbol{G}}_h^{(L+1)}(y)\boldsymbol{B}_h^{(L+1)}. \tag{5.85}$$

$$\boldsymbol{H}_x^{(m)}(y) = \tilde{\boldsymbol{S}}_h^{(m)}\left(\tilde{\boldsymbol{F}}_h^{(m)}(y)\boldsymbol{A}_h^{(m)} + \tilde{\boldsymbol{G}}_h^{(m)}(y)\boldsymbol{B}_h^{(m)}\right). \tag{5.86}$$

Matching E_x fields at the bottom interface,

$$\boldsymbol{E}_x^{(0)}(0) = \tilde{\boldsymbol{S}}_e^{(0)}\boldsymbol{A}_e^{(0)} = \tilde{\boldsymbol{S}}_e^{(1)}\boldsymbol{A}_e^{(1)} = \boldsymbol{E}_x^{(1)}(0). \tag{5.87}$$

At the top interface,

$$\boldsymbol{E}_x^{(L+1)}(y_L) = \tilde{\boldsymbol{S}}_e^{(L+1)}\boldsymbol{B}_e^{(L+1)} = \tilde{\boldsymbol{S}}_e^{(L)}\left(\tilde{\boldsymbol{F}}_e^{(L)}(y_L)\boldsymbol{A}_e^{(L)} + \tilde{\boldsymbol{G}}_e^{(L)}(y_L)\boldsymbol{B}_e^{(L)}\right) = \boldsymbol{E}_x^{(L)}(y_L). \tag{5.88}$$

At other interfaces between layers,

$$\boldsymbol{E}_x^{(m)}(y_m) = \tilde{\boldsymbol{S}}_e^{(m)}\left(\tilde{\boldsymbol{F}}_e^{(m)}(y_m)\boldsymbol{A}_e^{(m)} + \tilde{\boldsymbol{G}}_e^{(m)}(y_m)\boldsymbol{B}_e^{(m)}\right)$$

$$= \tilde{\boldsymbol{S}}_e^{(m+1)}\left(\tilde{\boldsymbol{F}}_e^{(m+1)}(y_m)\boldsymbol{A}_e^{(m+1)} + \tilde{\boldsymbol{G}}_e^{(m+1)}(y_m)\boldsymbol{B}_e^{(m+1)}\right) = \boldsymbol{E}_x^{(m+1)}(y_m). \tag{5.89}$$

(5.87)–(5.89) provide $L + 1$ equations for the unknown amplitudes $\boldsymbol{A}$ and $\boldsymbol{B}$ in each layer. There are an additional $L + 1$ equations found by matching the $\boldsymbol{H}_x$ fields at each interface.

$$\boldsymbol{H}_x^{(0)}(0) = \tilde{\boldsymbol{S}}_h^{(0)}\boldsymbol{A}_h^{(0)} = \tilde{\boldsymbol{S}}_h^{(1)}\boldsymbol{A}_h^{(1)} = \boldsymbol{H}_x^{(1)}(0). \tag{5.90}$$

$$\boldsymbol{H}_x^{(L+1)}(y_L) = \tilde{\boldsymbol{S}}_h^{(L+1)}\boldsymbol{B}_h^{(L+1)} = \tilde{\boldsymbol{S}}_h^{(L)}\left(\tilde{\boldsymbol{F}}_h^{(L)}(y_L)\boldsymbol{A}_h^{(L)} + \tilde{\boldsymbol{G}}_h^{(L)}(y_L)\boldsymbol{B}_h^{(L)}\right) = \boldsymbol{H}_x^{(L)}(y_L). \tag{5.91}$$

$$\boldsymbol{H}_x^{(m)}(y_m) = \tilde{\boldsymbol{S}}_h^{(m)}\left(\tilde{\boldsymbol{F}}_h^{(m)}(y_m)\boldsymbol{A}_h^{(m)} + \tilde{\boldsymbol{G}}_h^{(m)}(y_m)\boldsymbol{B}_h^{(m)}\right)$$

$$= \tilde{\boldsymbol{S}}_h^{(m+1)}\left(\tilde{\boldsymbol{F}}_h^{(m+1)}(y_m)\boldsymbol{A}_h^{(m+1)} + \tilde{\boldsymbol{G}}_e^{(m+1)}(y_m)\boldsymbol{B}_h^{(m+1)}\right) = \boldsymbol{H}_x^{(m+1)}(y_m). \tag{5.92}$$

The boundary conditions for $\boldsymbol{E}_z$ at the bottom interface are found from (5.60) and (5.73).

$$\begin{aligned}\boldsymbol{E}_z^{(0)}(0) &= i\left(\frac{\tilde{\boldsymbol{P}}_h}{\Delta x^2} - k_0^2\,\tilde{\epsilon}_h^{(0)}\right)^{-1}\left(k_z\frac{\tilde{\boldsymbol{D}}^t}{\Delta x}\tilde{\boldsymbol{S}}_e^{(0)}\,\boldsymbol{A}_e^{(0)} + k_0\eta_0\tilde{\boldsymbol{S}}_h^{(0)}\,\tilde{\boldsymbol{k}}_{yh}^{(0)}\,\boldsymbol{A}_h^{(0)}\right)\\ &= i\left(\frac{\tilde{\boldsymbol{P}}_h}{\Delta x^2} - k_0^2\,\tilde{\epsilon}_h^{(1)}\right)^{-1}\left(k_z\frac{\tilde{\boldsymbol{D}}^t}{\Delta x}\tilde{\boldsymbol{S}}_e^{(1)}\,\boldsymbol{A}_e^{(1)} + k_0\eta_0\tilde{\boldsymbol{S}}_h^{(1)}\,\tilde{\boldsymbol{k}}_{yh}^{(1)}\,\boldsymbol{B}_h^{(1)}\right) = \boldsymbol{E}_z^{(1)}(0)\end{aligned} \tag{5.93}$$

At the top interface, from (5.66) and (5.73),

$$\begin{aligned}\boldsymbol{E}_z^{(L+1)}(y_L) &= i\left(\frac{\tilde{\boldsymbol{P}}_h}{\Delta x^2} - k_0^2\,\tilde{\epsilon}_h^{(L+1)}\right)^{-1}\left(\frac{k_z}{\Delta x}\tilde{\boldsymbol{D}}^t\,\tilde{\boldsymbol{S}}_e^{(L+1)}\,\boldsymbol{B}_e^{(L+1)} - k_0\eta_0\tilde{\boldsymbol{S}}_h^{(L+1)}\,\tilde{\boldsymbol{k}}_{yh}^{(L+1)}\,\boldsymbol{B}_h^{(L+1)}\right)\\ &= i\left(\frac{\tilde{\boldsymbol{P}}_h}{\Delta x^2} - k_0^2\,\tilde{\epsilon}_h^{(L)}\right)^{-1}\left[\frac{k_z}{\Delta x}\tilde{\boldsymbol{D}}^t\,\tilde{\boldsymbol{S}}_e^{(L)}\left(\tilde{\boldsymbol{F}}_e^{(L)}(y_L)\boldsymbol{A}_e^{(L)} + \tilde{\boldsymbol{G}}_e^{(L)}(y_L)\boldsymbol{B}_e^{(L)}\right)\right.\\ &\quad\left. + k_0\eta_0\tilde{\boldsymbol{S}}_h^{(L)}\,\tilde{\boldsymbol{k}}_{yh}^{(L)}\left(\tilde{\boldsymbol{G}}_h^{(L)}(y)\boldsymbol{A}_h^{(L)} + \tilde{\boldsymbol{F}}_h^{(L)}(y)\boldsymbol{B}_h^{(L)}\right)\right] = \boldsymbol{E}_z^{(L)}(y_L).\end{aligned} \tag{5.94}$$

For intermediate interfaces from (5.73),

$$\begin{aligned}\boldsymbol{E}_z^{(m)}(y_m) &= i\left(\frac{\tilde{\boldsymbol{P}}_h}{\Delta x^2} - k_0^2\tilde{\epsilon}_h\right)^{-1}\left[\frac{k_z}{\Delta x}\tilde{\boldsymbol{D}}^t\,\tilde{\boldsymbol{S}}_e^{(m)}\left(\tilde{\boldsymbol{F}}_e^{(m)}(y_m)\boldsymbol{A}_e^{(m)} + \tilde{\boldsymbol{G}}_e^{(m)}(y_m)\boldsymbol{B}_e^{(m)}\right)\right.\\ &\quad\left. + k_0\eta_0\tilde{\boldsymbol{S}}_h^{(m)}\,\tilde{\boldsymbol{k}}_{yh}^{(m)}\left(\tilde{\boldsymbol{G}}_h^{(m)}(y_m)\boldsymbol{A}_h^{(m)} + \tilde{\boldsymbol{F}}_h^{(m)}(y_m)\boldsymbol{B}_h^{(m)}\right)\right]\\ &= i\left(\frac{\tilde{\boldsymbol{P}}_h}{\Delta x^2} - k_0^2\tilde{\epsilon}_h\right)^{-1}\left[\frac{k_z}{\Delta x}\tilde{\boldsymbol{D}}^t\,\tilde{\boldsymbol{S}}_e^{(m+1)}\left(\tilde{\boldsymbol{F}}_e^{(m+1)}(y_m)\boldsymbol{A}_e^{(m+1)} + \tilde{\boldsymbol{G}}_e^{(m+1)}(y_m)\boldsymbol{B}_e^{(m+1)}\right)\right.\\ &\quad\left. + k_0\eta_0\tilde{\boldsymbol{S}}_h^{(m+1)}\,\tilde{\boldsymbol{k}}_{yh}^{(m+1)}\left(\tilde{\boldsymbol{G}}_h^{(m+1)}(y)\boldsymbol{A}_h^{(m+1)} + \tilde{\boldsymbol{F}}_h^{(m+1)}(y)\boldsymbol{B}_h^{(m+1)}\right)\right] = \boldsymbol{E}_z^{(m+1)}(y_m).\end{aligned} \tag{5.95}$$

The boundary conditions for $\boldsymbol{H}_z$ at the bottom interface are found from (5.61) and (5.74).

$$\begin{aligned}\boldsymbol{H}_z^{(0)}(0) &= i\left(-\frac{\tilde{\boldsymbol{D}}}{\Delta x}\tilde{\epsilon}_h^{(0)-1}\frac{\tilde{\boldsymbol{D}}^t}{\Delta x} + k_0^2\tilde{\boldsymbol{I}}_e\right)^{-1}\left(\frac{k_0}{\eta_0}\tilde{\boldsymbol{S}}_e^{(0)}\,\tilde{\boldsymbol{k}}_{ye}^{(0)}\,\boldsymbol{A}_e^{(0)} + k_z\frac{\tilde{\boldsymbol{D}}}{\Delta x}\tilde{\epsilon}_h^{(0)-1}\,\tilde{\boldsymbol{S}}_h^{(0)}\,\boldsymbol{A}_h^{(0)}\right)\\ &= i\left(-\frac{\tilde{\boldsymbol{D}}}{\Delta x}\tilde{\epsilon}_h^{(1)-1}\frac{\tilde{\boldsymbol{D}}^t}{\Delta x} + k_0^2\,\tilde{\boldsymbol{I}}_e\right)^{-1}\left(\frac{k_0}{\eta_0}\tilde{\boldsymbol{S}}_e^{(1)}\,\tilde{\boldsymbol{k}}_{ye}^{(1)}\,\boldsymbol{B}_e^{(1)} + k_z\frac{\tilde{\boldsymbol{D}}}{\Delta x}\tilde{\epsilon}_h^{(1)-1}\,\tilde{\boldsymbol{S}}_h^{(1)}\,\boldsymbol{A}_h^{(1)}\right) = \boldsymbol{H}_z^{(1)}(0).\end{aligned} \tag{5.96}$$

For the top interface, from (5.65) and (5.74),

$$\begin{aligned}
\boldsymbol{H}_z^{(L+1)}(y_L) &= i\left(-\frac{\tilde{\boldsymbol{D}}}{\Delta x}\tilde{\boldsymbol{\epsilon}}_h^{(L+1)-1}\frac{\tilde{\boldsymbol{D}}^t}{\Delta x}+k_0^2\tilde{\boldsymbol{I}}_e\right)^{-1}\left(-\frac{k_0}{\eta_0}\tilde{\boldsymbol{S}}_e^{(L+1)}\tilde{\boldsymbol{k}}_{ye}^{(L+1)}\boldsymbol{B}_e^{(L+1)}\right.\\
&\left.+k_z\frac{\tilde{\boldsymbol{D}}}{\Delta x}\tilde{\boldsymbol{\epsilon}}_h^{(L+1)-1}\tilde{\boldsymbol{S}}_h^{(L+1)}\boldsymbol{B}_h^{(L+1)}\right)\\
&= i\left(-\frac{\tilde{\boldsymbol{D}}}{\Delta x}\tilde{\boldsymbol{\epsilon}}_h^{(L)-1}\frac{\tilde{\boldsymbol{D}}^t}{\Delta x}+k_0^2\tilde{\boldsymbol{I}}_e\right)^{-1}\left[\frac{k_0}{\eta_0}\tilde{\boldsymbol{S}}_e^{(L)}\tilde{\boldsymbol{k}}_{ye}^{(L)}\left(\tilde{\boldsymbol{G}}_e^{(L)}(y_L)\boldsymbol{A}_e^{(L)}+\tilde{\boldsymbol{F}}_e^{(L)}(y_L)\boldsymbol{B}_e^{(L)}\right)\right.\\
&\left.+k_z\frac{\tilde{\boldsymbol{D}}}{\Delta x}\tilde{\boldsymbol{\epsilon}}_h^{(L)-1}\tilde{\boldsymbol{S}}_h^{(L)}\left(\tilde{\boldsymbol{F}}_h^{(L)}(y_L)\boldsymbol{A}_h^{(L)}+\tilde{\boldsymbol{G}}_h^{(L)}(y_L)\boldsymbol{B}_h^{(L)}\right)\right]=\boldsymbol{H}_z^{(L)}(y_L).
\end{aligned} \tag{5.97}$$

For intermediate interfaces from (5.74),

$$\begin{aligned}
\boldsymbol{H}_z^{(m)}(y_m) &= i\left(-\frac{\tilde{\boldsymbol{D}}}{\Delta x}\tilde{\boldsymbol{\epsilon}}_h^{(m)-1}\frac{\tilde{\boldsymbol{D}}^t}{\Delta x}+k_0^2\tilde{\boldsymbol{I}}_e\right)^{-1}\left[\frac{k_0}{\eta_0}\tilde{\boldsymbol{S}}_e^{(m)}\tilde{\boldsymbol{k}}_{ye}^{(m)}\left(\tilde{\boldsymbol{G}}_e^{(m)}(y_m)\boldsymbol{A}_e^{(m)}+\tilde{\boldsymbol{F}}_e^{(m)}(y_m)\boldsymbol{B}_e^{(m)}\right)\right.\\
&\left.+k_z\frac{\tilde{\boldsymbol{D}}}{\Delta x}\tilde{\boldsymbol{\epsilon}}_h^{(m)-1}\tilde{\boldsymbol{S}}_h^{(m)}\left(\tilde{\boldsymbol{F}}_h^{(m)}(y_m)\boldsymbol{A}_h^{(m)}+\tilde{\boldsymbol{G}}_h^{(m)}(y_m)\boldsymbol{B}_h^{(m)}\right)\right]\\
&= i\left(-\frac{\tilde{\boldsymbol{D}}}{\Delta x}\tilde{\boldsymbol{\epsilon}}_h^{(m+1)-1}\frac{\tilde{\boldsymbol{D}}^t}{\Delta x}+k_0^2\tilde{\boldsymbol{I}}_e\right)^{-1}\left[\frac{k_0}{\eta_0}\tilde{\boldsymbol{S}}_e^{(m)}\tilde{\boldsymbol{k}}_{ye}^{(m)}\left(\tilde{\boldsymbol{G}}_e^{(m+1)}(y_m)\boldsymbol{A}_e^{(m+1)}+\tilde{\boldsymbol{F}}_e^{(m+1)}(y_m)\boldsymbol{B}_e^{(m+1)}\right)\right.\\
&\left.+k_z\frac{\tilde{\boldsymbol{D}}}{\Delta x}\tilde{\boldsymbol{\epsilon}}_h^{(m+1)-1}\tilde{\boldsymbol{S}}_h^{(m+1)}\left(\tilde{\boldsymbol{F}}_h^{(m+1)}(y_m)\boldsymbol{A}_h^{(m+1)}+\tilde{\boldsymbol{G}}_h^{(m+1)}(y_m)\boldsymbol{B}_h^{(m+1)}\right)\right]=\boldsymbol{H}_z^{(m+1)}(y_m).
\end{aligned} \tag{5.98}$$

(5.93)–(5.98) provide an additional $2(L + 1)$ boundary equations, for a total of $4(L + 1)$ which are sufficient to solve for the mode index and the field amplitudes. It is convenient to rewrite these equations as

$$\tilde{\boldsymbol{n}}_1^{(0)}(0)\boldsymbol{A}_e^{(0)} = \tilde{\boldsymbol{n}}_1^{(1)}(0)\boldsymbol{A}_e^{(1)}, \tag{5.99}$$

$$\tilde{\boldsymbol{n}}_2^{(L+1)}(y_L)\boldsymbol{B}_e^{(L+1)} = \tilde{\boldsymbol{n}}_1^{(L)}(y_L)\boldsymbol{A}_e^{(L)} + \tilde{\boldsymbol{n}}_2^{(L)}(y_L)\boldsymbol{B}_e^{(L)}, \tag{5.100}$$

$$\tilde{\boldsymbol{n}}_1^{(m)}(y_m)\boldsymbol{A}_e^{(m)} + \tilde{\boldsymbol{n}}_2^{(m)}(y_m)\boldsymbol{B}_e^{(m)} = \tilde{\boldsymbol{n}}_1^{(m+1)}(y_m)\boldsymbol{A}_e^{(m+1)} + \tilde{\boldsymbol{n}}_2^{(m+1)}(y_m)\boldsymbol{B}_e^{(m+1)}, \tag{5.101}$$

$$\tilde{\boldsymbol{n}}_3^{(0)}(0)\boldsymbol{A}_h^{(0)} = \tilde{\boldsymbol{n}}_3^{(1)}(0)\boldsymbol{A}_h^{(1)}, \tag{5.102}$$

$$\tilde{\boldsymbol{n}}_4^{(L+1)}(y_L)\boldsymbol{B}_h^{(L+1)} = \tilde{\boldsymbol{n}}_3^{(L)}(y_L)\boldsymbol{A}_h^{(L)} + \tilde{\boldsymbol{n}}_4^{(L)}(y_L)\boldsymbol{B}_h^{(L)}, \tag{5.103}$$

$$\tilde{\boldsymbol{n}}_3^{(m)}(y_m)\boldsymbol{A}_h^{(m)} + \tilde{\boldsymbol{n}}_4^{(m)}(y_m)\boldsymbol{B}_h^{(m)} = \tilde{\boldsymbol{n}}_3^{(m+1)}(y_m)\boldsymbol{A}_h^{(m+1)} + \tilde{\boldsymbol{n}}_4^{(m+1)}(y_m)\boldsymbol{B}_h^{(m+1)}, \tag{5.104}$$

$$\tilde{\boldsymbol{n}}_5^{(0)}(0)\boldsymbol{A}_e^{(0)} + \tilde{\boldsymbol{n}}_8^{(0)}(0)\boldsymbol{A}_h^{(0)} = \tilde{\boldsymbol{n}}_5^{(1)}(0)\boldsymbol{A}_e^{(1)} + \tilde{\boldsymbol{n}}_8^{(1)}(0)\boldsymbol{B}_h^{1}, \tag{5.105}$$

$$\begin{aligned}\tilde{\boldsymbol{n}}_6^{(L+1)}(y_L)\boldsymbol{B}_e^{L+1} - \tilde{\boldsymbol{n}}_7^{(L+1)}(y_L)\boldsymbol{B}_h^{L+1}\\ = \tilde{\boldsymbol{n}}_5^{(L)}(y_L)\boldsymbol{A}_e^{(L)} + \tilde{\boldsymbol{n}}_6^{(L)}(y_L)\boldsymbol{B}_e^{L} + \tilde{\boldsymbol{n}}_7^{(L)}(y_L)\boldsymbol{A}_h^{(L)} + \tilde{\boldsymbol{n}}_8^{(L)}(y_L)\boldsymbol{B}_h^{L},\end{aligned} \tag{5.106}$$

$$\begin{aligned}\tilde{\boldsymbol{n}}_5^{(m)}(y_m)\boldsymbol{A}_e^{(m)} + \tilde{\boldsymbol{n}}_6^{(m)}(y_m)\boldsymbol{B}_e^{(m)} + \tilde{\boldsymbol{n}}_7^{(m)}(y_m)\boldsymbol{A}_h^{(m)} + \tilde{\boldsymbol{n}}_8^{(m)}(y_m)\boldsymbol{B}_h^{(m)}\\ = \tilde{\boldsymbol{n}}_5^{(m+1)}(y_m)\boldsymbol{A}_e^{(m+1)} + \tilde{\boldsymbol{n}}_6^{(m+1)}(y_m)\boldsymbol{B}_e^{(m+1)} + \tilde{\boldsymbol{n}}_7^{(m+1)}(y_m)\boldsymbol{A}_h^{(m+1)} + \tilde{\boldsymbol{n}}_8^{(m+1)}(y_m)\boldsymbol{B}_h^{(m+1)},\end{aligned} \tag{5.107}$$

$$\tilde{\boldsymbol{n}}_{10}^{(0)}(0)\boldsymbol{A}_e^{(0)} + \tilde{\boldsymbol{n}}_{11}^{(0)}(0)\boldsymbol{A}_h^{(0)} = \tilde{\boldsymbol{n}}_{11}^{(1)}(0)\boldsymbol{A}_h^{(1)} + \tilde{\boldsymbol{n}}_{10}^{(1)}(0)\boldsymbol{B}_e^{(1)}, \tag{5.108}$$

$$\begin{aligned}-\tilde{\boldsymbol{n}}_9^{(L+1)}(y_L)\boldsymbol{B}_e^{(L+1)} + \tilde{\boldsymbol{n}}_{12}^{(L+1)}(y_L)\boldsymbol{B}_h^{(L+1)}\\ = \tilde{\boldsymbol{n}}_9^{(L)}(y_L)\boldsymbol{A}_e^{(L)} + \tilde{\boldsymbol{n}}_{10}^{(L)}(y_L)\boldsymbol{B}_e^{(L)} + \tilde{\boldsymbol{n}}_{11}^{(L)}(y_L)\boldsymbol{A}_h^{(L)} + \tilde{\boldsymbol{n}}_{12}^{(L)}(y_L)\boldsymbol{B}_h^{(L)},\end{aligned} \tag{5.109}$$

and

$$\begin{aligned}\tilde{\boldsymbol{n}}_9^{(m)}(y_m)\boldsymbol{A}_e^{(m)} + \tilde{\boldsymbol{n}}_{10}^{(m)}(y_m)\boldsymbol{B}_e^{(m)} + \tilde{\boldsymbol{n}}_{11}^{(m)}(y_m)\boldsymbol{A}_h^{(m)} + \tilde{\boldsymbol{n}}_{12}^{(m)}(y_m)\boldsymbol{B}_h^{(m)}\\ = \tilde{\boldsymbol{n}}_9^{(m+1)}(y_m)\boldsymbol{A}_e^{(m+1)} + \tilde{\boldsymbol{n}}_{10}^{(m+1)}(y_m)\boldsymbol{B}_e^{(m+1)} + \tilde{\boldsymbol{n}}_{11}^{(m+1)}(y_m)\boldsymbol{A}_h^{(m+1)} + \tilde{\boldsymbol{n}}_{12}^{(m+1)}(y_m)\boldsymbol{B}_h^{(m+1)}\end{aligned} \tag{5.110}$$

where

$$\tilde{\boldsymbol{n}}_1^{(m)}(y) \equiv \tilde{\boldsymbol{S}}_e^{(m)}\tilde{\boldsymbol{F}}_e^{(m)}(y), \tag{5.111}$$

$$\tilde{\boldsymbol{n}}_2^{(m)}(y) \equiv \tilde{\boldsymbol{S}}_e^{(m)}\tilde{\boldsymbol{G}}_e^{(m)}(y), \tag{5.112}$$

$$\tilde{\boldsymbol{n}}_3^{(m)}(y) \equiv \tilde{\boldsymbol{S}}_h^{(m)}\tilde{\boldsymbol{F}}_h^{(m)}(y), \tag{5.113}$$

$$\tilde{\boldsymbol{n}}_4^{(m)}(y) \equiv \tilde{\boldsymbol{S}}_h^{(m)}\tilde{\boldsymbol{G}}_h^{(m)}(y), \tag{5.114}$$

$$\tilde{\boldsymbol{n}}_5^{(m)}(y) \equiv k_z\left(\frac{\tilde{\boldsymbol{P}}_h}{\Delta\mathrm{x}^2} - k_0^2\tilde{\epsilon}_h^{(m)}\right)^{-1}\frac{\tilde{\boldsymbol{D}}^t}{\Delta\mathrm{x}}\tilde{\boldsymbol{S}}_e^{(m)}\tilde{\boldsymbol{F}}_e^{(m)}(y), \tag{5.115}$$

$$\tilde{\boldsymbol{n}}_6^{(m)}(y) \equiv k_z \left(\frac{\tilde{\boldsymbol{P}}_h}{\Delta x^2} - k_0^2 \tilde{\boldsymbol{\epsilon}}_h^{(m)} \right)^{-1} \frac{\tilde{\boldsymbol{D}}^t}{\Delta x} \tilde{\boldsymbol{S}}_e^{(m)} \tilde{\boldsymbol{G}}_e^{(m)}(y), \tag{5.116}$$

$$\tilde{\boldsymbol{n}}_7^{(m)}(y) \equiv k_0 \eta_0 \left(\frac{\tilde{\boldsymbol{P}}_h}{\Delta x^2} - k_0^2 \tilde{\boldsymbol{\epsilon}}_h^{(m)} \right)^{-1} \tilde{\boldsymbol{S}}_h^{(m)} \tilde{\boldsymbol{k}}_{yh}^{(m)} \tilde{\boldsymbol{G}}_h^{(m)}(y), \tag{5.117}$$

$$\tilde{\boldsymbol{n}}_8^{(m)}(y) \equiv k_0 \eta_0 \left(\frac{\tilde{\boldsymbol{P}}_h}{\Delta x^2} - k_0^2 \tilde{\boldsymbol{\epsilon}}_h^{(m)} \right)^{-1} \tilde{\boldsymbol{S}}_h^{(m)} \tilde{\boldsymbol{k}}_{yh}^{(m)} \tilde{\boldsymbol{F}}_h^{(m)}(y), \tag{5.118}$$

$$\tilde{\boldsymbol{n}}_9^{(m)}(y) \equiv \frac{k_0}{\eta_0} \left(-\frac{\tilde{\boldsymbol{D}}}{\Delta x} \tilde{\boldsymbol{\epsilon}}_h^{(m)-1} \frac{\tilde{\boldsymbol{D}}^t}{\Delta x} + k_0^2 \tilde{\boldsymbol{I}}_e \right)^{-1} \tilde{\boldsymbol{S}}_e^{(m)} \tilde{\boldsymbol{k}}_{ye}^{(m)} \tilde{\boldsymbol{G}}_e^{(m)}(y_m), \tag{5.119}$$

$$\tilde{\boldsymbol{n}}_{10}^{(m)}(y) \equiv \frac{k_0}{\eta_0} \left(-\frac{\tilde{\boldsymbol{D}}}{\Delta x} \tilde{\boldsymbol{\epsilon}}_h^{(m)-1} \frac{\tilde{\boldsymbol{D}}^t}{\Delta x} + k_0^2 \tilde{\boldsymbol{I}}_e \right)^{-1} \tilde{\boldsymbol{S}}_e^{(m)} \tilde{\boldsymbol{k}}_{ye}^{(m)} \tilde{\boldsymbol{F}}_e^{(m)}(y_m), \tag{5.120}$$

$$\tilde{\boldsymbol{n}}_{11}^{(m)}(y) \equiv k_z \left(-\frac{\tilde{\boldsymbol{D}}}{\Delta x} \tilde{\boldsymbol{\epsilon}}_h^{(m)-1} \frac{\tilde{\boldsymbol{D}}^t}{\Delta x} + k_0^2 \tilde{\boldsymbol{I}}_e \right)^{-1} \frac{\tilde{\boldsymbol{D}}}{\Delta x} \tilde{\boldsymbol{\epsilon}}_h^{(m)-1} \tilde{\boldsymbol{S}}_h^{(m)} \tilde{\boldsymbol{F}}_h^{(m)}(y_m), \tag{5.121}$$

and

$$\tilde{\boldsymbol{n}}_{12}^{(m)}(y) \equiv k_z \left(-\frac{\tilde{\boldsymbol{D}}}{\Delta x} \tilde{\boldsymbol{\epsilon}}_h^{(m)-1} \frac{\tilde{\boldsymbol{D}}^t}{\Delta x} + k_0^2 \tilde{\boldsymbol{I}}_e \right)^{-1} \frac{\tilde{\boldsymbol{D}}}{\Delta x} \tilde{\boldsymbol{\epsilon}}_h^{(m)-1} \tilde{\boldsymbol{S}}_h^{(m)} \tilde{\boldsymbol{G}}_h^{(m)}(y_m). \tag{5.122}$$

5.6 MULTILAYER ITERATIVE SOLUTION FOR THE WAVEVECTOR

The four boundary condition equations for intermediate layers are (5.89), (5.92), (5.95), and (5.98). In matrix format,

$$\tilde{\boldsymbol{M}}^{(m)}(y_m) \begin{bmatrix} \boldsymbol{A}_e^{(m)} \\ \boldsymbol{A}_h^{(m)} \\ \boldsymbol{B}_e^{(m)} \\ \boldsymbol{B}_h^{(m)} \end{bmatrix} = \tilde{\boldsymbol{M}}^{(m+1)}(y_m) \begin{bmatrix} \boldsymbol{A}_e^{(m+1)} \\ \boldsymbol{A}_h^{(m+1)} \\ \boldsymbol{B}_e^{(m+1)} \\ \boldsymbol{B}_h^{(m+1)} \end{bmatrix} \tag{5.123}$$

where

$$\tilde{\boldsymbol{M}}^{(m)}(y) \equiv \begin{bmatrix} \tilde{\boldsymbol{n}}_1^{(m)}(y) & \tilde{\boldsymbol{0}} & \tilde{\boldsymbol{n}}_2^{(m)}(y) & \tilde{\boldsymbol{0}} \\ \tilde{\boldsymbol{0}} & \tilde{\boldsymbol{n}}_3^{(m)}(y) & \tilde{\boldsymbol{0}} & \tilde{\boldsymbol{n}}_4^{(m)}(y) \\ \tilde{\boldsymbol{n}}_5^{(m)}(y) & \tilde{\boldsymbol{n}}_7^{(m)}(y) & \tilde{\boldsymbol{n}}_6^{(m)}(y) & \tilde{\boldsymbol{n}}_8^{(m)}(y) \\ \tilde{\boldsymbol{n}}_9^{(m)}(y) & \tilde{\boldsymbol{n}}_{11}^{(m)}(y) & \tilde{\boldsymbol{n}}_{10}^{(m)}(y) & \tilde{\boldsymbol{n}}_{12}^{(m)}(y) \end{bmatrix}. \tag{5.124}$$

$\tilde{\boldsymbol{M}}^{(m)}$ is a square matrix of rank $4N - 2$. The boundary conditions at the top surface, (5.88), (5.91), (5.94), and (5.97), can be written as the matrix equation

$$\tilde{\boldsymbol{M}}^{(L+1)}(y_L)\begin{bmatrix}\boldsymbol{B}_e^{(L+1)}\\ \boldsymbol{B}_h^{(L+1)}\end{bmatrix}=\tilde{\boldsymbol{M}}^{(L)}(y_L)\begin{bmatrix}\boldsymbol{A}_e^{(L)}\\ \boldsymbol{A}_h^{(L)}\\ \boldsymbol{B}_e^{(L)}\\ \boldsymbol{B}_h^{(L)}\end{bmatrix} \tag{5.125}$$

where

$$\tilde{\boldsymbol{M}}^{(L+1)}(y)\equiv\begin{bmatrix}\tilde{\boldsymbol{n}}_2^{(L+1)}(y_L) & \tilde{0}\\ \tilde{0} & \tilde{\boldsymbol{n}}_4^{(L+1)}(y_L)\\ \tilde{\boldsymbol{n}}_6^{(L+1)}(y_L) & -\tilde{\boldsymbol{n}}_7^{(L+1)}(y_L)\\ -\tilde{\boldsymbol{n}}_9^{(L+1)}(y_L) & \tilde{\boldsymbol{n}}_{12}^{(L+1)}(y_L)\end{bmatrix}. \tag{5.126}$$

$\tilde{\boldsymbol{M}}^{(L+1)}$ is rectangular with $4N-2$ rows and $2N-1$ columns. Finally, at the bottom surface, (5.87), (5.90), (5.93), and (5.96) can be written as the matrix equation

$$\tilde{\boldsymbol{M}}^{(0)}(0)\begin{bmatrix}\boldsymbol{A}_e^{(0)}\\ \boldsymbol{A}_h^{(0)}\end{bmatrix}=\tilde{\boldsymbol{M}}^{(1)}(0)\begin{bmatrix}\boldsymbol{A}_e^{(1)}\\ \boldsymbol{A}_h^{(1)}\\ \boldsymbol{B}_e^{(1)}\\ \boldsymbol{B}_h^{(1)}\end{bmatrix} \tag{5.127}$$

where

$$\tilde{\boldsymbol{M}}^{(0)}(y)\equiv\begin{bmatrix}\tilde{\boldsymbol{n}}_1^{(0)}(0) & \tilde{0}\\ \tilde{0} & \tilde{\boldsymbol{n}}_3^{(0)}(0)\\ \tilde{\boldsymbol{n}}_5^{(0)}(0) & \tilde{\boldsymbol{n}}_8^{(0)}(0)\\ \tilde{\boldsymbol{n}}_{10}^{(0)}(0) & \tilde{\boldsymbol{n}}_{11}^{(0)}(0)\end{bmatrix}. \tag{5.128}$$

(5.123), (5.125), and (5.127) may be combined into a single matrix equation. It is preferable to multiply matrix equations for the intermediate layers beginning with the bottom layer. This is because matrix inversion is required and the matrices at the bottom surface of each layer are simpler than those at the top surface due to all $\tilde{\boldsymbol{G}}$ matrices being zero and all $\tilde{\boldsymbol{F}}$ matrices being identity matrices at the bottom surfaces. As a result, the matrix inversion is more accurate as is the entire calculation, enabling more lines to be included before the accuracy of the result suffers. The first step is inverting (5.127),

$$\begin{bmatrix}\boldsymbol{A}_e^{(1)}\\ \boldsymbol{A}_h^{(1)}\\ \boldsymbol{B}_e^{(1)}\\ \boldsymbol{B}_h^{(1)}\end{bmatrix}=\left(\tilde{\boldsymbol{M}}^{(1)}(0)\right)^{-1}\tilde{\boldsymbol{M}}^{(0)}(0)\begin{bmatrix}\boldsymbol{A}_e^{(0)}\\ \boldsymbol{A}_h^{(0)}\end{bmatrix}. \tag{5.129}$$

Continuing the same process for each intermediate layer gives finally

$$\tilde{\boldsymbol{M}}^{(L)}\left(y_L\right)\begin{bmatrix}\boldsymbol{A}_e^{(L)}\\ \boldsymbol{A}_h^{(L)}\\ \boldsymbol{B}_e^{(L)}\\ \boldsymbol{B}_h^{(L)}\end{bmatrix}=\tilde{\boldsymbol{R}}\begin{bmatrix}\boldsymbol{A}_e^{(0)}\\ \boldsymbol{A}_h^{(0)}\end{bmatrix}=\tilde{\boldsymbol{M}}^{(L+1)}\left(y_L\right)\begin{bmatrix}\boldsymbol{B}_e^{(L+1)}\\ \boldsymbol{B}_h^{(L+1)}\end{bmatrix} \tag{5.130}$$

where

$$\tilde{\boldsymbol{R}}\equiv\tilde{\boldsymbol{M}}^{(L)}\left(y_L\right)\left(\tilde{\boldsymbol{M}}^{(L)}\left(y_{L-1}\right)\right)^{-1}\ldots\left(\tilde{\boldsymbol{M}}^{(2)}\left(y_1\right)\right)^{-1}\tilde{\boldsymbol{M}}^{(1)}\left(y_1\right)\left(\tilde{\boldsymbol{M}}^{(1)}(0)\right)^{-1}\tilde{\boldsymbol{M}}^{(0)}(0). \tag{5.131}$$

Also,

$$\tilde{\boldsymbol{M}}^{(L+1)}\left(y_L\right)\equiv\begin{bmatrix}\tilde{\boldsymbol{n}}_2^{(L+1)}\left(y_L\right) & \tilde{0}\\ \tilde{0} & \tilde{\boldsymbol{n}}_4^{(L+1)}\left(y_L\right)\\ \tilde{\boldsymbol{n}}_6^{(L+1)}\left(y_L\right) & -\tilde{\boldsymbol{n}}_7^{(L+1)}\left(y_L\right)\\ -\tilde{\boldsymbol{n}}_9^{(L+1)}\left(y_L\right) & \tilde{\boldsymbol{n}}_{12}^{(L+1)}\left(y_L\right)\end{bmatrix}. \tag{5.132}$$

Therefore, a new square matrix $\tilde{\boldsymbol{Q}}$ can be created that is filled with the elements of $\tilde{\boldsymbol{R}}$ and $\tilde{\boldsymbol{M}}^{(L+1)}$.

$$\tilde{\boldsymbol{Q}}\left(k_z\right)\equiv\begin{bmatrix}r_{11} & r_{21} & -m_{11} & -m_{21}\\ r_{12} & r_{22} & -m_{12} & -m_{22}\\ r_{13} & r_{23} & -m_{13} & -m_{23}\\ r_{14} & r_{24} & -m_{14} & -m_{24}\end{bmatrix}. \tag{5.133}$$

Thus,

$$\tilde{\boldsymbol{Q}}\left(k_z\right)\begin{bmatrix}\boldsymbol{A}_e^{(0)}\\ \boldsymbol{A}_h^{(0)}\\ \boldsymbol{B}_e^{(L+1)}\\ \boldsymbol{B}_h^{(L+1)}\end{bmatrix}=0. \tag{5.134}$$

$\tilde{\boldsymbol{Q}}$ is a function of the wavevector k_z. (5.134) is a system of linear equations that can only be satisfied for those values of the wavevector k_z for which the determinant of $\tilde{\boldsymbol{Q}}$ is zero.

$$\left|\tilde{\boldsymbol{Q}}\right|=0. \tag{5.135}$$

5.7 AMPLITUDE VECTORS

Once the wavevector(s) for the waveguide has been determined, the $\boldsymbol{A}$ and $\boldsymbol{B}$ amplitude vectors for both electric and magnetic field lines are calculated. The system of linear equations in (5.134) is not independent when the determinant of $\tilde{\boldsymbol{Q}}$ is zero, so one of the $4N-2$ rows, such as the last row, can be dropped to form a new square $\tilde{\boldsymbol{q}}$ matrix with $4N-3$ rows and $4N-3$ columns in which the first column of $\tilde{\boldsymbol{Q}}$ has also been removed to form a new vector $\mathbf{v}_2$. The first element of the vector amplitude $\boldsymbol{A}_e^{(0)}$ of layer 0 is normalized to 1 and the remaining vector amplitudes are solved relative to that amplitude. The matrix equation is

$$\tilde{\boldsymbol{q}}_1 \begin{bmatrix} A_{e2}^{(0)} \\ \vdots \\ A_{eN}^{(0)} \\ A_{h1}^{(0)} \\ \vdots \\ A_{h,N-1}^{(0)} \\ B_{e1}^{(L+1)} \\ \vdots \\ B_{eN}^{(L+1)} \\ B_{h1}^{(L+1)} \\ \vdots \\ A_{h,N-1}^{(L+1)} \end{bmatrix} \equiv \begin{bmatrix} Q_{12} & Q_{13} & Q_{14} & Q_{15} & \cdots & Q_{1,4N-3} \\ Q_{22} & Q_{23} & Q_{24} & Q_{25} & \cdots & Q_{2,4N-3} \\ Q_{32} & Q_{33} & Q_{34} & Q_{35} & \cdots & Q_{3,4N-3} \\ Q_{42} & Q_{43} & Q_{44} & Q_{45} & \cdots & Q_{4,4N-3} \\ Q_{52} & Q_{53} & Q_{54} & Q_{55} & \cdots & Q_{5,4N-3} \\ \vdots & \vdots & \vdots & \vdots & \ddots & \vdots \\ Q_{4N-3,2} & Q_{4N-3,3} & Q_{4N-3,4} & Q_{4N-3,5} & \cdots & Q_{4N-3,4N-3} \end{bmatrix} \begin{bmatrix} A_{e2}^{(0)} \\ \vdots \\ A_{eN}^{(0)} \\ A_{h1}^{(0)} \\ \vdots \\ A_{h,N-1}^{(0)} \\ B_{e1}^{(L+1)} \\ \vdots \\ B_{eN}^{(L+1)} \\ B_{h1}^{(L+1)} \\ \vdots \\ A_{h,N-1}^{(L+1)} \end{bmatrix}$$

$$= \begin{bmatrix} Q_{11} \\ Q_{21} \\ Q_{31} \\ Q_{41} \\ Q_{51} \\ Q_{61} \\ \vdots \\ Q_{4N-3,1} \end{bmatrix} \equiv \mathbf{v}_2 \tag{5.136}$$

The $\tilde{\boldsymbol{q}}$ matrix can then be inverted and multiplied by the vector $\mathbf{v}_2$ to obtain the complete vector amplitudes for the top and bottom claddings, $\boldsymbol{A}_e^{(0)}$, $\boldsymbol{A}_h^{(0)}$, $\boldsymbol{B}_e^{(L+1)}$, and $\boldsymbol{B}_h^{(L+1)}$. The amplitude vectors are still required for all intermediate layers. The amplitude vectors for the first layer above the bottom cladding are obtained from (5.129),

$$\begin{bmatrix} \boldsymbol{A}_e^{(1)} \\ \boldsymbol{A}_h^{(1)} \\ \boldsymbol{B}_e^{(1)} \\ \boldsymbol{B}_h^{(1)} \end{bmatrix} = \left(\tilde{\boldsymbol{M}}^{(1)}(0)\right)^{-1} \tilde{\boldsymbol{M}}^{(0)}(0) \begin{bmatrix} \boldsymbol{A}_e^{(0)} \\ \boldsymbol{A}_h^{(0)} \end{bmatrix} \tag{5.137}$$

and then all the remaining amplitude vectors are obtained by repeated application of (5.123).

$$\begin{bmatrix} \boldsymbol{A}_e^{(m+1)} \\ \boldsymbol{A}_h^{(m+1)} \\ \boldsymbol{B}_e^{(m+1)} \\ \boldsymbol{B}_h^{(m+1)} \end{bmatrix} = \left(\tilde{\boldsymbol{M}}^{(m+1)}(y_m)\right)^{-1} \tilde{\boldsymbol{M}}^{(m)}(y_m) \begin{bmatrix} \boldsymbol{A}_e^{(m)} \\ \boldsymbol{A}_h^{(m)} \\ \boldsymbol{B}_e^{(m)} \\ \boldsymbol{B}_h^{(m)} \end{bmatrix}. \tag{5.138}$$

After calculating all the amplitude vectors, the fields for any y-position along each line are obtained from (5.45)–(5.48), (5.75)–(5.80), and (5.81)–(5.86).

5.8 BILATERAL SYMMETRY

The problem can be greatly simplified for symmetric waveguides in order to improve both accuracy and speed of the calculation. There are two types of symmetry to consider. A waveguide may be symmetric with respect to reflection across a vertical axis or across a horizontal axis. In the former case, the symmetry axis is aligned with the y-axis and only the fields in the positive x half-space are calculated. In the latter case the symmetry axis is aligned with the x-axis and only the fields in the positive y half space are calculated. Of course, waveguides may also have symmetry across both axes, in which case the fields only need to be computed in the first quadrant.

5.8.1 Bilateral Symmetry Across the y-axis

a. $\boldsymbol{E_y}\big|_{x=0} = \mathbf{0}$ **and** $\partial \boldsymbol{E_x} / \partial \mathbf{x}\big|_{x=0} = \mathbf{0}$

Consider a waveguide with bilateral symmetry across the y-axis with the boundary conditions at $x = 0$ of $\partial E_x/\partial x = 0$ and $E_y = E_z = 0$. If an x_h line is placed at $x = 0$ as shown in Figure 5.4, then the boundary condition forces the tangential E_y and E_z fields to be zero at the $x = 0$ boundary of the computation space.

There is an electric wall at $x = 0$ with Dirichlet boundary conditions for the F_h fields E_y, E_z, and H_x and Neumann boundary conditions for the F_e fields E_x, H_y, and H_z. The $\tilde{\boldsymbol{D}}$ matrix multiplies the N-1 magnetic x_h lines to perform partial differentiation with respect to x. For the first line of field F_h at $x_{h,1}$,

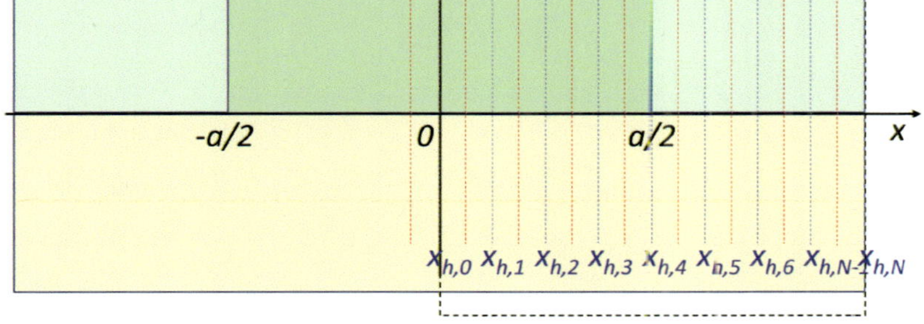

FIGURE 5.4 Waveguide with bilateral symmetry about the y-axis discretized for modes for which E_y and E_z vanish at $x = 0$. At $x = 0$ there is a magnetic line, x_{h0}, and an electric wall with Dirichlet boundary conditions for the F_h fields, and Neumann boundary conditions for the F_e fields. At the outer boundary $x_{h,N}$ there is also a magnetic line and an electric wall with the same boundary conditions. As in the original calculation that did not make use of symmetry, there are N electric lines and N-1 magnetic lines within the computation space not including the boundaries.

$$\frac{\partial F_{\text{h}1}}{\partial x} = \frac{F_{\text{h}1} - F_{\text{h}0}}{\Delta \text{x}} = \frac{F_{\text{h}1}}{\Delta \text{x}}. \tag{5.139}$$

At the final line for field F_h at $x_{h,\,N-1}$,

$$\frac{\partial F_{h,N-1}}{\partial x} = \frac{F_{h,N} - F_{h,N-1}}{\Delta \text{x}} = -\frac{F_{h,N-1}}{\Delta \text{x}}. \tag{5.140}$$

For all other lines in between,

$$\frac{\partial F_{h,n}}{\partial x} = \frac{F_{h,n} - F_{h,n-1}}{\Delta \text{x}}. \tag{5.141}$$

The matrix equation that combines (5.139)–(5.141) is

$$\frac{\partial \boldsymbol{F}_h}{\partial x} = \frac{\tilde{\boldsymbol{D}}}{\Delta \text{x}} \begin{bmatrix} F_{\text{h}1} \\ F_{\text{h}2} \\ \vdots \\ F_{h,N-2} \\ F_{h.N-1} \end{bmatrix} = \frac{1}{\Delta \text{x}} \begin{bmatrix} F_{\text{h}1} \\ F_{h,2} - F_{h,1} \\ \vdots \\ F_{h,N-1} - F_{h,N-2} \\ -F_{h,N-1} \end{bmatrix} = \frac{1}{\Delta \text{x}} \begin{bmatrix} 1 & 0 & 0 & 0 & \cdots & 0 \\ -1 & 1 & 0 & 0 & \cdots & 0 \\ \vdots & \vdots & \vdots & \ddots & \ddots & 0 \\ 0 & 0 & 0 & 0 & -1 & 1 \\ 0 & 0 & 0 & 0 & 0 & -1 \end{bmatrix} \begin{bmatrix} F_{\text{h}1} \\ F_{\text{h}2} \\ \vdots \\ F_{h.N-2} \\ F_{h.N-1} \end{bmatrix}. \tag{5.142}$$

Therefore, the $\tilde{\boldsymbol{D}}$ matrix has N rows and $N - 1$ columns,

$$\tilde{\boldsymbol{D}} \equiv \begin{bmatrix} 1 & 0 & 0 & 0 & \cdots & 0 \\ -1 & 1 & 0 & 0 & \cdots & 0 \\ \vdots & \vdots & \vdots & \ddots & \ddots & 0 \\ 0 & 0 & 0 & 0 & -1 & 1 \\ 0 & 0 & 0 & 0 & 0 & -1 \end{bmatrix} \tag{5.143}$$

which is the same matrix as used previously in (5.32) for the general waveguide case. The $\tilde{\boldsymbol{D}}^t$ matrix multiplies the electric x_e lines to perform partial differentiation with respect to x. At $x_{\text{e},1}$,

$$\frac{\partial F_{\text{e}1}}{\partial x} = \frac{F_{\text{e}2} - F_{\text{e}1}}{\Delta \text{x}}. \tag{5.144}$$

At $x_{\text{e, N}-1}$,

$$\frac{\partial F_{e,N-1}}{\partial x} = \frac{F_{e,N} - F_{e,N-1}}{\Delta \text{x}}. \tag{5.145}$$

Therefore,

$$\frac{\partial \boldsymbol{F}_e}{\partial x} = -\frac{\tilde{\boldsymbol{D}}^t}{\Delta \text{x}} \begin{bmatrix} F_{\text{e}1} \\ F_{\text{e}2} \\ \vdots \\ F_{e,N-1} \\ F_{e.N} \end{bmatrix} = -\frac{1}{\Delta \text{x}} \begin{bmatrix} F_{\text{e}2} - F_{\text{e}1} \\ F_{\text{e}3} - F_{\text{e}2} \\ \vdots \\ F_{e,N-1} - F_{e,N-2} \\ F_{e,N} - F_{e,N-1} \end{bmatrix} = -\frac{1}{\Delta \text{x}} \begin{bmatrix} 1 & -1 & 0 & 0 & \cdots & 0 \\ 0 & 1 & -1 & 0 & \cdots & 0 \\ \vdots & \vdots & \ddots & \ddots & \ddots & 0 \\ 0 & 0 & 0 & 1 & -1 & 0 \\ 0 & 0 & 0 & 0 & 1 & -1 \end{bmatrix} \begin{bmatrix} F_{\text{e}1} \\ F_{\text{e}2} \\ \vdots \\ F_{e.N-1} \\ F_{e.N} \end{bmatrix} \tag{5.146}$$

and

$$\tilde{\boldsymbol{D}}^t \equiv \begin{bmatrix} 1 & -1 & 0 & 0 & \cdots & 0 \\ 0 & 1 & -1 & 0 & \cdots & 0 \\ \vdots & \vdots & \ddots & \ddots & \ddots & 0 \\ 0 & 0 & 0 & 1 & -1 & 0 \\ 0 & 0 & 0 & 0 & 1 & -1 \end{bmatrix} \tag{5.147}$$

which is a matrix of $N - 1$ rows and N columns, the same as used previously in (5.36). As before, $\tilde{\boldsymbol{D}}^t$ is the transpose of $\tilde{\boldsymbol{D}}$, and

$$\tilde{\boldsymbol{P}}_h \equiv \tilde{\boldsymbol{D}}^t \tilde{\boldsymbol{D}} = \begin{bmatrix} 2 & -1 & 0 & \cdots & 0 \\ -1 & 2 & -1 & \cdots & 0 \\ 0 & -1 & 2 & \cdots & 0 \\ \vdots & \vdots & \vdots & \ddots & -1 \\ 0 & 0 & 0 & -1 & 2 \end{bmatrix}. \tag{5.148}$$

With reference to Figure 5.4, the dielectric matrices must also be modified from that of the nonsymmetric case. There are $N - 1$ ϵ_h lines. The dielectric matrix within layer m is

$$\tilde{\epsilon}_h^{(m)} \equiv \begin{bmatrix} n_{core}^2 & 0 & 0 & 0 & \cdots & 0 \\ 0 & n_{core}^2 & 0 & 0 & \cdots & 0 \\ 0 & 0 & \frac{1}{2}\left(n_{core}^2 + n_{clad}^2\right) & 0 & \cdots & 0 \\ 0 & 0 & 0 & n_{clad}^2 & \cdots & 0 \\ \vdots & \vdots & \vdots & \vdots & \ddots & 0 \\ 0 & 0 & 0 & 0 & 0 & n_{clad}^2 \end{bmatrix} \tag{5.149}$$

which is a square, diagonal matrix of rank $N - 1$. When an x_h field line falls on a border between two regions of different dielectric constants, the average dielectric constant value is used.

There are N x_e field lines. By design none of these lines lies on a border between regions. The dielectric matrix within layer m is

$$\tilde{\epsilon}_e^{(m)} \equiv \begin{bmatrix} n_{core}^2 & 0 & 0 & 0 & \cdots & 0 \\ 0 & n_{core}^2 & 0 & 0 & \cdots & 0 \\ 0 & 0 & n_{core}^2 & 0 & \cdots & 0 \\ 0 & 0 & 0 & n_{clad}^2 & \cdots & 0 \\ \vdots & \vdots & \vdots & \vdots & \ddots & 0 \\ 0 & 0 & 0 & 0 & 0 & n_{clad}^2 \end{bmatrix} \tag{5.150}$$

which is a square, diagonal matrix of rank N. After satisfying the boundary conditions and defining the dielectric matrices, the remaining procedure for calculating the mode indices and fields is the same as for the general case previously considered. See Example 1.

b. $\boldsymbol{E}_x\big|_{x=0} = \mathbf{0}$ **and** $\partial \boldsymbol{E}_y / \partial \boldsymbol{x}\big|_{x=0} = \mathbf{0}$

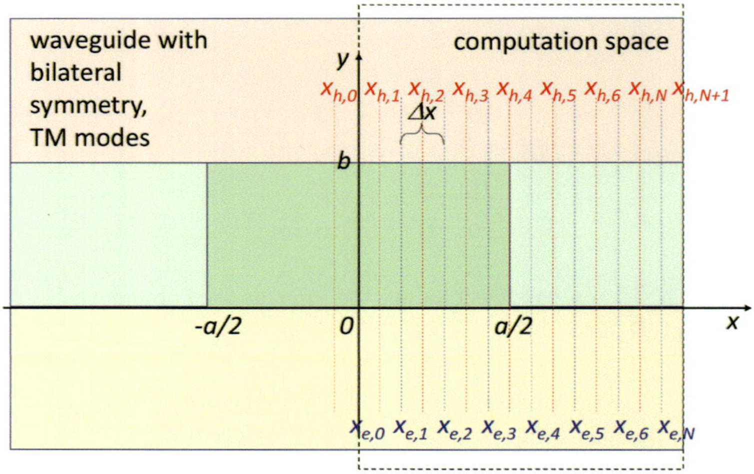

FIGURE 5.5 Waveguide with bilateral symmetry about the y-axis for quasi-TM mode analysis. At $x = 0$ there is an electric line, x_{e0}, and a magnetic wall with Dirichlet boundary conditions for F_h fields, and Neumann boundary conditions for F_e fields. With this discretization, there are N electric *and* magnetic lines within the computation space (not on the boundaries).

For the same waveguide symmetry there can also be the boundary conditions $E_x = 0$ and $\partial E_y/\partial x = \partial E_z/\partial x = 0$ at $x = 0$. In this case, an electric line must be placed at $x = 0$ as shown in Figure 5.5 where there is a magnetic wall with Dirichlet boundary conditions for the F_e fields E_x, H_y, and H_z, and Neumann boundary conditions for the F_h fields, E_y, E_z, and H_x. At $x_{e,1}$,

$$\frac{\partial F_{e1}}{\partial x} = \frac{F_{e,1} - F_{e,0}}{\Delta x} = \frac{F_{e,1}}{\Delta x}. \tag{5.151}$$

At $x_{e,N-1}$,

$$\frac{\partial F_{e,N}}{\partial x} = \frac{F_{e,N} - F_{e,N-1}}{\Delta x} \tag{5.152}$$

and the remaining lines are given by

$$\frac{\partial F_{e,n}}{\partial x} = \frac{F_{e,n} - F_{e,n-1}}{\Delta x}. \tag{5.153}$$

(5.151)–(5.153) may be combined into a matrix equation

$$\frac{\partial \boldsymbol{F}_e(y)}{\partial x} = \left(\frac{1}{\Delta x}\right)\begin{bmatrix} F_{e1}(y) \\ F_{e2}(y) - F_{e1}(y) \\ \vdots \\ F_{e,N-1}(y) - F_{e,N-2}(y) \\ F_{e,N}(y) - F_{e,N-1}(y) \end{bmatrix} = \left(\frac{1}{\Delta x}\right)\begin{bmatrix} 1 & 0 & \cdots & 0 & 0 \\ -1 & 1 & \cdots & 0 & 0 \\ \vdots & \ddots & \ddots & 0 & 0 \\ 0 & 0 & -1 & 1 & 0 \\ 0 & 0 & 0 & -1 & 1 \end{bmatrix}\begin{bmatrix} F_{e1}(y) \\ F_{e2}(y) \\ \vdots \\ F_{eN-1}(y) \\ F_{e,N}(y) \end{bmatrix}$$

$$= -\left(\frac{1}{\Delta x}\right)\tilde{\boldsymbol{D}}'\boldsymbol{F}_e(y) \tag{5.154}$$

where $\tilde{\boldsymbol{D}}^t$ is a square matrix of rank N,

$$\tilde{\boldsymbol{D}}^t \equiv \begin{bmatrix} -1 & 0 & \cdots & 0 & 0 \\ 1 & -1 & \cdots & 0 & 0 \\ \vdots & \ddots & \ddots & 0 & 0 \\ 0 & 0 & 1 & -1 & 0 \\ 0 & 0 & 0 & 1 & -1 \end{bmatrix}. \tag{5.155}$$

For the F_h fields, at the boundaries,

$$\frac{\partial F_{h,1}}{\partial x} = \frac{F_{h,2} - F_{h,1}}{\Delta \mathrm{x}} \tag{5.156}$$

and

$$\frac{\partial F_{h,N}}{\partial x} = \frac{F_{h,N+1} - F_{h,N}}{\Delta \mathrm{x}} = -\frac{F_{h,N}}{\Delta \mathrm{x}}, \tag{5.157}$$

and the partial derivatives of the remaining lines are given by

$$\frac{\partial F_{h,n}}{\partial x} = \frac{F_{h,n+1} - F_{h,n}}{\Delta \mathrm{x}}. \tag{5.158}$$

(5.156)–(5.158) may be combined into the matrix equation

$$\frac{\partial \boldsymbol{F}_h}{\partial x} = \left(\frac{1}{\Delta \mathrm{x}}\right) \begin{bmatrix} F_{h,2} - F_{h,1} \\ F_{\mathrm{h}3} - F_{\mathrm{h}2} \\ \vdots \\ F_{h,N} - F_{h,N-1} \\ -F_{h,N}(y) \end{bmatrix} = \left(\frac{1}{\Delta \mathrm{x}}\right) \begin{bmatrix} -1 & 1 & \cdots & 0 & 0 \\ 0 & -1 & \cdots & 0 & 0 \\ \vdots & \vdots & \ddots & 1 & 0 \\ 0 & 0 & 0 & -1 & 1 \\ 0 & 0 & 0 & 0 & -1 \end{bmatrix} \begin{bmatrix} F_{h,1}(y) \\ F_{h,2}(y) \\ \vdots \\ F_{h,N-1}(y) \\ F_{h,N}(y) \end{bmatrix}$$

$$= \left(\frac{1}{\Delta \mathrm{x}}\right) \tilde{\boldsymbol{D}} \begin{bmatrix} F_{h,1}(y) \\ F_{h,2}(y) \\ \vdots \\ F_{h,N-1}(y) \\ F_{h,N}(y) \end{bmatrix} = \left(\frac{1}{\Delta \mathrm{x}}\right) \tilde{\boldsymbol{D}} \boldsymbol{F}_h(y) \tag{5.159}$$

where $\tilde{\boldsymbol{D}}$ is also a square matrix of rank N,

$$\tilde{\boldsymbol{D}} \equiv \begin{bmatrix} -1 & 1 & \cdots & 0 & 0 \\ 0 & -1 & \cdots & 0 & 0 \\ \vdots & \vdots & \ddots & 1 & 0 \\ 0 & 0 & 0 & -1 & 1 \\ 0 & 0 & 0 & 0 & -1 \end{bmatrix} \tag{5.160}$$

and $\tilde{\boldsymbol{D}}^t$ is again the transpose of $\tilde{\boldsymbol{D}}$. In this case, $\tilde{\boldsymbol{P}}_h$ is a square matrix of rank N and the element for the upper left corner is 1.

$$\tilde{\boldsymbol{P}}_h \equiv \tilde{\boldsymbol{D}}^t \tilde{\boldsymbol{D}} = \begin{bmatrix} 1 & -1 & \cdots & & 0 \\ -1 & 2 & \cdots & 0 & 0 \\ \vdots & \vdots & \ddots & -1 & 0 \\ 0 & 0 & -1 & 2 & -1 \\ 0 & 0 & 0 & -1 & 2 \end{bmatrix}. \tag{5.161}$$

With reference to Figure (5.5) there are N ϵ_h lines within the computation space. The dielectric matrix for layer m is

$$\tilde{\epsilon}_h^{(m)} \equiv \begin{bmatrix} n_{core}^2 & 0 & 0 & 0 & \cdots & 0 \\ 0 & n_{core}^2 & 0 & 0 & \cdots & 0 \\ 0 & 0 & \frac{1}{2}\left(n_{core}^2 + n_{clad}^2\right) & 0 & \cdots & 0 \\ 0 & 0 & 0 & n_{clad}^2 & \cdots & 0 \\ \vdots & \vdots & \vdots & \vdots & \ddots & 0 \\ 0 & 0 & 0 & 0 & 0 & n_{clad}^2 \end{bmatrix} \tag{5.162}$$

which is a square, diagonal matrix of rank N. When an x_h field line falls on a border between two regions of different dielectric constants, the average dielectric constant value is used. There are also N x_e field lines within the computation space. By design none of these lines lies on a border between regions. The dielectric matrix within layer m is

$$\tilde{\epsilon}_e^{(m)} \equiv \begin{bmatrix} n_{core}^2 & 0 & 0 & 0 & \cdots & 0 \\ 0 & n_{core}^2 & 0 & 0 & \cdots & 0 \\ 0 & 0 & n_{clad}^2 & 0 & \cdots & 0 \\ 0 & 0 & 0 & n_{clad}^2 & \cdots & 0 \\ \vdots & \vdots & \vdots & \vdots & \ddots & 0 \\ 0 & 0 & 0 & 0 & 0 & n_{clad}^2 \end{bmatrix} \tag{5.163}$$

which is also a square, diagonal matrix of rank N.

Having satisfied boundary conditions and defined the dielectric matrices, the remaining procedure for calculating the mode indices and fields is the same as for the general case previously considered. See Example 1.

5.8.2 Bilateral Symmetry Across the x-axis

Consider a waveguide with bilateral symmetry across the x-axis. The lines are arranged in the same way as for the general case, placing magnetic lines at the outer x-axis boundaries and the boundaries between regions as shown in Figure 5.6. Due to the symmetry, the fields must either be zero at $y = 0$ or their partial derivatives with respect to y must be zero. The fields in the upper half plane still decay exponentially. For the bottom layer in the computation space, the layer that borders $y = 0$, only either the *cosh* or the *sinh* solutions are used, but not both together. In the first case for the bottom

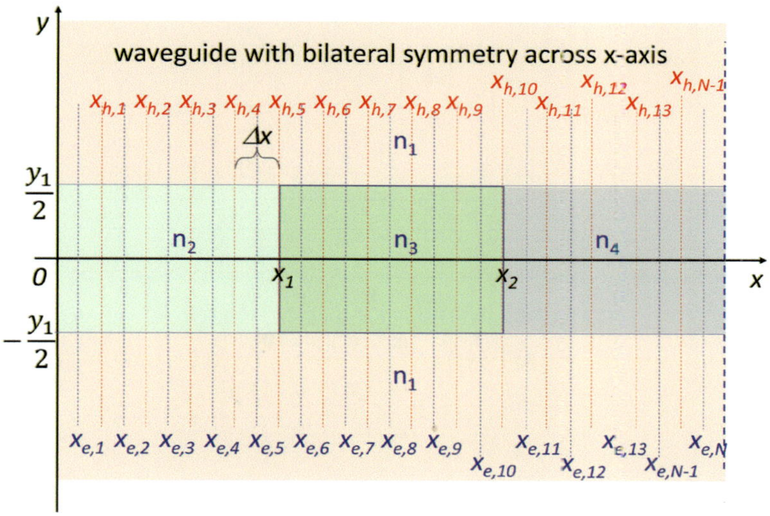

FIGURE 5.6 Waveguide with bilateral symmetry about the x-axis discretized for modes for which either E_x or E_y vanish at $y = 0$.

layer the *sinh* solutions are used for E_x, which guarantees that the E_x field component itself is zero at $y = 0$, and the *cosh* solutions are used for H_x, which guarantees that the partial derivative of the H_x field component with respect to y is zero at $y = 0$. Both *cosh* and *sinh* solutions are required as before for all other layers between the bottom layer and the top cladding. In the second symmetry case, these solutions are swapped so that in the bottom layer only the *cosh* solutions are used for E_x and only the *sinh* solutions are used for H_x.

a. $\boldsymbol{E_x}\big|_{y=0} = \mathbf{0}$ and $\partial \boldsymbol{H_x} / \partial \boldsymbol{y}\big|_{y=0} = \mathbf{0}$

The boundary condition equations (5.99)–(5.110) are the same as before with the exception of the equations for the first interface at $y = 0$, (5.99), (5.102), (5.105), and (5.108). These get replaced by equations for the new first interface at $y = y_1$.

$$\tilde{\boldsymbol{n}}_2^{(1)}(y_1)\boldsymbol{B}_e^{(1)} = \tilde{\boldsymbol{n}}_1^{(2)}(y_1)\boldsymbol{A}_e^{(2)} + \tilde{\boldsymbol{n}}_2^{(2)}(y_1)\boldsymbol{B}_e^{(2)}. \tag{5.164}$$

$$\tilde{\boldsymbol{n}}_3^{(1)}(y_1)\boldsymbol{A}_h^{(1)} = \tilde{\boldsymbol{n}}_3^{(2)}(y_1)\boldsymbol{A}_h^{(2)} + \tilde{\boldsymbol{n}}_4^{(2)}(y_1)\boldsymbol{B}_h^{(2)}. \tag{5.165}$$

$$\tilde{\boldsymbol{n}}_7^{(1)}(y_1)\boldsymbol{A}_h^{(1)} + \tilde{\boldsymbol{n}}_6^{(1)}(y_1)\boldsymbol{B}_e^{(1)} = \tilde{\boldsymbol{n}}_5^{(2)}(y_1)\boldsymbol{A}_e^{(2)} + \tilde{\boldsymbol{n}}_7^{(2)}(y_1)\boldsymbol{A}_h^{(2)} + \tilde{\boldsymbol{n}}_6^{(2)}(y_1)\boldsymbol{B}_e^{(2)} + \tilde{\boldsymbol{n}}_8^{(2)}(y_1)\boldsymbol{B}_h^{(2)}. \tag{5.166}$$

$$\tilde{\boldsymbol{n}}_{11}^{(1)}(y_1)\boldsymbol{A}_h^{(1)} + \tilde{\boldsymbol{n}}_{10}^{(1)}(y_1)\boldsymbol{B}_e^{(1)} = \tilde{\boldsymbol{n}}_9^{(2)}(y_1)\boldsymbol{A}_e^{(2)} + \tilde{\boldsymbol{n}}_{11}^{(2)}(y_1)\boldsymbol{A}_h^{(2)} + \tilde{\boldsymbol{n}}_{10}^{(2)}(y_1)\boldsymbol{B}_e^{(2)} + \tilde{\boldsymbol{n}}_{12}^{(2)}(y_1)\boldsymbol{B}_h^{(2)}. \tag{5.167}$$

Note that the $\boldsymbol{A}_e^{(1)}$ and $\boldsymbol{B}_h^{(1)}$ terms are missing from these equations. The matrix equations (5.123) and (5.125) remain the same. The matrix equation (5.127) for the bottom interface is now

$$\tilde{\boldsymbol{M}}^{(1)}(y_1)\begin{bmatrix}\boldsymbol{A}_h^{(1)}\\ \boldsymbol{B}_e^{(1)}\end{bmatrix}=\tilde{\boldsymbol{M}}^{(2)}(y_1)\begin{bmatrix}\boldsymbol{A}_e^{(2)}\\ \boldsymbol{A}_h^{(2)}\\ \boldsymbol{B}_e^{(2)}\\ \boldsymbol{B}_h^{(2)}\end{bmatrix} \tag{5.168}$$

where

$$\tilde{\boldsymbol{M}}^{(1)}(y_1)\equiv\begin{bmatrix}\boldsymbol{0} & \tilde{\boldsymbol{n}}_2^{(1)}(y_1)\\ \tilde{\boldsymbol{n}}_3^{(1)}(y_1) & \tilde{\boldsymbol{0}}\\ \tilde{\boldsymbol{n}}_7^{(1)}(y_1) & \tilde{\boldsymbol{n}}_6^{(1)}(y_1)\\ \tilde{\boldsymbol{n}}_{11}^{(1)}(y_1) & \tilde{\boldsymbol{n}}_{10}^{(1)}(y_1)\end{bmatrix} \tag{5.169}$$

and if layer 2 is not the top cladding,

$$\tilde{\boldsymbol{M}}^{(2)}(y_1)\equiv\begin{bmatrix}\tilde{\boldsymbol{n}}_1^{(2)}(y_1) & \tilde{\boldsymbol{0}} & \tilde{\boldsymbol{n}}_2^{(2)}(y_1) & \tilde{\boldsymbol{0}}\\ \tilde{\boldsymbol{0}} & \tilde{\boldsymbol{n}}_3^{(2)}(y_1) & \tilde{\boldsymbol{0}} & \tilde{\boldsymbol{n}}_4^{(2)}(y_1)\\ \tilde{\boldsymbol{n}}_5^{(2)}(y_1) & \tilde{\boldsymbol{n}}_7^{(2)}(y_1) & \tilde{\boldsymbol{n}}_6^{(2)}(y_1) & \tilde{\boldsymbol{n}}_8^{(2)}(y_1)\\ \tilde{\boldsymbol{n}}_9^{(2)}(y_1) & \tilde{\boldsymbol{n}}_{11}^{(2)}(y_1) & \tilde{\boldsymbol{n}}_{10}^{(2)}(y_1) & \tilde{\boldsymbol{n}}_{12}^{(2)}(y_1)\end{bmatrix}. \tag{5.170}$$

The matrix multiplication equation for the entire waveguide is

$$\tilde{\boldsymbol{R}}\begin{bmatrix}\boldsymbol{A}_h^{(1)}\\ \boldsymbol{B}_e^{(1)}\end{bmatrix}=\tilde{\boldsymbol{M}}^{(L+1)}(y_L)\begin{bmatrix}\boldsymbol{B}_e^{(L+1)}\\ \boldsymbol{B}_h^{(L+1)}\end{bmatrix} \tag{5.171}$$

where

$$\tilde{\boldsymbol{R}}\equiv\tilde{\boldsymbol{M}}^{(L)}(y_L)\left[\tilde{\boldsymbol{M}}^{(L)}(y_{L-1})\right]^{-1}\ldots\left[\tilde{\boldsymbol{M}}^{(2)}(y_1)\right]^{-1}\tilde{\boldsymbol{M}}^{(1)}(y_1) \tag{5.172}$$

and

$$\tilde{\boldsymbol{M}}^{(L+1)}(y)\equiv\begin{bmatrix}\tilde{\boldsymbol{n}}_2^{(L+1)}(y_L) & \tilde{\boldsymbol{0}}\\ \tilde{\boldsymbol{0}} & \tilde{\boldsymbol{n}}_4^{(L+1)}(y_L)\\ \tilde{\boldsymbol{n}}_6^{(L+1)}(y_L) & -\tilde{\boldsymbol{n}}_7^{(L+1)}(y_L)\\ -\tilde{\boldsymbol{n}}_9^{(L+1)}(y_L) & \tilde{\boldsymbol{n}}_{12}^{(L+1)}(y_L)\end{bmatrix}. \tag{5.173}$$

This is the same form for $\tilde{\boldsymbol{R}}$ as found previously in (5.131) for the general approach without symmetry, with the exception that the two matrices evaluated at $y = 0$, $\tilde{\boldsymbol{M}}^{(0)}(0)$ and $\tilde{\boldsymbol{M}}^{(1)}(0)$, are missing. Thus,

$$\tilde{Q}(k_z)\begin{bmatrix} A_h^{(1)} \\ B_e^{(1)} \\ B_e^{(L+1)} \\ B_h^{(L+1)} \end{bmatrix} = 0 \tag{5.174}$$

where the $\tilde{Q}$ matrix is composed of the elements from the new $\tilde{R}$ and $\tilde{M}^{(L+1)}(y_L)$ matrices,

$$\tilde{Q}(k_z) \equiv \begin{bmatrix} r_{11} & r_{21} & -m_{11} & -m_{21} \\ r_{12} & r_{22} & -m_{12} & -m_{22} \\ r_{13} & r_{23} & -m_{13} & -m_{23} \\ r_{14} & r_{24} & -m_{14} & -m_{24} \end{bmatrix}. \tag{5.175}$$

If there are only two layers, a bottom core and top cladding, then the $\tilde{R}$ matrix is particularly simple,

$$\tilde{R} = \tilde{M}^{(1)}(y_1) \tag{5.176}$$

and $\tilde{M}^{(2)}(y_1)$ is

$$\tilde{M}^{(2)}(y_1) \equiv \begin{bmatrix} \tilde{n}_2^{(2)}(y_1) & \tilde{0} \\ \tilde{0} & \tilde{n}_4^{(2)}(y_1) \\ \tilde{n}_6^{(2)}(y_1) & -\tilde{n}_7^{(2)}(y_1) \\ -\tilde{n}_9^{(2)}(y_1) & \tilde{n}_{12}^{(2)}(y_1) \end{bmatrix}. \tag{5.177}$$

(5.158) reduces to

$$\tilde{M}^{(1)}(y_1)\begin{bmatrix} A_h^{(1)} \\ B_e^{(1)} \end{bmatrix} = \tilde{M}^{(2)}(y_1)\begin{bmatrix} B_e^{(2)} \\ B_h^{(2)} \end{bmatrix}. \tag{5.178}$$

So

$$\tilde{Q}(k_z) \equiv \begin{bmatrix} \tilde{0}_{eh} & \tilde{n}_2^{(1)}(y_1) & -\tilde{n}_2^{(2)}(y_L) & \tilde{0}_{eh} \\ \tilde{n}_3^{(1)}(y_1) & \tilde{0}_{he} & \tilde{0}_{he} & -\tilde{n}_4^{(2)}(y_1) \\ \tilde{n}_7^{(1)}(y_1) & \tilde{n}_6^{(1)}(y_1) & -\tilde{n}_6^{(2)}(y_1) & +\tilde{n}_7^{(2)}(y_1) \\ \tilde{n}_{11}^{(1)}(y_1) & \tilde{n}_{10}^{(1)}(y_1) & +\tilde{n}_9^{(2)}(y_1) & -\tilde{n}_{12}^{(2)}(y_1) \end{bmatrix}. \tag{5.179}$$

See Example 1.

b. $\boldsymbol{H_x|_{y=0} = 0}$ **and** $\boldsymbol{\partial E_x / \partial y|_{y=0} = 0}$

The boundary condition equations (5.99)–(5.110) are the same as before with the exception of the equations for the first interface at $y = 0$, (5.99), (5.102), (5.105), and (5.108). These get replaced by equations for the new first interface at $y = y_l$.

$$\tilde{\boldsymbol{n}}_1^{(1)}(y_1)\boldsymbol{A}_e^{(1)} = \tilde{\boldsymbol{n}}_1^{(2)}(y_1)\boldsymbol{A}_e^{(2)} + \tilde{\boldsymbol{n}}_2^{(2)}(y_1)\boldsymbol{B}_e^{(2)}. \tag{5.180}$$

$$\tilde{\boldsymbol{n}}_4^{(1)}(y_1)\boldsymbol{B}_h^{(1)} = \tilde{\boldsymbol{n}}_3^{(2)}(y_1)\boldsymbol{A}_h^{(2)} + \tilde{\boldsymbol{n}}_4^{(2)}(y_1)\boldsymbol{B}_h^{(2)}. \tag{5.181}$$

$$\tilde{\boldsymbol{n}}_5^{(1)}(y_1)\boldsymbol{A}_e^{(1)} + \tilde{\boldsymbol{n}}_8^{(1)}(y_1)\boldsymbol{B}_h^{(1)} = \tilde{\boldsymbol{n}}_5^{(2)}(y_1)\boldsymbol{A}_e^{(2)} + \tilde{\boldsymbol{n}}_6^{(2)}(y_1)\boldsymbol{B}_e^{(2)} + \tilde{\boldsymbol{n}}_7^{(2)}(y_1)\boldsymbol{A}_h^{(2)} + \tilde{\boldsymbol{n}}_8^{(2)}(y_1)\boldsymbol{B}_h^{(2)}. \tag{5.182}$$

$$\tilde{\boldsymbol{n}}_9^{(1)}(y_1)\boldsymbol{A}_e^{(1)} + \tilde{\boldsymbol{n}}_{12}^{(1)}(y_1)\boldsymbol{B}_h^{(1)} = \tilde{\boldsymbol{n}}_9^{(2)}(y_1)\boldsymbol{A}_e^{(2)} + \tilde{\boldsymbol{n}}_{10}^{(2)}(y_1)\boldsymbol{B}_e^{(2)} + \tilde{\boldsymbol{n}}_{11}^{(2)}(y_1)\boldsymbol{A}_h^{(2)} + \tilde{\boldsymbol{n}}_{12}^{(2)}(y_1)\boldsymbol{B}_h^{(2)}. \tag{5.183}$$

Now the $\boldsymbol{A}_h^{(1)}$ and $\boldsymbol{B}_e^{(1)}$ terms are missing. The matrix equations (5.123) and (5.125) remain the same. The matrix equation (5.127) for the bottom interface is

$$\tilde{\boldsymbol{M}}^{(1)}(y_1)\begin{bmatrix}\boldsymbol{A}_e^{(1)}\\ \boldsymbol{B}_h^{(1)}\end{bmatrix} = \tilde{\boldsymbol{M}}^{(2)}(y_1)\begin{bmatrix}\boldsymbol{A}_e^{(2)}\\ \boldsymbol{A}_h^{(2)}\\ \boldsymbol{B}_e^{(2)}\\ \boldsymbol{B}_h^{(2)}\end{bmatrix} \tag{5.184}$$

where

$$\tilde{\boldsymbol{M}}^{(1)}(y_1) \equiv \begin{bmatrix}\tilde{\boldsymbol{n}}_1^{(1)}(y_1) & \tilde{\boldsymbol{0}}\\ \tilde{\boldsymbol{0}} & \tilde{\boldsymbol{n}}_4^{(1)}(y_1)\\ \tilde{\boldsymbol{n}}_5^{(1)}(y_1) & \tilde{\boldsymbol{n}}_8^{(1)}(y_1)\\ \tilde{\boldsymbol{n}}_9^{(1)}(y_1) & \tilde{\boldsymbol{n}}_{12}^{(1)}(y_1)\end{bmatrix} \tag{5.185}$$

and if layer 2 is not the top cladding,

$$\tilde{\boldsymbol{M}}^{(2)}(y_1) \equiv \begin{bmatrix}\tilde{\boldsymbol{n}}_1^{(2)}(y_1) & \tilde{\boldsymbol{0}}_{\mathrm{eh}} & \tilde{\boldsymbol{n}}_2^{(2)}(y_1) & \tilde{\boldsymbol{0}}_{\mathrm{eh}}\\ \tilde{\boldsymbol{0}}_{\mathrm{he}} & \tilde{\boldsymbol{n}}_3^{(2)}(y_1) & \tilde{\boldsymbol{0}}_{\mathrm{he}} & \tilde{\boldsymbol{n}}_4^{(2)}(y_1)\\ \tilde{\boldsymbol{n}}_5^{(2)}(y_1) & \tilde{\boldsymbol{n}}_7^{(2)}(y_1) & \tilde{\boldsymbol{n}}_6^{(2)}(y_1) & \tilde{\boldsymbol{n}}_8^{(2)}(y_1)\\ \tilde{\boldsymbol{n}}_9^{(2)}(y_1) & \tilde{\boldsymbol{n}}_{11}^{(2)}(y_1) & \tilde{\boldsymbol{n}}_{10}^{(2)}(y_1) & \tilde{\boldsymbol{n}}_{12}^{(2)}(y_1)\end{bmatrix}. \tag{5.186}$$

The matrix multiplication equation for the entire waveguide is

$$\tilde{\boldsymbol{R}}\begin{bmatrix}\boldsymbol{A}_e^{(1)}\\ \boldsymbol{B}_h^{(1)}\end{bmatrix} = \tilde{\boldsymbol{M}}^{(L+1)}(y_L)\begin{bmatrix}\boldsymbol{B}_e^{(L+1)}\\ \boldsymbol{B}_h^{(L+1)}\end{bmatrix} \tag{5.187}$$

where

$$\tilde{\boldsymbol{R}} \equiv \tilde{\boldsymbol{M}}^{(L)}(y_L)\left(\tilde{\boldsymbol{M}}^{(L)}(y_{L-1})\right)^{-1}\ldots\left(\tilde{\boldsymbol{M}}^{(2)}(y_1)\right)^{-1}\tilde{\boldsymbol{M}}^{(1)}(y_1) \tag{5.188}$$

and

$$\tilde{\boldsymbol{M}}^{(L+1)}(y) \equiv \begin{bmatrix} \tilde{\boldsymbol{n}}_2^{(L+1)}(y_L) & \tilde{\boldsymbol{0}} \\ \tilde{\boldsymbol{0}} & \tilde{\boldsymbol{n}}_4^{(L+1)}(y_L) \\ \tilde{\boldsymbol{n}}_6^{(L+1)}(y_L) & -\tilde{\boldsymbol{n}}_7^{(L+1)}(y_L) \\ -\tilde{\boldsymbol{n}}_9^{(L+1)}(y_L) & \tilde{\boldsymbol{n}}_{12}^{(L+1)}(y_L) \end{bmatrix}. \tag{5.189}$$

This is the same form for $\tilde{\boldsymbol{R}}$ as found previously in (5.131) for the general approach without symmetry, with the exception that the two matrices evaluated at $y = 0$, $\tilde{\boldsymbol{M}}^{(0)}(0)$ and $\tilde{\boldsymbol{M}}^{(1)}(0)$, are missing. Thus,

$$\tilde{\boldsymbol{Q}}(k_z) \begin{bmatrix} \boldsymbol{A}_e^{(1)} \\ \boldsymbol{B}_h^{(1)} \\ \boldsymbol{B}_e^{(L+1)} \\ \boldsymbol{B}_h^{(L+1)} \end{bmatrix} = 0 \tag{5.190}$$

where the $\tilde{\boldsymbol{Q}}$ matrix is composed of the elements from the new $\tilde{\boldsymbol{R}}$ and $\tilde{\boldsymbol{M}}^{(L+1)}(y_L)$ matrices,

$$\tilde{\boldsymbol{Q}}(\boldsymbol{k}_z) \equiv \begin{bmatrix} r_{11} & r_{21} & -m_{11} & -m_{21} \\ r_{12} & r_{22} & -m_{12} & -m_{22} \\ r_{13} & r_{23} & -m_{13} & -m_{23} \\ r_{14} & r_{24} & -m_{14} & -m_{24} \end{bmatrix}. \tag{5.191}$$

If there are only two layers, a bottom core and top cladding, then the $\tilde{\boldsymbol{R}}$ matrix is particularly simple,

$$\tilde{\boldsymbol{R}} = \tilde{\boldsymbol{M}}^{(1)}(y_1) \tag{5.192}$$

and $\tilde{\boldsymbol{M}}^{(2)}(y_1)$ is

$$\tilde{\boldsymbol{M}}^{(2)}(y_1) \equiv \begin{bmatrix} \tilde{\boldsymbol{n}}_2^{(2)}(y_1) & \tilde{\boldsymbol{0}} \\ \tilde{\boldsymbol{0}} & \tilde{\boldsymbol{n}}_4^{(2)}(y_1) \\ \tilde{\boldsymbol{n}}_6^{(2)}(y_1) & -\tilde{\boldsymbol{n}}_7^{(2)}(y_1) \\ -\tilde{\boldsymbol{n}}_9^{(2)}(y_1) & \tilde{\boldsymbol{n}}_{12}^{(2)}(y_1) \end{bmatrix}. \tag{5.193}$$

(5.170) reduces to

$$\tilde{\boldsymbol{M}}^{(1)}(y_1) \begin{bmatrix} \boldsymbol{A}_e^{(1)} \\ \boldsymbol{B}_h^{(1)} \end{bmatrix} = \tilde{\boldsymbol{M}}^{(2)}(y_1) \begin{bmatrix} \boldsymbol{B}_e^{(2)} \\ \boldsymbol{B}_h^{(2)} \end{bmatrix}. \tag{5.194}$$

So

$$\tilde{Q}(k_z) \equiv \begin{bmatrix} \tilde{n}_1^{(1)}(y_1) & \tilde{0} & -\tilde{n}_2^{(2)}(y_1) & \tilde{0} \\ \tilde{0} & \tilde{n}_4^{(1)}(y_1) & \tilde{0} & -\tilde{n}_4^{(2)}(y_1) \\ \tilde{n}_5^{(1)}(y_1) & \tilde{n}_8^{(1)}(y_1) & -\tilde{n}_6^{(2)}(y_1) & +\tilde{n}_7^{(2)}(y_1) \\ \tilde{n}_9^{(1)}(y_1) & \tilde{n}_{12}^{(1)}(y_1) & +\tilde{n}_9^{(2)}(y_1) & -\tilde{n}_{12}^{(2)}(y_1) \end{bmatrix}. \quad (5.195)$$

See Example 1.

5.9 ACCURACY AND CONVERGENCE

The method of lines technique has been compared to a variety of other techniques and found to be highly accurate [2]. The convergence of the wavevectors calculated by the method of lines algorithm is monotonic and asymptotically approaches the continuous solution. Consider the waveguide shown in Figure 5.7(a) after [4]. The x-component of the electric field is calculated at the mid-point of the core, $y = 0.3$ μm, and plotted in Figure 5.7(b) after calculation using various numbers of lines. It is clearly seen that even with only 21 lines the electric field amplitude is very close to the converged value with 77 lines. Obviously the calculated wavevector or mode index will increase in accuracy as more lines are included. At least, this will be true until the accuracy is affected by the digital precision of the computer/software.

Consider the rib waveguide cross section in Figure 5.8. In this example, after Figures 5 and 6 in [4], the rib extends to positive infinity to reduce the number of layers required for the calculation. Also, the three bottom layers each consist of a single material region. The orange region with index 3.17 is InP. The green region with index 3.38 is InGaAsP. The blue region is air. The entire width of the calculation space is 14.4 μm, while the rib itself is only 2.4 μm wide. This ensures that the fields drop nearly to zero as they approach the lateral boundaries. The InP layer right below the rib has a thickness t, where $t = 0.0$, 0.2, or 0.4 μm. The convergence behavior for the HE_{00} mode of the waveguide is plotted in Figure 5.9 as the number of lines is increased and the data are given in Table 5.1.

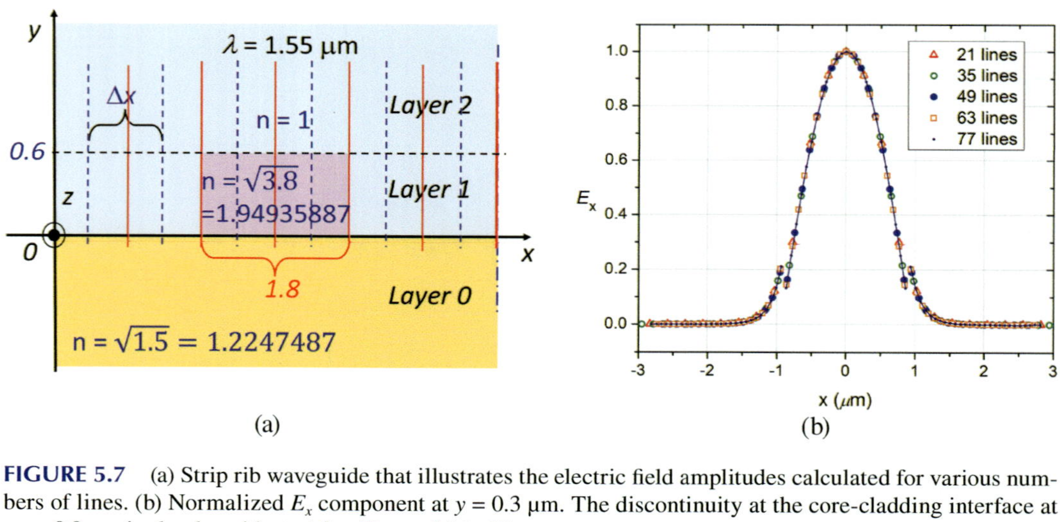

FIGURE 5.7 (a) Strip rib waveguide that illustrates the electric field amplitudes calculated for various numbers of lines. (b) Normalized E_x component at $y = 0.3$ μm. The discontinuity at the core-cladding interface at $x = \pm 0.9$ μm is clearly evident. After Figure 5.7 in [4].

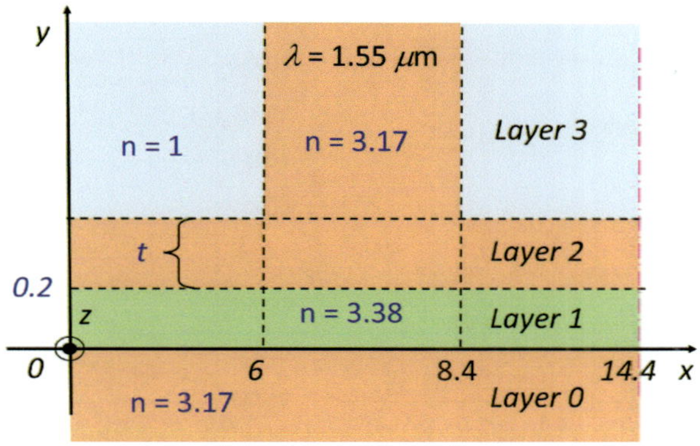

FIGURE 5.8 Four-layer waveguide after Figure 6 in [4].

TABLE 5.1
Data for Figure 5.8

Δx	0.0 μm	normalized	0.2 μm	normalized	0.4 μm	normalized
1.2	3.20011485	1.00146441	3.20170713	1.00086958	3.20343950	1.00034851
0.6	3.19747145	1.00063717	3.20011496	1.00037186	3.20281146	1.00015239
0.4	3.19662212	1.00037138	3.19963555	1.00022200	3.20262355	1.00009371
0.3	3.19624090	1.00025207	3.19941791	1.00015396	3.20253474	1.00006598
0.2	3.19590811	1.00014793	3.19922277	1.00009296	3.20245239	1.00001026
0.15	3.19576712	1.00010381	3.19913229	1.00006467	3.20241543	1.00003872
0.075	3.19560127	1.00005190	3.19902884	1.00003234	3.20236944	1.00001436

From Figure 5.9(b), it can be seen that the mode index converges very rapidly to the limiting value. In fact, the convergence is much more rapid that that shown in Figure 6 of [2] for the same waveguide. The reason for this discrepancy is not understood. Nevertheless, it is certainly true that a large number of very closely spaced lines are not required to obtain accurate mode indices.

A thicker InP layer right below the rib not surprisingly causes the mode index to increase, but the change in mode index as the number of lines increases is smaller, that is, the mode index converges more quickly to its true value as the thickness of the InP layer below the rib is increased. It is important to calculate the field profile to ensure that the boundaries are sufficiently separated from the high field regions. As seen in Figure 5.10, when the InP layer thickness below the rib is increased, the high field region spreads laterally. However, the boundary at $x = 7.2$ μm is still sufficiently far from the center of the waveguide and the high field region for accurate results.

5.10 PRECISION

As the number of lines within the computation space increases, the resolution increases as does the time to carry out the computation. Beyond a certain number of lines errors due to the precision of the arithmetic begin to limit the calculation. Therefore, it is generally necessary to program in double precision code in which floating point numbers occupy 64 bits and complex numbers 128 bits. The exponent range is –307 to +309. There are 16 digits of precision, and the range is ±0.1797693134862320D + 309 to ±0.2225073858507202D – 307. For calculation of the general waveguide with arbitrary symmetry, errors begin to become apparent usually around 50 lines.

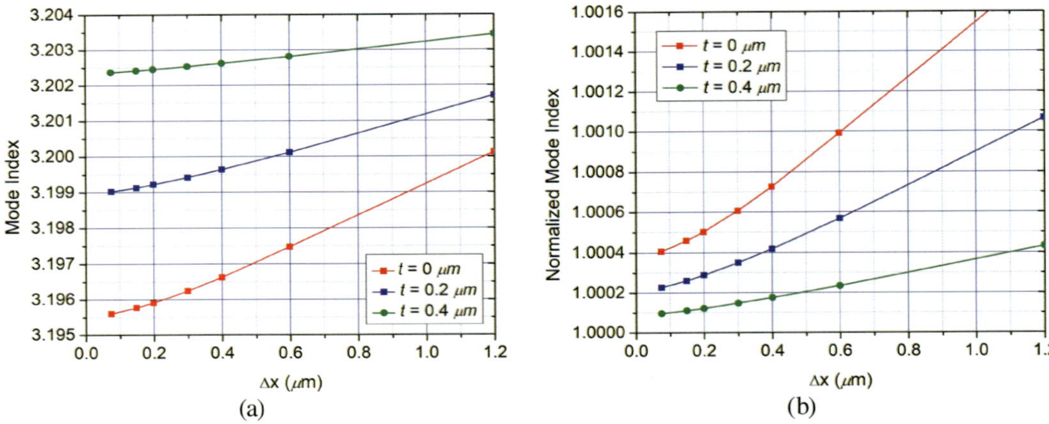

FIGURE 5.9 (a) Calculated mode indices for three InP layer thicknesses as a function of spacing between electric or magnetic lines. (b) Normalized mode indices for the three InP layer thicknesses as a function of spacing between electric or magnetic lines. The two points with the smallest spacing are extrapolated to zero spacing to determine the normalization constant.

After Figure 5.6 in [2].

When bilateral symmetry can be used, the number of lines can sometimes be increased up to ~90 or more before calculation errors become apparent in the wavevector though the field profiles often start to exhibit divergences much sooner. For the simple example in an earlier section (Figure 5.4) the number of lines was increased to 192 without any indication of calculation errors in the wavevector or mode indices due to loss of precision.

On the other hand, a simple strip waveguide was considered in Figure 5.1. The field profiles were calculated using bilateral symmetry with 48 lines and an outer x-axis boundary that was far from the core. However, many of the field profiles exhibited errors near the outer x-axis boundaries that were able to be cut out of the figures because they were sufficiently far from the core. However, the complete calculated field profiles with the entire x-axis in the computation space showing the divergent field regions are shown in Figure 5.11 for comparison to Figures 5.1(a) and (b).

5.11 PROGRAMMING TIPS

The algorithm for finding the wavevector is important for several reasons. It is the key factor in the time required to find solutions. The accuracy of the calculated wavevector value(s) or mode index has important consequences for the accuracy of the calculated fields as previously discussed. The mode index must lie between the largest core index value and the largest cladding index value. If the mode index drops below the largest cladding index, then the mode becomes leaky/radiative through that cladding. Generally, the algorithm should start from the largest core index and search downward with a mode index increment of, for example, $0.005 \cdot (n_{max} - n_{min})$. In some cases, for example, for very wide waveguides or waveguides in which core and cladding indices are nearly the same, this mode index increment is too coarse and it will miss mode solutions. The mode increment in the search algorithm should then be decreased to perhaps $0.0001 \cdot (n_{max} - n_{min})$. Of course, the smaller increment requires much more calculation time. Alternatively, the user can narrow the search range if the approximate mode index is already known.

It is usually wise to start with the largest core index and search downwards in mode index rather than start with the largest cladding index and search upwards when one is also varying parameters such as the wavelength or thickness of a layer. In that case, once a mode index solution is found,

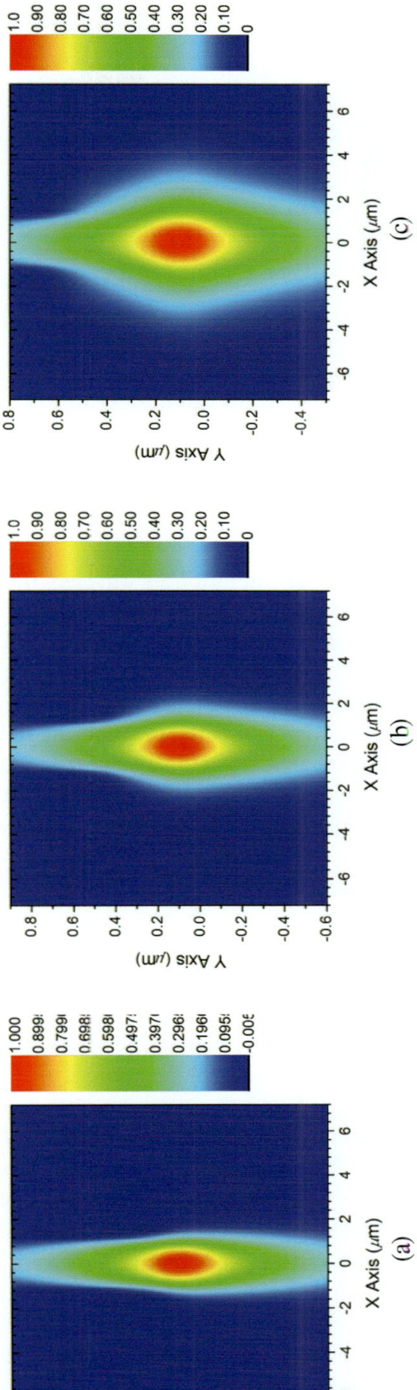

FIGURE 5.10 Contour plot of E_x for HE_{00} mode with (a) $t = 0$ μm, (b) t = 0.2 μm, and (c) t = 0.4 μm for the InP layer below the rib.

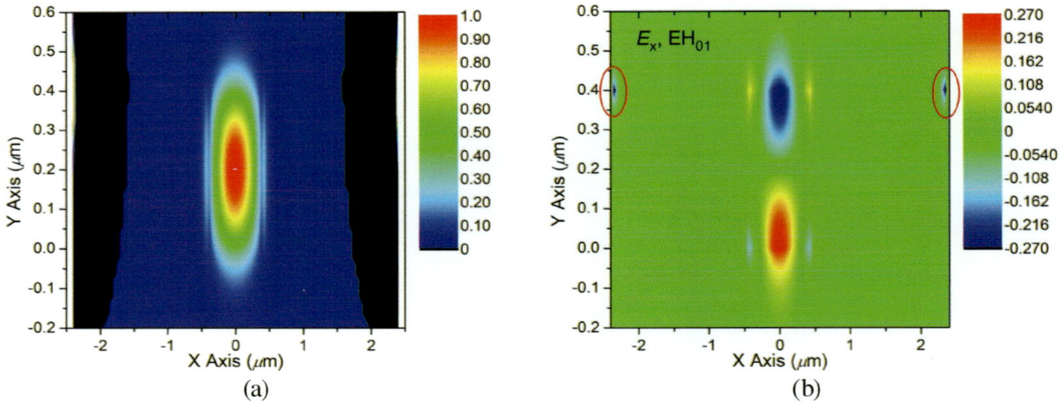

FIGURE 5.11 Calculation of the E_x field profiles for two modes of the strip waveguide in Figure 5.1. The black region which is where the calculation is diverging extends to a large area of the computation space for the HE_{00} mode in (a) and can be seen to be just starting to expand from two points on the outer edges of the computation space for the EH_{01} mode in (b).

it can become the starting point for the next iteration in wavelength or thickness, thereby greatly reducing the search time without missing modes with smaller indices.

As previously described, a mode index solution occurs when the determinant in (5.135) is zero. A search algorithm can look for a sign change between two iterations of the mode index trial value. Sometimes this occurs when the determinant is exponentially increasing towards $+\infty$ for one trial index and exponentially decreasing towards $-\infty$ for the neighboring trial index. The program, therefore must not identify these type of sign changes as indicative of true mode index solutions. It appears that a true solution always occurs when the real part of the determinant is zero, but this is not a sufficient condition. There are cases in which the real part of the determinant is zero, but the imaginary part is not.

A numerical library is useful for calculation of determinants. The IMSL routines return the exponent and mantissa of the determinant separately. This is advantageous because often for large numbers of lines the exponent is outside the range of double precision numbers. Because one is only looking for a wavevector that provides a zero in the determinant, it is a sign change as k_z is scanned that is important. The exponent is not particularly useful in finding the zero. A numerical library is also convenient for calculating eigenvectors, eigenvalues, and matrix inversions.

In the chapter on eigensystems in the standard book on numerical algorithms, [5] it is stated, "You have probably gathered by now that the solution of eigensystems is a fairly complicated business. It is. It is one of the few subjects covered in this book for which we do *not* recommend that you avoid canned routines." They go on to say, "We consider the problem of finding *all* eigenvectors of a non-symmetric matrix as lying beyond the scope of this book." Perhaps, therefore, it is not surprising that this seems to be one of the limiting factors in the accuracy of the method of lines calculations. For the initial program development, the eigenvalues and eigenvectors were calculated using the complex IMSL routine DEVCCG in the misguided assumption that keeping all arithmetic in complex double precision would make life simpler. However, the input matrices to this routine are composed of real numbers, so the IMSL routine DEVCRG may also be used and the calculated values are sometimes slightly different. In fact, when there is only one region within a layer, the input matrix is both real and symmetric. In that case, the DEVCSF routine can be used. It was found that the DEVCRG and DEVCSF routines also seem to avoid some of the fluctuating values around zero of the error function when searching for the mode index and wavevector that solve the determinant equation and that are sometimes seen when the DEVCCG routine is used with a very fine mode index increment. Therefore, the DEVCRG/DEVCSF routines should be used to improve the search routine speed and reliability. The one drawback with the IMSL routines is that they are limited to double precision arithmetic.

5.12 EXAMPLES

5.12.1 Bilateral Symmetry

The mode indices for the waveguide in Figure 5.1 required 1.2 minutes to calculate on a rather slow PC. This computation time can be substantially reduced by making use of the symmetry in the waveguide. For the half waveguide in Figure 5.12(a) designed for $E_y(x = 0) = 0$, the maximum number of regions is reduced from 3 to 2. There are four modes with indices

1. 1.802126628718: HE_{00}
2. 1.575598315668: EH_{10}
3. 1.345187492236: HE_{01}
4. 1.276650699235: HE_{20}.

The calculation time was 7 s. The field contour plots reproduce those in Figure 5.2 for their respective modes.

In order to make use of the symmetry calculation for $E_x(x = 0) = 0$, the Δx increment must be slightly changed. Whereas in the previous calculation for $E_y(x = 0) = 0$ an increment of 0.05 μm put the eighth electric line on the core boundary at 0.4 μm and the 24th electric line on the computation space boundary at 1.2 μm, for the $E_x(x = 0) = 0$ case, there must be 8.5 electric line spaces to reach the core boundary for this symmetry. If one chooses an x increment of 0.04706 μm, then the core boundary is at the correct position, and with 25 electric lines, the computation space boundary is also at 1.2 μm as before. Note that the precise location of the computation space boundary is not a hard and fast location but can be adjusted as necessary to enable the required number of electric lines within the core. Using this calculation symmetry, the other five modes are found for the Figure 5.1 waveguide with indices

1. 1.732840463583: EH_{00}
2. 1.609131394007: HE_{10}
3. 1.308949102942: EH_{20}
4. 1.112618867737: EH_{01}
5. 1.077136695569: HE_{11}

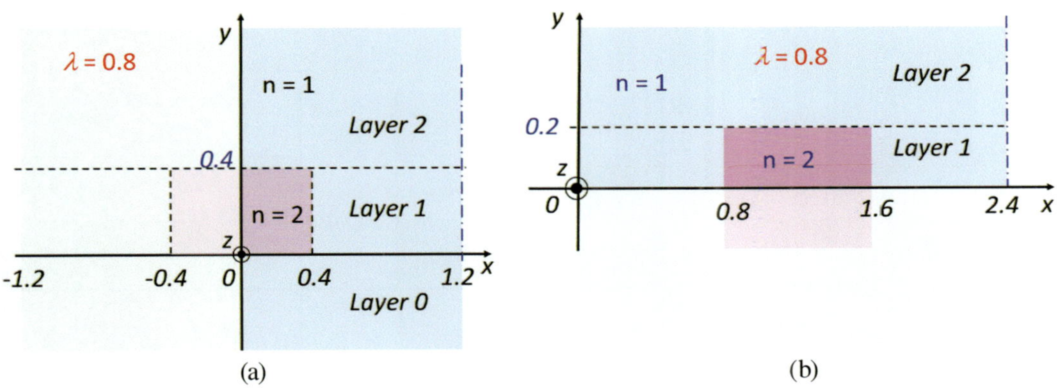

FIGURE 5.12 Waveguide design for bilateral symmetry calculation. (a) Symmetry across the y-axis. If $E_y(x = 0) = 0$, there are 24 electric lines with spacing 0.05 μm. If $E_x(x = 0) = 0$, there are 25 electric lines with spacing 0.04705882353 μm. (b) Symmetry across the x-axis.

These mode index values match those from Figure 5.2 to 4 or 5 significant figures. By reducing the x increment and increasing the x cutoff of the computation space, the agreement could be further improved. The calculation time was only 8 s.

Bilateral symmetry across the x-axis could also have been used instead in this example to reduce the computation time. For the half waveguide in Figure 5.12(b), the number of layers is reduced from 3 to 2. There are 48 lines with $\Delta x = 0.05$ μm. For $E_x(y = 0) = 0$, there are five modes with indices

```
1. 1.732771932076: EH00
2. 1.575598315663: EH10
3. 1.345187492231: HE01
4. 1.309261469107: EH20
5. 1.077827774307: HE11.
```

The calculation time to find these mode indices was 20 s.

The number of lines and x increment spacing do not need to be changed for the bilateral symmetry calculation with $E_y(y = 0) = 0$, but the waveguide parameter set must be modified to indicate the proper symmetry for the calculation. The results provide the other four mode indices. The calculation time was 24 s.

```
1. 1.802126628718: HE00
2. 1.609638433608: HE10
3. 1.276650699246: HE20
4. 1.112688570311: EH01
```

5.12.2 Rib Waveguide Dispersion as a Function of Wavelength or Wavevector

As a second example, consider the rib waveguide in Figure 5.13(a). This waveguide has been modeled in Figure 4 of [4], and Figure 2.25 of [2], and is also considered in [6]. There are many closely-spaced EH and HE modes in this bilateral symmetric waveguide. Therefore, this is an excellent waveguide to analyze by making use of its symmetry. Because there are fewer modes to discover when analyzing the waveguide by each type of symmetry separately, the mode index spacing between modes tends to be greater, making it easier for the search algorithm to identify all the modes. The two symmetric waveguide geometries for analysis are shown in Figure 5.13(b) and (c).

For the symmetric waveguide (b) for which $E_y(x = 0) = 0$, and (c) for which $E_x(x = 0) = 0$, the core boundary is at 0.6 μm and contains six electric lines. The analysis is carried out using 48 electric lines. In (b), $\Delta x = 3d/6 = 0.1\,\mu\text{m}$, and in (c), $\Delta x = 3d/6.5 = 0.0923\,\mu\text{m}$. The computation space

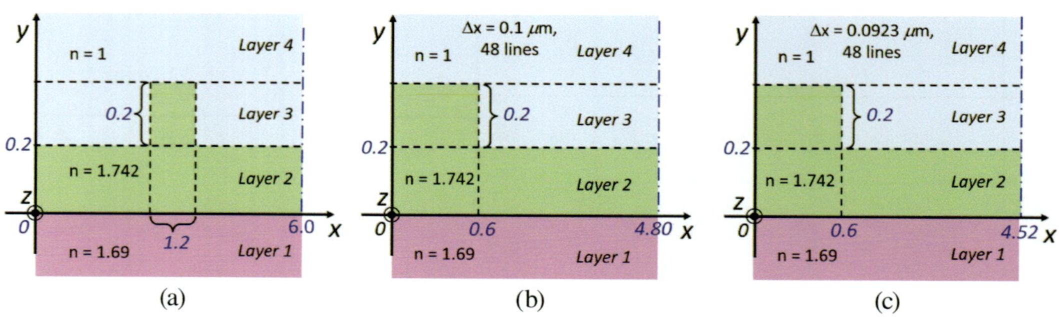

FIGURE 5.13 (a) Rib waveguide design. (b) Half waveguide design for E_y(x = 0) = 0. (c) Half waveguide design for E_x(x = 0) = 0.

boundary on the x-axis for (b) is at 48·0.1 = 4.80 μm, and for (c) is 48.5·0.0923 = 4.52 μm. To identify the modes, the fields are calculated at a wavelength of 0.2 μm. There are 17 mode solutions for $E_y(x = 0) = 0$ and 19 for $E_x(x = 0) = 0$. The mode indices for the first three mode indices for each symmetry are listed below and contour plots of the fields plotted in Figures 5.14 and 5.15.

	(b) $E_y(x=0) = 0$	(c) $E_x(x=0) = 0$
1.	1.728919696126	1.727894238272
2.	1.722952832477	1.723888950334
3.	1.717256916096	1.715089673406

After having identified the first three modes of each symmetry, the dispersion curves can be calculated as a function of wavelength or wavevector as shown in Figure 5.16.

Although the method of lines algorithm with 64 bit precision has no difficulty determining mode indices in this example, the calculation of the field profiles is problematic. The field profile plots in Figures 5.14 and 5.15 are truncated at ±1 μm along the x-axis even though 48 lines were used in the calculations for which the outer boundary is at ±~5 μm. If the entire calculation space is plotted, there are regions at the outer boundary where the field blows up. Generally, the first appearance of field divergence occurs at one of the boundaries between layers, perhaps indicating that the precision for calculation of the *sinh* and *cosh* terms is insufficient.

5.12.3 Photonic Integrated Circuit (PIC) Chip Strip Waveguide

Silicon strip or rectangular waveguides are commonly used in photonic integrated circuit (PIC) chips. The waveguide width varies but is typically between 0.4 and 0.6 μm. A common waveguide height is 0.22 μm (see Figure 3.14 in [7]). The Si waveguide core is cladded with SiO_2. Generally, the mode of interest is the quasi-TE_0 or HE_{00} mode because the quasi-TM_0 or EH_{00} mode extends substantially outside of the core and can be heavily damped by interaction with the lower Si substrate. In this example bilateral symmetry is used to calculate the wavelength-dependence of the HE_{00} mode for several waveguide widths. The geometry is shown in Figure 5.17(a).

The wavelength-dependence of the refractive index for Si is approximated analytically (p. 60 of [7]),

$$n_{Si}^2 = \epsilon_0 + \frac{\epsilon_L \omega_0^2}{\omega_0^2 - 2i\delta_0\left(\frac{2\pi c}{\lambda}\right) - \left(\frac{2\pi c}{\lambda}\right)^2} \tag{5.196}$$

where $\epsilon_0 = 7.9874$, $\epsilon_L = 3.6880$, $\omega_0 = 3.9328 \times 10^{15}$, and $\delta_0 = 0$. The variation of the mode index with wavelength and waveguide width is shown in Figure 5.17(b) for comparison to Figure 3.22 of [7].

5.12.4 PIC Rib Waveguide

Rib waveguides are also a standard waveguide structure in PIC foundries. A typical rib waveguide half cross section is shown in Figure 5.18(a) [7]. The variation with wavelength of the mode index for several different rib widths w (and proportionately scaling the width of the slab and of the computation space) is shown in Figure 5.18(b) for comparison to Figure 3.23(a) in [7].

5.12.5 Intersecting Mode Crossings

Consider the rectangular waveguide in Figure 5.19(a) that is symmetric across two orthogonal axes. The mode indices are graphed in Figure 5.19(b) as the core width is varied from 1 μm down to 0.01 μm. In this case, the quasi-TE and quasi-TM modes do not interact with each other (in reality, any deviation

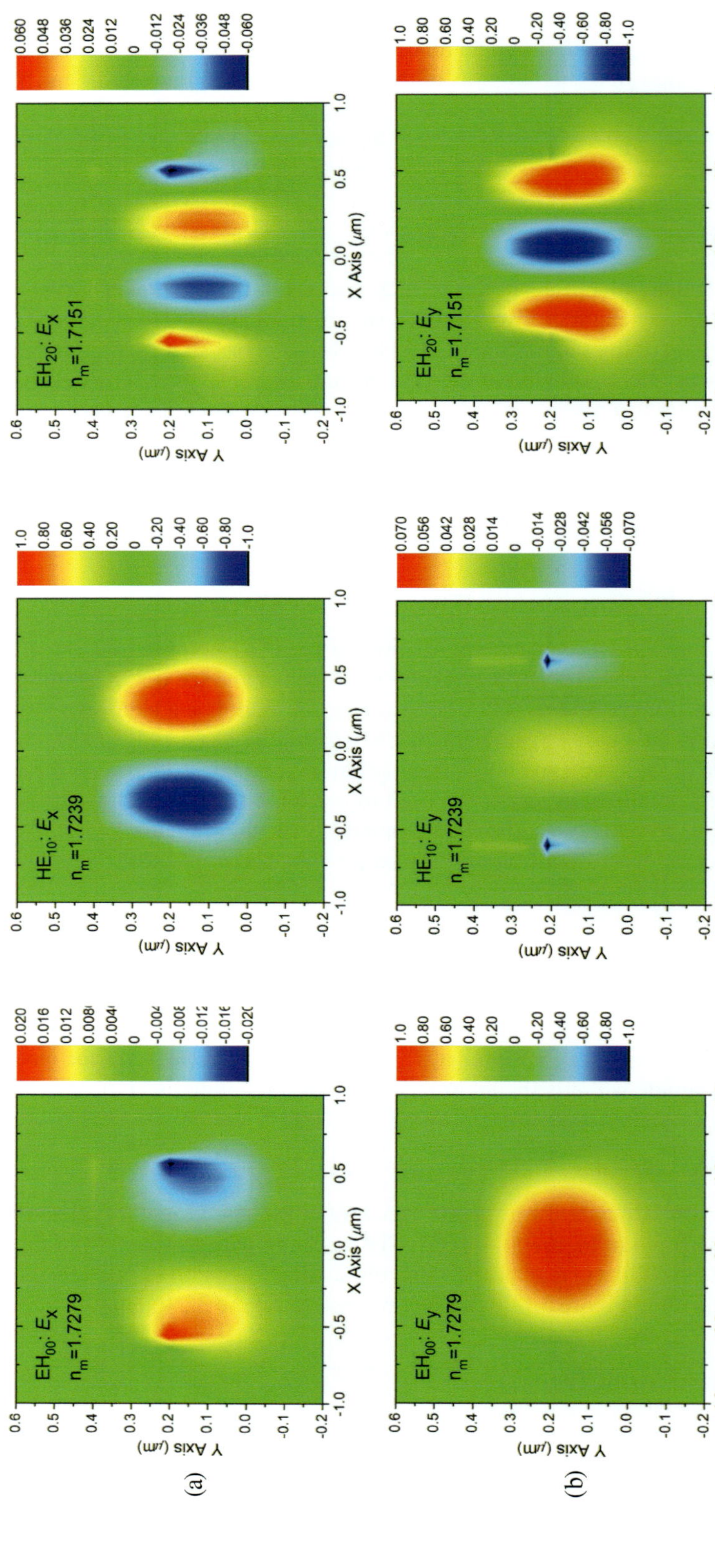

FIGURE 5.14 (a) E_x and (b) E_y contour plots for first three modes of waveguide in Figure 5.10(b) with $E_x(x = 0) = 0$.

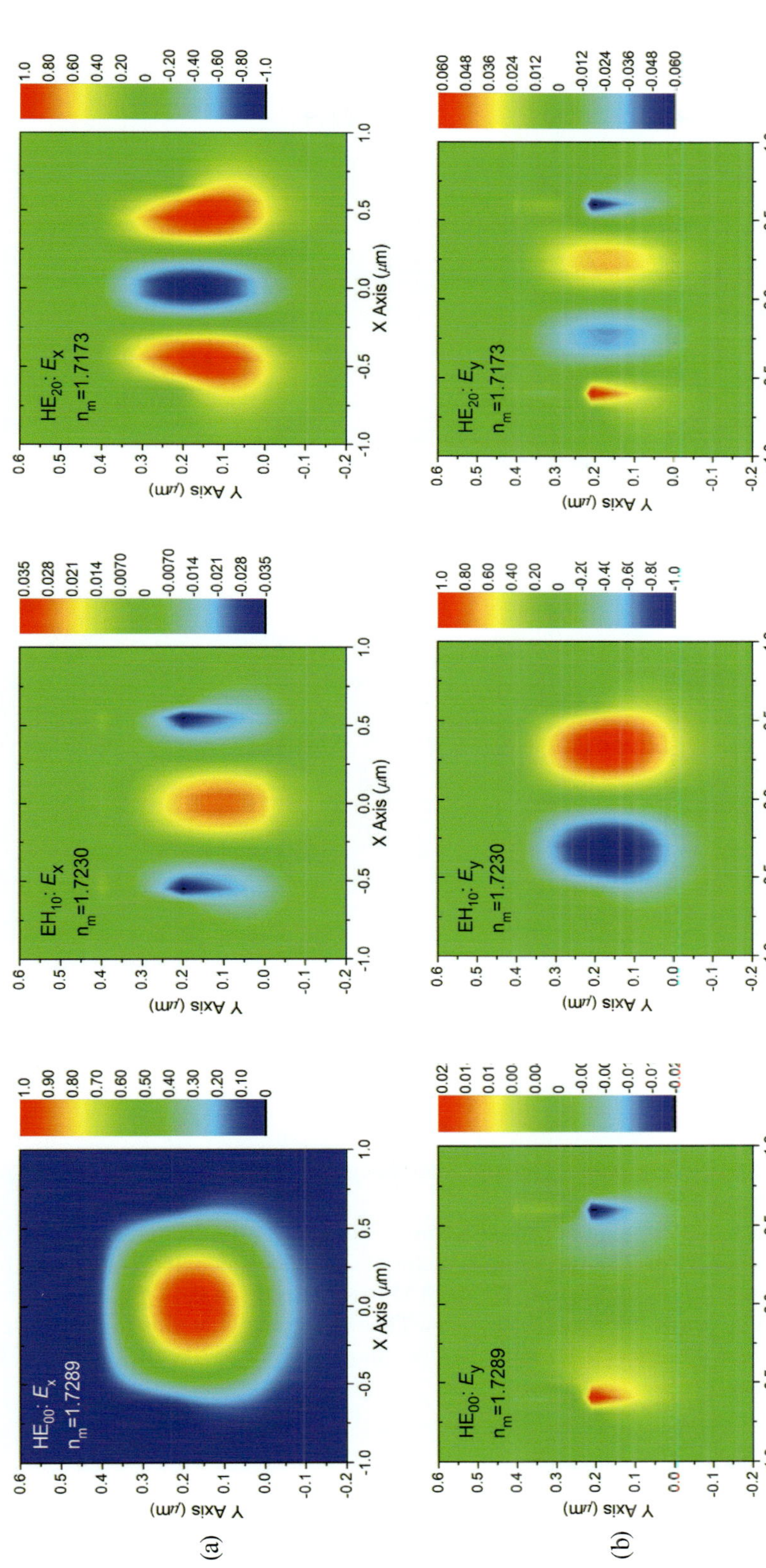

FIGURE 5.15 (a) E_x and (b) E_y contour plots for first three modes of waveguide in Figure 5.13(b) with $E_y(x = 0) = 0$. Compare to the references in the text. Note that the definitions for HE and EH modes are interchanged in the references.

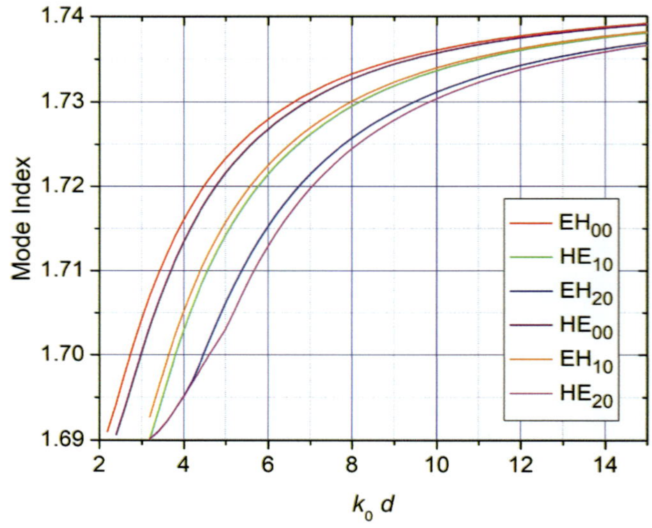

FIGURE 5.16 Dispersion curves of waveguide modes of Figures 5.14 and 5.15 as a function of wavevector.

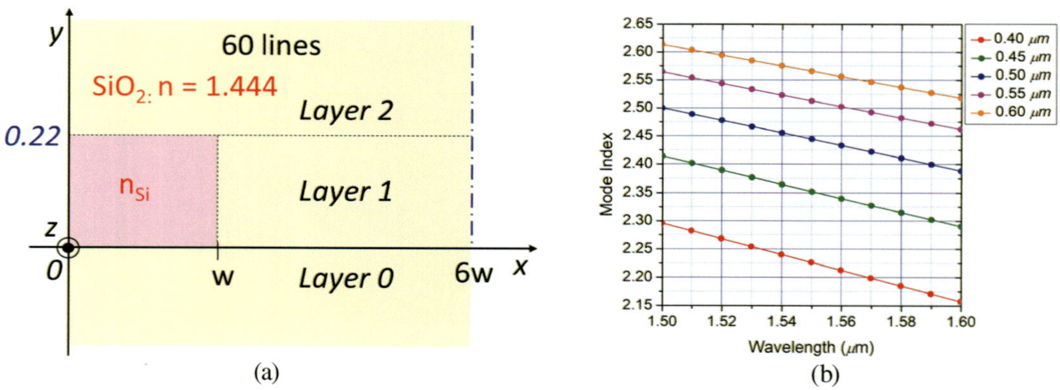

FIGURE 5.17 (a) PIC strip waveguide design. (b) Dispersion of mode index with wavelength for several different waveguide widths.

from the symmetry will cause interaction and interchange of energy between modes at points of intersection – see the following example).

The first TE/TM mode crossing occurs at a core width of 0.64 μm, which is slightly different than the value found in [7] of 0.68 μm using a vectorial 2D mode calculation (an approximation). The single-mode cutoff in [7] for the first TE and TM modes occurs at 0.44 μm, while the method of lines calculation indicates that the second quasi-TE mode actually begins to propagate at 0.42 μm. This illustrates the importance in some cases of using a technique like the method of lines for more accurate results.

5.12.6 Avoided Mode Crossings

Consider the strip waveguide in Figure 5.20 [8]. Because this is a symmetric waveguide, it is always advisable to make use of the symmetry to speed the calculation and improve the spatial resolution. It is also wise to make at least two individual calculations for the beginning width and the final

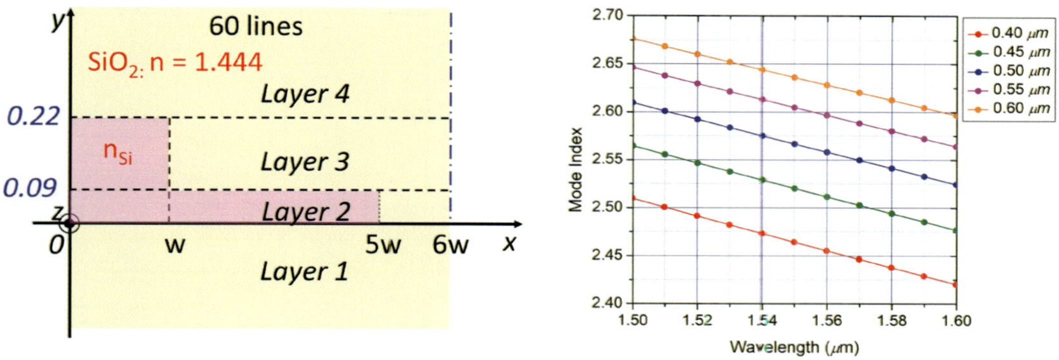

(a) (b)

FIGURE 5.18 (a) Typical rib waveguide design for PIC chip. (b) Variation of mode index with wavelength for several different values of the rib width w.

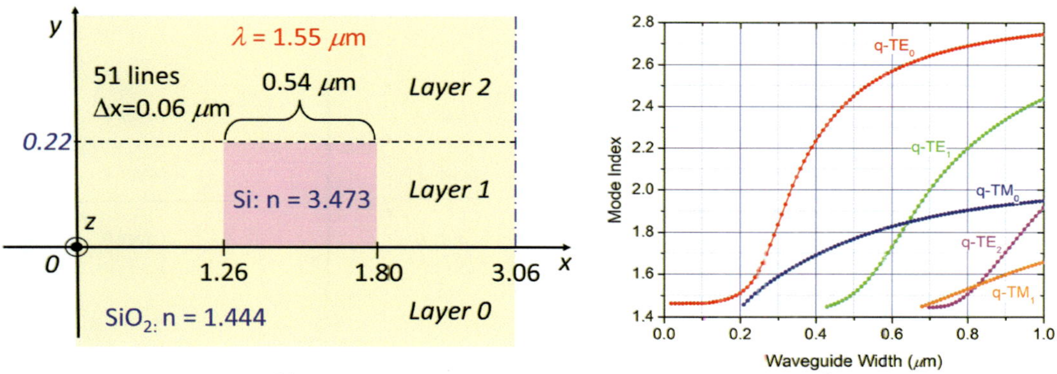

(a) (b)

FIGURE 5.19 (a) Si strip waveguide for PIC. (b) Dispersion of mode indices with waveguide width.

Compare to Figure 3.18(a) in [7].

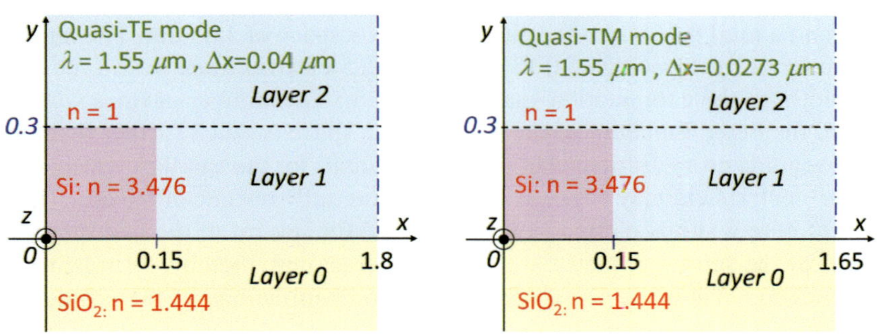

(a) (b)

FIGURE 5.20 (a) Waveguide design for $E_y(x = 0) = 0$ symmetry and (b) for $E_x(x = 0) = 0$ symmetry.

width of the core to ensure that the line spacing and computation space boundaries are suitable at both ends of the scan.

For the waveguide cross sections in Figure 5.20, 60 lines have been used. For the quasi-TE mode, the smallest core half-width 0.15 μm has a line spacing of 0.03 μm with five TM lines within the half-core, and a total half-width for the computation space of 1.8 μm. The largest

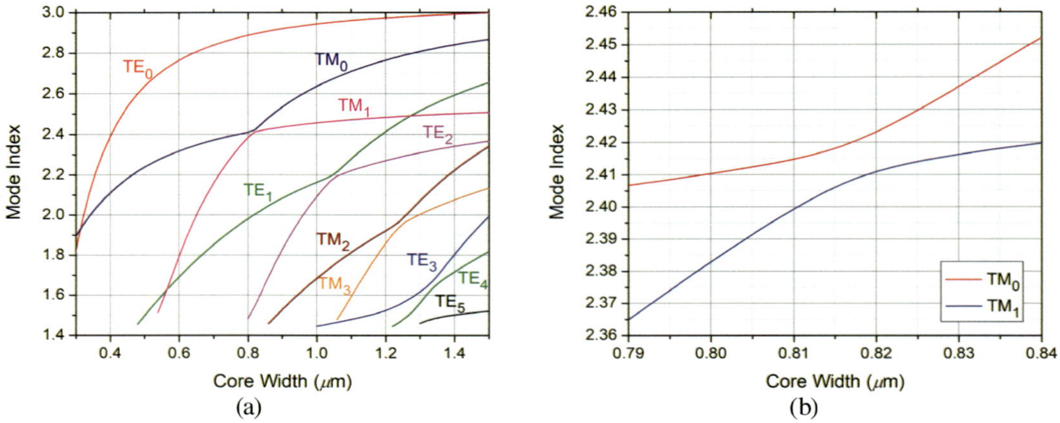

FIGURE 5.21 (a) Dispersion curves of the quasi-TE and quasi-TM modes for the waveguide in Figure 5.16. After Figure 5(a) in [8]. (b) Expanded scale for clarity of one of the avoided crossings.

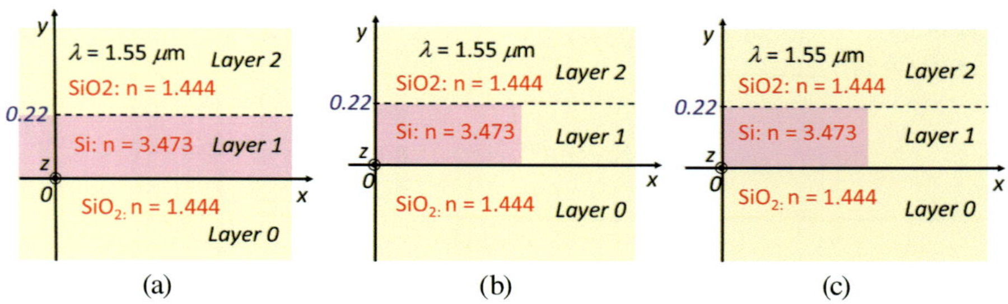

FIGURE 5.22 (a) Planar waveguide design. (b) Strip waveguide with $E_y(x = 0) = 0$ symmetry and (c) with $E_x(x = 0) = 0$ symmetry.

core half-width 0.75 μm has a line spacing of 0.15 μm, still with five electric lines within the half-core, and a total half-width for the computation space of 9 μm. For the quasi-TM mode, the smallest core half-width 0.15 μm has a line spacing of 0.0273 μm with 5.5 electric lines within the half-core and a total half-width for the computation space of 1.65 μm. The largest core half-width 0.75 μm has a line spacing of 0.136 μm, with 5.5 electric lines within the half-core, and a total half-width for the computation space of 8.18 μm. The number of lines within the core is acceptable and the outer boundary of the computation space is much further than necessary for the largest waveguide cross sections, but may be too small for the smallest waveguide cross sections when the electric field may begin to spread substantially outside of the core. The dispersion as a function of core width is plotted in Figure 5.21(a) for several of the first quasi-TE and -TM modes. Modes of the same symmetry are seen to interact and, thereby, form hybrid modes that avoid an intersection. The scale is expanded in (b) to examine one of the avoided crossings in greater detail.

5.12.7 Comparison of Strip Waveguide to Planar Waveguide

It is straightforward to analytically compute the TE and TM modes for a planar waveguide as described Chapter 4. It is interesting to compare the results from the planar calculation to that of the method of lines for a very wide strip waveguide. Consider the planar waveguide in Figure 5.22(a).

TABLE 5.2
Quasi-TE Mode Results

w/2	*D*	Δx	n_1	n_2	n_3
2	8	0.2	2.8383	2.7870	2.6867
4	16	0.4	2.8432	2.8059	2.7719
6	24	0.6	2.8441	2.8276	2.8126
8	32	0.8	2.8444	2.8351	2.8267
10	40	1.0	2.8446	2.8386	2.8333
20	40	1.0	2.8448	2.8442	2.8432

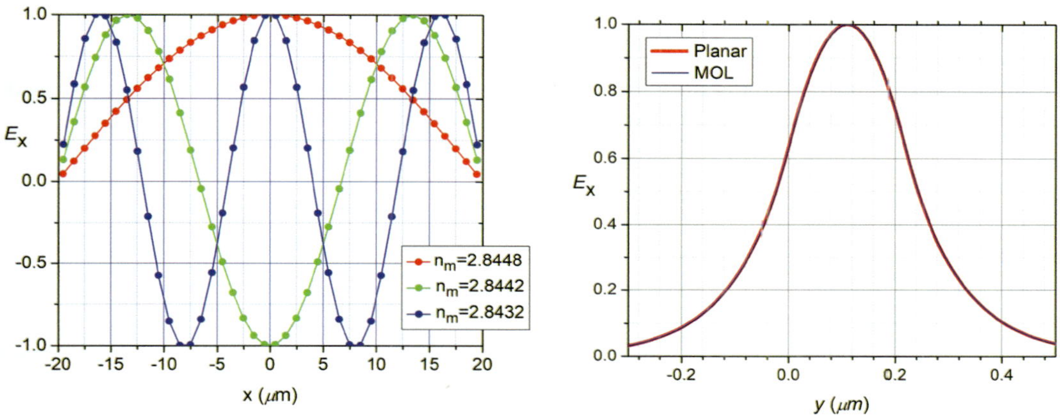

FIGURE 5.23 (a) E_x field component plotted along the x-axis for the first three waveguide modes. (b) Comparison of E_x field component along the y-axis at the center of the waveguide for the first mode with the same field component calculated for a planar waveguide.

There is one TE mode with mode index 2.8448 and one TM mode with mode index 2.0511 for the planar waveguide. This result can be compared to the mode indices for rectangular waveguides of various widths. The waveguide modes are calculated for the half-waveguide structures in Figure 5.22(b) and (c) with symmetries $E_y(x = 0) = 0$ and $E_x(x = 0) = 0$, respectively. Table 5.2 gives the calculated mode indices for the first three quasi-TE modes as the total width of the waveguide increases from 4 to 40 μm. As can be seen, the mode indices indeed quickly approach the planar waveguide result.

It is interesting to calculate the E_x field component at $x = 0$ along the y-axis, and at $y = 0.11$ (midpoint of the waveguide) along the x-axis (i.e., for each discrete line on the x-axis). Figure 5.23(a) shows the E_x field amplitude along the x-axis for the first three modes. The modes are mutually orthogonal. E_x has only one maximum for the first symmetric rectangular mode, two for the second, three for the third, etc. In Figure 5.23(b), $E_x(x = 0)$ for the first rectangular mode is compared to E_x for the planar waveguide TE_0 mode along the y-axis. To within a precision of four decimal places, there is no difference between the two curves. Indeed, no difference is found with any of the first three normalized rectangular waveguide modes along the y-axis.

One can also calculate the antisymmetric modes. Again, the mode indices are found to converge to the TE planar mode as the waveguide width increases. However, in this case E_x is always found to have a node at $x = 0$ and to have successively more maxima and minima as the waveguide width increases.

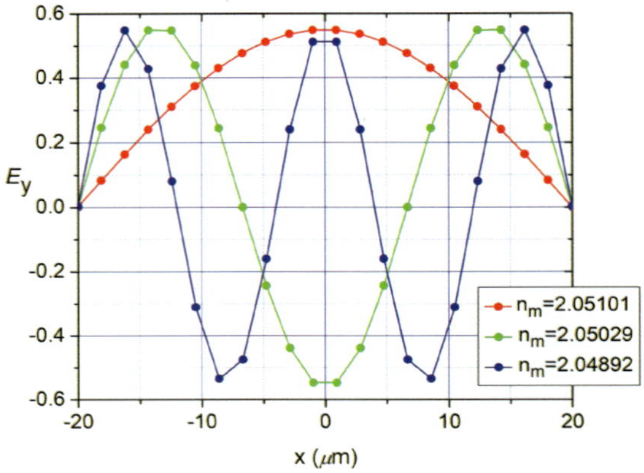

(a)

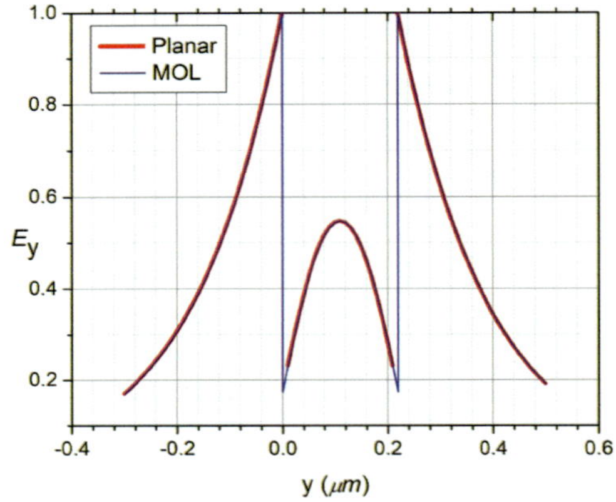

(b)

FIGURE 5.24 (a) Normalized $E_y(y)$ at $x = 0$ for TM_0 planar waveguide mode and first rectangular waveguide mode. (b) $E_y(x)$ at $y = 0.11$ μm for first three symmetric rectangular waveguide modes.

TABLE 5.3
Quasi-TM Mode Results

$w/2$	D	Δx	n_1	n_2	n_3
2	7.714	0.1905	2.0368	2.0157	1.9689
4	15.43	0.3810	2.0489	2.0314	1.9974
6	23.14	0.5714	2.0501	2.0422	2.0270
8	30.86	0.7619	2.0505	2.0375	2.0256
10	38.60	0.9524	2.0507	2.0424	2.0348
20	77.1	1.9048	2.0510	2.0503	2.0489

E_x is zero for the planar TM modes, while E_y is large. The results for the first three quasi-TM modes using 40 lines are given in Table 5.3. Again, as the width of the waveguide increases, the calculated mode indices approach that of the TM_0 planar waveguide mode.

It is interesting to calculate the E_y field component at $x = 0$ along the y-axis, and at $y = 0.11$ along the x-axis. The rectangular waveguide modes are also mutually orthogonal along the x-axis. E_y has only one maximum for the first symmetric rectangular mode, two for the second, three for the third, etc. as shown in Figure 5.24(a). One could also calculate the antisymmetric modes, and the mode indices are found to converge to the TM planar mode as the waveguide width increases. However, in this case E_y has a node at $x = 0$ and successively more maxima and minima as the waveguide width increases. In Figure 5.24(b), the field amplitudes, $E_y(x = 0)$ for the first rectangular waveguide mode is compared to E_y for the planar waveguide TM_0 mode along the y-axis. To within a precision of four decimal places, there is again no difference between the two curves or with any of the first three normalized rectangular waveguide modes.

5.12.8 GaAs Rib Waveguide

Another example of a rib waveguide is considered in [9] as shown in Figure 5.25. The effective index approximation method used in [9] results in an approximate mode index of 3.348 for a single

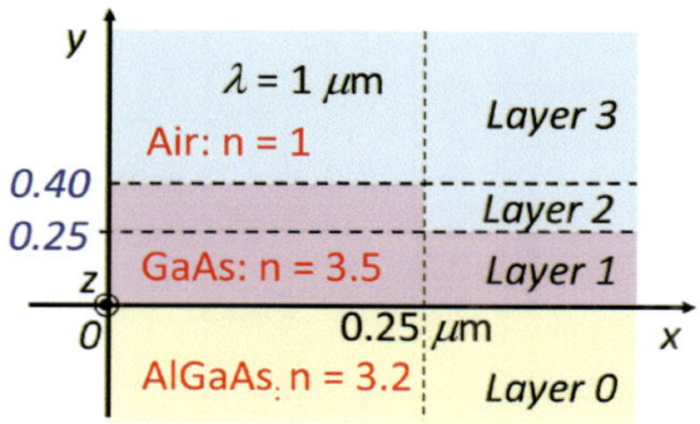

FIGURE 5.25 GaAs rib waveguide after [9]. Such a waveguide actually supports many modes.

TE mode. The method of lines technique using 50 lines with $\Delta x = 0.05$ μm actually identifies 13 total propagating modes with more accurate mode indices listed below.

	Ey(x=0)=0	Ex(x=0)=0
1.	3.332534208780	3.306863712319
2.	3.299215597060	3.299538490212
3.	3.283471774546	3.285802758220
4.	3.253882188473	3.258125842392
5.	3.247029266705	3.243166807756
6.	3.230341713258	3.216602159810
7.	3.211537892446	

5.12.9 Optimized Lens Coupling to Rectangular Waveguide

In order to optimize the coupling of light focused by a lens onto the end of a rectangular waveguide, the focused field amplitudes are calculated as described in Chapter 7 for light linearly polarized along the x-axis and for various lens numerical apertures over an area of ±10 μm along the x- and y- axes. The calculation increments are 0.05 μm along the x-axis and 0.11 μm along the y-axis for compatibility with the field calculation of the waveguide. Such a large area takes a fairly long time to evaluate and it is really only necessary for NA's < 0.6. The full width at half maximum (FWHM) for the focused fields as a function of numerical aperture is plotted in Figure 5.26(a). The values assume a Gaussian intensity apodization of the incident beam for which the FWHM is equal to the lens diameter. Therefore, the fraction of the incident beam captured by the lens is $f_c = 0.5$.

Let P_f be the total power in the focused spot that has been normalized to a peak electric field of 1 calculated as described in Chapter 7. Then the incident power P_{in} is given by

$$P_f = f_c\, P_{in}\,. \tag{5.197}$$

Now consider the SiN waveguide example in Figure 5.27(a). The E_x contour plot is shown for the quasi-TE$_0$ mode in Figure 5.27(b). E_x is plotted along the x-axis in Figure 5.26(b) at the y-center of the waveguide for comparison to that of the focused field for two different NAs. The electric field inside the waveguide is not Gaussian-like along the x-axis, but rather has two hot spots on the side walls of the core. It might be expected that matching the spot size of the focused beam to the FWHM intensity of the waveguide mode should give the best coupling efficiency. The FWHM of the waveguide mode appears to be ~2 μm, which occurs for the focused spot NA of ~0.45. However, the waveguide field amplitude decreases substantially within the core, so it is not completely obvious what lens NA will provide the best coupling efficiency.

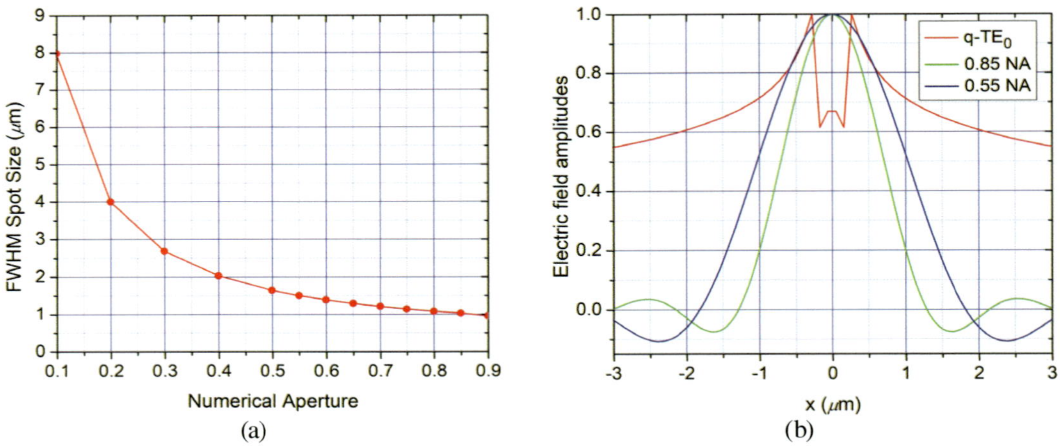

FIGURE 5.26 (a) Calculated FWHM focused spot size as a function of lens numerical aperture as described in Chapter 6 for an incident linearly polarized Gaussian beam that is apodized such that the FWHM of the incident beam occurs at the edge of the lens. (b) E_x for the SiN waveguide in Figure 5.25 at the center of the waveguide, $y = 0.11$ μm, along with E_x for focused spots with NA's of 0.85 and 0.55.

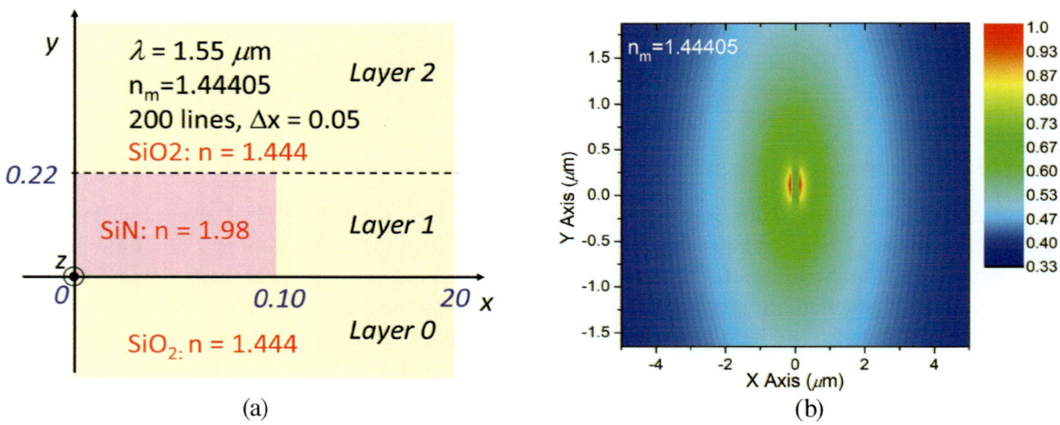

FIGURE 5.27 (a) Waveguide design. (b) E_x field profile for quasi-TE_0 waveguide mode.

The power in the waveguide is $P_w = 2.462 \times 10^{-14}$ W for the normalized electric field of the q-TE_0 mode. The power inside the waveguide scales as the square of the peak electric field, and the power in the focused spot also scales as the square of the peak focused field. Only the real part of the E_x component of the field is considered in this analysis as this is the dominant field component. As previously discussed, the overlap integral of the incident focused beam with the waveguide mode is an approximation that ignores any reflected light amplitude that could modify the total field at the entrance face of the waveguide. If the peak electric field outside the waveguide is normalized to 1, then the approximate optical power propagating inside the waveguide is given by the square of the overlap integral multiplied by the power in the waveguide mode P_w with an electric field amplitude of 1. The optical power transfer efficiency is the ratio of P_{out} to P_{in},

$$\epsilon = \frac{P_{out}}{P_{in}} \approx \frac{f_c \; P_w \, O^2}{P_f}. \tag{5.198}$$

TABLE 5.4
Coupling Efficiency Calculation

NA	FWHM of incident beam	Lens diameter	FWHM of focused spot	Total power in focused spot, P_f	Total incident power, $P_i = \frac{P_f}{f_c}$	Overlap, O	Efficiency, $\epsilon = O^2\left(\frac{P_w}{P_i}\right)$
0.9	10 mm	10 mm	0.954 μm	1.165×10^{-15} W	2.33×10^{-15} W	0.210	0.466
0.85	10	10	1.029	1.366×10^{-15}	2.732×10^{-15}	0.249	0.559
0.8	10	10	1.080	1.535×10^{-15}	3.07×10^{-15}	0.258	0.534
0.75	10	10	1.140	1.735×10^{-15}	3.470×10^{-15}	0.272	0.525
0.7	10	10	1.209	1.975×10^{-15}	3.950×10^{-15}	0.291	0.528
0.65	10	10	1.291	2.267×10^{-15}	4.534×10^{-15}	0.311	0.525
0.6	10	10	1.387	2.630×10^{-15}	5.260×10^{-15}	0.340	0.541
0.55	10	10	1.502	3.093×10^{-15}	6.186×10^{-15}	0.374	0.557
0.5	10	10	1.641	3.699×10^{-15}	7.398×10^{-15}	0.404	0.543
0.4	10	10	2.030	5.640×10^{-15}	1.280×10^{-14}	0.486	0.454
0.3	10	10	2.684	9.419×10^{-15}	1.884×10^{-14}	0.565	0.417
0.2	10	10	4.004	1.737×10^{-14}	3.474×10^{-14}	0.666	0.314
0.1	10	10	7.980	3.958×10^{-14}	7.916×10^{-14}	0.914	0.260

The coupling efficiency is calculated from (5.198) for NA's between 0.1 and 0.9 as shown in Table 5.4. The peak efficiency occurs at NAs of 0.85 and 0.55. The smaller NA would undoubtedly be preferable in practical applications. It is always wise to test the convergence of the overlap integration by repeating it with a smaller increment to see if similar results are obtained.

REFERENCES

1 R. E. Collin, *Field Theory of Guided Waves* (New York: McGraw-Hill, 1960).
2 R. Pregla, *Analysis of Electromagnetic Fields and Waves* (Chichester: John Wiley & Sons, Ltd., 2008).
3 E. A. J. Marcatili, "Dielectric rectangular waveguide and directional coupler for integrated optics," *Bell Syst. Tech. J.* **48** (1969) 2071.
4 U. Rogge and R. Pregla, "Method of lines for the analysis of strip-loaded optical waveguides," *J. Opt. Soc. Am. B.* **8** (1991) 459–463.
5 W. H. Press, S. A. Teukolsky, W. T. Vetterling, and B. P. Flannery, *Numerical Recipes in FORTRAN*, 2nd ed. (Cambridge: Cambridge University Press, 1992).
6 K. Yasuura, K. Shimohara, and T. Miyamoto, "Numerical analysis of a thin-film waveguide by mode-matching method," *J. Opt. Soc. Am.* **70** (1980) 183–191.
7 L. Chrostowski and M. Hochberg, *Silicon Photonics Design* (New York: Cambridge University Press, 2015).
8 W. J. Westerveld, S. M. Leinders, K. W. A. van Dongen, H. P. Urbach, and M. Yousefi, "Extension of Marcatili's analytical approach for rectangular silicon optical waveguides," *J. Lightwave Technol.* **30** (2012) 2388.
9 P. Yeh, *Optical Waves in Layered Media* (New York: John Wiley & Sons, 1988) 365.

6 Fiber Waveguides

6.1 INTRODUCTION

Cylindrical waveguides are important for a variety of applications but perhaps the most important is that of optical fiber. Depending on the indices and thicknesses of core and claddings, these waveguides support transverse electric (TE), transverse magnetic (TM), and hybrid modes usually denoted by "EH" or "HE." In this chapter the equations for calculating mode indices and fields are derived for multi-cladded fiber waveguides characterized by complex refractive indices at each wavelength.

Consider a fiber waveguide composed of multiple concentric cylinders as shown in Figure 6.1. Maxwell's equations in the absence of sources and assuming an $e^{-i\omega t}$ time dependence are

$$\mathrm{i}\omega\mu_0 \boldsymbol{H} = \nabla\times\boldsymbol{E}, \tag{6.1}$$

$$-\mathrm{i}\omega\epsilon\,\boldsymbol{E} = \nabla\times\boldsymbol{H}, \tag{6.2}$$

$$\nabla\cdot(\epsilon\,\boldsymbol{E}) = 0, \tag{6.3}$$

and

$$\nabla\cdot(\mu\boldsymbol{H}) = 0. \tag{6.4}$$

μ_0 is the vacuum permeability. ϵ is the medium permittivity. In cylindrical coordinates, (6.1) and (6.2) become

$$i\omega\mu_0\boldsymbol{H} = \hat{\boldsymbol{r}}\left(\frac{1}{r}\frac{\partial E_z}{\partial\phi} - \frac{\partial E_\phi}{\partial z}\right) + \hat{\boldsymbol{\phi}}\left(\frac{\partial E_r}{\partial z} - \frac{\partial E_z}{\partial r}\right) + \hat{\boldsymbol{z}}\frac{1}{r}\left(\frac{\partial(rE_\phi)}{\partial r} - \frac{\partial E_r}{\partial\phi}\right) \tag{6.5}$$

and

$$-i\omega\epsilon\,\boldsymbol{E} = \hat{\boldsymbol{r}}\left(\frac{1}{r}\frac{\partial H_z}{\partial\phi} - \frac{\partial H_\phi}{\partial z}\right) + \hat{\boldsymbol{\phi}}\left(\frac{\partial H_r}{\partial z} - \frac{\partial H_z}{\partial r}\right) + \hat{\boldsymbol{z}}\frac{1}{r}\left(\frac{\partial(rH_\phi)}{\partial r} - \frac{\partial H_r}{\partial\phi}\right). \tag{6.6}$$

Six scalar equations are obtained from (6.5) and (6.6),

$$i\eta_0 k_0 H_r = \frac{1}{r}\frac{\partial E_z}{\partial\phi} - \frac{\partial E_\phi}{\partial z}, \tag{6.7}$$

$$i\eta_0 k_0 H_\phi = \frac{\partial E_r}{\partial z} - \frac{\partial E_z}{\partial r}, \tag{6.8}$$

$$i\eta_0 k_0 H_z = \frac{1}{r}\left(\frac{\partial(rE_\phi)}{\partial r} - \frac{\partial E_r}{\partial\phi}\right), \tag{6.9}$$

DOI: 10.1201/9781003714231-6

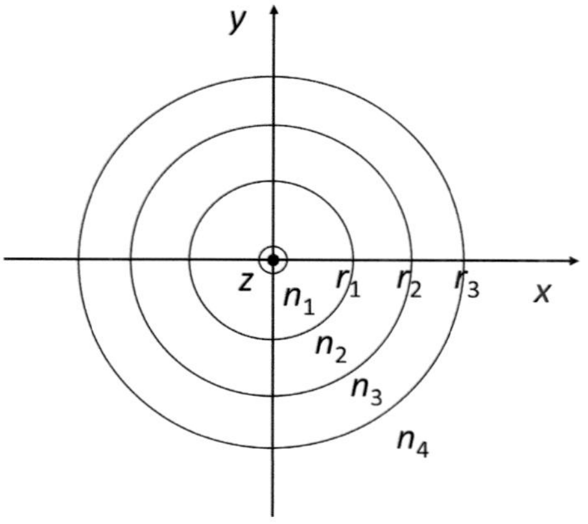

FIGURE 6.1 Cylindrical waveguide composed of a core and multiple claddings.

$$-i\frac{n^2k_0}{\eta_0}E_r = \frac{1}{r}\frac{\partial H_z}{\partial \phi} - \frac{\partial H_\phi}{\partial z}, \tag{6.10}$$

$$-i\frac{n^2k_0}{\eta_0}E_\phi = \frac{\partial H_r}{\partial z} - \frac{\partial H_z}{\partial r}, \tag{6.11}$$

$$-i\frac{n^2k_0}{\eta_0}E_z = \frac{1}{r}\left(\frac{\partial (rH_\phi)}{\partial r} - \frac{\partial H_r}{\partial \phi}\right), \tag{6.12}$$

where

$$\epsilon = n^2\epsilon_0, \tag{6.13}$$

$$\omega = ck_0 = \frac{k_0}{\sqrt{\epsilon_0\mu_0}}, \tag{6.14}$$

and

$$\eta_0 \equiv \sqrt{\frac{\mu_0}{\epsilon_0}}. \tag{6.15}$$

η_0 is the impedance of vacuum and n is the complex refractive index of the medium.

The E_r, E_φ, H_r, and H_φ components are expressed in terms of the longitudinal field components E_z and H_z. Assuming propagation along the z-axis with wavevector k_z (that is. the fields are proportional to $e^{ik_z z}$) then from (6.7) and (6.11),

$$h^2H_r = -\frac{in^2k_0}{\eta_0 r}\frac{\partial E_z}{\partial \phi} + ik_z\frac{\partial H_z}{\partial r} \tag{6.16}$$

and

$$h^2E_\phi = \frac{ik_z}{r}\frac{\partial E_z}{\partial \phi} - ik_0\eta_0\frac{\partial H_z}{\partial r}. \tag{6.17}$$

From (6.8) and (6.10),

$$h^2 H_\phi = \frac{in^2 k_0}{\eta_0}\frac{\partial E_z}{\partial r} + \frac{ik_z}{r}\frac{\partial H_z}{\partial \phi} \tag{6.18}$$

and

$$h^2 E_r = ik_z \frac{\partial E_z}{\partial r} + \frac{i\eta_0 k_0}{r}\frac{\partial H_z}{\partial \phi} \tag{6.19}$$

where

$$h^2 \equiv n^2 k_0^2 - k_z^2. \tag{6.20}$$

6.2 WAVE EQUATIONS

From (6.9) and (6.12),

$$i\eta_0 k_0 H_z = \frac{1}{r}E_\phi + \frac{\partial E_\phi}{\partial r} - \frac{1}{r}\frac{\partial E_r}{\partial \phi} \tag{6.21}$$

and

$$-i\frac{n^2 k_0}{Z_0}E_z = \frac{1}{r}H_\phi + \frac{\partial H_\phi}{\partial r} - \frac{1}{r}\frac{\partial H_r}{\partial \phi}. \tag{6.22}$$

Substituting for H_ϕ from (6.18), and for H_r from (6.16) into (6.22),

$$\frac{\partial^2 E_z}{\partial r^2} + \frac{1}{r^2}\frac{\partial^2 E_z}{\partial \phi^2} + \frac{1}{r}\frac{\partial E_z}{\partial r} = -h^2 E_z. \tag{6.23}$$

Substituting for E_ϕ from (6.17) and E_r from (6.19) into (6.21),

$$\frac{\partial^2 H_z}{\partial r^2} + \frac{1}{r^2}\frac{\partial^2 H_z}{\partial \phi^2} + \frac{1}{r}\frac{\partial H_z}{\partial r} = -h^2 H_z. \tag{6.24}$$

These partial differential equations can be solved by the separation of variables technique. Let

$$E_z = R(r)Q(\phi)\mathrm{e}^{ik_z z} \tag{6.25}$$

where R is only a function of r, and Q is only a function of ϕ. Substituting (6.25) into (6.23) and dividing through by RQ,

$$\frac{1}{R(r)}\frac{\partial^2 R(r)}{\partial r^2} + \frac{1}{rR(r)}\frac{\partial R(r)}{\partial r} + \frac{1}{Q(\phi)r^2}\frac{\partial^2 Q(\phi)}{\partial \phi^2} + h^2 = 0. \tag{6.26}$$

This equation can be separated into

$$\frac{d^2 Q(\phi)}{\mathrm{d}\phi^2} + m^2 Q(\phi) = 0 \tag{6.27}$$

and

$$\frac{\partial^2 R(r)}{\partial r^2} + \frac{1}{r}\frac{\partial R(r)}{\partial r} + \left(h^2 - \frac{m^2}{r^2}\right)R(r) = 0 \tag{6.28}$$

where m is a constant. The equation is of the same form for H_z. The solution of (6.27) is

$$Q(\phi) = \cos\left[m(\phi + \phi_0)\right] \tag{6.29}$$

where ϕ_0 is a constant. To ensure that Q is single-valued, $2\pi\ m$ must be an integer. Furthermore, the constant angle ϕ_0 can be omitted by simply choosing the orientation of the xy axes to make this constant equal to zero.

The solutions to (6.28) are

$$R(r) = A_m J_m(hr) + B_m Y_m(hr) \tag{6.30}$$

where J_m is a Bessel function of the first kind, and Y_m is a Bessel function of the second kind. At $r = 0$ inside the core, Y_m diverges. Therefore, only J_m can be used as a solution within the core. In the outer medium $J_m(hr)$ and $Y_m(hr)$ both diverge as $r \rightarrow \infty$. The Hankel functions are two independent linear combinations of the Bessel functions and are also solutions to (6.28).

$$H_m^{(1)}(h_{L+1}r) = J_m(h_{L+1}r) + iY_m(h_{L+1}r) \tag{6.31}$$

and

$$H_m^{(2)}(h_{L+1}r) = J_m(h_{L+1}r) - iY_m(h_{L+1}r) \tag{6.32}$$

where $H_m^{(1)}$ is a Hankel function of the first kind, and $H_m^{(2)}$ is a Hankel function of the second kind. The Hankel function of the first kind asymptotically goes to zero as $r \rightarrow \infty$ while the Hankel function of the second kind diverges. Therefore, the solution to (6.28) in the outer medium is a $H_m^{(1)}$ function. For cylindrical shells between the core and outer medium, both Bessel functions J_m and Y_m must be used to completely describe the fields.

6.3 ELECTRIC AND MAGNETIC FIELDS

6.3.1 Core Longitudinal Fields

The E_z and H_z fields inside the core from (6.25), (6.29), and (6.30) are

$$E_z^{(0)} = e^{ik_z z} \sum_{m=1}^{\infty} A_m^{(0)} J_m(h_0 r) \cos(m\phi + \phi_0) \tag{6.33}$$

and

$$H_z^{(0)} = e^{ik_z z} \sum_{m=1}^{\infty} C_m^{(0)} J_m(h_0 r) \sin(m\phi + \phi_0). \tag{6.34}$$

The superscript (0) represents the core region, numbering the different refractive index regions in the waveguide from inside to outside.

6.3.2 Cylindrical Shell Longitudinal Fields

For L concentric cylindrical shells around the core, the $E_z^{(l)}$ and $H_z^{(l)}$ field components for the lth shell are

$$E_z^{(l)} = \mathrm{e}^{ik_z z}\sum\nolimits_{m=1}^{\infty}\left[A_m^{(l)}J_m\left(h_l r\right)+B_m^{(l)}Y_m\left(h_l r\right)\right]\cos\left(m\phi+\phi_0\right) \tag{6.35}$$

and

$$H_z^{(l)} = \mathrm{e}^{ik_z z}\sum\nolimits_{m=1}^{\infty}\left[C_m^{(l)}J_m\left(h_l r\right)+D_m^{(l)}Y_m\left(h_l r\right)\right]\sin\left(m\phi+\phi_0\right). \tag{6.36}$$

6.3.3 Outer Medium Longitudinal Fields

The E_z and H_z fields in the outer medium are

$$E_z^{(L+1)} = \mathrm{e}^{ik_z z}\sum\nolimits_{m=1}^{\infty}A_m^{(L+1)}H_m^{(1)}\left(h_{L+1}r\right)\cos\left(m\phi+\phi_0\right) \tag{6.37}$$

and

$$H_z^{(L+1)} = \mathrm{e}^{ik_z z}\sum\nolimits_{m=1}^{\infty}C_m^{(L+1)}H_m^{(1)}\left(h_{L+1}r\right)\sin\left(m\phi+\phi_0\right). \tag{6.38}$$

6.4 BOUNDARY CONDITIONS

At the inner boundary, E_z must be continuous, so from (6.33) and (6.35),

$$\begin{aligned} E_z\left(z,r_1,\phi\right) &= \mathrm{e}^{ik_z z}\sum_{m=1}^{\infty}A_m^{(0)}J_m\left(h_0 r_1\right)\cos\left(m\phi+\phi_0\right) \\ &= \mathrm{e}^{ik_z z}\sum\nolimits_{m=1}^{\infty}\left[A_m^{(1)}J_m\left(h_1 r_1\right)+B_m^{(1)}Y_m\left(h_1 r_1\right)\right]\cos\left(m\phi+\phi_0\right) \end{aligned} \tag{6.39}$$

where the mode field amplitude in the core for each mode $A_m^{(0)} = 1$. Matching term-by-term,

$$A_m^{(0)}J_m\left(h_0 r_1\right) = A_m^{(1)}J_m\left(h_1 r_1\right)+B_m^{(1)}Y_m\left(h_1 r_1\right). \tag{6.40}$$

Similarly, at boundaries between cylinders, from (6.35),

$$A_m^{(l)}J_m\left(h_l r_{l+1}\right)+B_m^{(l)}Y_m\left(h_l r_{l+1}\right) = A_m^{(l+1)}J_m\left(h_{l+1}r_{l+1}\right)+B_m^{(l+1)}Y_m\left(h_{l+1}r_{l+1}\right). \tag{6.41}$$

Finally, at the outer boundary, from (6.35) and (6.37),

$$A_m^{(L+1)}H_m^{(1)}\left(h_{L+1}r_{L+1}\right) = A_m^{(L)}J_m\left(h_L r_{L+1}\right)+B_m^{(L)}Y_m\left(h_L r_{L+1}\right) \tag{6.42}$$

At each boundary H_z must also be continuous resulting in the boundary conditions,

$$C_m^{(0)}J_m\left(h_0 r_1\right) = C_m^{(1)}J_m\left(h_1 r_1\right)+D_m^{(1)}Y_m\left(h_1 r_1\right), \tag{6.43}$$

$$C_m^{(l)}J_m\left(h_l r_{l+1}\right)+D_m^{(l)}Y_m\left(h_l r_{l+1}\right) = C_m^{(l+1)}J_m\left(h_{l+1}r_{l+1}\right)+D_m^{(l+1)}Y_m\left(h_{l+1}r_{l+1}\right), \tag{6.44}$$

and

$$C_m^{(L+1)} H_m^{(1)}\left(h_{L+1} r_{L+1}\right) = C_m^{(L)} J_m\left(h_L r_{L+1}\right) + D_m^{(L)} Y_m\left(h_L r_{L+1}\right). \tag{6.45}$$

E_ϕ is found from (6.17).

$$E_\phi = \frac{ik_z}{h^2 r}\frac{\partial E_z}{\partial \phi} - \frac{ik_0}{h^2}\eta_0 \frac{\partial H_z}{\partial r}. \tag{6.46}$$

Matching boundary conditions at the core-to-first cylinder interface, cylinder-to-cylinder interfaces, and cylinder-to-outer medium interface,

$$\frac{mk_z}{r_1} J_m\left(h_0 r_1\right) A_m^{(0)} + k_0 h_0 \eta_0 J_m'\left(h_0 r_1\right) C_m^{(0)}$$

$$= \frac{mk_z}{r_1}\left[A_m^{(1)} J_m\left(h_1 r_1\right) + B_m^{(1)} Y_m\left(h_1 r_1\right)\right] + k_0 h_1 \eta_0 \left[C_m^{(1)} J_m'\left(h_1 r_1\right) + D_m^{(1)} Y_m'\left(h_1 r_1\right)\right], \tag{6.47}$$

$$\frac{mk_z}{r_{l+1}}\left[A_m^{(l)} J_m\left(h_l r_{l+1}\right) + B_m^{(l)} Y_m\left(h_l r_{l+1}\right)\right] + k_0 h_{l+1} \eta_0 \left[C_m^{(l)} J_m'\left(h_l r_{l+1}\right) + D_m^{(l)} Y_m'\left(h_l r_{l+1}\right)\right]$$

$$= \frac{mk_z}{r_{l+1}}\left[A_m^{(l+1)} J_m\left(h_{l+1} r_{l+1}\right) + B_m^{(l+1)} Y_m\left(h_{l+1} r_{l+1}\right)\right]$$

$$+ k_0 h_1 \eta_0 \left[C_m^{(l+1)} J_m'\left(h_{l+1} r_{l+1}\right) + D_m^{(l+1)} Y_m'\left(h_{l+1} r_{l+1}\right)\right], \tag{6.48}$$

and

$$\frac{mk_z}{r_{L+1}}\left[A_m^{(L)} J_m\left(h_L r_{L+1}\right) + B_m^{(L)} Y_m\left(h_L r_{L+1}\right)\right] + k_0 h_{L+1} \eta_0 \left[C_m^{(L)} J_m'\left(h_L r_{L+1}\right) + D_m^{(L)} Y_m'\left(h_L r_{L+1}\right)\right]$$

$$= \frac{mk_z}{r_{L+1}} A_m^{(L+1)} H_m^{(1)}\left(h_{L+1} r_{L+1}\right) + k_0 h_{L+1} \eta_0 C_m^{(L+1)} H_m'^{(1)}\left(h_{L+1} r_{L+1}\right). \tag{6.49}$$

Finally, H_ϕ is found from (6.18),

$$H_\phi = \frac{in^2 k_0}{h^2 \eta_0}\frac{\partial E_z}{\partial r} + \frac{ik_z}{h^2 r}\frac{\partial H_z}{\partial \phi}. \tag{6.50}$$

Continuity of H_ϕ at the interfaces between the core, shells, and outer medium requires

$$\frac{n_0^2 k_0}{h_0 \eta_0} J_m'\left(h_0 r_1\right) A_m^{(0)} + \frac{mk_z}{h_0^2 r_1} J_m\left(h_1 r_1\right) C_m^{(0)}$$

$$= \frac{n_1^2 k_0}{h_1 \eta_0}\left[A_m^{(1)} J_m'\left(h_1 r_1\right) + B_m^{(1)} Y_m'\left(h_1 r_1\right)\right] + \frac{mk_z}{h_1^2 r_1}\left[C_m^{(1)} J_m\left(h_1 r_1\right) + D_m^{(1)} Y_m\left(h_1 r_1\right)\right], \tag{6.51}$$

$$\frac{n_l^2 k_0}{h_l \eta_0}\left[A_m^{(l)} J_m'\left(h_l r_{l+1}\right)+B_m^{(l)} Y_m'\left(h_l r_{l+1}\right)\right]+\frac{m k_z}{h_l^2 r_{l+1}}\left[C_m^{(l)} J_m\left(h_l r\right)+D_m^{(l)} Y_m\left(h_l r\right)\right]$$

$$=\frac{n_{l+1}^2 k_0}{h_{l+1}\eta_0}\left[A_m^{(l+1)} J_m'\left(h_{l+1} r_{l+1}\right)+B_m^{(l+1)} Y_m'\left(h_{l+1} r_{l+1}\right)\right]$$

$$+\frac{m k_z}{h_{l+1}^2 r_{l+1}}\left[C_m^{(l+1)} J_m\left(h_{l+1} r_{l+1}\right)+D_m^{(l+1)} Y_m\left(h_{l+1} r_{l+1}\right)\right], \tag{6.52}$$

and

$$\frac{n_L^2 k_0}{h_L \eta_0}\left[A_m^{(L)} J_m'\left(h_L r_{L+1}\right)+B_m^{(L)} Y_m'\left(h_L r_{L+1}\right)\right]+\frac{m k_z}{h_L^2 r_{L+1}}\left[C_m^{(L)} J_m\left(h_L r_{L+1}\right)+D_m^{(L)} Y_m\left(h_L r_{L+1}\right)\right]$$

$$=\frac{n_L^2 k_0}{h_{L+1}\eta_0} A_m^{(L+1)} H_m'^{(1)}\left(h_{L+1} r_{L+1}\right)+\frac{m k_z}{h_{L+1}^2 r_{L+1}} C_m^{(L+1)} H_m^{(1)}\left(h_{L+1} r_{L+1}\right). \tag{6.53}$$

6.5 MATRIX EQUATIONS

These boundary conditions can be written as matrix equations.

$$\begin{bmatrix} J_m\left(h_0 r_1\right) & 0 \\ 0 & J_m\left(h_0 r_1\right) \end{bmatrix}\begin{bmatrix} A_m^{(0)} \\ C_m^{(0)} \end{bmatrix}=\begin{bmatrix} J_m\left(h_1 r_1\right) & Y_m\left(h_1 r_1\right) & 0 & 0 \\ 0 & 0 & J_m\left(h_1 r_1\right) & Y_m\left(h_1 r_1\right) \end{bmatrix}\begin{bmatrix} A_m^{(1)} \\ B_m^{(1)} \\ C_m^{(1)} \\ D_m^{(1)} \end{bmatrix}. \tag{6.54}$$

$$\begin{bmatrix} \frac{m k_z}{h_0^2 r_1} J_m\left(h_0 r_1\right) & \frac{k_0 \eta_0}{h_0} J_m'\left(h_0 r_1\right) \\ \frac{n_0^2 k_0}{h_0 \eta_0} J_m'\left(h_0 r_1\right) & \frac{m k_z}{h_0^2 r_1} J_m\left(h_0 r_1\right) \end{bmatrix}\begin{bmatrix} A_m^{(0)} \\ C_m^{(0)} \end{bmatrix}$$

$$=\begin{bmatrix} \frac{m k_z}{h_1^2 r_1} J_m\left(h_1 r_1\right) & \frac{m k_z}{h_1^2 r_1} Y_m\left(h_1 r_1\right) & \frac{k_0 \eta_0}{h_1} J_m'\left(h_1 r_1\right) & \frac{k_0 \eta_0}{h_1} Y_m'\left(h_1 r_1\right) \\ \frac{n_1^2 k_0}{h_1 \eta_0} J_m'\left(h_1 r_1\right) & \frac{n_1^2 k_0}{h_1 \eta_0} Y_m'\left(h_1 r_1\right) & \frac{m k_z}{h_1^2 r_1} J_m\left(h_1 r_1\right) & \frac{m k_z}{h_1^2 r_1} Y_m\left(h_1 r_1\right) \end{bmatrix}\begin{bmatrix} A_m^{(1)} \\ B_m^{(1)} \\ C_m^{(1)} \\ D_m^{(1)} \end{bmatrix}. \tag{6.55}$$

$$\begin{bmatrix} H_m^{(1)}\left(h_{L+1} r_{L+1}\right) & 0 \\ 0 & H_m^{(1)}\left(h_{L+1} r_{L+1}\right) \end{bmatrix}\begin{bmatrix} A_m^{(L+1)} \\ C_m^{(L+1)} \end{bmatrix}$$

$$=\begin{bmatrix} J_m\left(h_L r_{L+1}\right) & Y_m\left(h_L r_{L+1}\right) & 0 & 0 \\ 0 & 0 & J_m\left(h_L r_{L+1}\right) & Y_m\left(h_L r_{L+1}\right) \end{bmatrix}\begin{bmatrix} A_m^{(L)} \\ B_m^{(L)} \\ C_m^{(L)} \\ D_m^{(L)} \end{bmatrix}. \tag{6.56}$$

$$\begin{bmatrix} \dfrac{mk_z}{r_{L+1}} H_m^{(1)}\left(h_{L+1} r_{L+1}\right) & k_0 h_{L+1} \eta_0 H_m'^{(1)}\left(h_{L+1} r_{L+1}\right) \\ \dfrac{n_L^2 k_0}{h_{L+1}\eta_0} H_m'^{(1)}\left(h_{L+1} r_{L+1}\right) & \dfrac{mk_z}{h_{L+1}^2 r_{L+1}} H_m^{(1)}\left(h_{L+1} r_{L+1}\right) \end{bmatrix} \begin{bmatrix} A_m^{(L+1)} \\ C_m^{(L+1)} \end{bmatrix} =$$

$$\begin{bmatrix} \dfrac{mk_z}{r_{L+1}} J_m\left(h_L r_{L+1}\right) & \dfrac{mk_z}{r_{L+1}} Y_m\left(h_L r_{L+1}\right) & k_0 h_{L+1} \eta_0 J_m'\left(h_L r_{L+1}\right) & k_0 h_{L+1} \eta_0 Y_m'\left(h_L r_{L+1}\right) \\ \dfrac{n_L^2 k_0}{h_L \eta_0} J_m'\left(h_L r_{L+1}\right) & \dfrac{n_L^2 k_0}{h_L \eta_0} Y_m'\left(h_L r_{L+1}\right) & \dfrac{mk_z}{h_L^2 r_{L+1}} J_m\left(h_L r_{L+1}\right) & \dfrac{mk_z}{h_L^2 r_{L+1}} Y_n\left(h_L r_{L+1}\right) \end{bmatrix} \begin{bmatrix} A_m^{(L)} \\ B_m^{(L)} \\ C_m^{(L)} \\ D_m^{(L)} \end{bmatrix}. \tag{6.57}$$

$$\begin{bmatrix} J_m\left(h_l r_{l+1}\right) & Y_m\left(h_l r_{l+1}\right) & 0 & 0 \\ 0 & 0 & J_m\left(h_l r_{l+1}\right) & Y_m\left(h_l r_{l+1}\right) \\ \dfrac{mk_z}{r_{l+1}} J_m\left(h_l r_{l+1}\right) & \dfrac{mk_z}{r_{l+1}} Y_m\left(h_l r_{l+1}\right) & k_0 h_{l+1} \eta_0 J_m'\left(h_l r_{l+1}\right) & k_0 h_{l+1} \eta_0 Y_m'\left(h_l r_{l+1}\right) \\ \dfrac{n_l^2 k_0}{h_l \eta_0} J_m'\left(h_l r_{l+1}\right) & \dfrac{n_l^2 k_0}{h_l \eta_0} Y_m'\left(h_l r_{l+1}\right) & \dfrac{mk_z}{h_l^2 r_{l+1}} J_m\left(h_l r\right) & \dfrac{mk_z}{h_l^2 r_{l+1}} Y_m\left(h_l r\right) \end{bmatrix} \begin{bmatrix} A_m^{(l)} \\ B_m^{(l)} \\ C_m^{(l)} \\ D_m^{(l)} \end{bmatrix} =$$

$$\begin{bmatrix} J_m\left(h_{l+1} r_{l+1}\right) & Y_m\left(h_{l+1} r_{l+1}\right) & 0 & 0 \\ 0 & 0 & J_m\left(h_{l+1} r_{l+1}\right) & Y_m\left(h_{l+1} r_{l+1}\right) \\ \dfrac{mk_z}{r_{l+1}} J_m\left(h_{l+1} r_{l+1}\right) & \dfrac{mk_z}{r_{l+1}} Y_m\left(h_{l+1} r_{l+1}\right) & k_0 h_l \eta_0 J_m'\left(h_{l+1} r_{l+1}\right) & k_0 h_l \eta_0 Y_m'\left(h_{l+1} r_{l+1}\right) \\ \dfrac{n_{l+1}^2 k_0}{h_{l+1} \eta_0} J_m'\left(h_{l+1} r_{l+1}\right) & \dfrac{n_{l+1}^2 k_0}{h_{l+1} \eta_0} Y_m'\left(h_{l+1} r_{l+1}\right) & \dfrac{mk_z}{h_{l+1}^2 r_{l+1}} J_m\left(h_{l+1} r_{l+1}\right) & \dfrac{mk_z}{h_{l+1}^2 r_{l+1}} Y_m\left(h_{l+1} r_{l+1}\right) \end{bmatrix} \begin{bmatrix} A_m^{(l+1)} \\ B_m^{(l+1)} \\ C_m^{(l+1)} \\ D_m^{(l+1)} \end{bmatrix} \tag{6.58}$$

Defining new matrices to simplify the notation,

$$\tilde{\boldsymbol{M}}_0 \begin{bmatrix} A_m^{(0)} \\ C_m^{(0)} \end{bmatrix} = \tilde{\boldsymbol{Q}}_1 \begin{bmatrix} A_m^{(1)} \\ B_m^{(1)} \\ C_m^{(1)} \\ D_m^{(1)} \end{bmatrix}, \tag{6.59}$$

$$\tilde{\boldsymbol{Q}}_{l+1} \begin{bmatrix} A_m^{(l+1)} \\ B_m^{(l+1)} \\ C_m^{(l+1)} \\ D_m^{(l+1)} \end{bmatrix} = \tilde{\boldsymbol{P}}_l \begin{bmatrix} A_m^{(l)} \\ B_m^{(l)} \\ C_m^{(l)} \\ D_m^{(l)} \end{bmatrix}, \tag{6.60}$$

and

$$\tilde{\boldsymbol{M}}_{L+1} \begin{bmatrix} A_m^{(L+1)} \\ C_m^{(L+1)} \end{bmatrix} = \tilde{\boldsymbol{P}}_L \begin{bmatrix} A_m^{(L)} \\ B_m^{(L)} \\ C_m^{(L)} \\ D_m^{(L)} \end{bmatrix}, \tag{6.61}$$

where

$$\tilde{\boldsymbol{M}}_0 \equiv \begin{bmatrix} J_m(h_0 r_1) & 0 \\ 0 & J_m(h_0 r_1) \\ \dfrac{mk_z}{h_0^2 r_1} J_m(h_0 r_1) & \dfrac{k_0 \eta_0}{h_0} J'_m(h_0 r_1) \\ \dfrac{n_0^2 k_0}{h_0 \eta_0} J'_m(h_0 r_1) & \dfrac{mk_z}{h_0^2 r_1} J_m(h_0 r_1) \end{bmatrix}, \tag{6.62}$$

$$\tilde{\boldsymbol{Q}}_l \equiv \begin{bmatrix} J_m(h_l r_l) & Y_m(h_l r_l) & 0 & 0 \\ 0 & 0 & J_m(h_l r_l) & Y_m(h_l r_l) \\ \dfrac{mk_z}{h_l^2 r_l} J_m(h_l r_l) & \dfrac{mk_z}{h_l^2 r_l} Y_m(h_l r_l) & \dfrac{k_0 \eta_0}{h_l} J'_m(h_l r_l) & \dfrac{k_0 \eta_0}{h_l} Y'_m(h_l r_l) \\ \dfrac{n_l^2 k_0}{h_l \eta_0} J'_m(h_l r_l) & \dfrac{n_l^2 k_0}{h_l \eta_0} Y'_m(h_l r_l) & \dfrac{mk_z}{h_l^2 r_l} J_m(h_l r_l) & \dfrac{mk_z}{h_l^2 r_l} Y_m(h_l r_l) \end{bmatrix}, \tag{6.63}$$

$$\tilde{\boldsymbol{P}}_l \equiv \begin{bmatrix} J_m(h_l r_{l+1}) & Y_m(h_l r_{l+1}) & 0 & 0 \\ 0 & 0 & J_m(h_l r_{l+1}) & Y_m(h_l r_{l+1}) \\ \dfrac{mk_z}{h_l^2 r_{l+1}} J_m(h_l r_{l+1}) & \dfrac{mk_z}{h_l^2 r_{l+1}} Y_m(h_l r_{l+1}) & \dfrac{k_0 \eta_0}{h_l} J'_m(h_l r_{l+1}) & \dfrac{k_0 \eta_0}{h_l} Y'_m(h_l r_{l+1}) \\ \dfrac{n_l^2 k_0}{h_l \eta_0} J'_m(h_l r_{l+1}) & \dfrac{n_l^2 k_0}{h_l \eta_0} Y'_m(h_l r_{l+1}) & \dfrac{mk_z}{h_l^2 r_{l+1}} J_m(h_l r_{l+1}) & \dfrac{mk_z}{h_l^2 r_{l+1}} Y_m(h_l r_{l+1}) \end{bmatrix}, \tag{6.64}$$

and

$$\tilde{\boldsymbol{M}}_{L+1} \equiv \begin{bmatrix} H_m^{(1)}(h_{L+1} r_{L+1}) & 0 \\ 0 & H_m^{(1)}(h_{L+1} r_{L+1}) \\ \dfrac{mk_z}{h_{L+1}^2 r_{L+1}} H_m^{(1)}(h_{L+1} r_{L+1}) & \dfrac{k_0 \eta_0}{h_{L+1}} H_m'^{(1)}(h_{L+1} r_{L+1}) \\ \dfrac{n_{L+1}^2 k_0}{h_{L+1} \eta_0} H_m'^{(1)}(h_{L+1} r_{L+1}) & \dfrac{mk_z}{h_{L+1}^2 r_{L+1}} H_m^{(1)}(h_{L+1} r_{L+1}) \end{bmatrix}. \tag{6.65}$$

For L shells,

$$\tilde{\boldsymbol{M}}_0 \begin{bmatrix} A_m^{(0)} \\ C_m^{(0)} \end{bmatrix} = \tilde{\boldsymbol{Q}}_1 \left(\tilde{\boldsymbol{P}}_1\right)^{-1} \tilde{\boldsymbol{Q}}_2 \left(\tilde{\boldsymbol{P}}_2\right)^{-1} \tilde{\boldsymbol{Q}}_3 \cdots \tilde{\boldsymbol{Q}}_L \left(\tilde{\boldsymbol{P}}_L\right)^{-1} \tilde{\boldsymbol{M}}_{L+1} \begin{bmatrix} A_m^{(L+1)} \\ C_m^{(L+1)} \end{bmatrix} \equiv \tilde{\boldsymbol{R}}_L \begin{bmatrix} A_m^{(L+1)} \\ C_m^{(L+1)} \end{bmatrix}, \tag{6.66}$$

$$\tilde{\boldsymbol{M}}_0 \begin{bmatrix} A_m^{(0)} \\ C_m^{(0)} \end{bmatrix} - \tilde{\boldsymbol{R}}_L \begin{bmatrix} A_m^{(L+1)} \\ C_m^{(L+1)} \end{bmatrix} = 0, \tag{6.67}$$

and

$$\tilde{S}_L \begin{bmatrix} A_m^{(0)} \\ C_m^{(0)} \\ A_m^{(L+1)} \\ C_m^{(L+1)} \end{bmatrix} \equiv \begin{bmatrix} M_{11} & M_{12} & -R_{11} & -R_{12} \\ M_{21} & M_{22} & -R_{21} & -R_{22} \\ M_{31} & M_{32} & -R_{31} & -R_{32} \\ M_{41} & M_{42} & -R_{41} & -R_{42} \end{bmatrix} \begin{bmatrix} A_m^{(0)} \\ C_m^{(0)} \\ A_m^{(L+1)} \\ C_m^{(L+1)} \end{bmatrix} = 0. \tag{6.68}$$

This equation is valid even for $L = 0$ (no cylinder shells). For a solution to these four simultaneous equations, k_z must be chosen to make the determinant $\left|\tilde{S}_L\right| = 0$.

6.6 SPECIAL CASE: NO SHELLS

If there are no shells, the longitudinal field components in the core and outer medium are

$$E_z^{(0)}(r,\phi) = \mathrm{e}^{ik_z z} \sum\nolimits_{m=1}^{\infty} A_m^{(0)} J_m(h_0 r)\cos(m\phi + \phi_0), \tag{6.69}$$

$$H_z^{(0)}(r,\phi) = \mathrm{e}^{ik_z z} \sum\nolimits_{m=1}^{\infty} C_m^{(0)} J_m(h_0 r)\sin(m\phi + \phi_0), \tag{6.70}$$

$$E_z^{(1)}(r,\phi) = \mathrm{e}^{ik_z z} \sum\nolimits_{m=1}^{\infty} A_m^{(1)} H_m^{(1)}(h_1 r)\cos(m\phi + \phi_0), \tag{6.71}$$

and

$$H_z^{(1)}(r,\phi) = \mathrm{e}^{ik_z z} \sum\nolimits_{m=1}^{\infty} C_m^{(1)} H_m^{(1)}(h_1 r)\sin(m\phi + \phi_0), \tag{6.72}$$

and the tangential field components are

$$E_\phi^{(0)}(r,\phi) = -\mathrm{e}^{ik_z z} \sum\nolimits_{m=1}^{\infty} \left[\frac{imk_z}{h_0^2 r} A_m^{(0)} J_m(h_0 r) + \frac{ik_0\eta_0}{h_0} C_m^{(0)} J_m'(h_0 r)\right] \sin(m\phi + \phi_0), \tag{6.73}$$

$$H_\phi^{(0)}(r,\phi) = \mathrm{e}^{ik_z z} \sum\nolimits_{m=1}^{\infty} \left[\frac{in_0^2 k_0}{h_0\eta_0} A_m^{(0)} J_m'(h_0 r) + \frac{ik_z m}{h_0^2 r} C_m^{(0)} J_m(h_0 r)\right] \cos(m\phi + \phi_0), \tag{6.74}$$

$$E_\phi^{(1)}(r,\phi) = -\mathrm{e}^{ik_z z} \sum\nolimits_{m=1}^{\infty} \left[\frac{imk_z}{h_1^2 r} A_m^{(1)} H_m^{(1)}(h_1 r) + \eta_0 \frac{ik_0}{h_1} C_m^{(1)} H_m'^{(1)}(h_1 r)\right] \sin(m\phi + \phi_0), \tag{6.75}$$

and

$$H_\phi^{(1)}(r,\phi) = \mathrm{e}^{ik_z z} \sum\nolimits_{m=1}^{\infty} \left[\frac{in_1^2 k_0}{h_1\eta_0} A_m^{(1)} H_m'^{(1)}(h_1 r) + \frac{imk_z}{h_1^2 r} C_m^{(1)} H_m^{(1)}(h_1 r)\right] \cos(m\phi + \phi_0). \tag{6.76}$$

Matching E_z and H_z term-by-term at the boundary,

$$A_m^{(1)} = \frac{J_m(h_0 r)}{H_m^{(1)}(h_1 r)} A_m^{(0)}, \tag{6.77}$$

and

$$C_m^{(1)} = \frac{J_m(h_0 r_1)}{H_m^{(1)}(h_1 r_1)} C_m^{(0)}. \tag{6.78}$$

Matching E_ϕ and H_ϕ at boundary,

$$\frac{mk_z}{h_0^2 r_1} A_m^{(0)} J_m(h_0 r_1) + \frac{k_0 \eta_0}{h_0} C_m^{(0)} J'_m(h_0 r_1) = \frac{mk_z}{h_1^2 r_1} A_m^{(1)} H_m^{(1)}(h_1 r_1) + \frac{k_0 \eta_0}{h_1} C_m^{(1)} H_m'^{(1)}(h_1 r_1), \tag{6.79}$$

and

$$\frac{n_0^2 k_0}{h_0 \eta_0} A_m^{(0)} J'_m(h_0 r_1) + \frac{k_z m}{h_0^2 r_1} C_m^{(0)} J_m(h_0 r_1) = \frac{n_1^2 k_0}{h_1 \eta_0} A_m^{(1)} H_m'^{(1)}(h_1 r_1) + \frac{mk_z}{h_1^2 r_1} C_m^{(1)} H_m^{(1)}(h_1 r_1). \tag{6.80}$$

Substitute for $A_m^{(1)}$ and $C_m^{(1)}$ using (6.77) and (6.78).

$$\begin{aligned} &\frac{mk_z}{r_1} H_m^{(1)}(h_1 r_1)\left[\frac{1}{h_0^2} J_m(h_0 r) - \frac{1}{h_1^2} J_m(h_0 r)\right] A_m^{(0)} \\ &= k_0 \eta_0 \left[\frac{1}{h_1} J_m(h_0 r_1) H_m'^{(1)}(h_1 r) - \frac{1}{h_0} J'_m(h_0 r) H_m^{(1)}(h_1 r_1)\right] C_m^{(0)}. \end{aligned} \tag{6.81}$$

$$\begin{aligned} &\frac{k_0}{\eta_0}\left[\frac{n_0^2}{h_0} J'_m(h_0 r) H_m^{(1)}(h_1 r) - \frac{n_1^2}{h_1} H_m'^{(1)}(h_1 r) J_m(h_0 r)\right] A_m^{(0)} \\ &= \frac{mk_z}{r_1} H_m^{(1)}(h_1 r)\left[\frac{1}{h_1^2} J_m(h_0 r_1) - \frac{1}{h_0^2} J_m(h_0 r)\right] C_m^{(0)}. \end{aligned} \tag{6.82}$$

Solving (6.81) and (6.82) for $C_m^{(0)}$ and equating,

$$\left[\frac{n_1^2}{h_1}\frac{H_m'^{(1)}(h_1 r_1)}{H_m^{(1)}(h_1 r_1)} - \frac{n_0^2}{h_0}\frac{J'_m(h_0 r_1)}{J_m(h_0 r_1)}\right]\left[\frac{1}{h_0}\frac{J'_m(h_0 r_1)}{J_m(h_0 r_1)} - \frac{1}{h_1}\frac{H_m'^{(1)}(h_1 r_1)}{H_m^{(1)}(h_1 r_1)}\right] + \left(\frac{mk_z}{k_0 r_1}\right)^2 \left(\frac{1}{h_1^2} - \frac{1}{h_0^2}\right)^2 = 0. \tag{6.83}$$

The derivatives of the Bessel and Hankel functions can be calculated from the identities,

$$J'_m(x) = \frac{1}{2}\left[J_{m-1}(x) - J_{m+1}(x)\right] \tag{6.84}$$

and

$$H'_m(x) = \frac{1}{2}\left[H_{m-1}(x) - H_{m+1}(x)\right]. \tag{6.85}$$

For $m = 0$,

$$J'_m(x) = -J_1(x) \tag{6.86}$$

and

$$H'_m(x) = -H_1(x). \tag{6.87}$$

(6.83) reduces to

$$\left(\frac{n_0^2}{h_0}\mathfrak{I}_m - \frac{n_1^2}{h_1}\mathcal{H}_m\right)\left(\frac{1}{h_0}\mathfrak{I}_m - \frac{1}{h_1}\mathcal{H}_m\right) = \left(\frac{2mk_z}{k_0 r_1}\right)^2\left(\frac{1}{h_1^2} - \frac{1}{h_0^2}\right)^2 \tag{6.88}$$

where

$$\mathfrak{I}_m \equiv \frac{J_{m-1}(h_0 r_1) - J_{m+1}(h_0 r_1)}{J_m(h_0 r)} \tag{6.89}$$

and

$$\mathcal{H}_m \equiv \frac{H_{m-1}^{(1)}(h_1 r_1) - H_{m+1}^{(1)}(h_1 r_1)}{H_m^{(1)}(h_1 r)}. \tag{6.90}$$

There can be multiple values of k_z that satisfy (6.88). If each solution for a specific value of m is designated by an index value v, then each waveguide mode is specified by two indices, m and v. m is the number of zeroes around the 360° azimuth, and v is the number of zeroes in the radial direction.

If $m = 0$ then from (6.83) it must be true that either

$$\frac{n_1^2}{h_1}\frac{H_m^{\prime(1)}(h_1 r_1)}{H_m^{(1)}(h_1 r_1)} - \frac{n_0^2}{h_0}\frac{J_m'(h_0 r_1)}{J_m(h_0 r_1)} = 0 \tag{6.91}$$

or

$$\frac{1}{h_0}\frac{J_m'(h_0 r_1)}{J_m(h_0 r_1)} - \frac{1}{h_1}\frac{H_m^{\prime(1)}(h_1 r_1)}{H_m^{(1)}(h_1 r_1)} = 0. \tag{6.92}$$

If (6.91) is true, then from (6.94), $C_m^{(0)} = 0$, and $A_m^{(0)}$ is arbitrary and can be set to 1. The values of k_z that satisfy (6.91) are TM modes because by (6.34) the longitudinal component of the H field, H_z, is zero. If (6.92) is true then $A_m^{(0)} = 0$ and $C_m^{(0)}$ is arbitrary and can be set to 1. The values of k_z that satisfy (6.92) are TE modes because by (6.33) the longitudinal component of the E field, E_z, is zero.

If $m \neq 0$, then the values for k_z must be found that satisfy (6.83). $A_m^{(0)}$ can be arbitrarily set equal to 1 and then $C_m^{(0)}$ is determined from (6.81) or (6.82).

$$C_m^{(0)} = \frac{mk_z\left(\frac{1}{h_0^2} - \frac{1}{h_1^2}\right)}{k_0\eta_0 r_1\left(\frac{1}{h_1}\frac{H_m^{\prime(1)}(h_1 r)}{H_m^{(1)}(h_1 r_1)} - \frac{1}{h_0}\frac{J_m'(h_0 r)}{J_m(h_0 r)}\right)} A_m^{(0)} = \frac{k_0 r_1\left(\frac{n_0^2}{h_0}\frac{J_m'(h_0 r)}{J_m(h_0 r)} - \frac{n_1^2}{h_1}\frac{H_m^{\prime(1)}(h_1 r)}{H_m^{(1)}(h_1 r)}\right)}{m\eta_0 k_z\left(\frac{1}{h_1^2} - \frac{1}{h_0^2}\right)} A_m^{(0)}. \tag{6.93}$$

An important point to note is that for $m \geq 1$, there are both E_z and H_z components for each mode. There are no longer strictly transverse TE and TM modes. These new modes are hybrid modes and are denoted as EH or HE modes. Following the proposal of [1], the modes are labeled HE modes if the real part of the ratio $C_m^{(0)} / A_m^{(0)}$ is greater than zero, and EH modes otherwise.

6.7 SPECIAL CASE: ONE SHELL

When there is a single cylindrical shell between the core and outer medium, the longitudinal fields are

$$E_{m,z}^{(0)} = A_m^{(0)} J_m(h_0 r)\cos(m\phi + \phi_0), \tag{6.94}$$

$$E_{m,z}^{(1)} = \left[A_m^{(1)} J_m(h_1 r) + B_m^{(1)} Y_m(h_1 r)\right]\cos(m\phi + \phi_0), \tag{6.95}$$

$$E_{m,z}^{(2)} = A_m^{(2)} H_m^{(1)}(h_2 r)\cos(m\phi + \phi_0), \tag{6.96}$$

$$H_{m,z}^{(0)} = C_m^{(0)} J_m(h_0 r)\sin(m\phi + \phi_0), \tag{6.97}$$

$$H_{m,z}^{(1)} = \left[C_m^{(1)} J_m(h_1 r) + D_m^{(1)} Y_m(h_1 r)\right]\sin(m\phi + \phi_0), \tag{6.98}$$

and

$$H_{m,z}^{(2)} = C_m^{(2)} H_m^{(1)}(h_2 r)\sin(m\phi + \phi_0). \tag{6.99}$$

Matching boundary conditions gives

$$A_m^{(0)} J_m(h_0 r_1) = A_m^{(1)} J_m(h_1 r_1) + B_m^{(1)} Y_m(h_1 r_1), \tag{6.100}$$

$$A_m^{(1)} J_m(h_1 r_2) + B_m^{(1)} Y_m(h_1 r_2) = A_m^{(2)} H_m^{(1)}(h_2 r_2), \tag{6.101}$$

$$C_m^{(0)} J_m(h_0 r_1) = C_m^{(1)} J_m(h_1 r_1) + D_m^{(1)} Y_m(h_1 r_1), \tag{6.102}$$

and

$$C_m^{(1)} J_m(h_1 r_2) + D_m^{(1)} Y_m(h_1 r_2) = C_m^{(2)} H_m^{(1)}(h_2 r_2). \tag{6.103}$$

Solving for $A_m^{(1)}$, $B_m^{(1)}$, $C_m^{(1)}$, and $D_m^{(1)}$,

$$A_m^{(1)} = \frac{J_m(h_0 r_1) Y_m(h_1 r_2) A_m^{(0)} - Y_m(h_1 r_1) H_m^{(1)}(h_2 r_2) A_m^{(2)}}{J_m(h_1 r_1) Y_m(h_1 r_2) - J_m(h_1 r_2) Y_m(h_1 r_1)} = \alpha_1 A_m^{(0)} + \beta_1 A_m^{(2)}, \tag{6.104}$$

$$B_m^{(1)} = -\frac{J_m(h_0 r_1) J_m(h_1 r_2) A_m^{(0)} - J_m(h_1 r_1) H_m^{(1)}(h_2 r_2) A_m^{(2)}}{J_m(h_1 r_1) Y_m(h_1 r_2)] - J_m(h_1 r_2) Y_m(h_1 r_1)} = \alpha_2 A_m^{(0)} + \beta_2 A_m^{(2)}, \tag{6.105}$$

$$C_m^{(1)} = \frac{J_m(h_0 r_1) Y_m(h_1 r_2) C_m^{(0)} - Y_m(h_1 r_1) H_m^{(1)}(h_2 r_2) C_m^{(2)}}{J_m(h_1 r_1) Y_m(h_1 r_2) - J_m(h_1 r_2) Y_m(h_1 r_1)} = \alpha_3 C_m^{(0)} + \beta_3 C_m^{(2)}, \tag{6.106}$$

and

$$D_m^{(1)} = -\frac{J_m(h_0 r_1) J_m(h_1 r_2) C_m^{(0)} - J_m(h_1 r_1) H_m^{(1)}(h_2 r_2) C_m^{(2)}}{J_m(h_1 r_1) Y_m(h_1 r_2) - J_m(h_1 r_2) Y_m(h_1 r_1)} = \alpha_4 C_m^{(0)} + \beta_4 C_m^{(2)} \tag{6.107}$$

where

$$\alpha_1 = \frac{J_m(h_0 r_1) Y_m(h_1 r_2)}{J_m(h_1 r_1) Y_m(h_1 r_2) - J_m(h_1 r_2) Y_m(h_1 r_1)} \tag{6.108}$$

$$\alpha_2 = -\frac{J_m(h_0 r_1) J_m(h_1 r_2)}{J_m(h_1 r_1) Y_m(h_1 r_2) - J_m(h_1 r_2) Y_m(h_1 r_1)}, \tag{6.109}$$

$$\alpha_3 = \frac{J_m(h_0 r_1) Y_m(h_1 r_2)}{J_m(h_1 r_1) Y_m(h_1 r_2) - J_m(h_1 r_2) Y_m(h_1 r_1)} \tag{6.110}$$

$$\alpha_4 = -\frac{J_m(h_0 r_1) J_m(h_1 r_2)}{J_m(h_1 r_1) Y_m(h_1 r_2) - J_m(h_1 r_2) Y_m(h_1 r_1)}, \tag{6.111}$$

$$\beta_1 = -\frac{Y_m(h_1 r_1) H_m^{(1)}(h_2 r_2)}{J_m(h_1 r_1) Y_m(h_1 r_2) - J_m(h_1 r_2) Y_m(h_1 r_1)}, \tag{6.112}$$

$$\beta_2 = \frac{J_m(h_1 r_1) H_m^{(1)}(h_2 r_2)}{J_m(h_1 r_1) Y_m(h_1 r_2) - J_m(h_1 r_2) Y_m(h_1 r_1)} \tag{6.113}$$

$$\beta_3 = -\frac{Y_m(h_1 r_1) H_m^{(1)}(h_2 r_2)}{J_m(h_1 r_1) Y_m(h_1 r_2) - J_m(h_1 r_2) Y_m(h_1 r_1)}, \tag{6.114}$$

and

$$\beta_4 = \frac{J_m(h_1 r_1) H_m^{(1)}(h_2 r_2)}{J_m(h_1 r_1) Y_m(h_1 r_2) - J_m(h_1 r_2) Y_m(h_1 r_1)}. \tag{6.115}$$

Similarly, the azimuthal fields are given by

$$E_{m,\phi}^{(0)} = -\frac{imk_z}{h_0^2 r} A_m^{(0)} J_m(h_0 r)\sin(m\phi+\phi_0) - \frac{ik_0\eta_0}{h_0} C_m^{(0)} J_m'(h_0 r)\sin(m\phi+\phi_0), \tag{6.116}$$

$$E_{m,\phi}^{(1)} = -\frac{imk_z}{h_1^2 r}\left[A_m^{(1)} J_m(h_1 r) + B_m^{(1)} Y_m(h_1 r)\right]\sin(m\phi-\phi_0)$$

$$-\frac{ik_0\eta_0}{h_1}\left[C_m^{(1)} J_m'(h_1 r) + D_m^{(1)} Y_m'(h_1 r)\right]\sin(m\phi+\phi_0), \tag{6.117}$$

$$E_{m,\phi}^{(2)} = -\frac{imk_z}{h_2^2 r} A_m^{(2)} H_m^{(1)}(h_2 r)\sin(m\phi+\phi_0) - \frac{ik_0\eta_0}{h_2} C_m^{(2)} H_m'^{(1)}(h_2 r)\sin(m\phi+\phi_0), \tag{6.118}$$

$$H_{m,\phi}^{(0)} = \frac{in_0^2 k_0}{h_0\eta_0} A_m^{(0)} J_m'(h_0 r)\cos(m\phi+\phi_0) + \frac{ik_z}{h_0^2 r} m C_m^{(0)} J_m(h_0 r)\cos(m\phi+\phi_0), \tag{6.119}$$

$$\begin{aligned} H_{m,\phi}^{(1)} &= \frac{in_1^2 k_0}{h_1\eta_0}\left[A_m^{(1)} J_m'(h_1 r) + B_m^{(1)} Y_m'(h_1 r)\right]\cos(m\phi+\phi_0) \\ &+ \frac{ik_z}{h_1^2 r} m\left[C_m^{(1)} J_m(h_1 r) + D_m^{(1)} Y_m(h_1 r)\right]\cos(m\phi+\phi_0), \end{aligned} \tag{6.120}$$

and

$$H_{m,\phi}^{(2)} = \frac{in_2^2 k_0}{h_2 \eta_0} A_m^{(2)} H_m'^{(1)}(h_2 r)\cos(m\phi + \phi_0) + \frac{ik_z}{h_2^2 r} m C_m^{(2)} H_m^{(1)}(h_2 r)\cos(m\phi + \phi_0). \quad (6.121)$$

Matching boundary conditions,

$$\begin{aligned} &\frac{mk_z}{h_0^2 r_1} J_m(h_0 r_1) A_m^{(0)} + \frac{k_0 \eta_0}{h_0} J_m'(h_0 r_1) C_m^{(0)} \\ &= \frac{mk_z}{h_1^2 r_1} J_m(h_1 r_1) A_m^{(1)} + \frac{mk_z}{h_1^2 r_1} Y_m(h_1 r_1) B_m^{(1)} + \frac{k_0 \eta_0}{h_1} J_m'(h_1 r_1) C_m^{(1)} + \frac{k_0 \eta_0}{h_1} Y_m'(h_1 r_1) D_m^{(1)}, \end{aligned} \quad (6.122)$$

$$\begin{aligned} &\frac{mk_z}{h_2^2 r_2} H_m^{(1)}(h_2 r_2) A_m^{(2)} + \frac{k_0 \eta_0}{h_2} H_m'^{(1)}(h_2 r_2) C_m^{(2)} \\ &= \frac{mk_z}{h_1^2 r_2} J_m(h_1 r_2) A_m^{(1)} + \frac{mk_z}{h_1^2 r_2} Y_m(h_1 r_2) B_m^{(1)} + \frac{k_0 \eta_0}{h_1} J_m'(h_1 r_2) C_m^{(1)} + \frac{k_0 \eta_0}{h_1} Y_m'(h_1 r_2) D_m^{(1)}, \end{aligned} \quad (6.123)$$

$$\begin{aligned} &\frac{n_0^2 k_0}{h_0 \eta_0} J_m'(h_0 r_1) A_m^{(0)} + \frac{mk_z}{h_0^2 r_1} J_m(h_0 r_1) C_m^{(0)} \\ &= \frac{n_1^2 k_0}{h_1 \eta_0} J_m'(h_1 r_1) A_m^{(1)} + \frac{n_1^2 k_0}{h_1 \eta_0} Y_m'(h_1 r_1) B_m^{(1)} + \frac{mk_z}{h_1^2 r_1} J_m(h_1 r_1) C_m^{(1)} + \frac{mk_z}{h_1^2 r_1} Y_m(h_1 r_1) D_m^{(1)}, \end{aligned} \quad (6.124)$$

and

$$\begin{aligned} &\frac{n_2^2 k_0}{h_2 \eta_0} H_m'^{(1)}(h_2 r_2) A_m^{(2)} + \frac{mk_z}{h_2^2 r_2} H_m^{(1)}(h_2 r_2) C_m^{(2)} \\ &= \frac{n_1^2 k_0}{h_1 \eta_0} J_m'(h_1 r_2) A_m^{(1)} + \frac{n_1^2 k_0}{h_1 \eta_0} Y_m'(h_1 r_2) B_m^{(1)} + \frac{mk_z}{h_1^2 r_2} J_m(h_1 r_2) C_m^{(1)} + \frac{mk_z}{h_1^2 r_2} Y_m(h_1 r_2) D_m^{(1)}. \end{aligned} \quad (6.125)$$

A matrix equation is obtained by substituting (6.104)–(6.107) into (6.122)–(6.125),

$$\begin{bmatrix} a_1 & b_1 & -c_1 & -d_1 \\ a_2 & b_2 & -c_2 & -d_2 \\ a_3 & b_3 & -c_3 & -d_3 \\ a_4 & b_4 & -c_4 & -d_4 \end{bmatrix} \begin{bmatrix} A_m^{(0)} \\ C_m^{(0)} \\ A_m^{(2)} \\ C_m^{(2)} \end{bmatrix} = 0 \quad (6.126)$$

where

$$a_1 = \frac{mk_z}{h_0^2 r_1} J_m(h_0 r_1) - \frac{mk_z}{h_1^2 r_1} J_m(h_1 r_1)\alpha_1 - \frac{mk_z}{h_1^2 r_1} Y_m(h_1 r_1)\alpha_2, \quad (6.127)$$

$$a_2 = \frac{mk_z}{h_1^2 r_2} J_m(h_1 r_2)\alpha_1 + \frac{mk_z}{h_1^2 r_2} Y_m(h_1 r_2)\alpha_2, \quad (6.128)$$

$$a_3 = \frac{n_0^2 k_0}{h_0 \eta_0} J'_m(h_0 r_1) - \frac{n_1^2 k_0}{h_1 \eta_0} J'_m(h_1 r_1)\alpha_1 - \frac{n_1^2 k_0}{h_1 \eta_0} Y'_m(h_1 r_1)\alpha_2, \quad (6.129)$$

$$a_4 = \frac{n_1^2 k_0}{h_1 \eta_0} J'_m(h_1 r_2)\alpha_1 + \frac{n_1^2 k_0}{h_1 \eta_0} Y'_m(h_1 r_2)\alpha_2, \quad (6.130)$$

$$b_1 = \frac{k_0 \eta_0}{h_0} J'_m(h_0 r_1) - \frac{k_0 \eta_0}{h_1} J'_m(h_1 r_1)\alpha_3 - \frac{k_0 \eta_0}{h_1} Y'_m(h_1 r_1)\alpha_4, \quad (6.131)$$

$$b_2 = \frac{k_0 \eta_0}{h_1} J'_m(h_1 r_2)\alpha_3 + \frac{k_0 \eta_0}{h_1} Y'_m(h_1 r_2)\alpha_4, \quad (6.132)$$

$$b_3 = \frac{m k_z}{h_0^2 r_1} J_m(h_0 r_1) - \frac{m k_z}{h_1^2 r_1} J_m(h_1 r_1)\alpha_3 - \frac{m k_z}{h_1^2 r_1} Y_m(h_1 r_1)\alpha_4, \quad (6.133)$$

$$b_4 = \frac{m k_z}{h_1^2 r_2} J_m(h_1 r_2)\alpha_3 + \frac{m k_z}{h_1^2 r_2} Y_m(h_1 r_2)\alpha_4, \quad (6.134)$$

$$c_1 = \frac{m k_z}{h_1^2 r_1} J_m(h_1 r_1)\beta_1 + \frac{m k_z}{h_1^2 r_1} Y_m(h_1 r_1)\beta_2, \quad (6.135)$$

$$c_2 = \frac{m k_z}{h_2^2 r_2} H_m^{(1)}(h_2 r_2) - \frac{m k_z}{h_1^2 r_2} J_m(h_1 r_2)\beta_1 - \frac{m k_z}{h_1^2 r_2} Y_m(h_1 r_2)\beta_2, \quad (6.136)$$

$$c_3 = \frac{n_1^2 k_0}{h_1 \eta_0} J'_m(h_1 r_1)\beta_1 + \frac{n_1^2 k_0}{h_1 \eta_0} Y'_m(h_1 r_1)\beta_2, \quad (6.137)$$

$$c_4 = \frac{n_2^2 k_0}{h_2 \eta_0} H_m'^{(1)}(h_2 r_2) - \frac{n_1^2 k_0}{h_1 \eta_0} J'_m(h_1 r_2)\beta_1 - \frac{n_1^2 k_0}{h_1 \eta_0} Y'_m(h_1 r_2)\beta_2, \quad (6.138)$$

$$d_1 = \frac{k_0 \eta_0}{h_1} J'_m(h_1 r_1)\beta_3 + \frac{k_0 \eta_0}{h_1} Y'_m(h_1 r_1)\beta_4, \quad (6.139)$$

$$d_2 = \frac{k_0 \eta_0}{h_2} H_m'^{(1)}(h_2 r_2) - \frac{k_0 \eta_0}{h_1} J'_m(h_1 r_2)\beta_3 - \frac{k_0 \eta_0}{h_1} Y'_m(h_1 r_2)\beta_4, \quad (6.140)$$

$$d_3 = \frac{m k_z}{h_1^2 r_1} J_m(h_1 r_1)\beta_3 + \frac{m k_z}{h_1^2 r_1} Y_m(h_1 r_1)\beta_4, \quad (6.141)$$

and

$$d_4 = \frac{m k_z}{h_2^2 r_2} H_m^{(1)}(h_2 r_2) - \frac{m k_z}{h_1^2 r_2} J_m(h_1 r_2)\beta_3 - \frac{m k_z}{h_1^2 r_2} Y_m(h_1 r_2)\beta_4. \quad (6.142)$$

Again, the procedure is to find values for k_z for which the determinant of the 4×4 matrix in (6.126) is zero.

6.8 FIELDS

Once values for the mode index, or equivalently, k_z, have been determined, the fields associated with each mode index can be found. The first step is to determine values for all $A_m^{(n)}$, $B_m^{(n)}$, $C_m^{(n)}$, and $D_m^{(n)}$. In terms of $A_m^{(0)}$, from (6.68)

$$\begin{bmatrix} C_m^{(0)} \\ A_m^{(L+1)} \\ C_m^{(L+1)} \end{bmatrix} = -A_m^{(0)} \begin{bmatrix} M_{12} & -R_{11} & -R_{12} \\ M_{22} & -R_{21} & -R_{22} \\ M_{32} & -R_{31} & -R_{32} \end{bmatrix}^{-1} \begin{bmatrix} M_{11} \\ M_{21} \\ M_{31} \end{bmatrix}. \tag{6.143}$$

If there is a degeneracy preventing inversion of the matrix, then a different set of three equations in (6.68) can be selected such as

$$\begin{bmatrix} C_m^{(0)} \\ A_m^{(L+1)} \\ C_m^{(L+1)} \end{bmatrix} = -A_m^{(0)} \begin{bmatrix} M_{22} & -R_{21} & -R_{22} \\ M_{32} & -R_{31} & -R_{32} \\ M_{42} & -R_{41} & -R_{42} \end{bmatrix}^{-1} \begin{bmatrix} M_{21} \\ M_{31} \\ M_{41} \end{bmatrix}. \tag{6.144}$$

After determining the A_m and C_m coefficients for the core and outer medium, then (6.59) or (6.60) can be applied successively for each shell to determine all remaining coefficients. In general, because the fields are larger inside the core than the cladding, the $A_m^{(0)}$ and $C_m^{(0)}$ coefficients for the core are generally more accurate than the $A_m^{(L+1)}$ and $C_m^{(L+1)}$ coefficients for the cladding, particularly when multiple cylinders around the core are present. Therefore, it is best to first solve for the inner cylinder coefficients from the core coefficients,

$$\begin{bmatrix} A_m^{(1)} \\ B_m^{(1)} \\ C_m^{(1)} \\ D_m^{(1)} \end{bmatrix} = \left(\tilde{\boldsymbol{Q}}_1\right)^{-1} \tilde{\boldsymbol{M}}_0 \begin{bmatrix} A_m^{(0)} \\ C_m^{(0)} \end{bmatrix} \tag{6.145}$$

and then solve for any remaining cylinder coefficients by working outwards,

$$\tilde{\boldsymbol{Q}}_{l+1} \begin{bmatrix} A_m^{(l+1)} \\ B_m^{(l+1)} \\ C_m^{(l+1)} \\ D_m^{(l+1)} \end{bmatrix} = \tilde{\boldsymbol{P}}_l \begin{bmatrix} A_m^{(l)} \\ B_m^{(l)} \\ C_m^{(l)} \\ D_m^{(l)} \end{bmatrix}. \tag{6.146}$$

The longitudinal field components are calculated from (6.33)–(6.38). The remaining field components are then obtained from (6.16)–(6.19).

6.9 LP MODES

If the refractive index difference between the core and cladding is small (< 0.01), which is usually the case for communication fibers, then the weakly guiding approximation can be applied in which the longitudinal field components are assumed to be negligible. These modes are the so-called linearly polarized modes or "LP" modes. The LP modes satisfy the scalar wave equation

TABLE 6.1
General Relationship Between LP and EH/HE Modes

LP-mode designation	EH/HE mode designation	# degenerate modes
LP_{0n}	$HE_{1n} \times 2$	2
LP_{1n}	TE_{0n}, TM_{0n}, $HE_{2n} \times 2$	4
LP_{mn}	$EH_{m-1,n} \times 2$, $HE_{m+1,n} \times 2$	4

TABLE 6.2
Examples of LP and EH/HE Mode Relationships

LP-mode designation	EH/HE mode designation	# degenerate modes
LP_{01}	$HE_{11} \times 2$	2
LP_{11}	TE_{01}, TM_{01}, $HE_{21} \times 2$	4
LP_{21}	$EH_{11} \times 2$, $HE_{31} \times 2$	4
LP_{02}	$HE_{12} \times 2$	2
LP_{31}	$EH_{21} \times 2$, $HE_{41} \times 2$	4
LP_{12}	TE_{02}, TM_{02}, $HE_{22} \times 2$	4
LP_{41}	$EH_{31} \times 2$, $HE_{51} \times 2$	4
LP_{22}	$EH_{12} \times 2$, $HE_{32} \times 2$	4
LP_{03}	$HE_{13} \times 2$	2
LP_{51}	$EH_{41} \times 2$, $HE_{61} \times 2$	4
LP_{32}	$EH_{22} \times 2$, $HE_{42} \times 2$	4
LP_{13}	TE_{03}, TM_{03}, $HE_{23} \times 2$	4

$$\left(\frac{\partial^2}{\partial r^2}+\frac{1}{r}\frac{\partial}{\partial r}+\frac{1}{r^2}\frac{\partial^2}{\partial \phi^2}+n_{ave}^2k_0^2-k_z^2\right)\psi_{mn}=0. \tag{6.147}$$

There is a correspondence between LP modes and EH/HE modes as shown in Tables 6.1 and 6.2 [2].

6.10 OVERLAP INTEGRALS FOR COUPLING LIGHT INTO FIBER

It is often desirable to calculate the amount of incident power focused onto the tip of a fiber that gets coupled into a particular mode of the fiber. The simplest method to estimate this power coupling is through the use of overlap integrals. Because the optical modes within a fiber form a complete orthogonal set of propagating modes, any propagating field profile within the fiber can be divided into this set of modes. The amplitude of each mode in the set is given by the overlap integral,

$$O \equiv \frac{\int E_1 \cdot E_2^* \mathrm{d}A}{\sqrt{\int |E_1|^2 \,\mathrm{d}A \cdot \int |E_2|^2 \,\mathrm{d}A}}. \tag{6.148}$$

In this equation E_1 represents the electric field of the propagating mode and E_2 represents the electric field of one of the modes of the fiber. If the two modes are identical then the overlap integral is 1. If the two modes are orthogonal, then the integral is 0. The amount of power in the normal mode scales as the square of the overlap integral.

Of course, knowing the electric field profile from a source on the outside at the end face of a fiber does not enable one to immediately determine the field profile on the inside of the fiber end face even though the transverse electric field must be continuous across the end face. That is because this approach neglects the reflected field. However, if the reflected field is assumed to be negligible, which should be a good approximation for most fiber modes, the overlap integral should provide a good estimate of power transfer.

For a Gaussian focused spot intensity, the incident field profile is

$$E_g = E_0 e^{-\ln 2\left(\frac{r}{r_{1/2}}\right)^2}. \tag{6.149}$$

The intensity of the field within the spot is proportional to the square of the electric field.

$$I_g = \left|E_g\right|^2 = I_0 e^{-2\ln 2\left(\frac{r}{r_{1/2}}\right)^2}. \tag{6.150}$$

$r_{1/2}$ is the radius at which the *electric field* drops to half its peak value. The amplitude at the center of the spot has been adjusted so that the integral of the square of the electric field over the *xy* plane is unity. The full width at half maximum (FWHM) of the *intensity* is

$$d_0 = \sqrt{2}\, r_{1/2}. \tag{6.151}$$

The intensity drops to its e^{-2} value for

$$r_e = r_{1/2}\sqrt{\frac{1}{\ln 2}} = \sqrt{\frac{1}{2\ln 2}}\, d_0. \tag{6.152}$$

The Airy function is the scalar approximation to the diffracted field of a focused spot, so it is a better approximation than the Gaussian spot. Consider a lens of diameter D and focal length f that is uniformly filled with light, as shown in Figure 6.2.

In the Airy scalar approximation, the radial dependence of the electric field in the focal plane of the lens with numerical aperture $\mathcal{N}$ can be approximated as

$$\frac{E(r)}{E_{r=0}} = 2\frac{J_1(kr\mathcal{N})}{(kr\mathcal{N})} \tag{6.153}$$

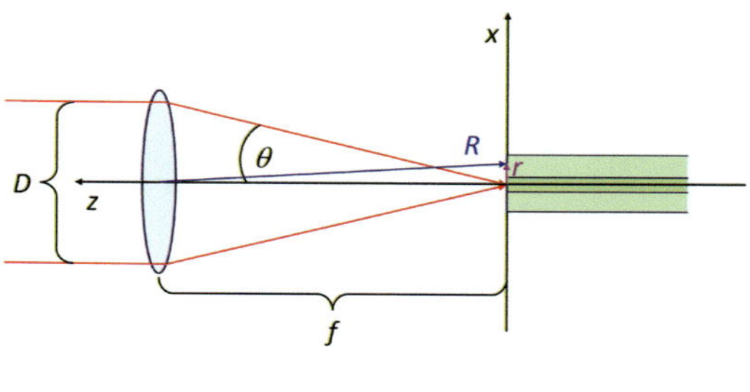

FIGURE 6.2 A lens with focal length f and diameter D focuses light onto the end face of a fiber. The vector $\boldsymbol{R}$ connects the center of the lens to the observation point on the end face of the fiber. The vector $\boldsymbol{r}$ lies in the end face and is the radius of the observation point.

where r is the radial distance from the center of the focused spot to a point in the focal plane and J_1 is a Bessel function of the first kind. Let

$$x \equiv kr\mathcal{N}. \tag{6.154}$$

$J_1(x)$ exhibits its first zero when $x = 3.83170597$. If r_0 is the radius of the first zero, then

$$kr_0\mathcal{N} = \left(\frac{2\pi r_0}{\lambda}\right)\mathcal{N} = 3.83170597. \tag{6.155}$$

The numerical aperture can be expressed in terms of the first zero,

$$\mathcal{N} = 3.83170597\left(\frac{\lambda}{2\pi r_0}\right) = 0.60983495\left(\frac{\lambda}{r_0}\right). \tag{6.156}$$

The first zero of the Airy function in the focal plane of the lens occurs at $x = 3.8317$. The radius at which the electric field drops to $1/\sqrt{2}$ of its value at the origin, corresponding to the half maximum in $|E|^2$ intensity, occurs at $x = 1.61634$ or

$$r_{1/2} = \frac{1.616339948}{3.83170597} r_0 = 0.42183298\, r_0. \tag{6.157}$$

The radius at which the electric field drops to its $1/e$ value, corresponding to the intensity dropping to its $1/e^2$ value is 2.5838393.

$$r_e = \frac{2.5838393}{3.83170597} r_0 = 0.67691876\, r_0. \tag{6.158}$$

The radial dependence of intensity is

$$I = \left[\frac{2J_1(kr\mathcal{N})}{(kr\mathcal{N})}\right]^2. \tag{6.159}$$

The integrated intensity is

$$\int_0^\infty |E|^2\, 2\pi r dr = 4\int_0^\infty \left|\frac{J_1(kr\mathcal{N})}{(kr\mathcal{N})}\right|^2 2\pi r dr = \frac{4\pi}{(k\mathcal{N})^2} \tag{6.160}$$

where J_1 is a Bessel function of the first kind. The half maximum of the $|E|^2$ intensity in the focused spot occurs at a radius of

$$r = \frac{1.61634}{k\mathcal{N}} = \frac{1.61634}{2\pi\mathcal{N}}\lambda = 0.25725\left(\frac{\lambda}{\mathcal{N}}\right) \tag{6.161}$$

and the first zero occurs at

$$r = \frac{3.8317}{k\mathcal{N}} = \frac{3.8317}{2\pi\mathcal{N}}\lambda = 0.6098\left(\frac{\lambda}{\mathcal{N}}\right). \tag{6.162}$$

6.11 FAR FIELD RADIATION

The far field radiation from a near field source can be calculated by a procedure due to Shelkunoff [3, 4]. Consider a closed region containing electromagnetic sources. If the fields on the surface are known, fictitious currents on this surface can be calculated that in turn will reproduce the fields outside the surface that would be generated by the internal sources. Although the fields on the surface may have both tangential and normal components, only the tangential components contribute to the far fields.

For a fiber waveguide that is truncated at the $z = 0$ plane with its end face normal to the z-axis as shown in Figure 6.3, the electric and magnetic surface currents are calculated across this plane. If $\boldsymbol{E}(x,y,z = 0)$ and $\boldsymbol{H}(x,y,z = 0)$ are the electric and magnetic fields, respectively, at the end face of the fiber, then the surface currents are

$$\boldsymbol{J}_s = \hat{\boldsymbol{n}} \times \boldsymbol{H} = H_y\, \hat{\boldsymbol{x}} - H_x\, \hat{\boldsymbol{y}} \tag{6.163}$$

$$\boldsymbol{M}_s = -\hat{\boldsymbol{n}} \times \boldsymbol{E} = -E_y\, \hat{\boldsymbol{x}} + E_x\, \hat{\boldsymbol{y}} \tag{6.164}$$

where $\hat{\boldsymbol{n}} = \hat{\boldsymbol{z}}$ is the unit normal to the end face. The vector potentials at the point of observation are calculated by integrating across the plane $z = 0$ at the end face of the fiber.

$$\boldsymbol{N}(x,y) \equiv \frac{\mu}{4\pi R} \int_{s'} \boldsymbol{J}_s(x',y') e^{+ik_0 \hat{\boldsymbol{R}}\cdot\boldsymbol{r}'} dx'dy' \tag{6.165}$$

$$\boldsymbol{L}(x,y) \equiv \frac{\epsilon}{4\pi R} \int_{s'} \boldsymbol{M}_s(x',y') e^{+ik_0 \hat{\boldsymbol{R}}\cdot\boldsymbol{r}'} dx'dy'. \tag{6.166}$$

$\boldsymbol{r}'$ is the vector to the source point on the end face and $\hat{\boldsymbol{R}}$ is the unit vector from the end face to the observation point. In principle, the surface s' in (6.165) and (6.166) extends to infinity in all directions, but in practice the integral only needs to be carried out where there is a substantial electric and magnetic field. If

$$\hat{\boldsymbol{R}} = x_o \hat{\boldsymbol{x}} + y_o \hat{\boldsymbol{y}} + z_o \hat{\boldsymbol{z}} \tag{6.167}$$

then

$$\hat{\boldsymbol{R}}\cdot\boldsymbol{r}' = x_o x' + y_o y'. \tag{6.168}$$

For each observation point (x_o, y_o, z_o) the integrals in (6.165) and (6.166) are summed numerically to determine the vector potentials at each observation point in the far field using the calculated electric and magnetic fields for a specific mode in the fiber.

The coordinate transforms are used to convert the vector potentials from Cartesian to polar coordinates,

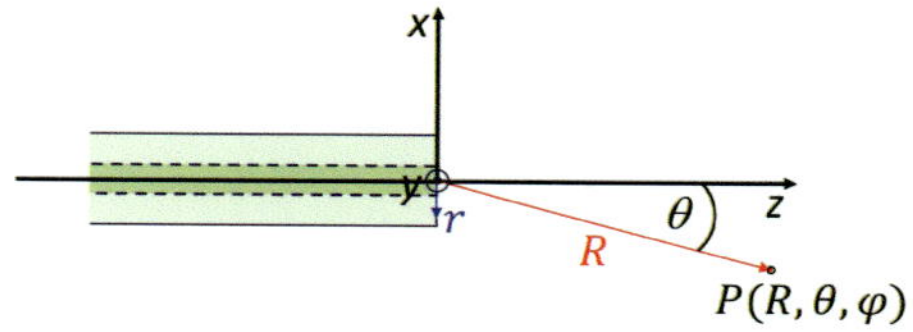

FIGURE 6.3 Diagram of far field radiation from end face of fiber.

$$\hat{\boldsymbol{r}} = \sin\theta\left(\cos\phi\,\hat{\boldsymbol{x}} + \sin\phi\,\hat{\boldsymbol{y}}\right) + \cos\theta\,\hat{\boldsymbol{z}}, \tag{6.169}$$

$$\hat{\boldsymbol{\theta}} = \cos\theta\left(\cos\phi\,\hat{\boldsymbol{x}} + \sin\phi\,\hat{\boldsymbol{y}}\right) - \sin\theta\,\hat{\boldsymbol{z}}, \tag{6.170}$$

and

$$\hat{\boldsymbol{\phi}} = -\sin\phi\,\hat{\boldsymbol{x}} + \cos\phi\,\hat{\boldsymbol{y}}. \tag{6.171}$$

Therefore,

$$\begin{aligned} N_\theta &= \boldsymbol{N}\cdot\hat{\boldsymbol{\theta}} = \boldsymbol{N}\cdot\left(\hat{\boldsymbol{x}}\cos\theta\cos\phi + \hat{\boldsymbol{y}}\cos\theta\sin\phi - \hat{\boldsymbol{z}}\sin\theta\right) \\ &= N_x\cos\theta\cos\phi + N_y\cos\theta\sin\phi - N_z\sin\theta \end{aligned} \tag{6.172}$$

and

$$N_\phi = \boldsymbol{N}\cdot\hat{\boldsymbol{\phi}} = \boldsymbol{N}\cdot\left(-\hat{\boldsymbol{x}}\sin\phi + \hat{\boldsymbol{y}}\cos\phi\right) = -N_x\sin\phi + N_y\cos\phi \tag{6.173}$$

and similarly for L_θ and L_ϕ. When the end face is perpendicular to the z-axis, there are no N_z or L_z components of the vector potentials. After calculating the Cartesian components of the vector potentials $\boldsymbol{N}$ and $\boldsymbol{L}$ from the summed integrations (6.165) and (6.166), then N_θ, N_ϕ, L_θ, and L_ϕ are calculated for each observation point from (6.172) and (6.173).

The electric and magnetic fields at any point outside the surface are given in terms of the vector potentials by

$$\boldsymbol{E} = i\omega\boldsymbol{N} + \frac{i\omega}{k^2}\nabla(\nabla\cdot\boldsymbol{N}) - \frac{1}{\epsilon}\nabla\times\boldsymbol{L} \tag{6.174}$$

and

$$\boldsymbol{H} = i\omega\boldsymbol{L} + \frac{i\omega}{k^2}\nabla(\nabla\cdot\boldsymbol{L}) + \frac{1}{\mu}\nabla\times\boldsymbol{N}. \tag{6.175}$$

However, these equations can be simplified in the far field by keeping only the terms that decay as $1/r^2$. The result is

$$E_\theta = i\frac{e^{ikR}}{2\lambda R}\left(\eta_0 N_\theta + L_\phi\right), \tag{6.176}$$

$$E_\phi = i\frac{e^{ikR}}{2\lambda R}\left(\eta_0 N_\phi - L_\theta\right), \tag{6.177}$$

$$H_\theta = -\frac{E_\phi}{\eta_0}, \tag{6.178}$$

and

$$H_\phi = \frac{E_\theta}{\eta_0}. \tag{6.179}$$

The fields in Cartesian coordinates at each observation point are found from the transforms (6.169)–(6.171).

$$E_x = \boldsymbol{E}\cdot\hat{\boldsymbol{x}} = \cos\phi\,\cos\theta E_\theta - \sin\phi E_\phi, \tag{6.180}$$

$$E_y = \boldsymbol{E}\cdot\hat{\boldsymbol{y}} = \sin\phi\,\cos\theta E_\theta + \cos\phi E_\phi, \tag{6.181}$$

$$H_x = \boldsymbol{H}\cdot\hat{\boldsymbol{x}} = -\frac{1}{\eta_0}\left(\cos\theta\cos\phi\, E_\phi + \sin\phi\, E_\theta\right), \tag{6.182}$$

and

$$H_y = \boldsymbol{H}\cdot\hat{\boldsymbol{y}} = -\frac{1}{\eta_0}\left(\cos\theta\sin\phi\, E_\phi - \cos\phi\, E_\theta\right). \tag{6.183}$$

The end face of a fiber is often polished at a slant angle α of 8° relative to the z-axis as shown in Figure 6.4 to prevent light from re-entering the fiber after reflecting from the end face. The unit normal to the end face surface is

$$\hat{\boldsymbol{n}} = \hat{\boldsymbol{x}}\sin\alpha + \hat{\boldsymbol{z}}\cos\alpha. \tag{6.184}$$

The surface currents in the plane of the end face now include a z-component.

$$\boldsymbol{J}_s = \hat{\boldsymbol{n}}\times\boldsymbol{H} = -H_y\cos\alpha\hat{\boldsymbol{x}} + \left(H_x\cos\alpha - H_z\sin\alpha\right)\hat{\boldsymbol{y}} + H_y\sin a\hat{\boldsymbol{z}} \tag{6.185}$$

and

$$\boldsymbol{M}_s = -\hat{\boldsymbol{n}}\times\boldsymbol{E} = E_y\cos\alpha\hat{\boldsymbol{x}} - \left(E_x\cos\alpha - E_z\sin\alpha\right)\hat{\boldsymbol{y}} - E_y\sin\alpha\hat{\boldsymbol{z}}. \tag{6.186}$$

The vector to the source point on the end face, $\boldsymbol{r}'$, is

$$\boldsymbol{r}' = x'\,\hat{\boldsymbol{x}} + y'\,\hat{\boldsymbol{y}} + z'\,\hat{\boldsymbol{z}} = x'\,\hat{\boldsymbol{x}} + y'\,\hat{\boldsymbol{y}} - x'\tan\alpha\,\hat{\boldsymbol{z}} \tag{6.187}$$

so

$$\hat{\boldsymbol{R}}\cdot\boldsymbol{r}' = \sin\theta\cos\phi x' + \sin\theta\sin\phi y' - x'\tan\alpha\cos\theta\, z'. \tag{6.188}$$

For a slanted end face, the fields must be calculated at the end face inside the cylinder for the correct z-position. The exponential term in (6.165) and (6.166) becomes

$$e^{i\left[x'\sin\theta\cos\phi + y'\sin\theta\sin\phi + z'(n_m - 1)k_0\right]} = e^{i\,k_0\left[x'\sin\theta\cos\phi + y'\sin\theta\sin\phi + x'(n_m - 1)\tan\alpha\cos\theta\right]}. \tag{6.189}$$

This approximation assumes that the mode within the fiber continues to propagate undisturbed after it reaches the leading edge of the slanted end face. Clearly this is incorrect and will lead to small errors in the far field result.

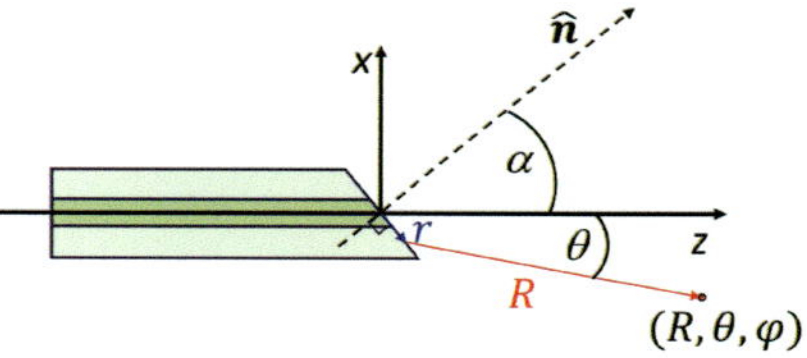

FIGURE 6.4 Far field calculation for fiber with angled end face.

When the end face is normal to the cylinder axis, the z-components of the fields do not enter into the calculation due to the dot product in (6.188) with $\alpha = 0$. When the end face is slanted, however, the z-components of the fields are retained in (6.188). However, there is a discontinuity of the component of the electric field normal to the surface of the end face. In order to make the component of the displacement field that is normal to the end face of the fiber at point (x,y) continuous across the boundary, the electric field components on the air side of the boundary (neglecting reflection at the surface) in terms of the field components calculated for the mode within the fiber at the same point are

$$E_{x,\text{air}}(x,y) = \left(\cos^2\alpha + n_m^2 \sin^2\alpha\right) E_{x,fib}(x,y) + \left(n_{fib}^2(x,y) - 1\right)\sin\alpha\cos\alpha E_{z,fib}(x,y) \quad (6.190)$$

$$E_{y,\text{air}}(x,y) = E_{y,\text{fib}}(x,y) \quad (6.191)$$

$$E_{z,\text{air}}(x,y) = \left(n_m^2 - 1\right)\sin\alpha\cos\alpha E_{x,fib}(x,y) + \left[\sin^2\alpha + n_{fib}^2(x,y)\cos^2\alpha\right] E_{z,fib}(x,y). \quad (6.192)$$

The vector potentials at the point of observation are calculated by integrating across the plane of the end face of the fiber.

$$\boldsymbol{N}(x,y) \equiv \frac{\mu}{4\pi R}\int_{s'} \boldsymbol{J}_s(x',y') e^{+ik_0 \hat{\boldsymbol{R}}\cdot\boldsymbol{r}'} \left(\frac{dx'}{\cos\alpha}\right) dy'. \quad (6.193)$$

$$\boldsymbol{L}(x,y) \equiv \frac{\epsilon}{4\pi R}\int_{s'} \boldsymbol{M}_s(x',y') e^{+ik_0 \hat{\boldsymbol{R}}\cdot\boldsymbol{r}'} \left(\frac{dx'}{\cos\alpha}\right) dy'. \quad (6.194)$$

Once the vector potentials have been calculated, the field equations (6.176)–(6.183) follow through as before.

When the fiber end face is slanted as in Figure 6.4, the radiated cone of light is deviated by an angle towards point P that can be estimated from Snell's law as shown in Figure 6.5.

$$\sin(\alpha + \xi) = n_m \sin\alpha. \quad (6.195)$$

The far field radiation example at the end of this chapter demonstrates the offset of the emitter spot in the far field for a slanted fiber end face.

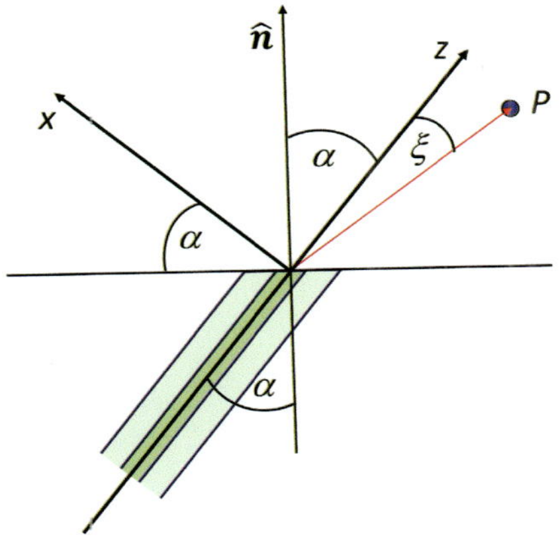

FIGURE 6.5 The waveguide mode propagates along z-axis with mode index n_m. At the end face the emitted beam is refracted by angle ξ with respect to the z-axis.

6.12 GRADED INDEX FIBERS

The formalism developed in this chapter can be used to calculate the modes and mode indices in graded index fiber. This fiber is manufactured such that the refractive index in the core is not constant but varies approximately according to the equation

$$n(r) = n_{core}\sqrt{1-2\Delta\left(\frac{r}{r_0}\right)^g} \tag{6.196}$$

where r_0 is the radius of the cladding, g is a parameter usually close to 2, and

$$\Delta = \frac{n_{core}^2 - n_{clad}^2}{2n_{core}^2}. \tag{6.197}$$

A graded index fiber is considered in the examples at the end of the chapter.

6.13 USEFUL EQUATIONS AND NOMENCLATURE

The V parameter is a measure of the difference between the refractive indices of core and cladding and is defined as

$$V = \frac{2\pi r_c}{\lambda}\sqrt{n_{co}^2 - n_{cl}^2}. \tag{6.198}$$

The mode index for all (guided) modes will lie between the largest core/shell index and the outer cladding index. As the difference between the core index and the cladding index decreases, that is, as V decreases, every mode except the HE_{11} mode reaches cutoff. For $V \leq 2.404$ only the fundamental HE_{11} mode can propagate so the waveguide is single mode. For larger values of V, it is multimode. The number of allowed modes is approximately $V^2/2$ which includes the polarization degeneracy.

The group velocity is defined as

$$\mathrm{v}_g \equiv \left(\frac{d\beta}{d\omega}\right)^{-1} = \frac{1}{\dfrac{d\left(\dfrac{n_m}{\lambda}\right)}{d\nu}} = \frac{1}{\dfrac{d\lambda}{d\nu}\dfrac{d\left(\dfrac{n_m}{\lambda}\right)}{d\lambda}} = -\frac{1}{\dfrac{\lambda^2}{c}\left(\dfrac{\lambda\dfrac{dn_m}{d\lambda} - n_m}{\lambda^2}\right)} = \frac{c}{\left(n_m - \lambda\dfrac{dn_m}{d\lambda}\right)} \tag{6.199}$$

where c is the speed of light in vacuum. The group velocity is the speed at which a light pulse composed of a range of wavelengths travels through a fiber. The group index can be defined in terms of the group velocity,

$$n_g \equiv \frac{c}{\mathrm{v}_g} = n_m - \lambda\frac{dn_m}{d\lambda}. \tag{6.200}$$

The group delay is the inverse of the group velocity,

$$\tau \equiv \frac{1}{\mathrm{v}_g} = \frac{n_g}{c} = \frac{n_m}{c} - \frac{\lambda}{c}\frac{dn_m}{d\lambda}. \tag{6.201}$$

Dispersion is proportional to the second derivative of the mode index and is also the derivative of the group delay with respect to wavelength.

$$D \equiv \frac{\mathrm{d}\tau}{\mathrm{d}\lambda} = \frac{d}{d\lambda}\left(\frac{n_m}{c} - \frac{\lambda}{c}\frac{dn_m}{d\lambda}\right) = \frac{1}{c}\left(\frac{dn_m}{d\lambda} - \frac{dn_m}{d\lambda} - \lambda\frac{d^2 n_m}{d\lambda^2}\right) = -\frac{\lambda}{c}\frac{d^2 n_m}{d\lambda^2}. \tag{6.202}$$

The group velocity dispersion parameter, β_2, is the second derivative of the mode wavevector with respect to angular frequency.

$$\beta_2 = \frac{d^2(n_m k_0)}{d\omega^2} = 2\pi \frac{d^2\left(\frac{n_m}{\lambda}\right)}{d\omega^2} = \frac{\lambda^3}{2\pi c^2}\left(\frac{d^2 n_m}{d\lambda^2}\right) = -\left(\frac{\lambda^2}{2\pi c}\right)\frac{d\tau}{d\lambda} = -\left(\frac{\lambda^2}{2\pi c}\right)D. \tag{6.203}$$

Dispersion is generally calculated in units of ps/km/nm and group velocity dispersion in units of fs^2/mm. A dispersion calculation for optical fiber is considered in the examples.

6.14 EXAMPLES

6.14.1 SMF-28 Single Mode Step Index Fiber

SMF-28 is a conventional single mode fiber for 1550 nm wavelength in the telecom C-band. The core diameter is 8.2 μm with a refractive index of 1.4513, and the cladding diameter is 125 μm with a refractive index of 1.44692 [5]. At a wavelength of 1.55 μm, assuming an infinitely thick cladding, the HE_{11} mode index is 1.4485518. The e^{-2} radius for the normalized Poynting vector occurs at a radius of 5.03.

Table 6.3 lists the refractive indices for several SMF-28 fiber varieties [6]. The HE_{11} mode index and e^{-2} mode radius are calculated at a wavelength of 1.55 μm assuming a core radius of 4.1 μm and an infinite cladding radius. The radial dependences of the fields and Poynting vector are plotted for standard SMF-28 fiber in Figure 6.6.

6.14.2 Simple Waveguide

Consider the fiber waveguide [7] in Figure 6.7. The mode indices at a wavelength of 1.55 μm for $m = 0$ to 4 are

TM_0 modes:
1. $n_m = 1.491257226078885$
2. $n_m = 1.472727935571726$

TE_0 modes:
1. $n_m = 1.491331904070452$
2. $n_m = 1.472787030163557$

EH/HE modes for $m = 1$:
1. $n_m = 1.496541292821156$ (HE_{11})
2. $n_m = 1.484532840593025$ (EH_{11})
3. $n_m = 1.482329903658630$ (HE_{12})

EH/HE modes for $m = 2$:
1. $n_m = 1.491278627243661$ (HE_{21})
2. $n_m = 1.476481802701236$ (EH_{21})
3. $n_m = 1.472716764599007$ (HE_{22})

EH/HE modes for $m = 3$:
1. $n_m = 1.484475681802761$ (HE_{31})

EH/HE modes for $m = 4$:
1. $n_m = 1.476357691424228$ (HE_{41})

The radial field profiles for these modes are plotted in Figures 6.8–6.14 and the mode index dispersion in Figure 6.15.

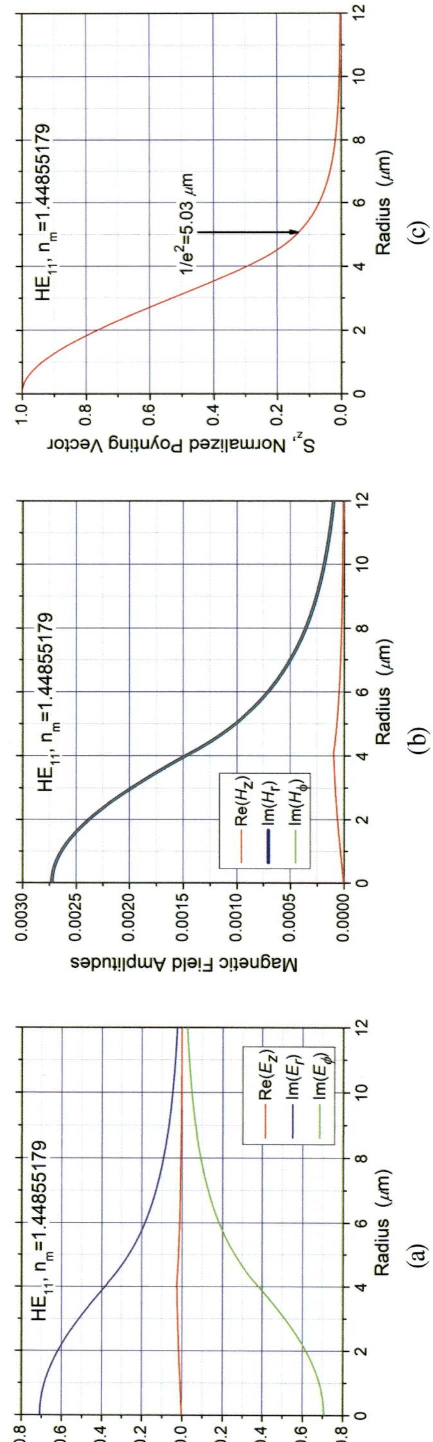

FIGURE 6.6 Amplitude of the (a) electric field, (b) magnetic field, and (c) Poynting vector for HE_{11} mode of SMF28 fiber as a function of fiber radius.

TABLE 6.3
Refractive Indices of SMF-26 Fiber Varieties

Fiber	Core Index	Cladding index	HE_{11} mode index	S_z: e^{-2} radius
SMF-28	1.45213	1.44692	1.44915636	4.76 μm
SMF-28e	1.4677	1.4624	1.46472241	4.73
SMF28e+	1.45173	1.44602	1.44863745	4.64
SMF28-e+LL	1.45223	1.44702	1.44925647	4.76
SMF-28 ULL	1.44525	1.44002	1.44226339	4.76

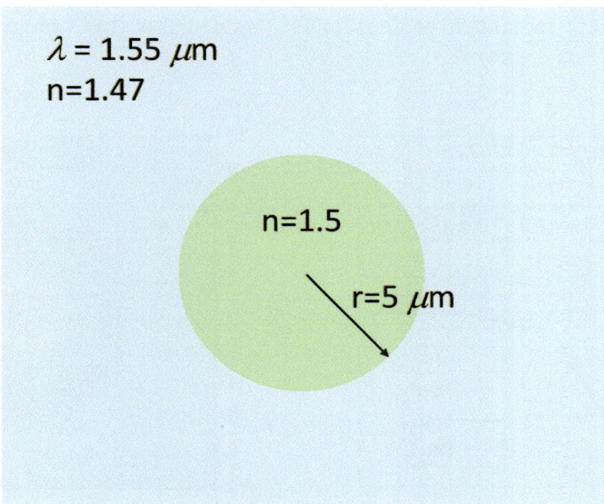

FIGURE 6.7 Fiber waveguide with core index 1.5 and cladding index 1.47 with a 5.0 μm radius at a wavelength of 1.55 μm.

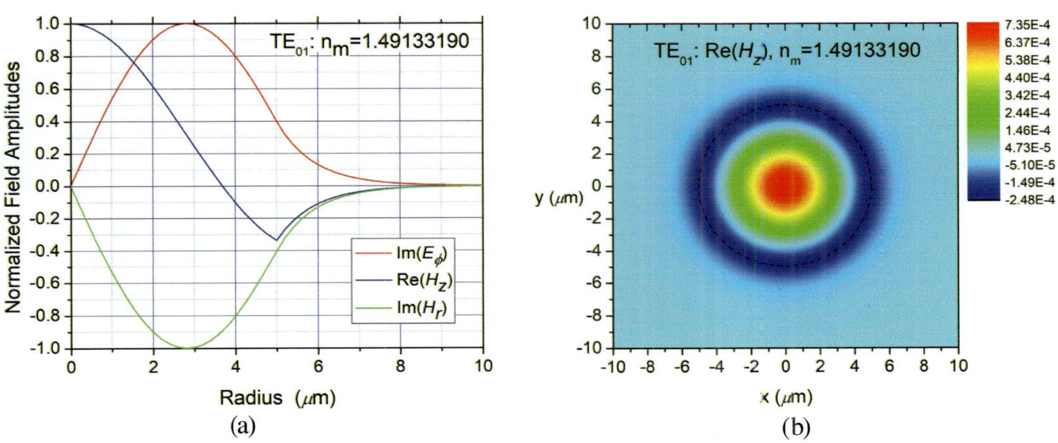

FIGURE 6.8 (a) Normalized radial field profiles for TE_{01} mode fields. (b) Contour plot of H_z field component.

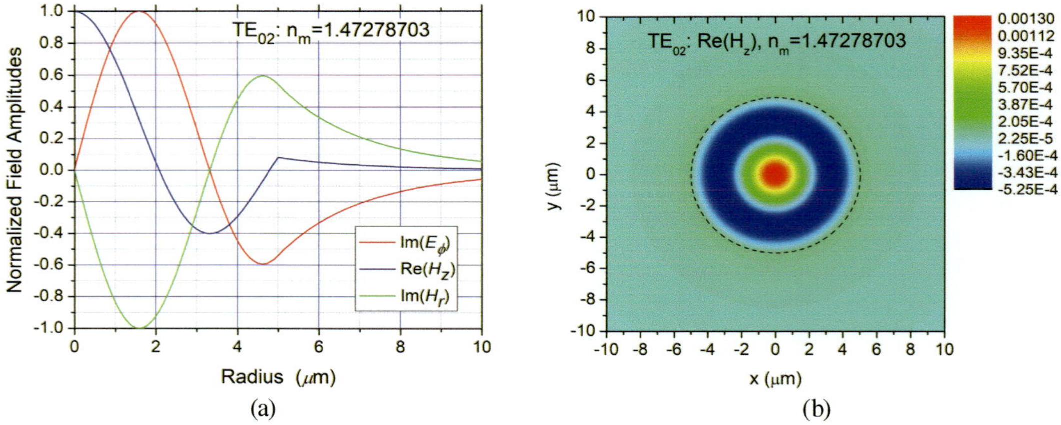

FIGURE 6.9 (a) Normalized radial field profiles for TE_{02} mode fields. (b) Contour plot of H_z field component.

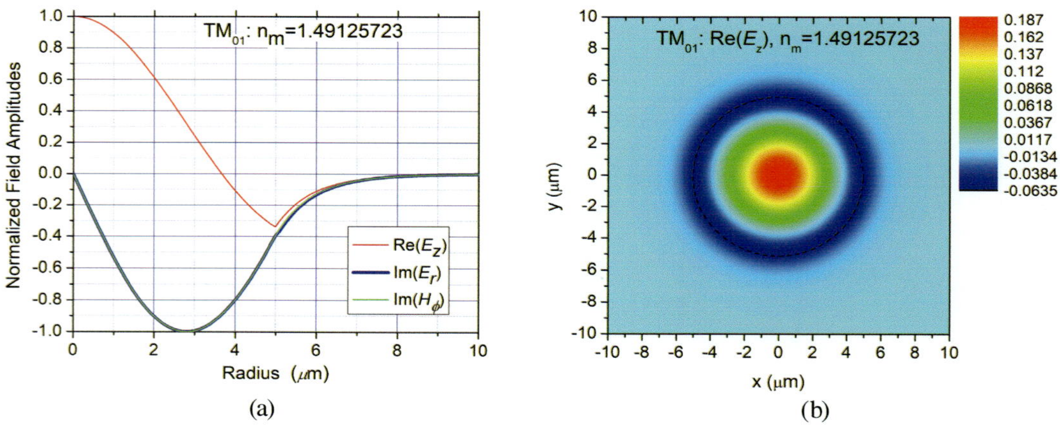

FIGURE 6.10 (a) Normalized radial profile for TM_{01} mode fields. (b) Contour plot of E_z field component.

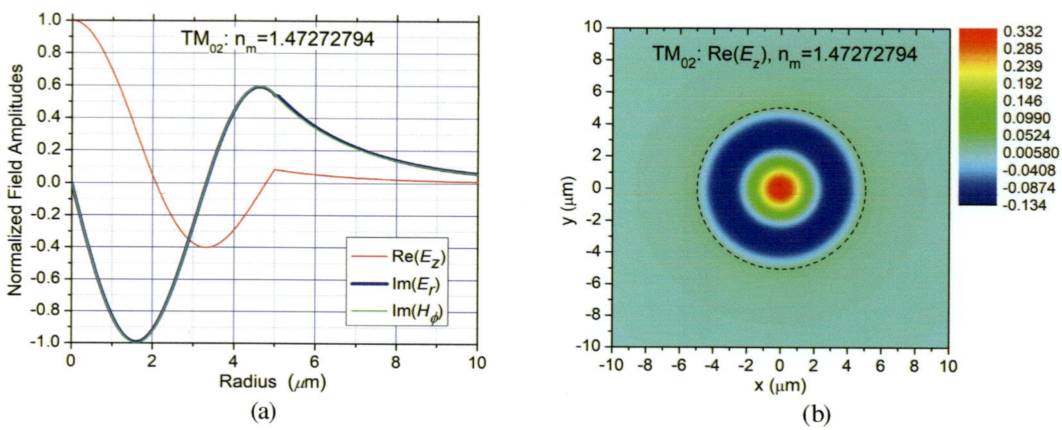

FIGURE 6.11 (a) Normalized radial profile for TM_{02} mode fields. (b) Contour plot of E_z field component.

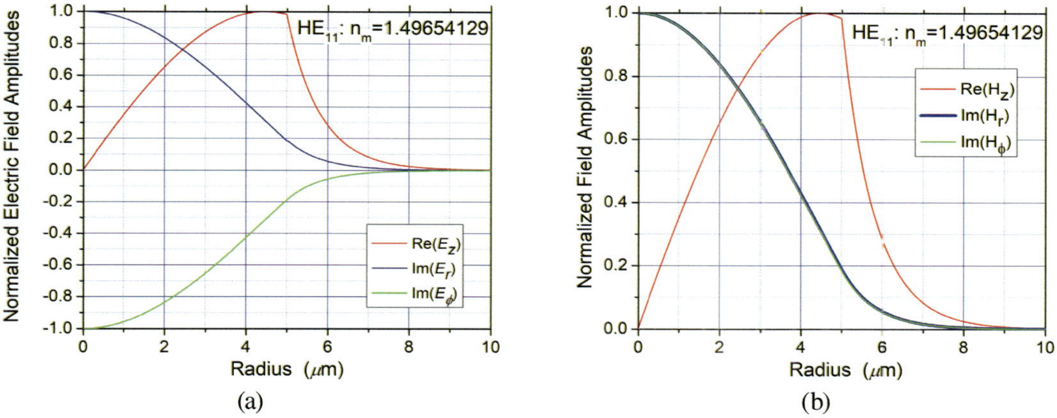

FIGURE 6.12 Normalized radial (a) electric and (b) magnetic field profiles for HE_{11} mode fields.

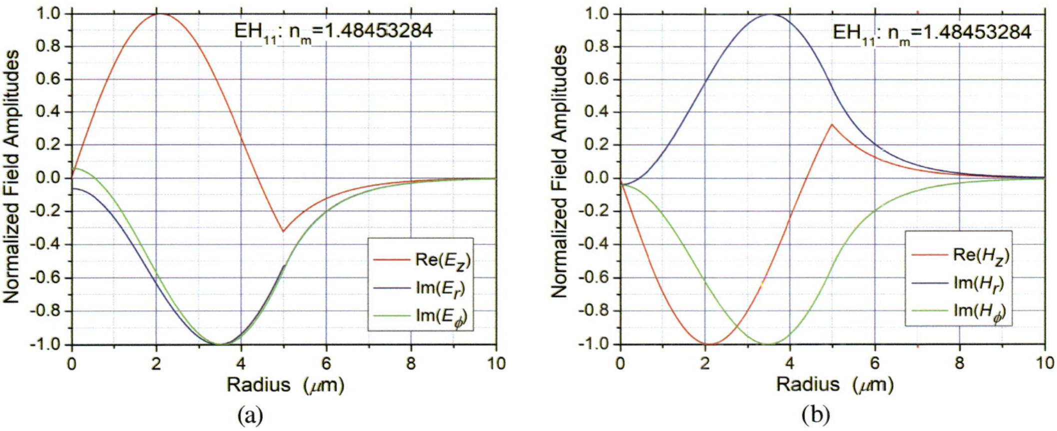

FIGURE 6.13 Normalized radial (a) electric and (b) magnetic field profiles for EH_{11} mode fields.

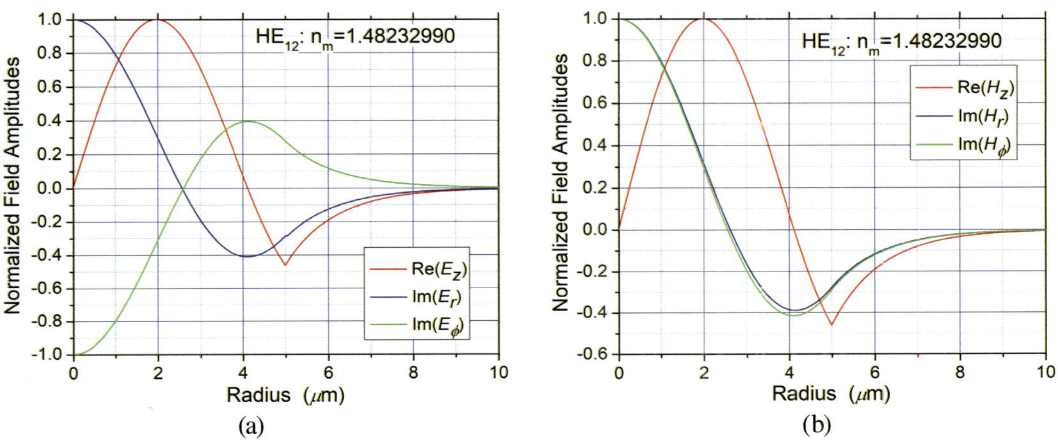

FIGURE 6.14 Normalized radial (a) electric and (b) magnetic field profiles for HE_{12} mode fields.

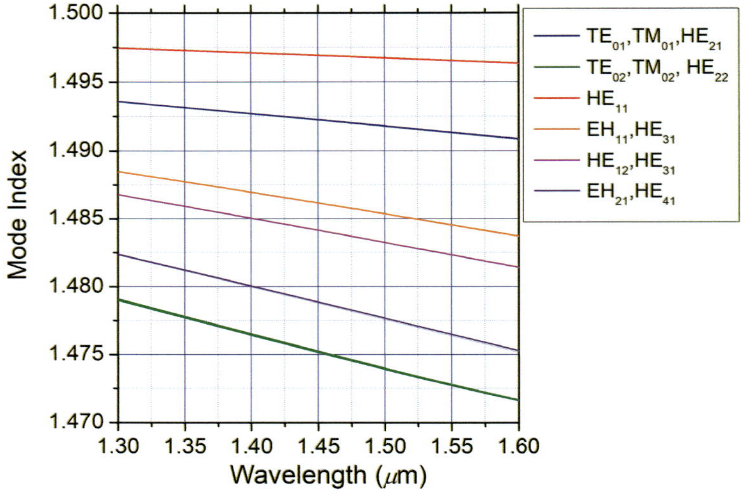

FIGURE 6.15 Dispersion curves for mode indices.

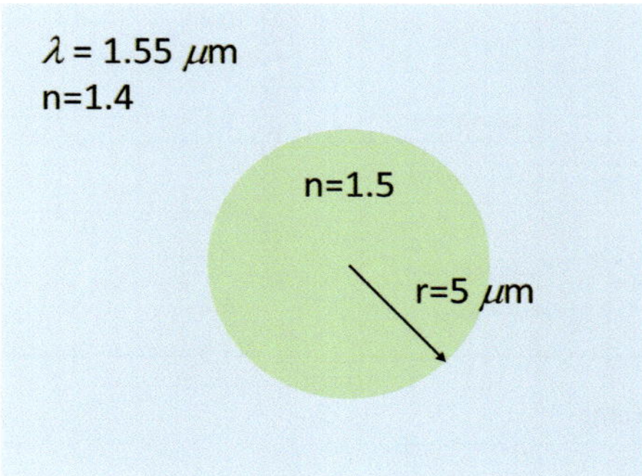

FIGURE 6.16 Simple waveguide to illustrate LP modes.

6.14.3 LP Mode Field Contour Plots

Consider the waveguide in Figure 6.16. The mode indices for the first several LP modes are given in Table 6.4 at the wavelength of 1.55 μm and dispersion as a function of radius and wavevector are plotted in Figure 6.17. The E_x-component of the field, which is the primary field polarization for the calculation, is shown in Figure 6.18 for several LP modes. However, these modes are degenerate with the modes polarized along the y-axis that have the same contour plots, but rotated by 90°. These other modes correspond to solving the mode equation with ϕ_0 in (6.29) set to 90°.

6.14.4 LP Mode Radial Field Plots

Consider the waveguide in Figure 6.19 [8]. Although the LP_{01} mode is well-confined to the core, the electric field for many of the higher order LP modes can extend well into the cladding as shown in

TABLE 6.4
LP Modes for Figure 6.17

LP-mode designation	EH/HE mode designation	Mode index
LP_{01}	$HE_{11} \times 2$	1.49601699
LP_{11}	TE_{01}, TM_{01}, $HE_{21} \times 2$	1.490
LP_{21}	$EH_{11} \times 2$, $HE_{31} \times 2$	1.482
LP_{02}	$HE_{12} \times 2$	1.47901664
LP_{31}	$EH_{21} \times 2$, $HE_{41} \times 2$	1.472
LP_{12}	TE_{02}, TM_{02}, $HE_{22} \times 2$	1.466
LP_{41}	$EH_{31} \times 2$, $HE_{51} \times 2$	1.460
LP_{22}	$EH_{12} \times 2$, $HE_{32} \times 2$	1.45
LP_{03}	$HE_{13} \times 2$	1.44864888
LP_{51}	$EH_{41} \times 2$, $HE_{61} \times 2$	1.447
LP_{32}	$EH_{22} \times 2$, $HE_{42} \times 2$	1.435
LP_{13}	TE_{03}, TM_{03}, $HE_{23} \times 2$	1.430
LP_{61}	$EH_{51} \times 2$	1.432
LP_{42}	$EH_{32} \times 2$, $HE_{52} \times 2$	1.416
LP_{71}	$EH_{61} \times 2$, $HE_{81} \times 2$	1.415
LP_{23}	$EH_{13} \times 2$, $HE_{33} \times 2$	1.409
LP_{04}	$HE_{14} \times 2$	1.407

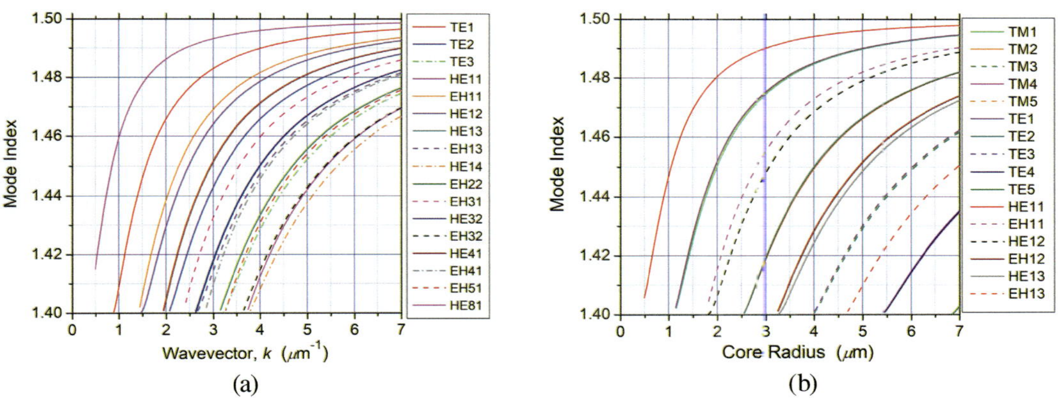

FIGURE 6.17 Mode dispersion as a function of (a) wavevector and (b) core radius.

Figure 6.20(a). The mode indices for several LP modes are plotted as a function of V parameter in Figure 6.20(b).

6.14.5 Graded Index Fiber

As discussed previously, graded index fiber can be modeled by using multiple concentric shells around the core with a radial refractive index variation given by (6.208). Consider the case of a fiber with 9 concentric cylindrical shells of equal thicknesses as shown in Figure 6.21(a). The radial variation of the refractive index is plotted for several different values of the parameter g in Figure 6.21(b). As g increases, the refractive index profile approaches that of a step index fiber.

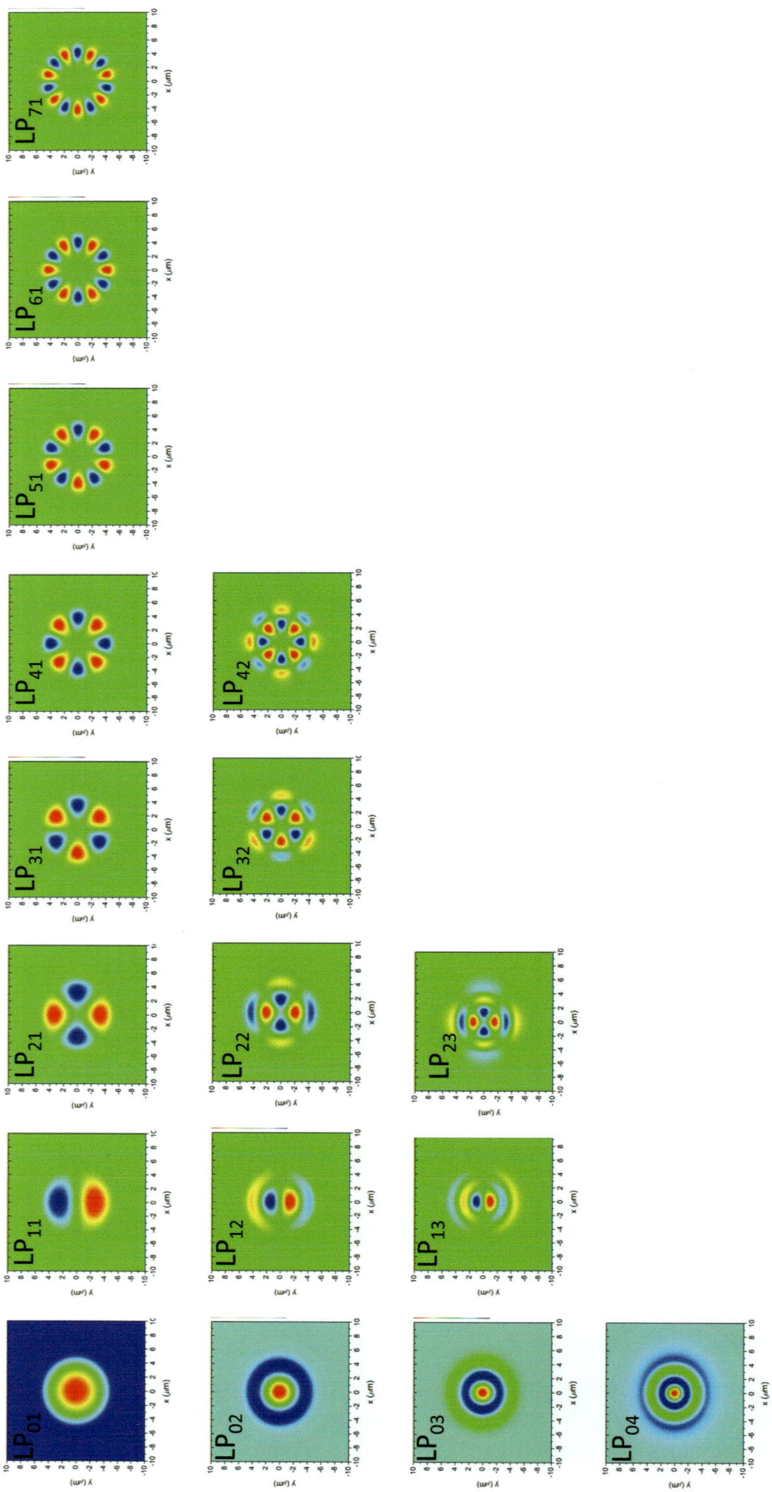

FIGURE 6.18 E_x electric field component for LP modes of the fiber waveguide in Figure 6.17. These modes are degenerate, so there is also a corresponding linearly polarized field plot for E_y that is rotated by 90°.

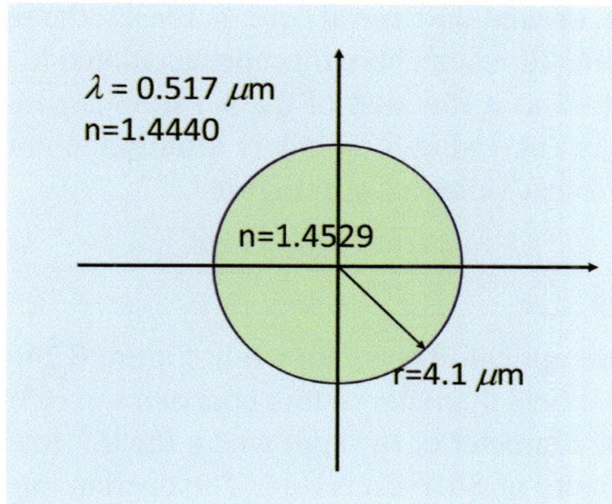

FIGURE 6.19 Example of radial dependence of electric field for LP modes.

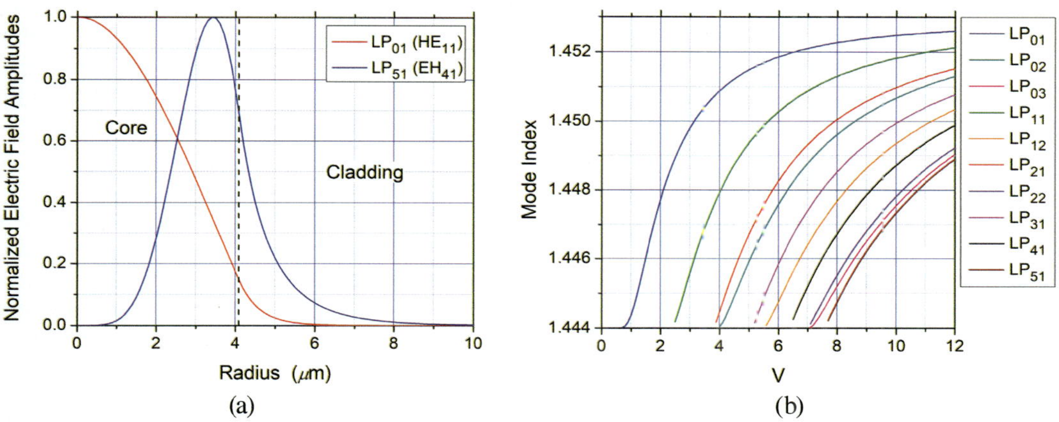

FIGURE 6.20 (a) Radial dependence of $|E|$ for two LP modes. Dashed line is radius of the core. (b) Mode indices as a function of V parameter after Figure 9 of [8].

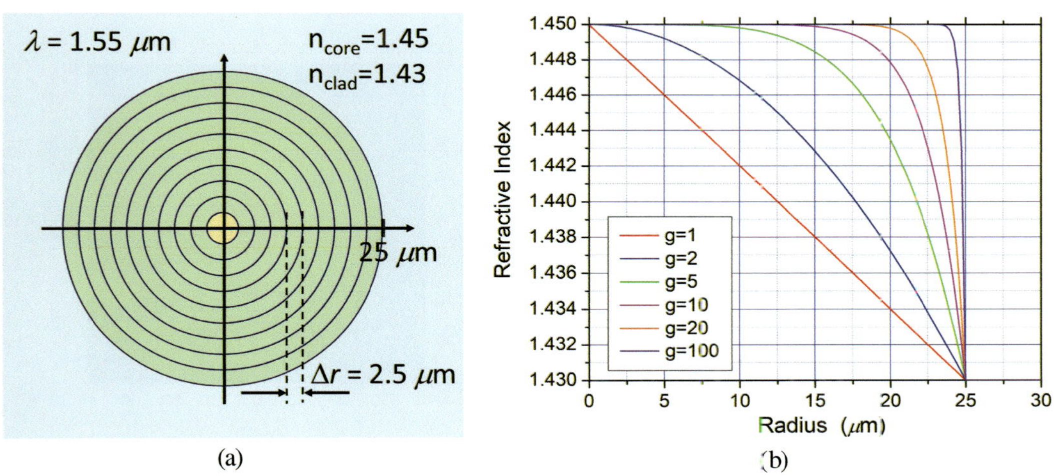

FIGURE 6.21 (a) Graded index fiber with nine shells. (b) Refractive index profile as a function of radius for various values of g according to (6.209).

As the number of cylinders around the central core increases, the required precision of the calculation also increases. Therefore, all results here must be calculated in quadruple (128-bit) precision. The HE_{11} mode index is plotted as a function of the g-parameter in Figure 6.22(a) and the mode profile for a step index fiber in (b). Using 9 cylinders around the core, the HE_{11} E_x field profile is calculated and plotted for different values of g in Figure 6.23.

6.14.6 Overlap Integrals

Consider SMF-28 single mode optical fiber as shown in Figure 6.24(a). The calculated HE_{11} mode index is 1.44915636. The mode field diameter of this fiber at a wavelength of 1.55 μm is specified [9] as 10.4 ± 0.8 μm, which is the diameter of the spot where the $|E|^2$ intensity drops to $1/e^2$ of its peak value, and the numerical aperture of SMF-28 is 0.14. The overlap integral calculation is performed in Cartesian coordinates and the contour plot of the electric field is calculated from −10 to +10 μm at intervals of 0.05 μm as shown in Figure 6.24(b).

6.14.6.1 Gaussian Focused Spot

The calculated amplitudes of the Gaussian electric field along the x-axis for several different values of intensity FWHM are compared to that of the SMF-28 fiber field in Figure 6.25(a). The overlap integral is plotted as a function of the FWHM of the spot size in Figure 6.25(b) and the peak value is 0.9972 which occurs for a FWHM intensity spot size of 5.9 μm corresponding to a e^{-2} full width spot size (mode diameter) of 10.02 μm, which is in good agreement with the specifications.

6.14.6.2 Airy Focused Spot

The calculated amplitude of the Airy electric field along the x-axis is compared to that of the fiber in Figure 6.26(a) for several Airy fields with different mode diameters. The Airy field with a mode diameter of 9.5 μm overlaps the HE_{11} field over much of its central radial variation, but the overlap integral actually peaks for an Airy field with a FWHM spot intensity of 5.69 μm as shown in Figure 6.26(b), corresponding to a first zero-crossing at 9.1 μm and a mode diameter of 12.0 μm. The numerical aperture for this modeled lens at the maximum overlap is 0.11. A numerical aperture at the fiber specification of 0.14 occurs for a FWHM spot size of only 5.7 μm or a mode diameter of 9.1 μm. The square of the optimum overlap integral, which approximates the optimized optical power transfer to the fiber, is 0.79 which is close to the power transfer for the ideal case.

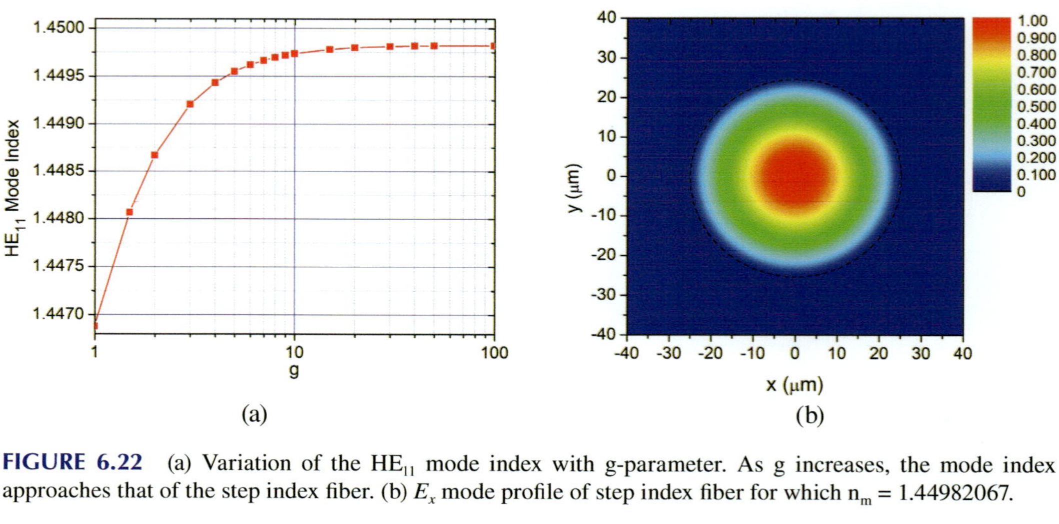

(a) (b)

FIGURE 6.22 (a) Variation of the HE_{11} mode index with g-parameter. As g increases, the mode index approaches that of the step index fiber. (b) E_x mode profile of step index fiber for which $n_m = 1.44982067$.

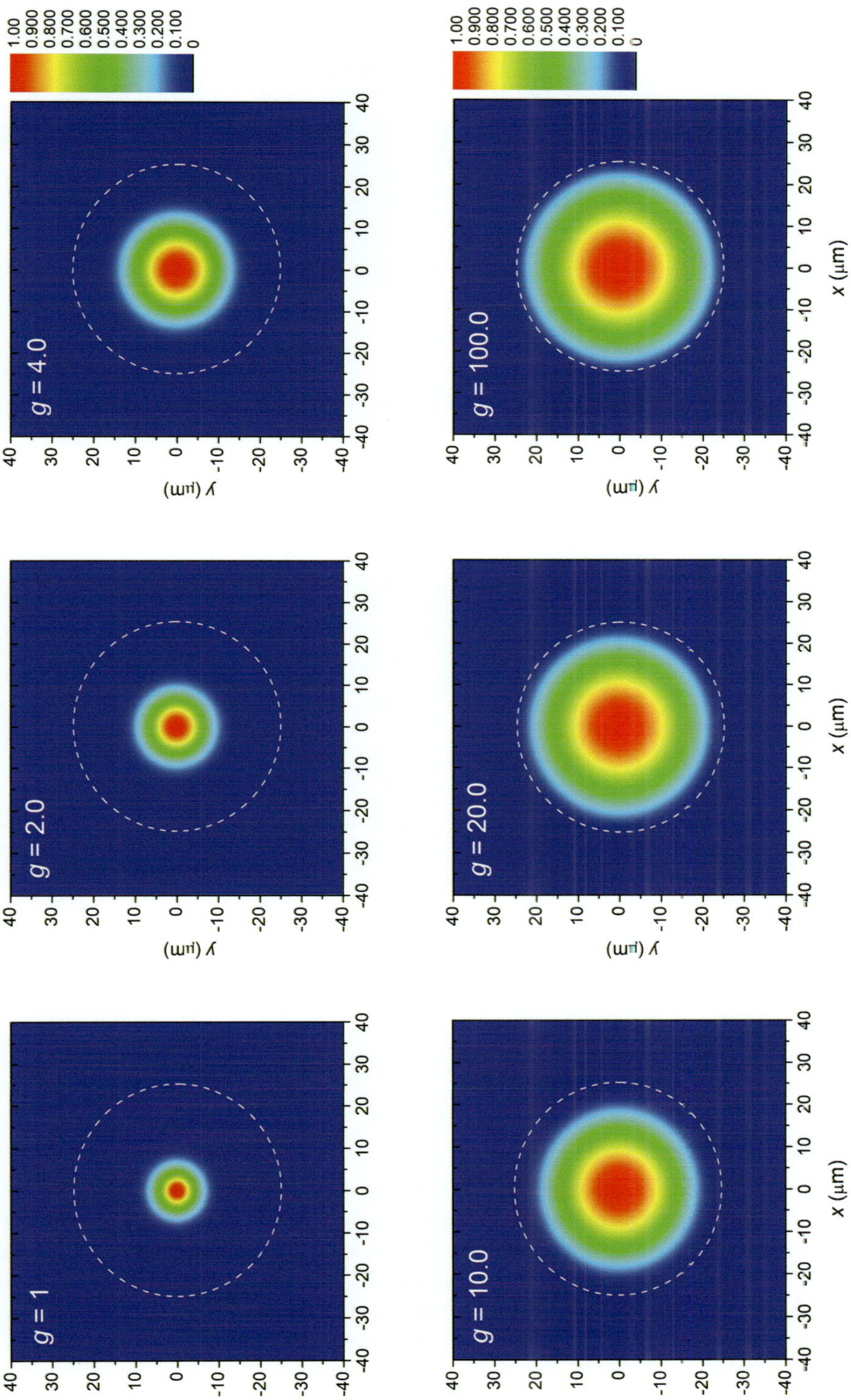

FIGURE 6.23 HE_{11} $Im(E_x)$ mode profile for several values of the g-parameter. The mode profile for g = 100 is almost indistinguishable from that of the step index fiber in Figure 6.22(b).

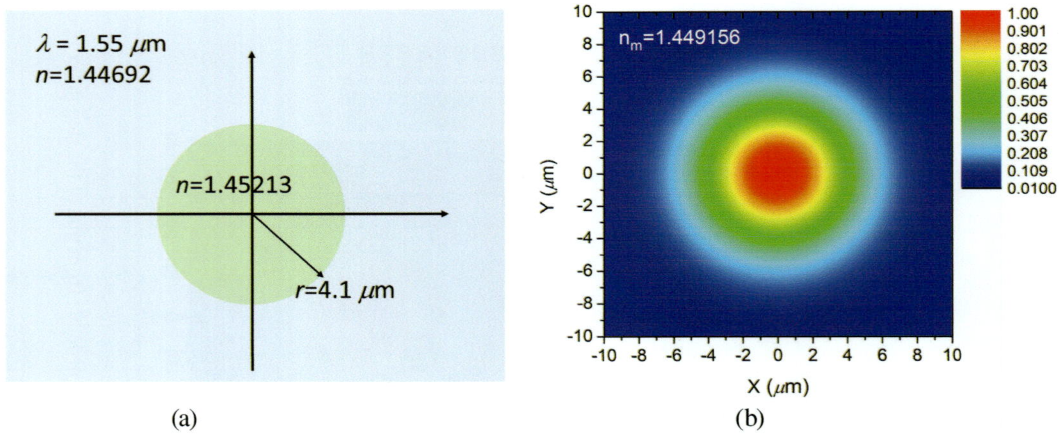

FIGURE 6.24 (a) SMF-28 fiber. (b) Contour plot of the electric field for the HE_{11} mode of SMF-28 optical fiber.

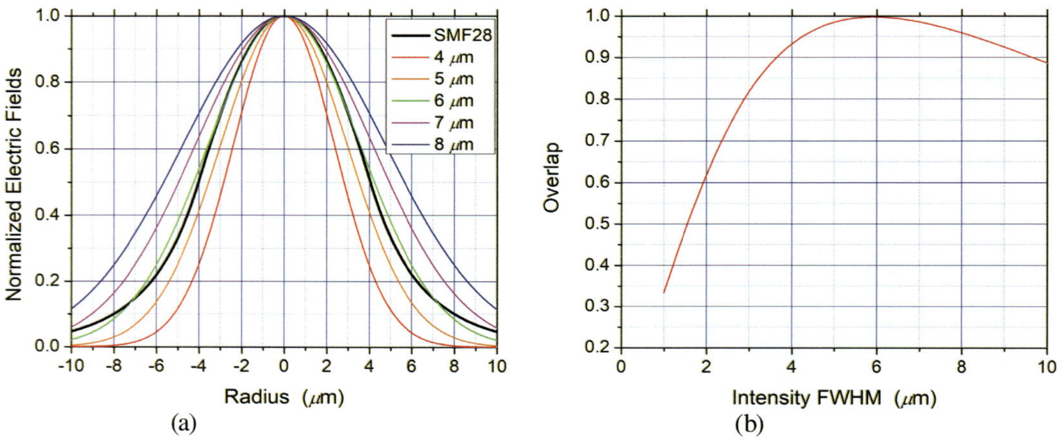

FIGURE 6.25 (a) Plot of the electric field along the x-axis for the fiber and the calculated Gaussian field for different values of the intensity FWHM. (b) Overlap integral of HE_{11} mode with Gaussian spots as a function of the FWHM intensity. The overlap is maximized for a FWHM radius of 5.9 μm at a value of 0.997. The overlap for the intensity is, therefore, 0.994.

6.14.7 Far Field Radiation

The far field radiation pattern is calculated for SMF-28 fiber using the parameters in Table 6.5 and plotted in Figure 6.27. The electric and magnetic fields are first calculated for the HE_{11} mode in the xy plane between ±10 μm in increments of 0.2 μm. The x-polarized electric field in the far field is calculated for three different slant angles of the end face: 0°, 8°, and 20°. The angular offset of the optical spot in the far field agrees well with the approximate value from Snell's law, and the estimated numerical aperture from the radiation angle (half angle at which the Poynting vector drops to 5% of its peak value) is also close to that of the fiber specification.

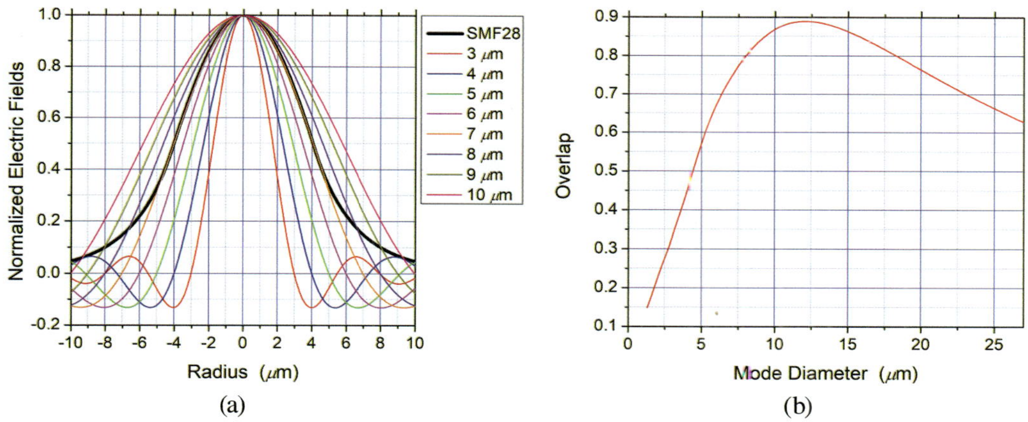

FIGURE 6.26 (a) Plot of the electric field along the *x*-axis of the SMF28 fiber end face and the calculated Airy focused field for different values of the mode diameter. (b) Overlap integral of HE_{11} mode with Airy functions of various mode diameters. The overlap is maximized for a FWHM spot size of 7.51 μm at a value of 0.889. The overlap for the *intensity* is, therefore, 0.790. This FWHM spot size corresponds to a first zero radius of 8.9 μm and a mode diameter of 12.1 μm.

TABLE 6.5
Far Field Radiation from SMF-28 Fiber

Slant angle	FWHM angle	Radiation angle	NA	Offset angle	Snell
0°	6.16°	6.52°	0.11	0°	0°
8°	6.18°	6.55°	0.11	3.58°	3.97°
20°	6.33°	6.71°	0.12	9.43°	10.6°

6.14.8 Multi-Shell Waveguide for Airy Mode Coupling

Recently a multimode waveguide has been designed to transport an Airy-like mode as one of its modes [10]. The refractive index as a function of radius is given by the equation,

$$r_0^2\, k_0^2 \left(n^2 - n_0^2\right) = -\frac{1}{\psi}\left(\frac{d^2\psi}{dR^2} + \frac{1}{R}\frac{d\psi}{dR}\right), \tag{6.204}$$

where $R = r/r_0$ is the radius r normalized by the radius r_0 of the first dark ring of the Airy function, n_0 is an effective index, and

$$\psi_m\left(R\right) = \sum_{i=1}^{3} A_{m,i}\left[J_0\left(P_{m,i}R\right) + C_{m,i}\, Y_0\left(P_{m,i}R\right)\right]. \tag{6.205}$$

m denotes a zero in the Airy function. The first dark ring occurs at $m = 1$. If the radius r lies between dark ring m and $m + 1$, then the function $\psi_m\left(r\right)$ is used to calculate the refractive index. For example, for radii less than the first dark ring, $\psi_0(r)$ is the appropriate function to use in (6.204). J_0 and Y_0 are Bessel functions of order 0 of the first and second kind, respectively. For this example, the values for $P_{m,i}$, $C_{m,i}$, and $A_{m,i}$ are given in Table 6.6.

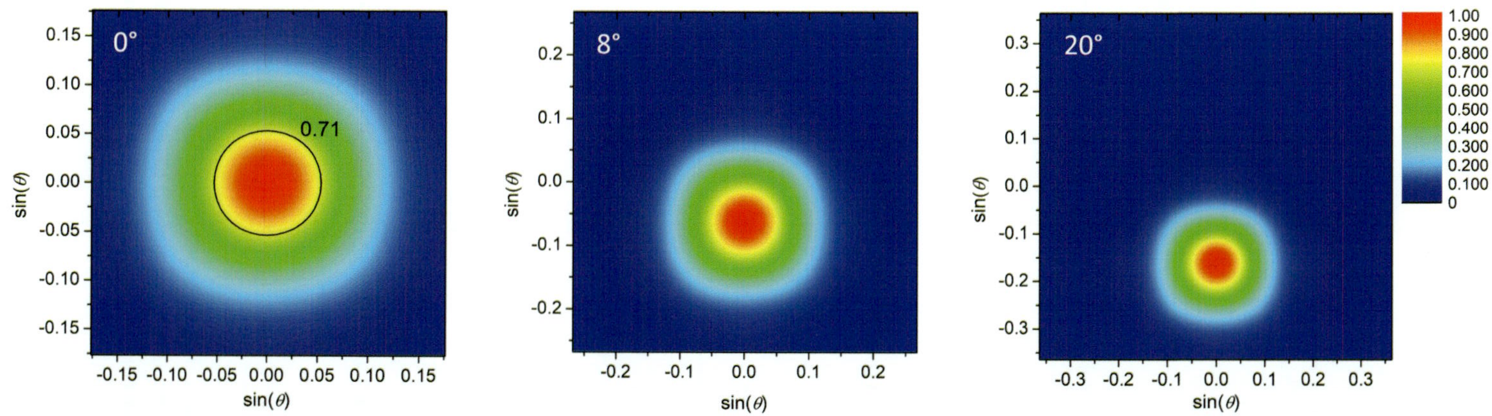

FIGURE 6.27 Contour plots of Im(E_x) in the far field as a function of sine of the polar angle for 0°, 8°, and 20° slant angle of the fiber end face. The black contour corresponds to $E_x = 0.71$, the FWHM of the$|E|^2$ intensity.

TABLE 6.6
Parameters for Computing Refractive Index [5]

m	i	$P_{m,i}$	$C_{m,i}$	$A_{m,i}$
0	1	2.4048	0	0.45734
	2	5.5201	0	0.063497
	3	8.6537	0	−0.020836
1	1	3.7637	5.0501	0.037719
	2	7.5527	−1.9797	0.018506
	3	11.336	−0.48985	−0.0036321
2	1	3.8053	8.3602	0.013937
	2	7.6205	−1.4872	−0.0086658
	3	11.434	−0.28498	−0.0016756
3	1	3.8177	11.577	0.0073579
	2	7.6406	−1.3281	0.0049240
	3	11.462	−0.20373	−0.00065756

Substituting (6.205) into (6.204),

$$k_0^2\left(n^2-n_{eff}^2\right)=-\frac{1}{\psi}\sum_{i=1}^{3}\left(A_{m,i}\left(\frac{P_{m,i}}{a}\right)^2\left[J_0''\left(\frac{P_{m,i}r}{a}\right)+C_{m,i}Y_0''\left(\frac{P_{m,i}r}{a}\right)\right]\right.$$
$$\left.+A_{m,i}\left(\frac{P_{m,i}}{ar}\right)\left[J_o'\left(\frac{P_{m,i}r}{a}\right)+C_{m,i}\,Y_0'\left(\frac{P_{m,i}r}{a}\right)\right]\right) \tag{6.206}$$

where

$$J_o'(x)\equiv\frac{d}{dx}J_0(x)=-J_1(x), \tag{6.207}$$

$$Y_0'(x)\equiv\frac{d}{dx}Y_0(x)=-Y_1(x), \tag{6.208}$$

$$J_0''(x)\equiv\frac{d^2}{dx^2}J_0(x)=-J_0(x)+\frac{1}{x}J_1(x)=\frac{-J_0(x)+J_2(x)}{2}, \tag{6.209}$$

and

$$Y_0''(x)\equiv\frac{d^2}{dx^2}Y_0(x)=-Y_0(x)+\frac{1}{x}Y_1(x)=\frac{-Y_0(x)+Y_2(x)}{2}. \tag{6.210}$$

(6.204) can be simplified and solved for the radial dependence of the refractive index.

$$n_m(r)=\sqrt{n_0^2+\frac{1}{(ak_0)^2}\left(\frac{\sum_{i=1}^{3}\left(\frac{P_{m,i}}{r_0}\right)^2A_{m,i}\left[J_0\left(\frac{P_{m,i}r}{r_0}\right)+C_{m,i}Y_0\left(\frac{P_{m,i}r}{r_0}\right)\right]}{\sum_{i=1}^{3}A_{m,i}\left[J_0\left(\frac{P_{m,i}r}{r_0}\right)+C_{m,i}\,Y_0\left(\frac{P_{m,i}r}{r_0}\right)\right]}-\upsilon_0\right)}. \tag{6.211}$$

The numerical value of the first dark ring of the Airy pattern, $r_0 = 3.8317$, is used as the normalization constant. $\lambda = 1.55$ μm. The core is divided into an inner core of radius 0.3375 μm and 33 concentric cylinder shells of thickness 0.675 μm around the core for a total diameter of 45.225 μm. The radial dependence of the refractive index is calculated according to (6.220) where $a = 6.75$ μm is the physical radius of the first dark ring of the Airy pattern in the fiber and $\upsilon_0 = 30$ is a normalization constant. $m = 0$ for $r < r_0$; $m = 1$ for $r < 2r_0$; $m = 2$ for $r < 2r_0$; $m = 3$ for $r < 3\,r_0$. $n_0 = 1.44$ is the refractive index for pure silica. The refractive index of each shell is the average value of the refractive index at the inner and outer edges of the shell calculated from (6.220) and the refractive index is a constant 1.423 for $r \geq 3r_0$. The calculated refractive index profile is shown in Figure 6.28.

Quad precision must be used to calculate the mode indices for this fiber geometry. The mode indices found for the first seven orders are shown below.

```
EH/HE modes for m = 0:
1.  n =  1.432053676561370  EH01
2.  n =  1.432039536577653  EH02
3.  n =  1.430263166205386  EH03
4.  n =  1.430238219760268  EH04
5.  n =  1.428387516238018  EH05
6.  n =  1.428349646569545  EH06
7.  n =  1.423116724151773  EH07
8.  n =  1.423104698400784  EH08

EH/HE modes for m = 1:
1.  n =  1.432196573729791  HE11
2.  n =  1.431653217743072  EH11
3.  n =  1.430488893337810  HE12
4.  n =  1.429674288336503  EH12
5.  n =  1.428611084013700  HE13
6.  n =  1.427756815888968  EH13
7.  n =  1.425967279083269  HE14
```

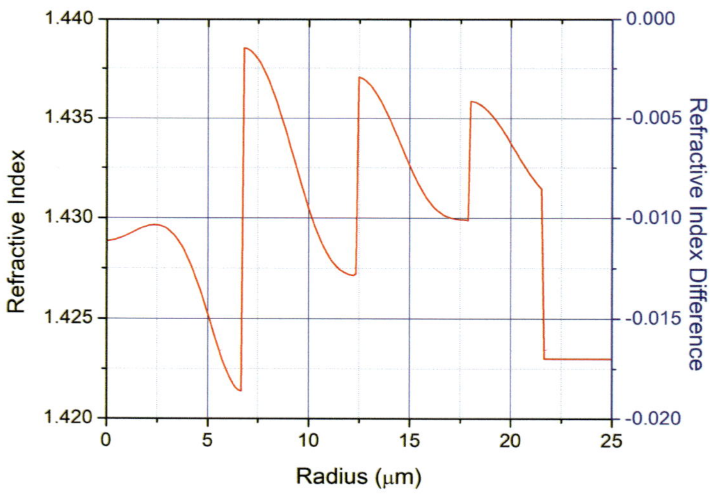

FIGURE 6.28 Plots of the absolute refractive index and the refractive index difference from 1.44 as a function of radius. After Figure 8(a) in [10].

```
EH/HE modes for m = 2:
1.  n =  1.432046923853217  HE21
2.  n =  1.431091372827402  EH21
3.  n =  1.430251799666687  HE22
4.  n =  1.428921514038095  EH22
5.  n =  1.428879285894407  EH23
6.  n =  1.428370336509287  HE23
7.  n =  1.426602703146555  EH24
8.  n =  1.423109760593846  HE24

EH/HE modes for m = 3:
1.  n =  1.431653473934228  HE31
2.  n =  1.430404471084068  EH31
3.  n =  1.429675431646686  HE32
4.  n =  1.428879285894407  EH32
5.  n =  1.428091873876636  EH33
6.  n =  1.427760140730341  HE33
7.  n =  1.424868991111945  EH34

EH/HE modes for m = 4:
1.  n =  1.431091149707293  HE41
2.  n =  1.429616248742088  EH41
3.  n =  1.428921779666032  EH42
4.  n =  1.428879285894407  EH43
5.  n =  1.427147753669215  EH44
6.  n =  1.426604479693264  HE42

EH/HE modes for m = 5:
1.  n =  1.430403600278512  HE51
2.  n =  1.428879285894407  EH51
3.  n =  1.428743848745530  EH52
4.  n =  1.428091577691295  EH53
5.  n =  1.426037281968422  EH54
6.  n =  1.424864974076770  EH55

EH/HE modes for m = 6:
1.  n =  1.429614549973799  HE61
2.  n =  1.428879285894407  EH61
3.  n =  1.427797124478801  EH62
4.  n =  1.427146769766563  EH63
5.  n =  1.424739328913272  EH64

EH/HE modes for m = 7:
1.  n =  1.428879285894407  EH71
2.  n =  1.428741182913327  EH72
3.  n =  1.426777632316869  EH73
4.  n =  1.426034695616892  EH74
5.  n =  1.423259485663194  EH75
```

Figure 6.29(a) shows the E_x field amplitude along the x-axis for the first four HE_{1n} modes, which are identical to the first four LP_{0n} modes, and (b) shows a 3D plot of the LP_{04} mode clearly demonstrating its Airy-like pattern. The E_x field contour plots in Figure 6.30 for these modes are linearly polarized along the x-axis.

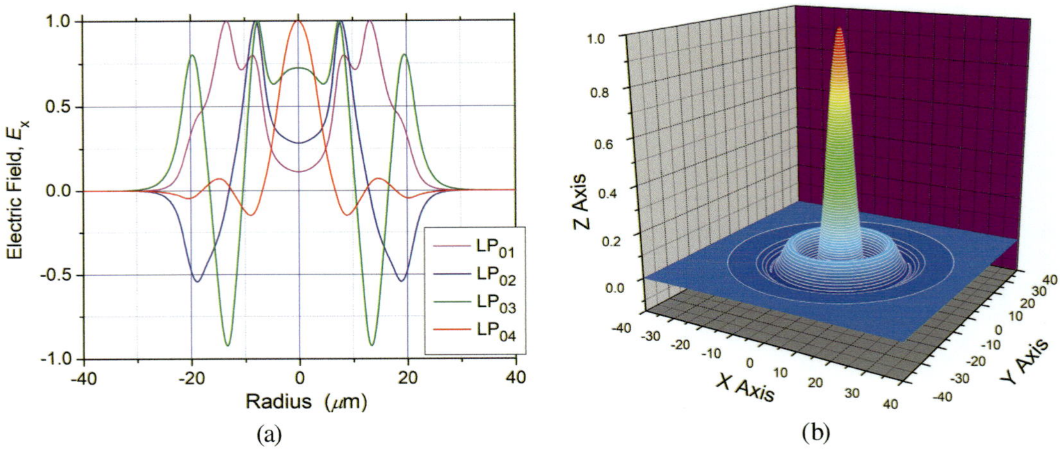

FIGURE 6.29 (a) Normalized radial $|E_x|$ field profile for first four LP modes (HE_{11} – HE_{14}). (b) 3D plot of LP_{04} mode.

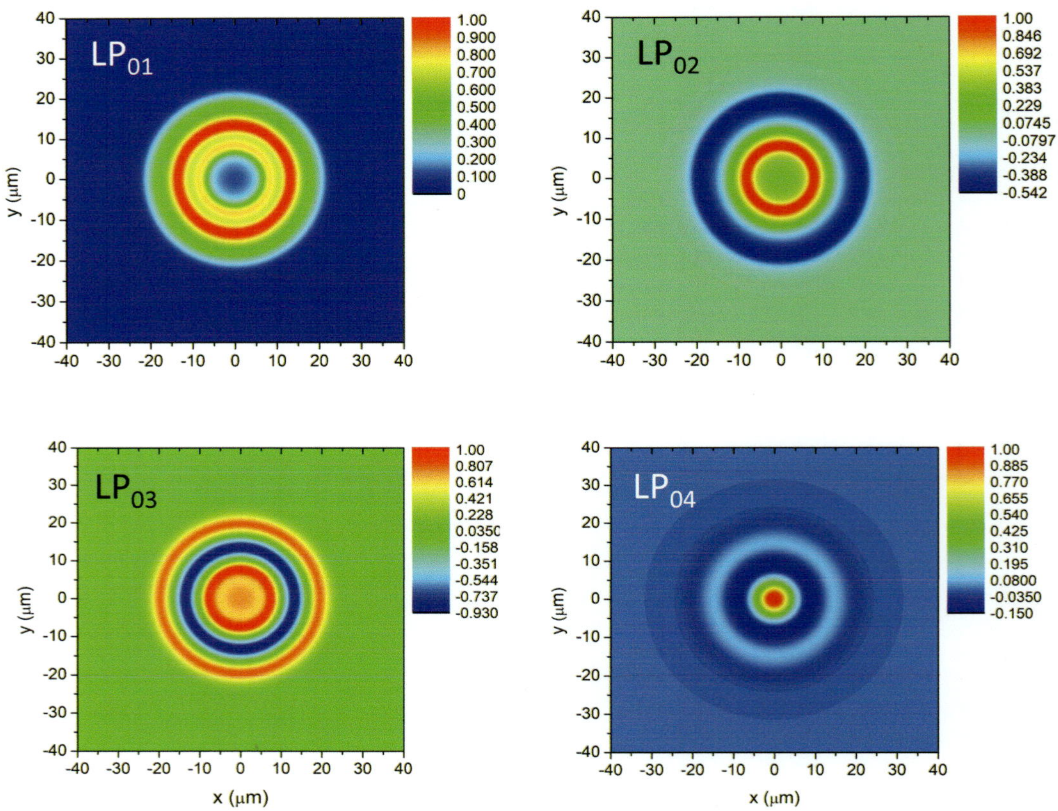

FIGURE 6.30 E_x field contour plots for the LP_{01} – LP_{04} modes at $\lambda = 1.55$ μm. LP_{04} (HE_{14}) is the desired Airy mode.

The overlap integral can also be calculated with the LP_{04} mode. The overlap is maximized at a value of 0.971 for a first dark ring radius of 6.76 μm, slightly larger than the design target radius. The comparison of the Airy function to the calculated electric field along the x-axis is plotted in Figure 6.31(a).

The dispersion of the first order mode indices with wavelength (assuming all refractive indices are constant) is shown in Figure 6.31(b). This is a difficult plot to generate because there are so many modes, many of which cross each other, and for short wavelengths there is insufficient precision even using quad precision. In cases like this, it is generally useful to first calculate the error function at the shortest wavelength over the complete mode index range to discover any precision problems. The error function at $\lambda = 0.6$ μm is shown in Figure 6.32. The loss in precision at large mode indices is apparent. The precision of the calculation limits the mode index calculation to a maximum wavevector of 9 for the lowest LP modes as shown in Figure 6.31(b).

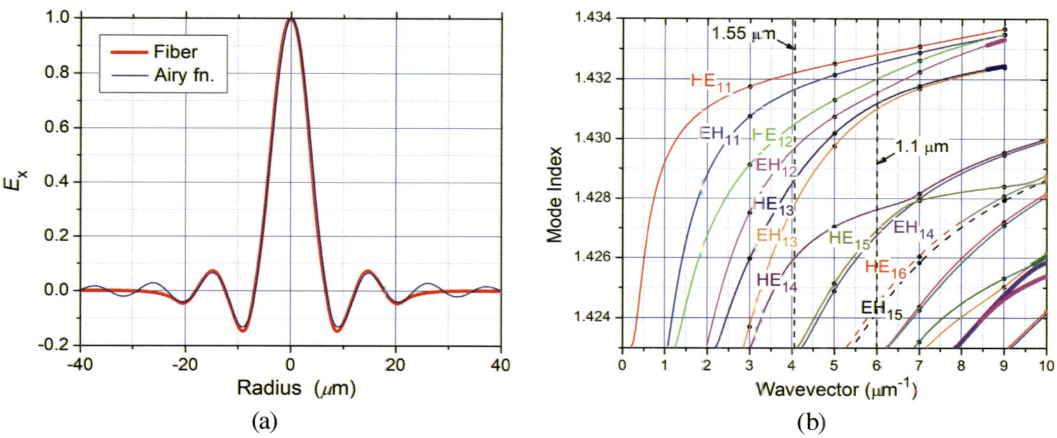

FIGURE 6.31 (a) Overlap of the calculated E_x electric field in the fiber for the LP_{04} mode (red) to that of the exact Airy function (blue). The overlap integral is 0.97. (b) 1st order mode dispersion plot. The design wavelength of 1.55 μm corresponds to a wavevector of 4.054.

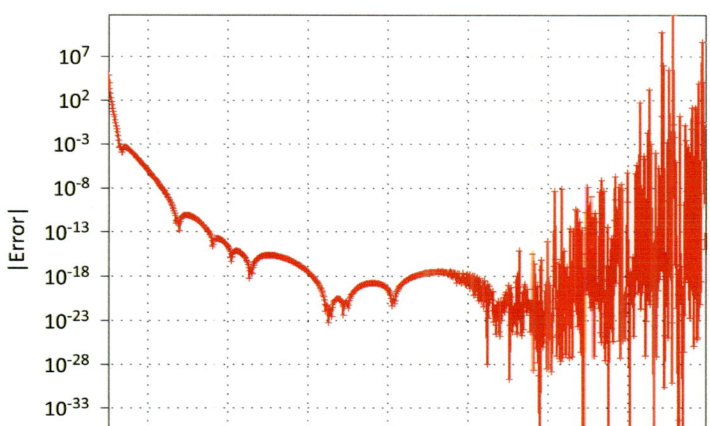

FIGURE 6.32 Absolute value of the error function at $\lambda = 0.6$ μm for the Airy function fiber of Figure 6.28 calculated in quad precision. For mode index values below ~1.432, a distinct kink in the error function is indicative of a solution.

Normally, dispersion curves should be calculated from short wavelengths to long wavelengths in order to not miss any orders. The search algorithm might initially search over the complete mode index range to find all modes within that range. It may then use the previously calculated value for each mode index as the starting point for the next wavelength iteration. This procedure works well except when there are mode crossings, in which case, the search algorithm fails to find one of the modes at the next iteration following a crossing. In that case, one must start the search at a shorter wavelength on the other side of the crossing and iterate towards longer wavelengths. Figure 6.31(b) with multiple mode crossings required searching over several limited wavevector ranges towards longer or shorter wavevectors to generate a complete set of dispersion curves.

The dispersion curves in Figure 6.31(b) are similar to those in Figure 7(b) of [10]. There is an avoided crossing between the LP_{04} (HE_{14}) and LP_{05} (HE_{15}) modes at a wavelength of ~1.1 μm (k≈6.8). For shorter wavelengths (higher frequencies) the LP_{04} mode loses its Airy function character while the LP_{05} mode picks it up as shown in Figure 6.33. For even shorter wavelengths there is another avoided crossing in Figure 6.31(b) at ~0.66 μm ($k = 9.5$) between the LP_{05} and LP_{06} modes which is not evident in the reference figure. A weak avoided crossing at k≈5.7 between the LP_{01} (HE_{11}) and LP_{02} (HE_{12}) modes in the reference figure does not appear in Figure 6.31(b), nor does a weak avoided crossing at k≈8.5 between the LP_{08} and LP_{09} modes. The avoided crossing in Figure 6.31(b) at $k \approx 9.5$ between the LP_{05} (HE_{15}) and LP_{06} (HE_{16}) modes does not appear in the reference figure.

6.14.9 W-Fibers

W-fibers are composed of a core and two claddings with a refractive index profile as shown in Figure 6.34(a). These fibers can be designed with single-mode anomalous dispersion, single-mode operation with large core size, a more tightly confined intensity within the core, and a non-zero cutoff frequency for the fundamental mode [11, 12]. In this example, the core and inner cladding radii are taken from Figure 9 of [11]. The outer cladding index n_3 is not specified in the reference but is chosen to be pure silica. The refractive index of the core and inner cladding are offset from that of pure silica by the amount specified in the figure caption. The dispersion for three fibers is calculated, that of a step index fiber and two W-fibers with parameters given in Figure 6.35. The results are shown in Figure 6.36. The dispersion for W-fiber (b) is nearly zero at the design wavelength of 1.55 μm. These results are slightly different from those in Figure 9 of [11]. This is probably due to somewhat different wavelength-dependent refractive indices in the calculations which were not explicitly specified in the reference. Also, the reference based all calculations on the weakly guiding scalar approximation for the LP modes rather than the exact vector calculation.

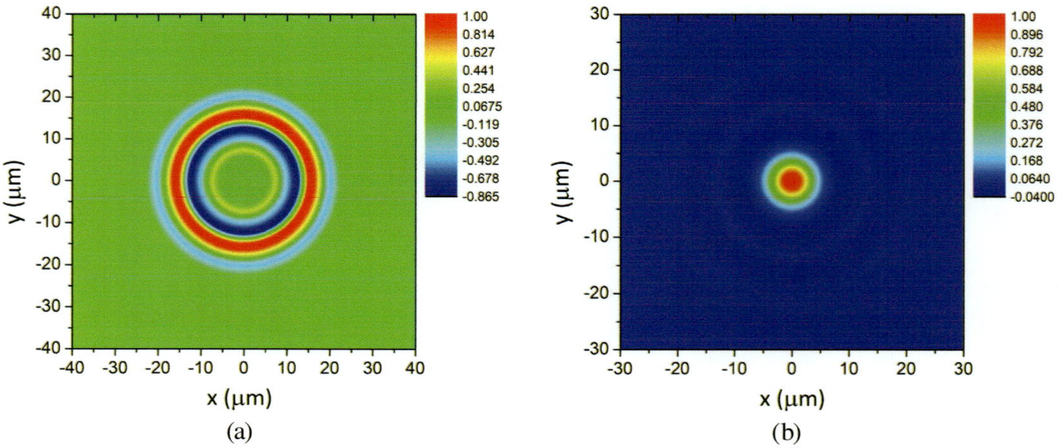

FIGURE 6.33 (a) Contour plot of Re(E_x) for the LP_{04} mode and (b) of Im(E_x) for the LP_{05} mode at a wavelength of 0.775 μm.

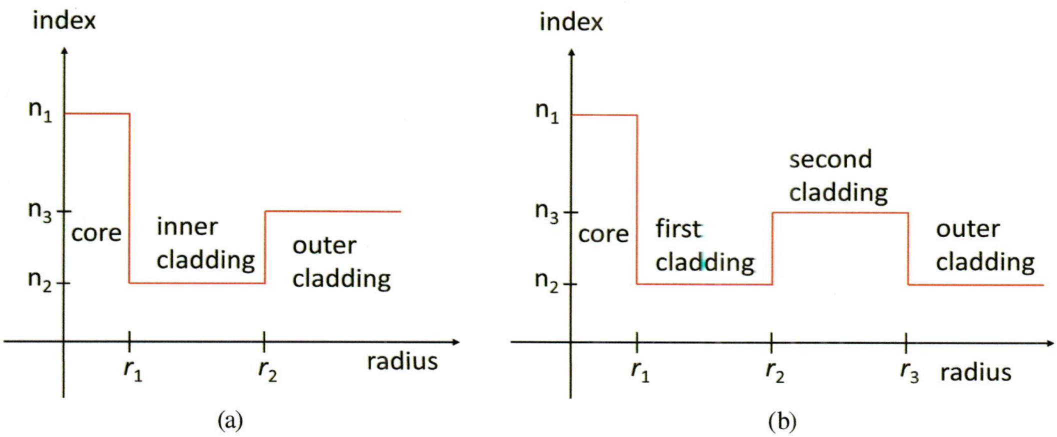

FIGURE 6.34 Refractive index profile for W-fiber (a) with one inner cladding, and (b) with two inner claddings.

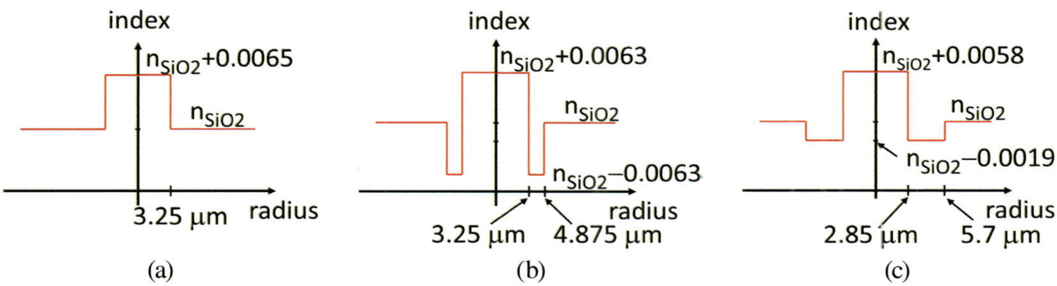

FIGURE 6.35 Refractive index profiles for three fibers. (a) Step index fiber. (b) W-fiber with narrow inner cladding. (c) W-fiber with a larger inner cladding but a smaller decrease in refractive index from pure silica.

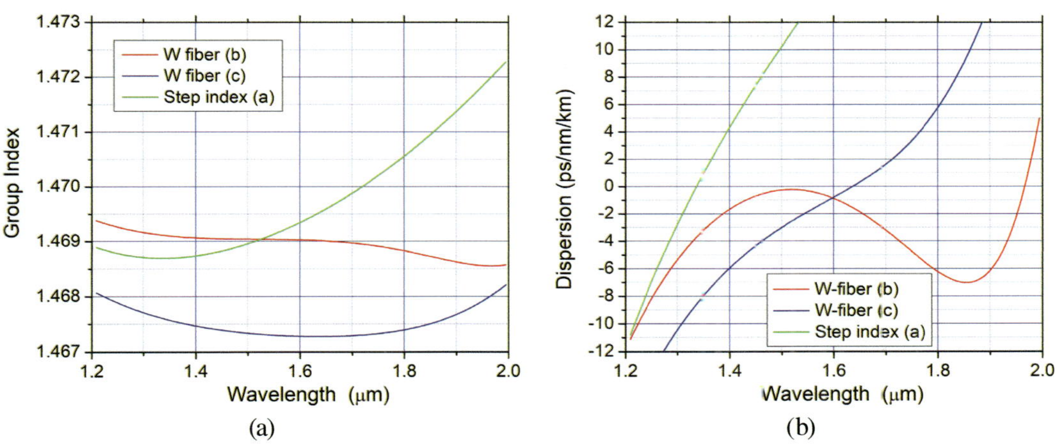

FIGURE 6.36 (a) Mode index and (b) dispersion as a function of wavelength for the three fibers in Figure 6.35.

The pure silica index file is a three term Sellmeier fit,

$$n^2 = 1 + \sum_{i=1}^{3} \frac{A_i \lambda^2}{\lambda^2 - \lambda_i^2}. \tag{6.212}$$

A fourth term can be included with $\lambda_4 = 0$ to provide a constant offset to the dielectric constant,

$$n^2 = n_{\mathrm{SiO_2}}^2 + A_4. \tag{6.213}$$

To model doped silica with small changes in refractive index, the Sellmeier coefficients for pure silica are selected and A_4 is then added to fit the desired refractive index at a specific wavelength. Using Figure 9 in [11] as a model with the outer cladding as pure silica, the index for the step index fiber core in plot (a) that matches the refractive index offset at λ = 1.55 μm is:

Sellmeier S_core

A_1	A_2	A_3	A_4
0.6961663	0.4079426	0.8974794	0.018815
0.0684043	0.1162414	9.8961610	0.000000

The index file for the core of the W-fiber in plot (b) is:

Sellmeier W_core_b

A_1	A_2	A_3	A_4
0.6961663	0.4079426	0.8974794	0.018234
0.0684043	0.1162414	9.8961610	0.000000

The index file for the inner cladding of the W-fiber in plot (b) is:

Sellmeier W_clad_b

A_1	A_2	A_3	A_4
0.6961663	0.4079426	0.8974794	-0.018155
0.0684043	0.1162414	9.8961610	0.000000

The index file for the core of the W-fiber in plot (c1) is:

Sellmeier W_core_c1

A_1	A_2	A_3	A_4
0.6961663	0.4079426	0.8974794	0.016784
0.0684043	0.1162414	9.8961610	0.000000

The index file for the inner cladding of the W-fiber in plot (c1) is:

Sellmeier W_clad

A_1	A_2	A_3	A_4
0.6961663	0.4079426	0.8974794	-0.005484
0.0684043	0.1162414	9.8961610	0.000000

Sometimes a second inner cladding is added to W-fiber as shown in Figure 6.34(b). If the refractive index profile for the W-fiber is modified to that in Figure 6.37(a), for example, then new modes appear. The mode dispersion is plotted in Figure 6.37(b).

It should be noted that the calculation should generally be run from smaller wavelengths to larger (or larger wavevectors to smaller) so that the search algorithm can locate all the modes at the first wavelength and can then subsequently just follow the modes to larger wavelengths. However, in this example there is a mode crossing at λ~1.35 μm between the HE_{22} mode and the EH_{21} mode, two

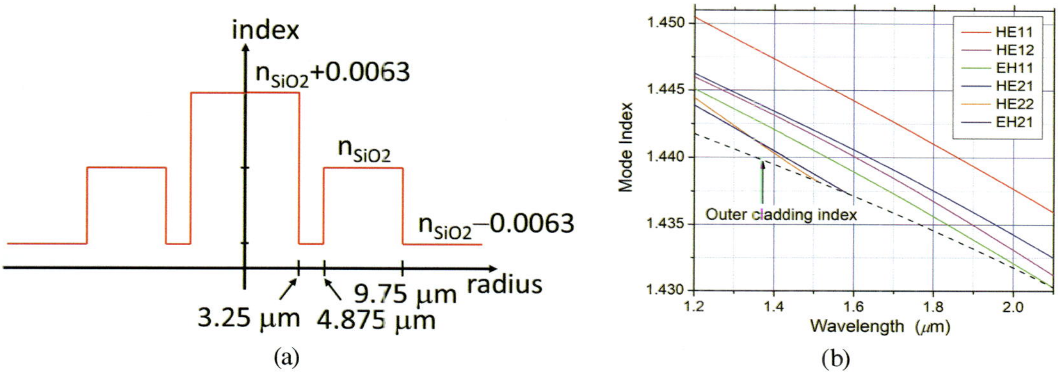

FIGURE 6.37 (a) W-fiber design with three claddings. (b) Mode dispersion. Note that the TE and TM modes essentially overlap the EH/HE modes and are not explicitly labelled. The HE_{22} and EH_{21} modes cross.

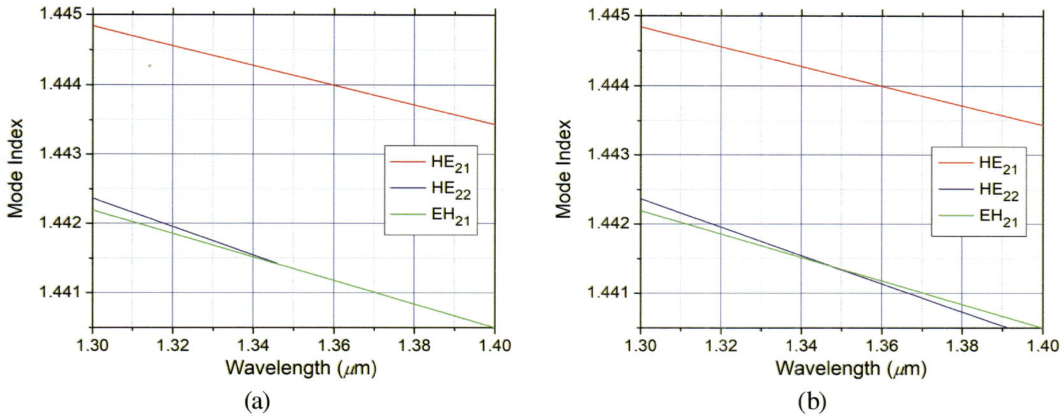

FIGURE 6.38 These plots are obtained by varying the wavelength from small to larger values. (a) The search increment is 0.001 μm. In this case, the search algorithm loses the HE_{22} mode at a mode crossing between two modes of the same order. (b) The search increment is 0.0001 μm. The search algorithm correctly finds both modes but incorrectly shows the modes as not crossing.

modes of the same order. Depending on the size of the search increment, the search algorithm may find only one or both of these modes as it continues to larger wavelengths as shown in Figure 6.38. If both modes are found, the algorithm may incorrectly switch the modes after the crossing, thinking that the crossing was an avoided crossing. If the search increment is too large, the search algorithm may only locate the mode with the larger index value, in this case the EH_{21} mode. To find the lower index mode (HE_{21}), the search should be rerun in the region of the crossing with a smaller search increment, or beginning after the crossing at $\lambda \sim 1.36$ μm.

REFERENCES

1 A. Safaai-Jazi and G. L. Yip, "Classification of hybrid modes in cylindrical dielectric optical waveguides," *Radio Sci.* **12** (1977) 603–609.
2 T. Okoshi, "Wave Theory of Uniform-Core Fibers," in *Optical Fibers* (New York: Academic Press, 1982) chapter 4, 63.
3 S. A. Shelkunoff, "A general radiation formula," *Proc. I.R. E.* **27** (1939) 660–666.

4 S. Ramo, J. R. Whinnery, and T. Van Duzer, *Fields and Waves in Communication Electronics* (New York: John Wiley & Sons, 1965) 661 ff.
5 T. Saktioto et al., "Birefringence and polarization mode dispersion phenomena of commercial optical fiber in telecommunication networks," *J. Phys. Conf. Ser.* **1655** (2020) 012160. https://doi.org/10.1088/1742-6596/1655/1/012160.
6 Saktioto et al., "Birefringence and polarization mode dispersion phenomena of commercial optical fiber in telecommunicaton networks," *J. Phys. Conf. Ser.* **1655** (2020) 012160.
7 E. N. Glytsis, "Cylindrical dielectric waveguides," *National Technical University of Athens*. http://users.ntua.gr/eglytsis/IO/Cylindrical_Waveguides_p.pdf
8 F. Bucholtz and J. M. Singley, "A brief introduction to core modes in optical fiber," *Naval Research Laboratory*. NRL/5650/MR-2021/2. (2021). https://apps.dtic.mil/sti/pdfs/AD1124366.pdf
9 *Corning® SMF-28™ Optical Fiber Product Information Data Sheet, PI1036* (April, 2002).
10 I. Gris-Sánchez, D. Van Ras, and T. A. Birks, "The Airy fiber: An optical fiber that guides light diffracted by a circular aperture," *Optica* **3** (2016) 270–276.
11 M. Monerie, "Propagation in doubly clad single-mode fibers," *IEEE J. Qu. Elect.* **QE-18** (1982) 535–542.
12 R. A. Sammut, "Range of monomode operation of W-fibres," *Opt. Qu. Elect.* **10** (1978) 509–514.

7 Focused Fields

7.1 INTRODUCTION

It is often useful to know the field amplitudes near the focal point of a lens as input to FDTD calculations or for calculating overlap integrals in order to optimize coupling into waveguides. There are a variety of scalar and vector approximations to the focal fields. This chapter describes the technique of Richards and Wolf [1–3] which provides a fully vectorial calculation for the electric and magnetic field amplitudes in the vicinity of the focal plane. Results for several different input polarization states and apodizations are also described. The procedure for calculating coupling efficiency into a waveguide using the overlap integral is explained.

7.2 ELECTRIC FIELD

7.2.1 Linear and Elliptical Polarization

This chapter follows the analysis of [3] (without the solid immersion lens) for calculating the fields near the focus of an aplanatic lens (without spherical or coma aberrations). A diagram of light rays passing through the lens is shown in Figure 7.1. The incident ray with x-amplitude a_x^0 and y-amplitude a_y^0 (generally complex numbers) is refracted at angle θ to the focal point. This ray has an azimuthal angle ϕ with respect to the x-axis. If the incident beam is linearly or elliptically polarized, it can be expressed in terms of s- (TE) and p- (TM) polarizations to make the calculation of the refracted ray simpler. Energy conservation of the refracted ray introduces an amplitude factor of $\sqrt{\cos\theta}$ as discussed in [2] and [3]. The field can then be converted into amplitudes in the primed Cartesian coordinate system in which the z' axis is parallel to the z axis and the y' axis is parallel to the s-polarization. Finally, the field amplitudes can be converted to the unprimed Cartesian coordinate system near the focal point.

According to the Richards-Wolf theory, the electric field at point $\mathbf{p} = (x_p, y_p, z_p)$ near the Gaussian focus is

$$\boldsymbol{E}(p) = \frac{i}{\lambda}\int_0^{\theta_m} \sin\theta d\theta \int_0^{2\pi} e^{-ik_0(\hat{s}\cdot p)}\, \boldsymbol{a}(\theta,\phi)\, d\phi \tag{7.1}$$

where θ_m is the angular semi-aperture of the lens and λ is the free space wavelength. The free space wavevector is

$$k_0 = \frac{2\pi}{\lambda}. \tag{7.2}$$

$\hat{\boldsymbol{a}}$ is the amplitude vector of the electric field. $\hat{\boldsymbol{s}}$ is the unit vector along the focused ray, [4] which is

$$\hat{\boldsymbol{s}} = -\sin\theta\cos\phi\,\hat{\boldsymbol{x}} - \sin\theta\sin\phi\,\hat{\boldsymbol{y}} + \cos\theta\,\hat{\boldsymbol{z}}. \tag{7.3}$$

The point $\boldsymbol{p}$ is

$$\boldsymbol{p} = (x_p, y_p, z_p) = (r_p\cos\phi_p, r_p\sin\phi_p, z_p) \tag{7.4}$$

DOI: 10.1201/9781003714231-7

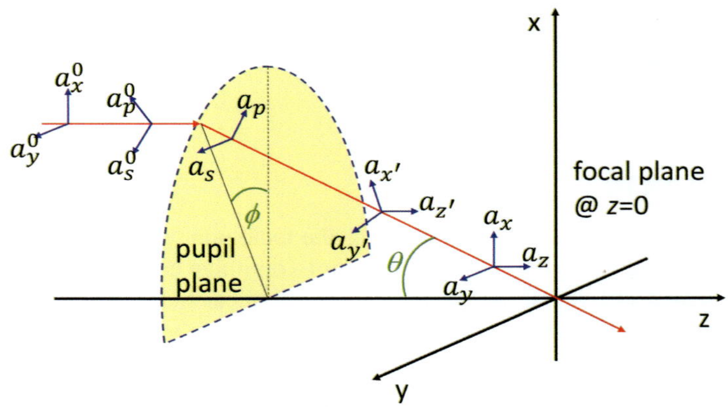

FIGURE 7.1 Diagram of light rays from lens to focused spot.

so

$$\hat{\boldsymbol{s}}\cdot\boldsymbol{p} = -r_p \sin\theta \cos\left(\phi-\phi_p\right) + z_p \cos\theta. \tag{7.5}$$

The s- and p-polarized components of the fields before the lens are related to Cartesian coordinates by

$$\begin{bmatrix} a_p^0 \\ a_s^0 \end{bmatrix} = \begin{bmatrix} \cos\phi & \sin\phi \\ -\sin\phi & \cos\phi \end{bmatrix} \begin{bmatrix} a_x^0 \\ a_y^0 \end{bmatrix}. \tag{7.6}$$

The amplitude vector after the lens is calculated from the amplitude vector before the lens, assuming no loss in transmission through the lens.

$$\begin{bmatrix} a_p \\ a_s \end{bmatrix} = \sqrt{\cos\theta} \begin{bmatrix} a_p^0 \\ a_s^0 \end{bmatrix}. \tag{7.7}$$

As previously mentioned, the factor $\sqrt{\cos\theta}$ accounts for power conservation. The x', y', z' coordinate system is defined such that z' is parallel to the z axis and y' is parallel to the s-polarization state.

$$\begin{bmatrix} a_x' \\ a_y' \\ a_z' \end{bmatrix} = \begin{bmatrix} \cos\theta & 0 \\ 0 & 1 \\ \sin\theta & 0 \end{bmatrix} \begin{bmatrix} a_p \\ a_s \end{bmatrix}. \tag{7.8}$$

The amplitude of the electric field in the focal plane Cartesian coordinate system is

$$\boldsymbol{a} = \begin{bmatrix} a_x \\ a_y \\ a_z \end{bmatrix} = \begin{bmatrix} \cos\phi & -\sin\phi & 0 \\ \sin\phi & \cos\phi & 0 \\ 0 & 0 & 1 \end{bmatrix} \begin{bmatrix} a_x' \\ a_y' \\ a_z' \end{bmatrix}. \tag{7.9}$$

Combining (6)–(9) gives

$$\begin{bmatrix} a_x \\ a_y \\ a_z \end{bmatrix} = \sqrt{\cos\theta}\begin{bmatrix} \cos\phi & -\sin\phi & 0 \\ \sin\phi & \cos\phi & 0 \\ 0 & 0 & 1 \end{bmatrix}\begin{bmatrix} \cos\theta & 0 \\ 0 & 1 \\ \sin\theta & 0 \end{bmatrix}\begin{bmatrix} \cos\phi & \sin\phi \\ -\sin\phi & \cos\phi \end{bmatrix}\begin{bmatrix} a_x^0 \\ a_y^0 \end{bmatrix} \tag{7.10}$$

$$= \sqrt{\cos\theta}\begin{bmatrix} \cos^2\phi\cos\theta + \sin^2\phi & \sin\phi\cos\phi\cos\theta - \sin\phi\cos\phi \\ \sin\phi\cos\phi\cos\theta - \sin\phi\cos\phi & \sin^2\phi\cos\theta + \cos^2\phi \\ \cos\phi\sin\theta & \sin\phi\sin\theta \end{bmatrix}\begin{bmatrix} a_x^0 \\ a_y^0 \end{bmatrix} \tag{7.11}$$

or

$$a_x = \frac{\sqrt{\cos\theta}}{2}\left\{\left[(1+\cos\theta)-(1-\cos\theta)\cos 2\phi\right]a_x^0 - (1-\cos\theta)\sin 2\phi a_y^0\right\}, \tag{7.12}$$

$$a_y = \frac{\sqrt{\cos\theta}}{2}\left\{-(1-\cos\theta)\sin 2\phi a_x^0 + \left[(1+\cos\theta)+(1-\cos\theta)\cos 2\phi\right]a_y^0\right\}, \tag{7.13}$$

and

$$a_z = \sqrt{\cos\theta}\sin\theta\left(\cos\phi a_x^0 + \sin\phi a_y^0\right). \tag{7.14}$$

Inserting (7.5) into (7.1),

$$\boldsymbol{E}(p) = \frac{i}{\lambda}\int_0^{\theta_m}\sin\theta d\theta\int_0^{2\pi} e^{\left[-ik\left(-r_p\sin\theta\cos\left(\phi-\phi_p\right)+z_p\cos\theta\right)\right]}\cdot\boldsymbol{a}(\theta,\phi)\,d\phi. \tag{7.15}$$

Two useful integral relations are

$$\int_0^{2\pi} e^{-ik(-r_p\sin\theta\cos\left(\phi-\phi_p\right)}\cdot\cos(n\phi)\,d\phi = 2\pi(-i)^n J_n\left(kr_p\sin\theta\right)\cos\left(n\phi_p\right) \tag{7.16}$$

$$\int_0^{2\pi} e^{-ik(-r_p\sin\theta\cos\left(\phi-\phi_p\right)}\cdot\sin(n\phi)\,d\phi = 2\pi(-i)^n J_n\left(kr_p\sin\theta\right)\sin\left(n\phi_p\right) \tag{7.17}$$

where J_n is a Bessel function of the first kind. The x component of the field is obtained by inserting (7.12) into (7.15),

$$E_x(p) = \frac{i}{2\lambda}\int_0^{\theta_m}\sqrt{\cos\theta}\sin\theta d\theta\int_0^{2\pi} e^{\left[-ik\left(-r_p\sin\theta\cos\left(\phi-\phi_p\right)+z_p\cos\theta\right)\right]}$$

$$\left\{\left[(1+\cos\theta)-(1-\cos\theta)\cos 2\phi\right]a_x^0 + (1-\cos\theta)\sin 2\phi a_y^0\right\}d\phi \tag{7.18}$$

$$= \frac{\pi i}{\lambda}\left\{a_x^0\int_0^{\theta_m}\sqrt{\cos\theta}(1+\cos\theta)\sin\theta J_0\left(kr_p\sin\theta\right)e^{-ikz_p\cos\theta}d\theta + \right.$$

$$\left.\left(a_x^0 \cos 2\phi_p + a_y^0 \sin 2\phi_p\right)\int_0^{\theta_m} \sqrt{\cos\theta}\,(1-\cos\theta)\sin\theta J_2\left(kr_p \sin\theta\right)e^{-ikz_p\cos\theta}\,d\theta\right\}. \tag{7.19}$$

Similarly, the y component is

$$E_y = \frac{\pi i}{\lambda}\left\{\left(a_x^0 \sin 2\phi_p - a_y^0 \cos 2\phi_p\right)\int_0^{\theta_m} \sqrt{\cos\theta}\,(1-\cos\theta)\sin\theta\, J_2\left(kr_p \sin\theta\right)e^{-ikz_p\cos\theta}\,d\theta\right.$$
$$\left.+a_y^0\int_0^{\theta_m} \sqrt{\cos\theta}\,(1+\cos\theta)\sin\theta\, J_0\left(kr_p \sin\theta\right)e^{-ikz_p\cos\theta}\right\}d\theta \tag{7.20}$$

and the z component is

$$E_z(p) = \frac{i}{\lambda}\int_0^{\theta_m} d\theta\, \sin\theta\sqrt{\cos\theta}\,\times$$
$$\int_0^{2\pi} e^{\left[-ik\left(-r_p\sin\theta\cos\left(\phi-\phi_p\right)+z_p\cos\theta\right)\right]}\left(\cos\phi\sin\theta a_x^0 + \sin\phi\sin\theta a_y^0\right)d\phi \tag{7.21}$$

$$= k_0\left(a_x^0 \cos\phi_p + a_y^0 \sin\phi_p\right)\int_0^{\theta_m} \sqrt{\cos\theta}\sin^2\theta\, J_1\left(kr_p \sin\theta\right)e^{-ikz_p\cos\theta}\,d\theta \tag{7.22}$$

Defining integrals,

$$i_1\left(x_p, y_p, z_p, \theta_m\right) \equiv \int_0^{\theta_m} \sqrt{\cos\theta}\,(1+\cos\theta)\sin\theta J_0\left(kr_p \sin\theta\right)e^{-ikz_p\cos\theta}\,d\theta, \tag{7.23}$$

$$i_2\left(x_p, y_p, z_p, \theta_m\right) \equiv \int_0^{\theta_m} \sqrt{\cos\theta}\,(1-\cos\theta)\sin\theta J_2\left(kr_p \sin\theta\right)e^{-ikz_p\cos\theta}\,d\theta, \tag{7.24}$$

and

$$i_3\left(x_p, y_p, z_p, \theta_m\right) \equiv \int_0^{\theta_m} \sqrt{\cos\theta}\sin^2\theta\, J_1\left(kr_p \sin\theta\right)e^{-ikz_p\cos\theta}\,d\theta, \tag{7.25}$$

the equations for the electric field components reduce to

$$E_x = \frac{\pi k_0 i}{2}\left[a_x^0 i_1\left(\theta_m\right) + \left(a_x^0 \cos 2\phi_p + a_y^0 \sin 2\phi_p\right)i_2\left(\theta_m\right)\right], \tag{7.26}$$

$$E_y = \frac{\pi k_0 i}{2}\left[\left(a_x^0 \sin 2\phi_p - a_y^0 \cos 2\phi_p\right)i_2\left(\theta_m\right) + a_y^0\, i_1\left(\theta_m\right)\right], \tag{7.27}$$

and

$$E_z(p) = k_0\left(a_x^0 \cos\phi_p + a_y^0 \sin\phi_p\right) i_3(\theta_m). \tag{7.28}$$

At the focal point, $z_p = r_p = 0$,

$$i_1^{(0)}\left(z_p, r_p = 0, \theta_m\right) \equiv \int_0^{\theta_m} \sqrt{\cos\theta}\,(1+\cos\theta)\sin\theta \mathrm{d}\theta = \frac{2}{15}\left[8 - \cos^{\frac{3}{2}}\theta_m\left(5 + 3\cos\theta_m\right)\right] \tag{7.29}$$

and

$$i_2^{(0)}\left(\theta_m, r_p = 0\right) = i_3^{(0)}\left(\theta_m, r_p = 0\right) = 0. \tag{7.30}$$

Therefore,

$$E_x\left(z_p, r_p = 0\right) = \frac{k_0 i}{2} a_x^0\, i_1(\theta_m), \tag{7.31}$$

$$E_y\left(z_p, r_p = 0\right) = \frac{k_0 i}{2} a_y^0\, i_1(\theta_m), \tag{7.32}$$

and

$$E_z\left(z_p, r_p = 0\right) = 0. \tag{7.33}$$

7.2.2 Radial and Azimuthal Polarization

The electric field profiles for radial and azimuthal polarizations are shown in Figure 7.2. These polarization states can be generated through appropriate waveplates or other means [5–13]. If the axis of the beam is incident on the center of the lens, then the focused spot profile has interesting characteristics.

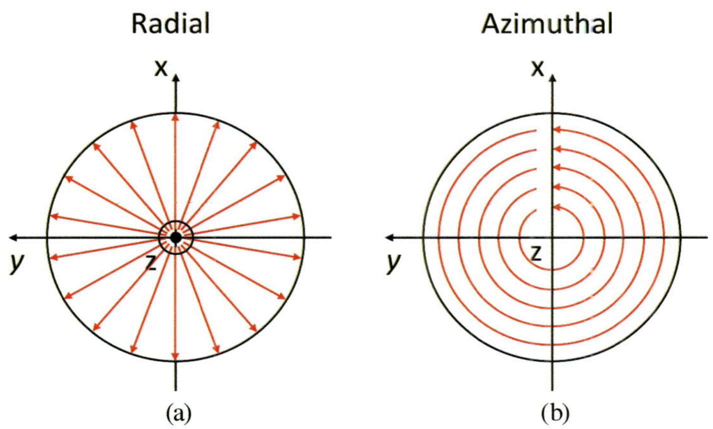

FIGURE 7.2 Electric field profiles for incident (a) radial and (b) azimuthal polarization states.

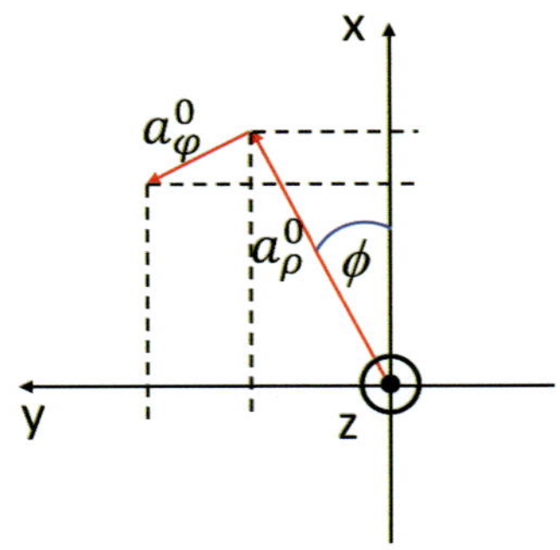

FIGURE 7.3 Relationship between Cartesian and polar coordinates.

As shown in Figure 7.3, the incident polarization in Cartesian coordinates can be expressed in terms of radial coordinates as

$$a_x^0 = a_\rho^0 \cos\phi - a_\phi^0 \sin\phi, \tag{7.34}$$

and

$$a_y^0 = a_\rho^0 \sin\phi + a_\phi^0 \cos\phi \tag{7.35}$$

or

$$\begin{bmatrix} a_x^0 \\ a_y^0 \end{bmatrix} = \begin{bmatrix} \cos\phi & -\sin\phi \\ \sin\phi & \cos\phi \end{bmatrix} \begin{bmatrix} a_\rho^0 \\ a_\phi^0 \end{bmatrix}. \tag{7.36}$$

Inserting (7.36) into (7.11),

$$\begin{bmatrix} a_x \\ a_y \\ a_z \end{bmatrix} = \sqrt{\cos\theta} \begin{bmatrix} \cos\phi\cos\theta & -\sin\phi \\ \sin\phi\cos\theta & \cos\phi \\ \sin\theta & 0 \end{bmatrix} \begin{bmatrix} a_\rho^0 \\ a_\phi^0 \end{bmatrix}. \tag{7.37}$$

Inserting (7.37) into (7.15), the electric fields are

$$E_x\left(x_p, y_p, z_p, \theta_m\right) = k_0\left[\cos\phi_p i_5\left(x_p, y_p, z_p, \theta_m\right) a_r^0 - \sin\phi_p i_4\left(x_p, y_p, z_p, \theta_m\right) a_\phi^0\right], \tag{7.38}$$

$$E_y\left(x_p, y_p, z_p, \theta_m\right) = k_0\left[\sin\phi_p i_5\left(x_p, y_p, z_p, \theta_m\right) a_r^0 + \cos\phi_p i_4\left(x_p, y_p, z_p, \theta_m\right) a_\phi^0\right], \tag{7.39}$$

and

$$E_z\left(x_p, y_p, z_p, \theta_m\right) = i k_0\, i_6\left(x_p, y_p, z_p, \theta_m\right) a_r^0 \tag{7.40}$$

where three new integrals are defined,

$$i_4\left(x_p, y_p, z_p, \theta_m\right) \equiv \int_0^{\theta_m} \sqrt{\cos\theta}\, \sin\theta\, e^{-ikz_p\cos\theta}\, J_1\left(kr_p \sin\theta\right) d\theta, \tag{7.41}$$

$$i_5\left(x_p,y_p,z_p,\theta_m\right)\equiv\int_0^{\theta_m}\sqrt{\cos\theta}\,\sin\theta\cos\theta\,e^{-ikz_p\cos\theta}\,J_1\left(kr_p\sin\theta\right)d\theta\,, \tag{7.42}$$

and

$$i_6\left(x_p,y_p,z_p,\theta_m\right)\equiv\int_0^{\theta_m}\sqrt{\cos\theta}\,\sin^2\theta\,e^{-ikz_p\cos\theta}\,J_0\left(kr_p\sin\theta\right)d\theta. \tag{7.43}$$

For z_p, $r_p = 0$, $J_0(k\,r_p\sin\theta) = 1$ and $J_1(k\,r_p\sin\theta) = 0$, so

$$E_x = E_y = 0 \tag{7.44}$$

and

$$E_z\left(z_p,r_p=0\right)=ik_0a_r^0\int_0^{\theta_m}\sqrt{\cos\theta}\sin^2\theta d\theta=\frac{ik_0a_r^0}{5}\left[4E\left(\frac{\theta_m}{2}\middle|2\right)-2\cos^{\frac{3}{2}}\theta_m\sin\theta_m\right] \tag{7.45}$$

where $E(\phi|m)$ is the elliptic integral of the second kind. The electric field at the focus is longitudinally polarized along the z-axis for incident radially polarized light. The magnetic field is similarly polarized for incident azimuthally polarized light. In cylindrical coordinates,

$$\begin{bmatrix}E_{r_p}\\E_{r_p}\end{bmatrix}=\begin{bmatrix}\cos\phi_p & \sin\phi_p\\-\sin\phi_p & \cos\phi_p\end{bmatrix}\begin{bmatrix}E_{x_p}\\E_{yp}\end{bmatrix}. \tag{7.46}$$

Therefore,

$$\boldsymbol{E}\left(r_p,\phi_p,z_p,\theta_m\right)=\left(\cos\phi_pE_{xp}+\sin\phi_pE_{yp}\right)\hat{\boldsymbol{r}}+\left(-\sin\phi_pE_{xp}+\cos\phi_pE_{yp}\right)\hat{\phi}+E_z\hat{\boldsymbol{z}} \tag{7.47}$$

Substituting from (7.38) and (7.39),

$$E_r\left(\rho_p,\phi_p,z_p,\theta_m\right)=k_0i_5\left(x_p,y_p,z_p,\theta_m\right)a_r^0\,, \tag{7.48}$$

and

$$E_\phi\left(\rho_p,\phi_p,z_p,\theta_m\right)=k_0i_4\left(x_p,y_p,z_p,\theta_m\right)a_\phi^0. \tag{7.49}$$

7.3 MAGNETIC FIELD

7.3.1 LINEAR AND ELLIPTICAL POLARIZATION

The unit vector along the ray is given by (7.3). The amplitude of the electric vector is given by (7.14)–(7.16). The magnetic field amplitude vector is then

$$\boldsymbol{b}=\frac{1}{\eta_0}\hat{\boldsymbol{s}}\times\boldsymbol{a}=\frac{1}{\eta_0}(\left(s_ya_z-s_za_y\right)\hat{\boldsymbol{x}}+\left(s_za_x-s_xa_z\right)\hat{\boldsymbol{y}}+\left(s_xa_y-s_ya_x\right)\hat{\boldsymbol{z}} \tag{7.50}$$

where η_0 is the impedance of free space. From (7.14)–(7.16),

$$b_x = \frac{\sqrt{\cos\theta}}{2\eta_0}\left\{-(1-\cos\theta)\sin 2\phi a_x^0 - \left[(1+\cos\theta)-(1-\cos\theta)\cos 2\phi\right]a_y^0\right\}, \quad (7.51)$$

$$b_y = \frac{\sqrt{\cos\theta}}{2\eta_0}\left\{(1+\cos\theta)+(1-\cos\theta)\cos 2\phi)a_x^0 + (1-\cos\theta)\sin 2\phi a_y^0\right\}, \quad (7.52)$$

and

$$b_z = \frac{\sqrt{\cos\theta}}{\eta_0}\left(\sin\phi\sin\theta a_x^0 - \cos\phi\sin\theta a_y^0\right). \quad (7.53)$$

The total magnetic field is then

$$\boldsymbol{H}(p) = \frac{i}{\lambda}\int_0^{\theta_m}\sin\theta d\theta\int_0^{2\pi} e^{\left[-ik\left(-r_p\sin\theta\cos(\phi-\phi_p)+z_p\cos\theta\right)\right]}\cdot\boldsymbol{b}(\theta,\phi)\,d\phi \quad (7.54)$$

and the field components are

$$H_x\left(x_p,y_p,z_p,\theta_m\right) = \frac{ik_0}{2\eta_0}\left[i_2\sin 2\phi_p a_x^0 - \left(i_1 + i_2\cos 2\phi_p\right)a_y^0\right], \quad (7.55)$$

$$H_y\left(x_p,y_p,z_p,\theta_m\right) = \frac{ik_0}{2\eta_0}\left[\left(i_1 - i_2\cos 2\phi_p\right)a_x^0 - i_2\sin 2\phi_p a_y^0\right], \quad (7.56)$$

and

$$H_z\left(x_p,y_p,z_p,\theta_m\right) = \frac{k_0}{\eta_0}i_3\left(\sin\phi_p a_x^0 - \cos\phi_p a_y^0\right). \quad (7.57)$$

7.3.2 Radial and Azimuthal Polarization

The magnetic field is easily calculated for radial and azimuthal polarization from (7.3) and (7.40).

$$\boldsymbol{b} = \frac{1}{\eta_0}\hat{\boldsymbol{s}}\times\left(a_r\hat{\boldsymbol{r}} + a_z\hat{\boldsymbol{z}}\right) = \frac{\sqrt{\cos\theta}}{\eta_0}\begin{bmatrix}\hat{\boldsymbol{x}} & \hat{\boldsymbol{y}} & \hat{\boldsymbol{z}}\\ -\sin\theta\cos\phi & -\sin\theta\sin\phi & \cos\theta\\ \cos\phi\cos\theta a_r^0 - \sin\phi a_\phi^0 & \sin\phi\cos\theta a_r^0 + \cos\phi a_\phi^0 & \sin\theta a_r^0\end{bmatrix} \quad (7.58)$$

$$= \frac{\sqrt{\cos\theta}}{\eta_0}\left[\left(-\sin\phi a_r^0 - \cos\phi\cos\theta a_\phi^0\right)\hat{\boldsymbol{x}} + \left(\cos\phi a_r^0 - \sin\phi\cos\theta a_\phi^0\right)\hat{\boldsymbol{y}} - \sin\theta a_\phi^0\hat{\boldsymbol{z}}\right]. \quad (7.59)$$

Using this result in (7.54) gives

$$H_x(z_p, r_p, \phi_p, \theta_m) = -\frac{k_0}{\eta_0}\left(\sin\phi_p i_4 a_r^0 + \cos\phi_p i_5 a_\phi^0\right), \tag{7.60}$$

$$H_y(z_p, r_p, \phi_p, \theta_m) = \frac{k_0}{\eta_0}\left(\cos\phi_p i_4 a_r^0 - \sin\phi_p i_5 a_\phi^0\right), \tag{7.61}$$

and

$$H_z(z_p, r_p, \phi_p, \theta_m) = -\frac{k_0 i}{\eta_0} i_6 a_\phi^0. \tag{7.62}$$

The conversion between Cartesian and radial coordinates is

$$\begin{bmatrix} H_{\rho_p} \\ H_{\phi_p} \end{bmatrix} = \begin{bmatrix} \cos\phi_p & \sin\phi_p \\ -\sin\phi_p & \cos\phi_p \end{bmatrix} \begin{bmatrix} H_{x_p} \\ H_{y_p} \end{bmatrix}. \tag{7.63}$$

In cylindrical coordinates,

$$H_{\rho_p}(z_p, r_p, \phi_p, \theta_m) = -\frac{k_0}{\eta_0} i_5 a_\phi^0, \tag{7.64}$$

$$H_{\phi_p}(z_p, r_p, \phi_p, \theta_m) = \frac{k_0}{\eta_0} i_4 a_r^0, \tag{7.65}$$

and

$$H_z(z_p, r_p, \phi_p, \theta_m) = -\frac{i k_0}{\eta_0} i_6 a_\phi^0. \tag{7.66}$$

At the focus for radial polarization there is only an E_z component, while for azimuthal polarization there is only an H_z component. If the incident polarization is radial, then the magnetic field in the focal plane is entirely azimuthal.

7.4 POYNTING VECTOR

7.4.1 Cartesian Coordinates

The Poynting vector is

$$\boldsymbol{S} \equiv \boldsymbol{E} \times \boldsymbol{H}^*. \tag{7.67}$$

The total power flow through a closed surface is

$$\boldsymbol{P} = \frac{1}{2}\mathrm{Re}\left(\int_A \boldsymbol{S} da\right). \tag{7.68}$$

The total power flow through the xy focal plane requires knowledge only of the S_z component of the Poynting vector.

$$P_z = \int_{-\infty}^{+\infty} dx_p \int_{-\infty}^{+\infty} dy_p S_z = \frac{1}{2}\mathrm{Re}\int_{-\infty}^{+\infty} dx_p \int_{-\infty}^{+\infty} dy_p \left(E_x H_y^* - E_y H_x^*\right). \tag{7.69}$$

7.4.2 Cylindrical Coordinates

In cylindrical coordinates the Poynting vector is

$$\boldsymbol{S} \equiv \boldsymbol{E} \times \boldsymbol{H}^* = \left(E_\phi H^*_z - E_z H^*_\phi\right)\hat{\boldsymbol{r}} + \left(E_z H^*_r - E_r H^*_z\right)\hat{\phi} + \left(E_r H^*_\phi - E_\phi H^*_r\right)\hat{\boldsymbol{z}}\,. \tag{7.70}$$

Combining (7.48), (7.49), (7.64), and (7.65), the power flow along the z-axis is

$$P_z = \frac{1}{2}\mathrm{Re}\int_A \left(E_r H^*_\phi - E_\phi H^*_r\right) dA = \frac{1}{2}\mathrm{Re}\int_0^\infty r_p dr_p \int_0^{2\pi} \left(E_r H^*_\phi - E_\phi H^*_r\right) d\phi \tag{7.71}$$

$$= \frac{\pi k_0^2}{\eta_0}\mathrm{Re}\left\{\left[\left(a_r^0\right)^2 + \left(a_\phi^0\right)^2\right]\int_0^\infty r i_4\left(r_p, \theta_m\right) i_5\left(r_p, \theta_m\right) dr\right\} \tag{7.72}$$

7.5 TILTED IMAGE PLANE

Sometimes it may be desirable to tilt the image plane so that reflected light does not get re-imaged to the same spot as the light emitter. Consider the previous results now calculated along an image plane at an angle α with respect to the xy plane as shown in Figure 7.4. No changes are necessary in the equations, but the calculation of the electric field is now carried out on the tilted plane. If the x' axis is tilted in the xz plane by angle α as shown, then the transform equations are

$$x_0 = x' \cos\alpha + z' \sin\alpha, \tag{7.73}$$

$$y_0 = y', \tag{7.74}$$

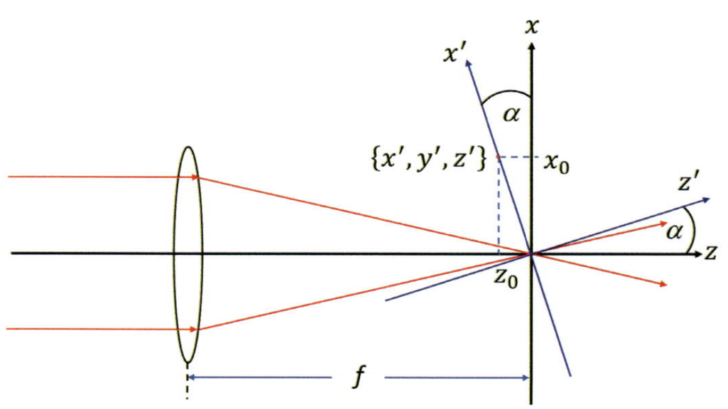

FIGURE 7.4 Image plane tilted by angle α from xy plane.

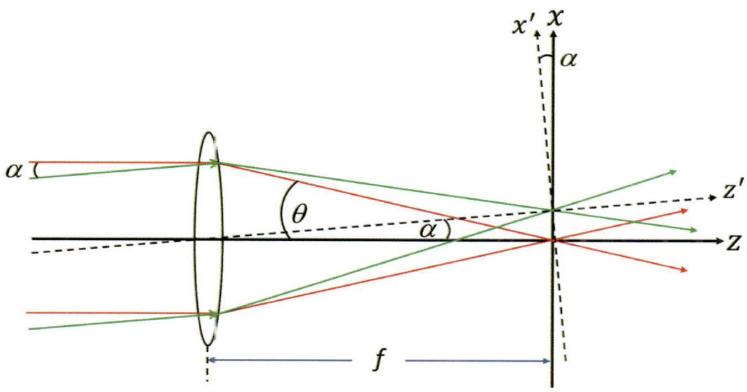

FIGURE 7.5 The incident collimated beam is shifted by a small angle α in the xz plane from normal. The focal point occurs at the origin of the $x'z'$ coordinate system which is offset from the original xz coordinate system. The focal length of the lens is f and its numerical aperture is $\sin\theta$.

and

$$z_0 = -x'\sin\alpha + z'\cos\alpha. \tag{7.75}$$

7.6 TILTED INCIDENT BEAM

If the incident collimated beam is tilted by a small angle α with respect to the optic axis as shown in Figure 7.5, then the focal point is shifted.

To first order, the position of the focal point in the xz coordinate system is ($f\sin\theta$, 0) where f is the focal length of the lens. The fields are calculated in the $x'z'$ coordinate system and then translated into the xz coordinate system using the transform equations

$$x = f\sin\alpha + x'\cos\alpha + z'\sin\alpha, \tag{7.76}$$

and

$$z = -x'\sin\alpha + z'\cos\alpha. \tag{7.77}$$

The inverse transforms are

$$x' = (\cos\alpha)x - (\sin\alpha)z - (\sin\alpha\cos\alpha)f, \tag{7.78}$$

and

$$z' = (\sin\alpha)x + (\cos\alpha)z - (\sin^2\alpha)f. \tag{7.79}$$

7.7 RADIAL VARIATION OF INCIDENT FIELD INTENSITY

The analysis in Section 1 was carried out for uniform illumination of the lens, a constant incident amplitude $|a(\theta,\phi)|$ across the lens. This situation is approximated by an overfilled lens with a relatively spatially uniform illumination. This type of illumination creates the smallest focused spot, but much of the incident optical power is not collected by the lens. More typically the lens is illuminated with a beam whose intensity varies radially across the lens with a Gaussian profile that is abruptly truncated at the edge of the lens. In this case, the amplitude factor $a(\theta,\phi)$ must be multiplied by an additional factor that varies with radius across the lens. For a lens with focal length *f*, the radius is a function of θ,

$$r = f\sin\theta. \tag{7.80}$$

If the intensity profile is Gaussian with a half width at half maximum of r_0, then

$$I(r) = I_0\, e^{-\ln(2)\left(\frac{r}{r_0}\right)^2}. \tag{7.81}$$

The electric field amplitude varies as the square root of the intensity. At radius *r* on the lens, the field amplitude is

$$E(r) = E_0\, e^{-\frac{\ln(2)}{2}\left(\frac{r}{r_0}\right)^2}. \tag{7.82}$$

Including this factor in the incident electric field amplitude (7.15) gives

$$\boldsymbol{E}(p) = \frac{i}{\lambda}\int_0^{\theta_m} e^{-\frac{\ln(2)}{2}\left(\frac{f\sin\theta}{r_0}\right)^2} \sin\theta d\theta \int_0^{2\pi} e^{\left[-ik\left(-r_p\sin\theta\cos(\phi-\phi_p)+z_p\cos\theta\right)\right]} \cdot \boldsymbol{a}(\theta,\phi)\,d\phi. \tag{7.83}$$

In order to normalize the field amplitude, the integration can be carried out at the focal point, $r_p = 0$,

$$i_1^{(0)}\left(r_p = 0, \theta_m\right) \equiv \int_0^{\theta_m} e^{-\frac{\ln(2)}{2}\left(\frac{f\sin\theta}{r_0}\right)^2} \sqrt{\cos\theta}\,(1+\cos\theta)\sin\theta d\theta. \tag{7.84}$$

This integral can be solved in closed form with Gamma functions, but the result is very messy. Therefore, it is often simpler in software to choose a very small value for r_p and then calculate the integral numerically.

The fraction of the incident power that is collected by the lens of radius *R* is

$$f_c = \frac{2\pi\int_0^R I_0 e^{-\ln(2)\left(\frac{r}{r_0}\right)^2} r dr}{2\pi\int_0^\infty I_0 e^{-\ln(2)\left(\frac{r}{r_0}\right)^2} r dr} = 1 - 2^{-\left(\frac{R^2}{r_0^2}\right)}. \tag{7.85}$$

Interestingly and perhaps somewhat surprisingly, when the lens diameter is equal to the FWHM of the incident Gaussian intensity beam, $R = r_0$, the lens collects half of the incident light.

7.8 OVERLAP INTEGRAL FOR COUPLING EFFICIENCY

The overlap integral *O* of two functions $F(x,y)$ and $G(x,y)$ over an area *A* is defined by

$$O = \frac{\iint_A F(x,y)G^*(x,y)\,dx\,dy}{\sqrt{\iint_A |F(x,y)|^2\,dx\,dy}\sqrt{\iint_A |G(x,y)|^2\,dx\,dy}}. \tag{7.86}$$

where G^* is the complex conjugate of G. When $F = G$, the overlap integral is 1. If the two functions are orthogonal, the overlap is 0.

Overlap integrals are useful when one wants to expand a function in a series of orthogonal functions, for example, Fourier analysis. If one has a complete set of orthogonal functions, then one can expand any other function in terms of a sum of those functions. The waveguide modes are mutually orthogonal, but they do not form a complete set. Any arbitrary field profile inside a waveguide can be expanded in terms of waveguide modes *plus* leaky modes. The leaky modes will be an infinite set of modes that approximate plane waves well outside of the core. However, usually one is not interested in the leaky modes and only wishes to ascertain amplitudes of the propagating modes inside the waveguide. Therefore, the overlap integral of a given field profile with each propagating mode provides the amplitude of that mode in the waveguide.

There is a caveat. This approach only works strictly for an unknown field profile *within* a waveguide. The field profile in the focal plane from a lens that is focused onto the end of a waveguide is the profile of the incident wave *outside* the waveguide. There will also presumably be a reflected wave. The transverse electric field component of the sum of incident and reflected waves is continuous across the end of the waveguide though the normal components of the total field are not, so there is always an approximation involved when using an overlap integral calculation with external field amplitudes to estimate the internal field amplitudes of the waveguide modes. However, the overlap integral is useful for optimizing the matching lens parameters, for example, even though the calculated coupling efficiency will not in general be completely accurate.

An example can make this clearer. Consider a plane wave normally incident from vacuum onto a substrate with refractive index n. From the Fresnel relations the reflected intensity is

$$R = \frac{(n-1)^2}{(n+1)^2} \tag{7.87}$$

and the transmitted intensity is

$$T = 1 - \frac{(n-1)^2}{(n+1)^2} = \frac{4n}{(n+1)^2}. \tag{7.88}$$

The amplitude of the transmitted field is

$$t = \sqrt{1 - \frac{(n-1)^2}{(n+1)^2}} = \frac{2\sqrt{n}}{n+1}. \tag{7.89}$$

In the case of an air/glass interface with $n = 1.5$, (7.89) gives $t = 0.98$ because the reflectance is only 4% and makes only a small change in the transmitted field. However, using the definition of (7.86) gives $t = 1$ because both incident and transmitted waves are plane waves, that is, their overlap is perfect. In this case, the error from the overlap integral is only 2%, which is a very good approximation to the transmitted field amplitude. If the reflection is larger, however, or if there are large normal components in the focused incident field due to a lens with a large numerical aperture, the accuracy of the overlap integral for estimating the mode amplitudes will correspondingly decrease.

7.9 EXAMPLES

7.9.1 Fields Located Behind the Focal Plane Along the x-axis for Linear Polarization

Consider a uniformly illuminated lens at a wavelength of 1.55 μm with a numerical aperture of 0.5. The incident beam is linearly polarized along the x-axis. The fields are calculated 0.2 μm behind

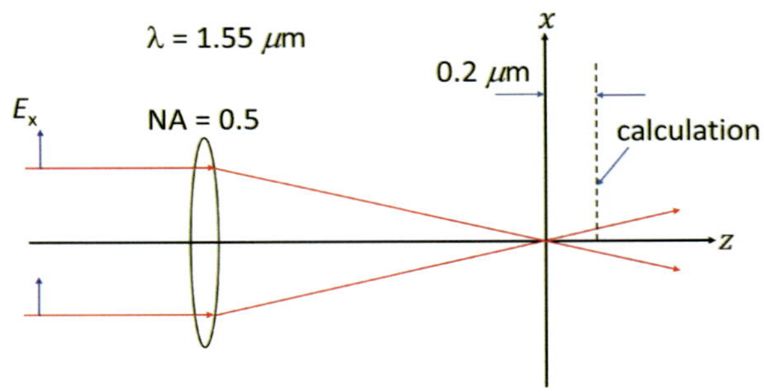

FIGURE 7.6 Electric field along line 0.2 μm behind the focal plane for incident linear polarization.

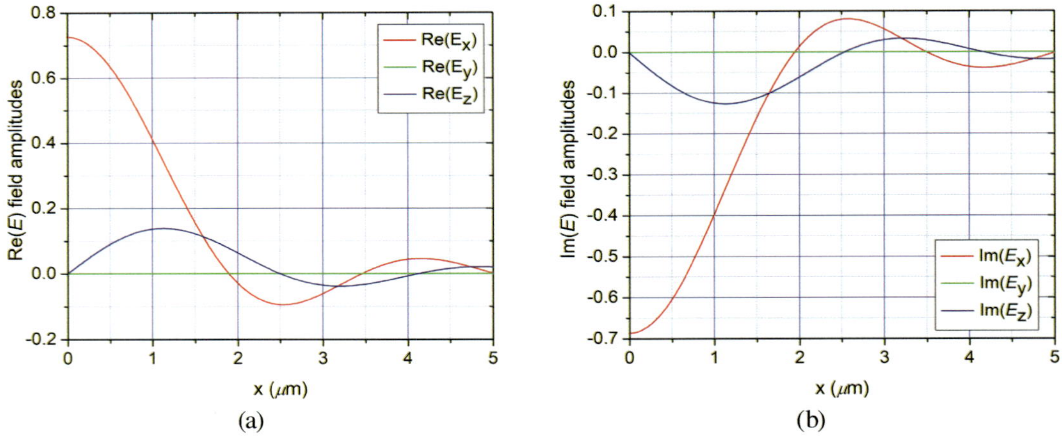

FIGURE 7.7 (a) Real and (b) imaginary parts of the electric field along line 2 μm behind the focal plane for incident linear polarization.

the focal plane away from the lens along the x-axis as shown in Figure 7.6. The real and imaginary parts of the electric field components are shown in Figure 7.7.

7.9.2 Contour Plot for Linear Polarization in the Focal Plane

For the previous example, the electric field is now calculated across the entire ($z = 0$) focal plane. The result is shown in Figure 7.8 for all three components of the electric field. Note that at the focus $E_y = E_z = 0$.

7.9.3 Circular Polarization in the Focal Plane

Consider a uniformly illuminated lens in Figure 7.9(a) at a wavelength of 1.55 μm with a numerical aperture of 0.5. The incident beam is circularly polarized (equal amplitudes of s- and p-polarization with a relative π phase shift). The real and imaginary components of the electric fields in the focal plane along the x-axis are shown in Figure 7.9(b).

7.9.4 Incident Radial Polarization

Consider the same optical setup as Figure 7.6, but the incident beam is radially polarized, $\boldsymbol{E} = E_0\,\hat{\boldsymbol{r}}$. Contour plots of the $\hat{\boldsymbol{r}}$ and $\hat{z}$ components in the focal plane are shown in Figure 7.10. It is

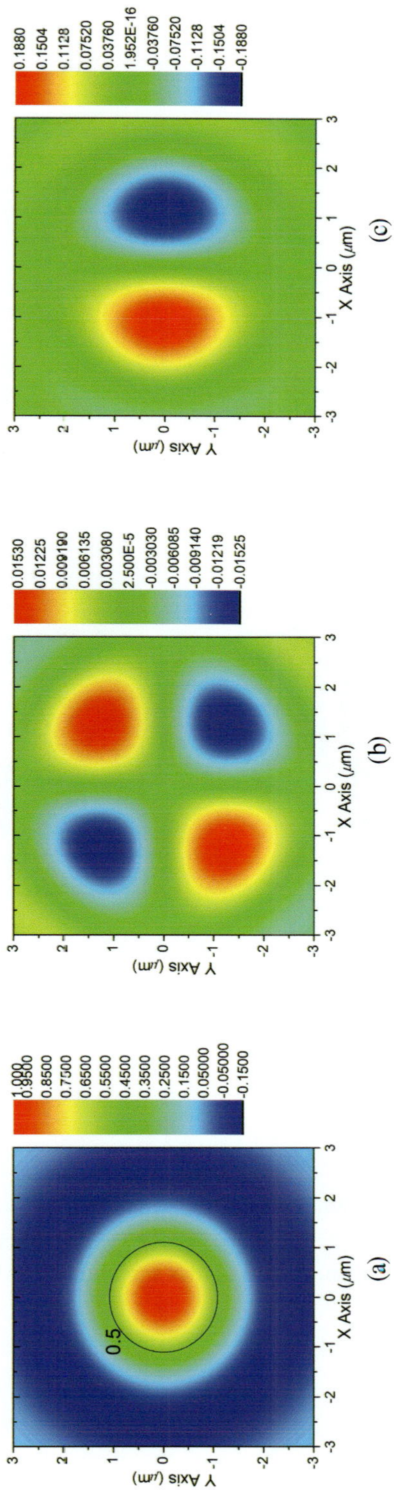

FIGURE 7.8 (a) Re(E_x) and (b) Re(E_y), and (c) Im(E_z) amplitudes in the focal plane for example in Figure 7.6.

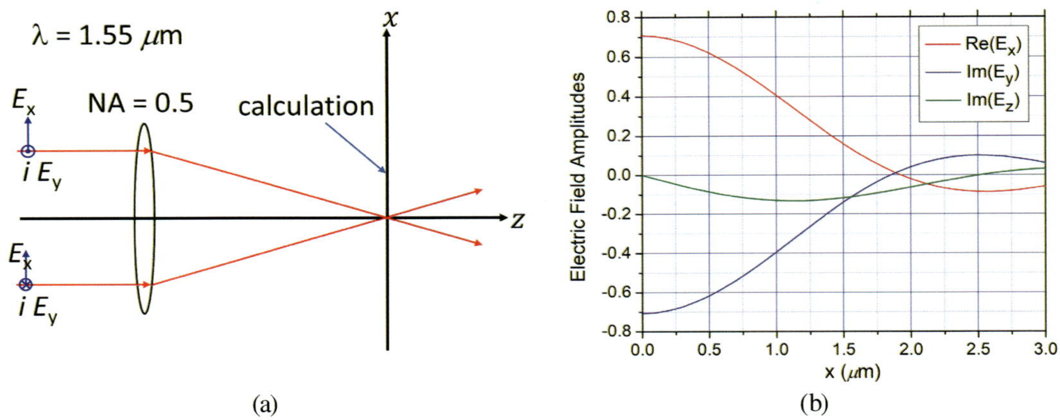

FIGURE 7.9 (a) Electric field along the x-axis for incident circular polarization. (b) Real, and imaginary parts of the electric field along line the x-axis for incident circular polarization.

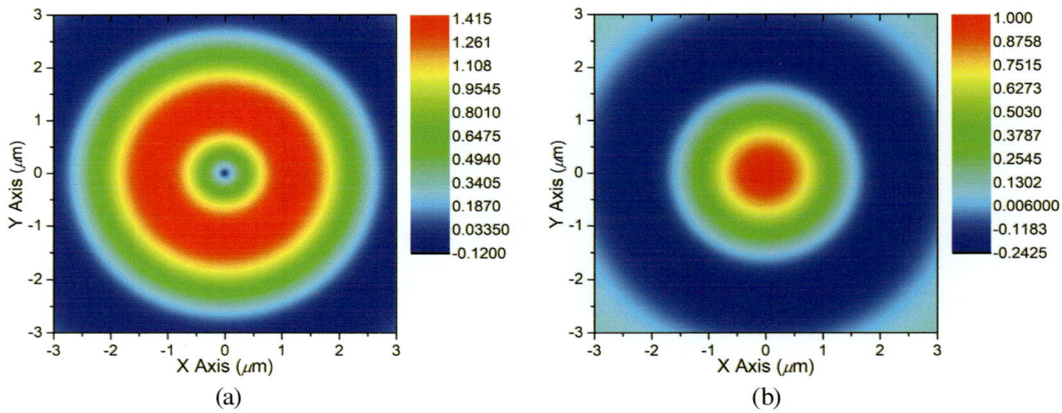

FIGURE 7.10 (a) Re(E_r) and (b) Im(E_z) for the optical setup in Figure 7.6 with incident radial polarization.

particularly interesting to notice that when incident radial polarization is focused, at the focal point there is a null in E_r and a maximum in E_z, that is, at the focus the entire electric field is directed along the direction of propagation of the beam.

7.9.5 Incident Azimuthal Polarization

Again, consider the same optical setup as Figure 7.6, but the incident beam is azimuthally polarized, $\boldsymbol{E} = E_0\hat{\boldsymbol{\phi}}$. A contour plot of E_ϕ is shown in Figure 7.11. At the focal point there is a null in E_ϕ . Unlike the case of radial polarization, $E_z = 0$ across the entire focal plane. Therefore, the electric field at the focal point is zero.

7.9.6 Tilted Image Plane

Consider a uniformly-illuminated lens that is linearly polarized along the x-axis at a wavelength of 1.55 μm with a numerical aperture 0.5 for which the image plane is tilted by 40° in the xz plane from the focal plane as shown in Figure 7.12. Contour plots of the field components are shown in Figure 7.13.

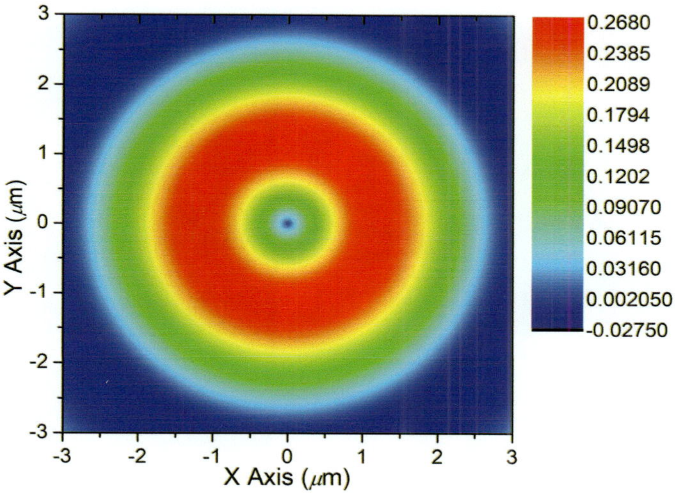

FIGURE 7.11 Re(E_ϕ) for the optical setup in Figure 7.6 with incident azimuthal polarization.

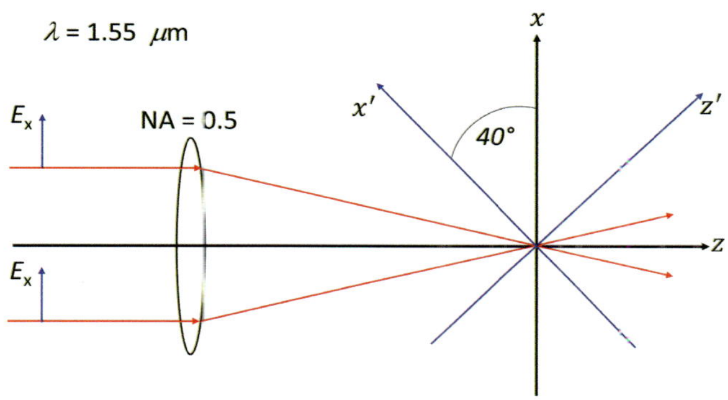

FIGURE 7.12 Focused spot in a plane that is tilted around the y-axis by 40°.

7.9.7 Tilted Incident Beam

Consider a uniformly illuminated lens with linearly polarized light, a numerical aperture 0.67, and a focal length of 11.25 mm at a wavelength of 1.55 μm as shown in Figure 7.14. The incident beam is tilted by 0.1528° with respect to the normal to the lens. This angular tilt causes a shift of $f \sin\alpha =$ 30 μm in the position of the focal point along the x-axis. The field profile is calculated in the $z = 0$ focal plane as shown in Figure 7.15.

7.9.8 Comparison to Scalar Airy Field Approximation

A scalar approximation to the focused field in the focal plane is given by the Airy pattern, [14]

$$E(r) = 2E_0 \frac{J_1(k_0 \text{NA}\, r_p)}{k_0 \text{NA}\, r_p}. \tag{7.90}$$

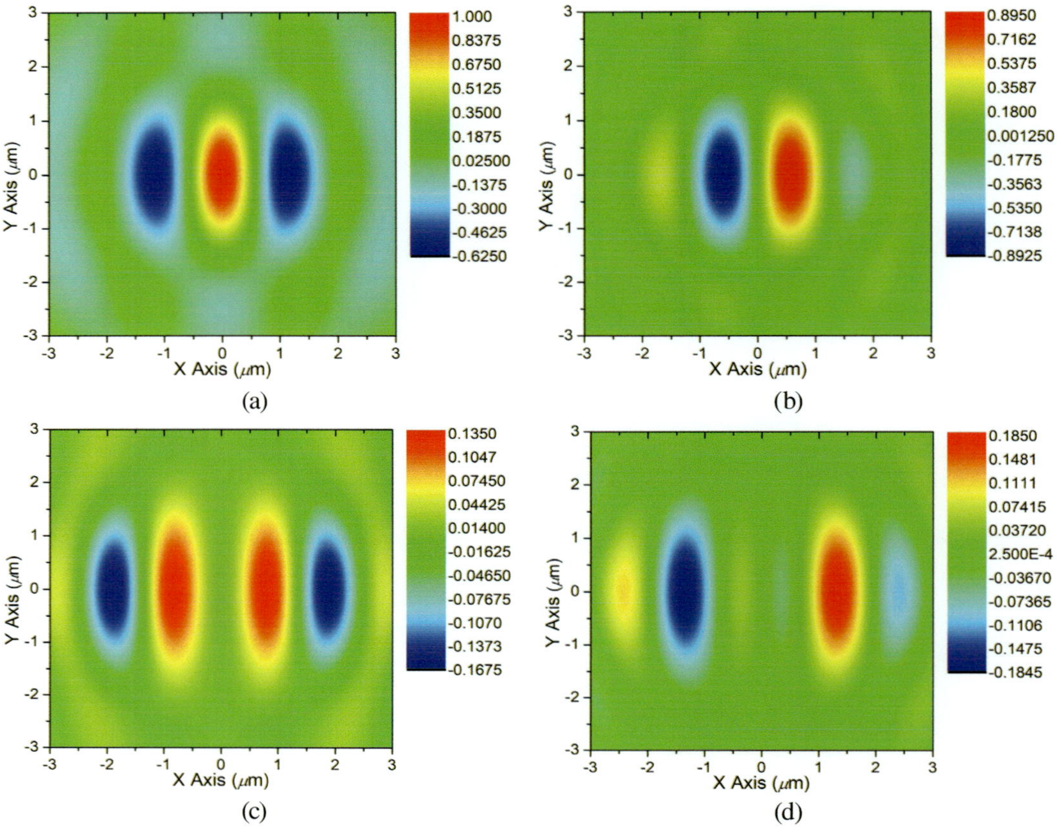

FIGURE 7.13 (a) Re(E_x), (b) Im(E_x), (c) Im(E_z), (d) Re(E_z) amplitudes for the example in Figure 7.6 in a plane tilted by 40° around the y-axis from the focal plane.

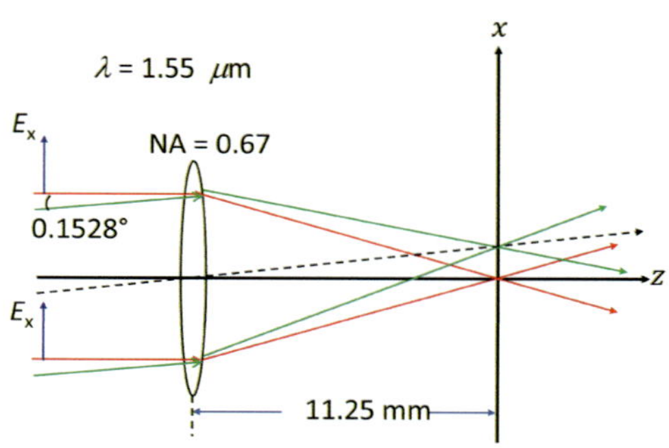

FIGURE 7.14 Example of tilted incident beam which displaces the focal point.

The Airy approximation neglects polarization effects and, therefore, is less accurate for lenses with large numerical apertures. For lenses with numerical apertures ≤ 0.5, however, the Airy approximation is actually very close to the full Richards-Wolf vector result for linearly polarized light. Consider again the first example in Figure 7.6 with a wavelength of 1.55 μm, numerical aperture of 0.5, a uniformly illuminated lens, and x-axis polarization. A comparison of the calculated field

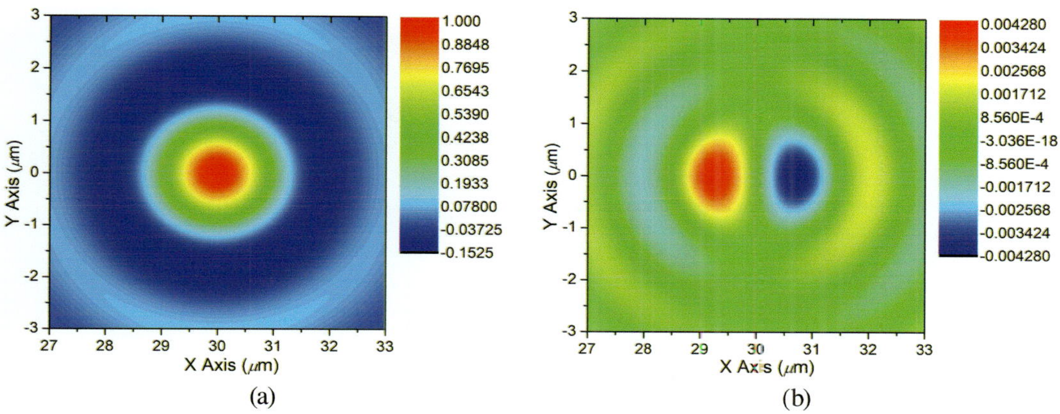

FIGURE 7.15 (a) Re(E_x) and (b) Im(E_x) for focused spot which is shifted by 30 μm along x-axis due to tilted input beam. Note the displacement of the *x*-axis.

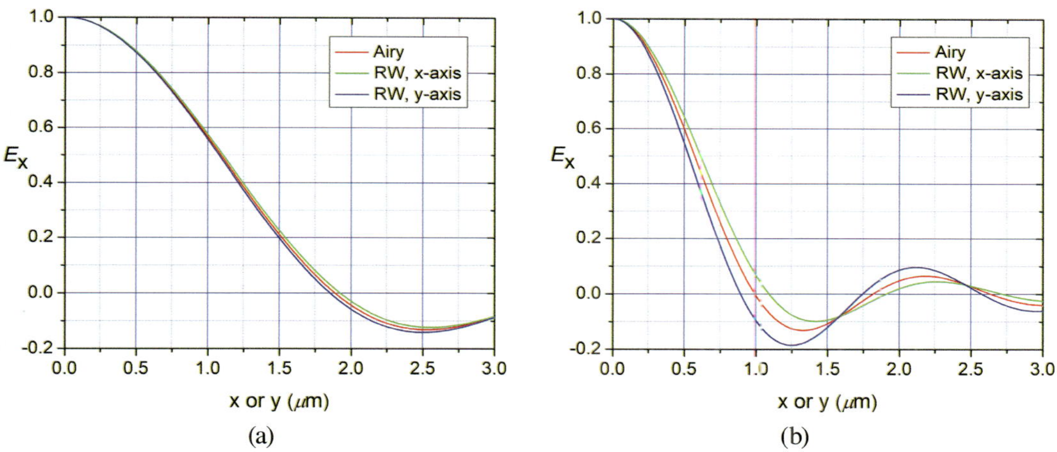

FIGURE 7.16 Airy approximation is compared to the vector calculation field amplitudes along *x*- and *y*-axes for a lens with (a) 0.5 NA, and (b) 0.95 NA. The actual focused spot is slightly elongated along the polarization direction compared to the scalar Airy approximation. The spot size, however, is smaller with larger NA's.

amplitudes in the focal plane is shown in Figure 7.16(a). A similar comparison for a lens with 0.95 NA is shown in Figure 7.16(b).

7.9.9 Comparison of Uniform Incident Beam to Gaussian Incident Beam

The focused spot profile for a lens that is illuminated uniformly is compared to one illuminated with a Gaussian beam. The parameters are given in Figure 7.17. The incident beam is polarized along the *x*-axis. The focal length of the lens is 10 mm, with a numerical aperture of 0.5. The diameter of the lens is

$$d_l = 2f \tan\left[\sin^{-1}\left(\mathrm{NA}\right)\right] = 11.547\,\mathrm{mm}. \tag{7.91}$$

The full width at half maximum for the incident beam is set to the edge of the lens, that is, $d_{1/2} = d_l$. The fields are calculated in the focal plane. The E_x field amplitude is plotted along the *x*- and *y*-axes in Figure 7.18. Clearly, for the apodized beam, the focused spot size is larger.

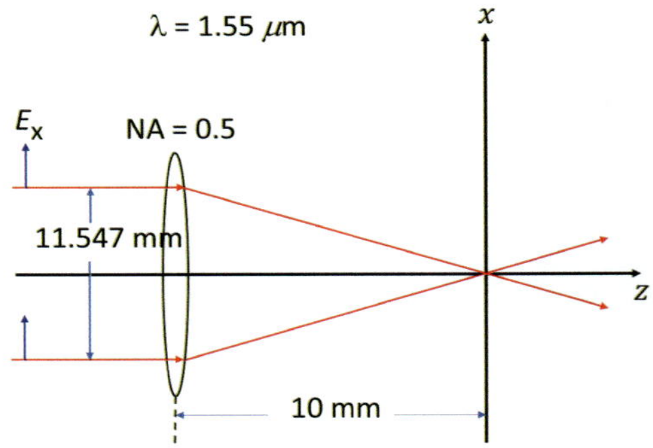

FIGURE 7.17 A low NA lens is illuminated either uniformly or by a beam with a Gaussian intensity for which the full width at half maximum is the lens diameter.

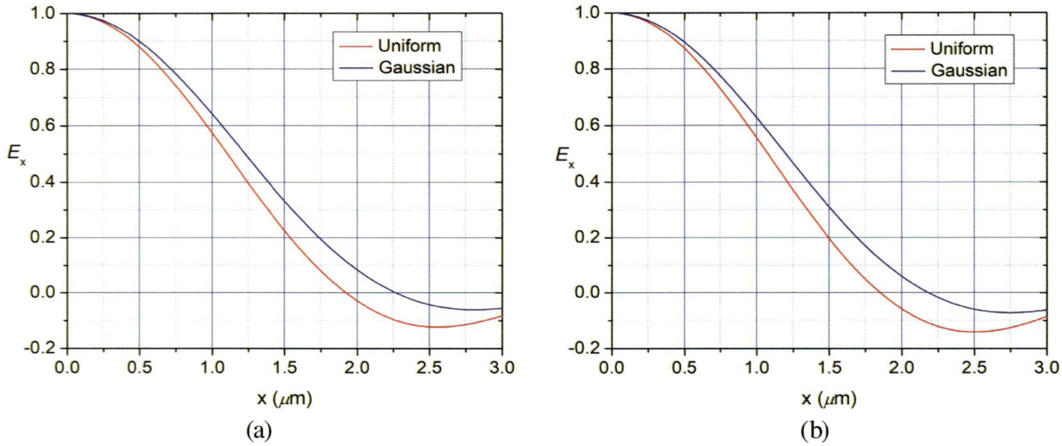

FIGURE 7.18 E_x field amplitude along (a) x-axis and (b) y-axis for a uniform intensity lens illumination and a Gaussian apodized intensity for which the FWHM of the incident beam intensity occurs at the edge of the lens.

7.9.10 Optimized Lens Coupling to Rectangular Waveguide

The optimum lens design for edge coupling focused light into a waveguide in the most efficient manner can be determined through an electric field overlap calculation. The electric field profile inside the waveguide is first calculated using the equations described in Chapter 5 and then the overlap integrals are calculated for a variety of lens numerical apertures.

Consider a 0.3 µm wide, 0.22 µm high SiN strip waveguide embedded in SiO_2. There is one quasi-TE mode with a mode index of 1.4451 at a wavelength of 1.55 µm. The waveguide geometry and the E_x field profile for this mode are shown in Figure 7.19. The peak value for E_x occurs at the edges of the waveguide, not at the center. The calculated FWHM for $|E_x|^2$ of this mode for which $E_x = 1/\sqrt{2}$ is 0.728 µm. The total power within the calculated region ±10 µm along the x-axis and ±2 µm along the y-axis is 8.82×10^{-15} W.

In this example, the focused field amplitudes are calculated and stored for light linearly polarized along the x-axis and for various lens numerical apertures over an area of ±6 µm along the x- and y- axes. Such a large area takes a fairly long time to evaluate, and it is really only necessary for NAs < 0.6. The full width at half maximum (FWHM) for the focused fields

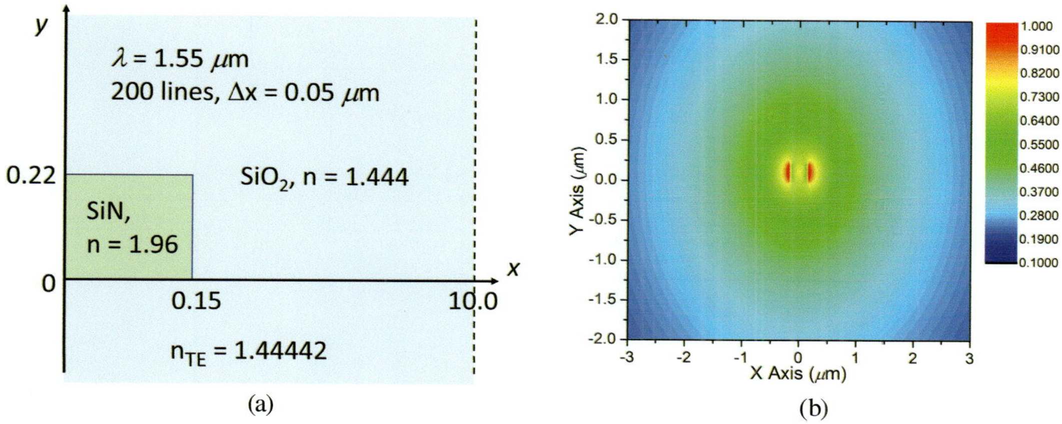

FIGURE 7.19 (a) Waveguide design. (b) E_x field profile.

as a function of numerical aperture is plotted in Figure 7.20. The values assume a Gaussian intensity apodization of the incident beam for which the FWHM is equal to the lens diameter. Therefore, the fraction of the incident beam captured by the lens is $f_c = 0.5$. It might be expected that matching the spot size of the focused beam to the FWHM intensity of the waveguide mode should give the best coupling efficiency in which case a lens with an extremely large numerical aperture to obtain a focused spot size of 0.78 μm would be required. This does indeed turn out to be true in this case.

Let P_f be the total power in the focused spot that has been normalized to a peak electric field of 1. Then the incident power P_{in} is given by

$$P_f = f_c P_{in}. \tag{7.92}$$

Let the peak electric field calculated for the waveguide from the method of lines be normalized to 1 and the power inside the waveguide for this electric field amplitude be P_w. The power inside the waveguide scales as the square of the peak electric field, and the power in the focused spot also scales as the square of the peak focused field. Only the real part of the E_x component of the field is considered in this analysis. As previously discussed, the overlap integral of the incident focused beam with the waveguide mode is an approximation that ignores any reflected light amplitude that could modify the total field at the entrance face of the waveguide. If the peak electric field outside the waveguide is normalized to 1, then the approximate optical power propagating inside the waveguide is given by the square of the overlap integral multiplied by the power in the waveguide mode P_w with an electric field amplitude of 1. From the method of lines this is found to be $P_w = 6.402 \times 10^{-15}$ W for the TE_0 waveguide mode of Figure 7.19. The optical power transfer efficiency is the ratio of P_{out} to P_{in},

$$\epsilon = \frac{P_{out}}{P_{in}} \approx \frac{f_c P_w O^2}{P_f}. \tag{7.93}$$

The coupling efficiency is calculated from (7.93) for NAs between 0.1 and 0.9 as shown in Table 7.1. The peak efficiency is seen to continually increase as the NA of the lens increases, though the overlap integral appears to saturate at a value of 0.77 once the NA has reached 0.8. This saturation effect might be an artifact of the step size. It is always wise to test the convergence of the overlap integration by repeating it with a smaller increment to see if similar results are obtained.

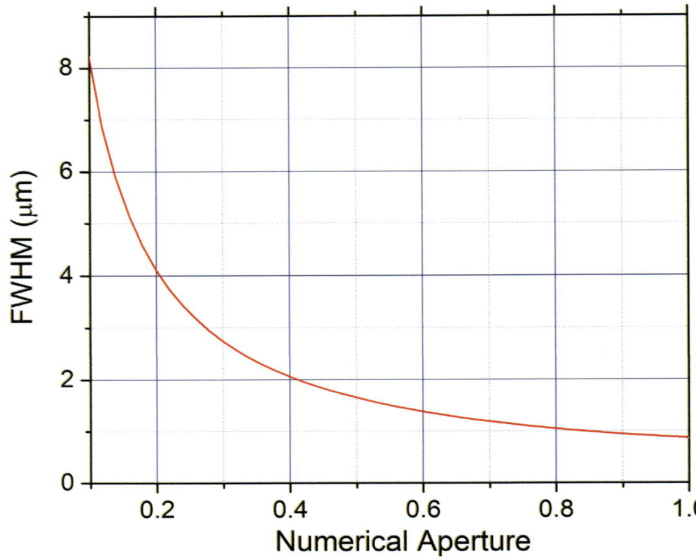

FIGURE 7.20 Calculated FWHM focused spot size as a function of lens numerical aperture for an incident linearly polarized Gaussian beam that is apodized such that the FWHM of the incident beam occurs at the edge of the lens.

TABLE 7.1
Coupling Efficiency Parameters

NA	FWHM of incident beam	Lens diameter	FWHM of focused spot	Total power in focused spot, P_f	Total incident power, $P_i = \frac{P_f}{f_c}$	Overlap, O	Efficiency, $\epsilon = O^2\left(\frac{P_w}{P_i}\right)$
0.9	10 mm	10 mm	1.288 μm	2.24×10^{-15} W	4.48×10^{-15} W	0.770	0.847
0.8	10	10	1.339	2.41×10^{-15}	4.82×10^{-15}	0.770	0.787
0.7	10	10	1.424	2.72×10^{-15}	5.44×10^{-15}	0.767	0.692
0.6	10	10	1.562	3.30×10^{-15}	6.60×10^{-15}	0.760	0.560
0.5	10	10	1.779	4.35×10^{-15}	8.70×10^{-15}	0.767	0.433
0.4	10	10	2.135	6.38×10^{-15}	1.28×10^{-14}	0.825	0.340
0.3	10	10	2.760	1.08×10^{-14}	2.16×10^{-14}	0.915	0.248
0.2	10	10	4.052	2.28×10^{-14}	4.56×10^{-14}	0.903	0.114
0.1	10	10	8.004	7.96×10^{-14}	1.59×10^{-13}	0.621	0.016

REFERENCES

1 E. Wolf, "Electromagnetic diffraction in optical systems: I. An integral representation of the image field," *Proc. R. Soc. Lond. Ser. A*. **253** (1959) 349–357.

2 B. Richards and E. Wolf, "Electromagnetic diffraction in optical systems: II. Structure of the image field in an aplanatic system," *Proc. R. Soc. Lond. Ser. A*. **253** (1959) 358–379.

3 I. Ichimura, S. Hayashi, and G. S. Kino, "High-density optical recording using a solid immersion lens," *Appl. Opt*. **36** (1997) 4339–4348.

4 Eq. (8) of [3] is incorrect.

5 S. C. Tidwell, D. H. Ford, and W. D. Kimura, "Generating radially polarized beams interferometrically," *Appl. Opt.* **29** (1990) 2234–2239.

6 T. Grosjean, A. Sabac, and D. Courjon, "A versatile and stable device allowing the efficient generation of beams with radial, azimuthal or hybrid polarizations," *Opt. Commun.* **252** (2005) 12–21.

7 X.-L. Wang, J. Ding, W.-J. Ni, C.-S. Guo, and H.-T. Wang, "Generation of arbitrary vector beams with a spatial light modulator and a common path interferometric arrangement," *Opt. Lett.* **32** (2007) 3549–3551.

8 K. C. Toussaint, Jr., S. Park, J. E. Jureller, and N. F. Scherer, "Generation of optical vector beams with a diffractive optical element interferometer," *Opt. Lett.* **30** (21) (2005) 2846–2848.

9 C. Maurer, A. Jesacher, S. Fürhapter, S. Bernet, and M. Ritsch-Marte, "Tailoring of arbitrary optical vector beams," *New J. Phys.* **9** (2007) 78.

10 Z. Bomzon, V. Kleiner, and E. Hasman, "Formation of radially and azimuthally polarized light using space-variant subwavelength metal stripe gratings," *Appl. Phys. Lett.* **79** (2001) 1587–1589.

11 K. J. Moh, X.-C. Yuan, J. Bu, R. E. Burge, and B. Z. Gao, "Generating radial or azimuthal polarization by axial sampling of circularly polarized vortex beams," *Appl. Opt.* **46** (2007) 7544–7551.

12 M. Stalder, and M. Schadt, "Linearly polarized light with axial symmetry generated by liquid-crystal polarization converters," *Opt. Lett.* **21** (1996) 1948–1950.

13 J. A. Davis, D. E. McNamara, D. M. Cottrell, and T. Sonehara, "Two-dimensional polarization encoding with a phase-only liquid-crystal spatial light modulator," *Appl. Opt.* **39** (2000) 1549–1554.

14 A. B. Marchant, *Optical Recording* (Reading, MA: Addison-Wesley Publishing Co., 1990) 105.

8 Mie Scattering

8.1 INTRODUCTION

The theory of light scattering from spherical particles was developed and published by Gustav Mie in 1908 [1]. Several others independently developed the theory at around the same time, including Ludvig Lorenz [2] and Peter Debye [3]. There is an enormous amount of literature available freely on the internet about this theory. Therefore, in this chapter only a simplified analysis is presented based primarily on chapter 13 of [4]. As in [4], the analysis is performed with Maxwell's equations in cgs units and assumes an $e^{-i\omega t}$ time-dependence. Excellent detailed discussions of light scattering from particles in general as well as derivations of Mie theory in particular are found in [5] and [6], upon which some parts of this chapter are also based. A FORTRAN program is also provided in an appendix of [6] for making scattering calculations. Other published algorithms and programs are found in, for example, [7–14]. The derivation in [4] is extended in this chapter to include spherical particles with multiple concentric shells, and the derivation in [6] for spheres and nanoshells is extended to include electric and magnetic field calculations.

The early formulations of Mie theory were not directly tailored to efficient and accurate computer algorithms, so much of the more recent research has been on how to implement these equations in a manner that enables fast calculations with finite precision machines. For example, Bessel functions must be calculated to higher orders. The standard upward-recursion formula becomes unstable for large orders. Kattawar and Plass [15] described a downward recursion approach to overcome this problem. A detailed discussion of computer algorithm optimization is given by Wiscombe [9, 16]. A further improvement to Wiscombe's algorithm is given by Du [17] that enables Mie calculations of raindrops as large as 1 mm.

8.2 WAVE EQUATION IN SPHERICAL COORDINATES

Maxwell's curl equations are

$$\nabla \times E = \left(\frac{i\omega}{c}\right)\boldsymbol{H} = ik_0\boldsymbol{H} \tag{8.1}$$

and

$$\nabla \times H = -\left(\frac{i\omega\epsilon}{c}\right)\boldsymbol{E} = -i\epsilon\, k_0\boldsymbol{E} \tag{8.2}$$

where ϵ is the dielectric constant, c is the speed of light, ω is the angular frequency, k_0 is the free space wavevector, $\boldsymbol{E}$ is the electric field, and $\boldsymbol{H}$ is the magnetic field.

The incident plane wave propagates in the $+z$ direction as shown in Figure 8.1. It is linearly polarized along the x direction, $E_i\,\hat{\boldsymbol{x}}$. The outer medium has dielectric constant ϵ_{N+1}. The sphere has dielectric constant ϵ_0 and it may be surrounded by N concentric shells (not shown) with dielectric constants ϵ_i, $1 \le i \le N$. The radius of each sphere/shell is r_i. The fields are observed at point P. In polar coordinates this point is defined by the radius, R, the polar angle measured from the z-axis, θ, and the azimuthal angle measured from the x-axis, ϕ.

The incident normalized field components are

$$E_x = e^{ik_1 z}, \tag{8.3}$$

DOI: 10.1201/9781003714231-8

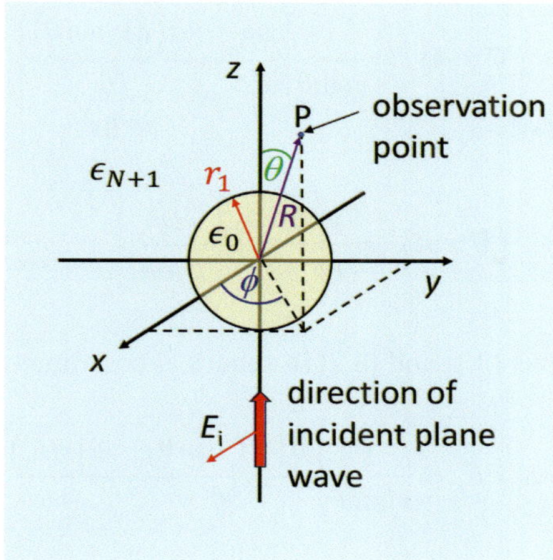

FIGURE 8.1 Coordinate system for light incident on a spherical particle.

$$H_y = i\epsilon_1 e^{ik_1 z} \tag{8.4}$$

and

$$E_y = E_z = H_x = H_z = 0. \tag{8.5}$$

The solution of Maxwell's equations is most easily handled in spherical coordinates where

$$x = r\sin\theta\cos\phi\,, \tag{8.6}$$

$$y = r\sin\theta\sin\phi\,, \tag{8.7}$$

and

$$z = r\cos\theta. \tag{8.8}$$

The transformation of a vector from Cartesian coordinates to spherical coordinates is

$$A_r = A_x\sin\theta\cos\phi + A_y\sin\theta\sin\phi + A_z\cos\theta\,, \tag{8.9}$$

$$A_\theta = A_x\sin\theta\cos\phi + A_y\cos\theta\sin\phi - A_z\sin\theta\,, \tag{8.10}$$

and

$$A_\phi = -A_x\sin\phi + A_y\cos\phi. \tag{8.11}$$

The curl of a vector in spherical coordinates is

$$(\nabla\times\boldsymbol{A})_r = \frac{1}{r^2\sin\theta}\left(\frac{\partial(rA_\phi\sin\theta)}{\partial\theta} - \frac{\partial(rA_\theta)}{\partial\phi}\right), \tag{8.12}$$

$$(\nabla \times A)_\theta = \frac{1}{r\sin\theta}\left(\frac{\partial A_r}{\partial \phi} - \frac{\partial(rA_\phi \sin\theta)}{\partial r}\right) \tag{8.13}$$

and

$$(\nabla \times A)_\phi = \frac{1}{r}\left(\frac{\partial(rA_\theta)}{\partial r} - \frac{\partial A_r}{\partial \theta}\right). \tag{8.14}$$

Therefore, Maxwell's equations (8.1) and (8.2) in spherical coordinates become

$$-ik_0\epsilon E_r = \frac{1}{r^2\sin\theta}\left(\frac{\partial(rH_\phi \sin\theta)}{\partial\theta} - \frac{\partial(rH_\theta)}{\partial\phi}\right), \tag{8.15}$$

$$-ik_0\epsilon E_\theta = \frac{1}{r\sin\theta}\left(\frac{\partial H_r}{\partial \phi} - \frac{\partial(rH_\phi \sin\theta)}{\partial r}\right), \tag{8.16}$$

$$-ik_0\epsilon E_\phi = \frac{1}{r}\left(\frac{\partial(rH_\theta)}{\partial r} - \frac{\partial H_r}{\partial \theta}\right), \tag{8.17}$$

$$ik_0 H_r = \frac{1}{r^2\sin\theta}\left(\frac{\partial(rE_\phi \sin\theta)}{\partial\theta} - \frac{\partial(rE_\theta)}{\partial\phi}\right), \tag{8.18}$$

$$ik_0 H_\theta = \frac{1}{r\sin\theta}\left(\frac{\partial E_r}{\partial \phi} - \frac{\partial(rE_\phi \sin\theta)}{\partial r}\right), \tag{8.19}$$

and

$$ik_0 H_\phi = \frac{1}{r}\left(\frac{\partial(rE_\theta)}{\partial r} - \frac{\partial E_r}{\partial \theta}\right). \tag{8.20}$$

At each interface between shells, the θ and ϕ components of the fields must be continuous.

Solving these equations results in two linearly independent fields that are denoted ($E^{(e)}$, $H^{(e)}$) and ($E^{(m)}$, $H^{(m)}$) for which

$$E_r^{(e)} = E_r \quad \text{and} \quad H_r^{(e)} = 0\,, \tag{8.21}$$

which is the electric wave, and

$$E_r^{(m)} = 0 \quad \text{and} \quad H_r^{(m)} = H_r\,, \tag{8.22}$$

which is the magnetic wave. For $H_r = 0$, (8.16) and (8.17) become

$$E_\theta^{(e)} = -\frac{i}{k_0\epsilon r}\left(\frac{\partial\left(rH_\phi^{(e)}\right)}{\partial r}\right), \tag{8.23}$$

and

$$E_\phi^{(e)} = \frac{i}{k_0 \epsilon r}\left(\frac{\partial\left(rH_\theta^{(e)}\right)}{\partial r}\right). \tag{8.24}$$

By substituting (8.23) and (8.24) into (8.19) and (8.20), two wave equations for the electric wave solution are obtained,

$$\left(\frac{\partial^2}{\partial r^2} + \epsilon k_0^2\right)\left(rH_\theta^{(e)}\right) = -\left(\frac{ik_0\epsilon}{\sin\theta}\right)\frac{\partial E_r^{(e)}}{\partial\phi}, \tag{8.25}$$

and

$$\left(\frac{\partial^2}{\partial r^2} + \epsilon k_0^2\right)\left(rH_\phi^{(e)}\right) = ik_0\epsilon\frac{\partial E_r^{(e)}}{\partial\theta}. \tag{8.26}$$

Similarly, for the magnetic wave with $E_r = 0$, (8.19) and (8.20) become

$$H_\theta^{(m)} = \frac{i}{k_0 r\sin\theta}\left(\frac{\partial\left(rE_\phi^{(m)}\sin\theta\right)}{\partial r}\right), \tag{8.27}$$

and

$$H_\phi^{(m)} = -\frac{i}{k_0 r}\left(\frac{\partial\left(rE_\theta^{(m)}\right)}{\partial r}\right). \tag{8.28}$$

Substituting (8.27) and (8.28) into (8.16) and (8.17) gives two wave equations for the magnetic wave,

$$\left(\frac{\partial^2}{\partial r^2} + k_0^2\epsilon\right)\left(rE_\theta^{(m)}\right) = \frac{ik_0}{\sin\theta}\frac{\partial H_r^{(m)}}{\partial\phi}, \tag{8.29}$$

and

$$\left(\frac{\partial^2}{\partial r^2} + k_0^2\epsilon\right)\left(rE_\phi^{(m)}\right) = -ik_0\frac{\partial H_r^{(m)}}{\partial\theta}. \tag{8.30}$$

8.3 SCALAR POTENTIALS

Scalar Debye potentials are defined from which the electric and magnetic wave solutions may be derived. For the electric wave, let

$$U \equiv \frac{\partial\left(r\Pi^{(e)}\right)}{\partial r} \tag{8.31}$$

where

$$E_{\phi}^{(e)} = \frac{1}{r\sin\theta}\frac{\partial U}{\partial\phi} \tag{8.32}$$

and

$$E_{\theta}^{(e)} = \frac{1}{r}\frac{\partial U}{\partial\theta} \tag{8.33}$$

so

$$E_{\phi}^{(e)} = \frac{1}{r\sin\theta}\frac{\partial^2\left(r\Pi^{(e)}\right)}{\partial r\,\partial\phi} \tag{8.34}$$

and

$$E_{\theta}^{(e)} = \frac{1}{r}\frac{\partial^2\left(r\Pi^{(e)}\right)}{\partial r\,\partial\theta}. \tag{8.35}$$

It is also true that

$$H_{\phi}^{(e)} = i\epsilon k_0\frac{\partial\Pi^{(e)}}{\partial\theta} = \frac{i\epsilon k_0}{r}\frac{\partial\left(r\Pi^{(e)}\right)}{\partial\theta} \tag{8.36}$$

and

$$H_{\theta}^{(e)} = -\frac{i\epsilon k_0}{\sin\theta}\frac{\partial\Pi^{(e)}}{\partial\phi} = -\frac{i\epsilon k_0}{r\sin\theta}\frac{\partial\left(r\Pi^{(e)}\right)}{\partial\phi}. \tag{8.37}$$

Substituting (8.36) and (8.37) into (8.23) and (8.24) immediately reduces to (8.34) and (8.35), respectively. Substituting from (8.36) and (8.37) into (8.15) gives

$$E_{r}^{(e)} = -\frac{1}{r\sin\theta}\left[\frac{\partial}{\partial\theta}\left(\sin\theta\frac{\partial\Pi^{(e)}}{\partial\theta}\right) + \frac{1}{\sin\theta}\frac{\partial^2\Pi^{(e)}}{\partial\phi^2}\right]. \tag{8.38}$$

Substituting from (8.36)–(8.38) into the wave equations (8.25) and (8.26) gives

$$\frac{\partial}{\partial\phi}\left\{\frac{1}{r\sin\theta}\left[\frac{\partial}{\partial\theta}\left(\sin\theta\frac{\partial\Pi^{(e)}}{\partial\theta}\right) + \frac{1}{\sin\theta}\frac{\partial^2\Pi^{(e)}}{\partial\phi^2}\right] + \left(\frac{\partial^2}{\partial r^2} + \epsilon k_0^2\right)\left(r\Pi^{(e)}\right)\right\} = 0, \tag{8.39}$$

and

$$\frac{\partial}{\partial\theta}\left\{\frac{1}{r\sin\theta}\left[\frac{\partial}{\partial\theta}\left(\sin\theta\frac{\partial\Pi^{(e)}}{\partial\theta}\right) + \frac{1}{\sin\theta}\frac{\partial^2\Pi^{(e)}}{\partial\phi^2}\right] + \left(\frac{\partial^2}{\partial r^2} + \epsilon k_0^2\right)\left(r\Pi^{(e)}\right)\right\} = 0. \tag{8.40}$$

The expressions in curly brackets in (8.39) and (8.40) are identical. The two independent equations may be satisfied by setting this expression to zero.

$$\frac{1}{r}\frac{\partial^2}{\partial r^2}\left(r\Pi^{(e)}\right)+\frac{1}{r^2\sin\theta}\frac{\partial}{\partial\theta}\left[\sin\theta\frac{\partial\Pi^{(e)}}{\partial\theta}\right]+\frac{1}{r^2\sin^2\theta}\frac{\partial^2\Pi^{(e)}}{\partial\phi^2}+\frac{\epsilon k_0^2}{r}\left(r\Pi^{(e)}\right)=0 \tag{8.41}$$

or

$$-\frac{1}{r\sin\theta}\left[\frac{\partial}{\partial\theta}\left(\sin\theta\frac{\partial\Pi^{(e)}}{\partial\theta}\right)+\frac{1}{\sin\theta}\frac{\partial^2\Pi^{(e)}}{\partial\phi^2}\right]=\frac{\partial^2}{\partial r^2}\left(r\Pi^{(e)}\right)+\epsilon k_0^2\left(r\Pi^{(e)}\right). \tag{8.42}$$

Substituting into (8.38),

$$E_r^{(e)}=\frac{\partial^2}{\partial r^2}\left(r\Pi^{(e)}\right)+\epsilon k_0^2\left(r\Pi^{(e)}\right). \tag{8.43}$$

Following the same procedure, the magnetic wave is derived from a magnetic potential, $\Pi^{(m)}$. The complete solution is the sum of both electric and magnetic waves.

$$E_r=E_r^{(e)}+E_r^{(m)}=\frac{\partial^2}{\partial r^2}\left(r\Pi^{(e)}\right)+\epsilon k_0^2 r\Pi^{(e)} \tag{8.44}$$

$$E_\theta=E_\theta^{(e)}+E_\theta^{(m)}=\frac{1}{r}\frac{\partial^2\left(r\Pi^{(e)}\right)}{\partial r\,\partial\theta}+\frac{ik_0}{r\sin\theta}\frac{\partial\left(r\Pi^{(m)}\right)}{\partial\phi}, \tag{8.45}$$

$$E_\phi=E_\phi^{(e)}+E_\phi^{(m)}=\frac{1}{r\sin\theta}\frac{\partial^2\left(r\Pi^{(e)}\right)}{\partial r\partial\phi}-\frac{ik_0}{r}\frac{\partial\left(r\Pi^{(m)}\right)}{\partial\theta}, \tag{8.46}$$

$$H_r=H_r^{(e)}+H_r^{(m)}=\frac{\partial^2}{\partial r^2}\left(r\Pi^{(m)}\right)+\epsilon k_0^2 r\Pi^{(m)}, \tag{8.47}$$

$$H_\theta=H_\theta^{(e)}+H_\theta^{(m)}=-\frac{i\epsilon k_0}{r\sin\theta}\frac{\partial\left(r\Pi^{(e)}\right)}{\partial\phi}+\frac{1}{r}\frac{\partial^2\left(r\Pi^{(m)}\right)}{\partial r\,\partial\theta}, \tag{8.48}$$

and

$$H_\phi=H_\phi^{(e)}+H_\phi^{(m)}=\frac{i\epsilon k_0}{r}\frac{\partial\left(r\Pi^{(e)}\right)}{\partial\theta}+\frac{1}{r\sin\theta}\frac{\partial^2\left(r\Pi^{(m)}\right)}{\partial r\,\partial\phi}. \tag{8.49}$$

In spherical coordinates,

$$\nabla^2\psi=\frac{1}{r}\frac{\partial^2\psi}{\partial r^2}+\frac{1}{r^2\sin\theta}\frac{\partial}{\partial\theta}\left(\sin\theta\frac{\partial\psi}{\partial\theta}\right)+\frac{1}{r^2\sin^2\theta}\frac{\partial^2\psi}{\partial\phi^2} \tag{8.50}$$

so (8.41) is simply the wave equation for both $\Pi^{(e)}$ and $\Pi^{(m)}$,

$$\nabla^2\Pi + \epsilon k_0^2\Pi = 0. \tag{8.51}$$

8.4 SOLUTION OF THE WAVE EQUATION

By inspection of (8.45), (8.46), (8.48), and (8.49), it must be true that continuity of the tangential field components, E_θ, E_ϕ, H_θ, and H_ϕ, across each interface at a fixed r requires continuity of $\epsilon k_0 r\Pi^{(e)}$, $k_0 r \Pi^{(m)}$, $d(r\,\Pi^{(e)})/dr$, and $d(r\,\Pi^{(m)})/dr$ across each interface. A separable solution of the form

$$\Pi = \mathrm{R}(r)\Theta(\theta)\Phi(\phi) \tag{8.52}$$

satisfies the wave equation. Substituting (8.52) into (8.41) gives

$$\left\{\frac{r^2}{\left[r\,\mathrm{R}(r)\right]}\frac{\partial^2}{\partial r^2}\left[r\,\mathrm{R}(r)\right] + \epsilon k_0^2 r^2\right\} + \frac{1}{\Theta(\theta)\sin\theta}\frac{\partial}{\partial\theta}\left(\sin\theta\frac{\partial\Theta(\theta)}{\partial\theta}\right) + \frac{1}{\Phi(\phi)\sin^2\theta}\frac{\partial^2\Phi(\phi)}{\partial\phi^2} = 0. \tag{8.53}$$

The three terms on the left-hand side of (8.53) must each be constant and their sum be zero. Let the radial term equal the unknown constant α.

$$\frac{r^2}{\left[rR(r)\right]}\frac{d^2}{\mathrm{d}r^2}\left[rR(r)\right] + \epsilon k_0^2 r^2 = \alpha, \tag{8.54}$$

and

$$\frac{d^2}{dr^2}\left[rR(r)\right] + \left(\epsilon k_0^2 - \frac{\alpha}{r^2}\right)\left[rR(r)\right] = 0. \tag{8.55}$$

In addition, let the Φ term equal the unknown constant $-\beta$,

$$\frac{1}{\Phi(\phi)}\frac{d^2\Phi(\phi)}{d\phi^2} = -\beta, \tag{8.56}$$

so

$$\frac{d^2\Phi(\phi)}{d\phi^2} + \beta\Phi(\phi) = 0. \tag{8.57}$$

Then substituting into (8.53) gives

$$\alpha + \frac{1}{\Theta(\theta)\sin\theta}\frac{d}{d\theta}\left(\sin\theta\frac{d\Theta(\theta)}{d\theta}\right) - \frac{\beta}{\sin^2\theta} = 0, \tag{8.58}$$

and

$$\frac{1}{\sin\theta}\frac{d}{d\theta}\left(\sin\theta\frac{d\Theta(\theta)}{d\theta}\right) + \left(\alpha - \frac{\beta}{\sin^2\theta}\right)\Theta(\theta) = 0. \tag{8.59}$$

The partial differential equations have all become ordinary differential equations by the separation of variables technique. The solution to (8.57) is

$$\Phi(\phi) = a\cos\left(\sqrt{\beta}\phi\right) + b\sin\left(\sqrt{\beta}\phi\right). \tag{8.60}$$

In order for Φ to remain single-valued as ϕ passes through 2π sr,

$$\beta = m^2 \tag{8.61}$$

where m is an integer. Therefore, (8.60) can be rewritten

$$\Phi(\phi) = a\cos(m\phi) + b\sin(m\phi). \tag{8.62}$$

(8.59) is the spherical harmonic equation. It too must be constrained to be single-valued as θ passes through 2π sr. The condition for this is

$$a = l(l+1) \tag{8.63}$$

where l is an integer and $l > |m|$. Let

$$\xi \equiv \cos\theta \tag{8.64}$$

so

$$d\xi = -\sin\theta\ d\theta. \tag{8.65}$$

Substituting for α, β, and θ into (8.59),

$$\frac{d}{d\xi}\left(\sqrt{1-\xi^2}\,\frac{d\Theta(\xi)}{d\xi}\right) + \left(l(l+1) - \frac{m^2}{\sqrt{1-\xi^2}}\right)\Theta(\xi) = 0. \tag{8.66}$$

The solutions of (8.66) are the associated Legendre functions.

$$\Theta(\xi) = P_l^{(m)}(\xi) = P_l^{(m)}(\cos\theta). \tag{8.67}$$

For each l there are $2l + 1$ functions for which

$$m = -l, -l+1, \cdots, l-1, l. \tag{8.68}$$

Finally, for (8.55) let

$$-\sqrt{\epsilon}\,k_0 r \equiv \rho \tag{8.69}$$

so

$$R(r) \equiv \frac{1}{\sqrt{\rho}} Z(\rho), \tag{8.70}$$

$$rR(r) \equiv -\frac{\sqrt{\rho}}{\sqrt{\epsilon}\,k_0} Z(\rho) \tag{8.71}$$

$$\frac{d}{dr}\left[r\,\mathrm{R}(r)\right]=\frac{d}{d\rho}\left[\sqrt{\rho}\mathrm{Z}(\rho)\right]=\frac{1}{2\sqrt{\rho}}\mathrm{Z}(\rho)+\sqrt{\rho}\frac{dZ(\rho)}{d\rho}, \tag{8.72}$$

and

$$\frac{d^2}{dr^2}\left[rR(r)\right]=-\sqrt{\epsilon}\,k_0\frac{d}{d\rho}\left(\frac{1}{2\sqrt{\rho}}Z(\rho)+\sqrt{\rho}\frac{dZ(\rho)}{d\rho}\right) \tag{8.73}$$

$$=-\sqrt{\epsilon\,\rho}\,k_0\left(-\frac{1}{4\rho^2}Z(\rho)+\frac{1}{\rho}\frac{dZ(\rho)}{d\rho}+\frac{d^2Z(\rho)}{d\rho^2}\right). \tag{8.74}$$

Then (8.55) becomes

$$\frac{d^2Z(\rho)}{d\rho^2}+\frac{1}{\rho}\frac{dZ(\rho)}{d\rho}+\left(1-\frac{\left(l+\frac{1}{2}\right)^2}{\rho^2}\right)\mathrm{Z}(\rho)=0 \tag{8.75}$$

which is the Bessel equation with solution the cylindrical function $Z_{l+\frac{1}{2}}(\rho)$. So,

$$R=\frac{1}{\sqrt{\sqrt{\epsilon}\,k_0 r}}Z_{l+\frac{1}{2}}\left(\sqrt{\epsilon}\,k_0 r\right). \tag{8.76}$$

Each cylindrical function can be written as a linear combination of a Bessel function, $J_{l+\frac{1}{2}}(\rho)$, and a Neumann function, $N_{l+\frac{1}{2}}(\rho)$. The Riccati-Bessel functions are defined as

$$\psi_l(\rho)\equiv\sqrt{\frac{\pi\rho}{2}}J_{l+\frac{1}{2}}(\rho) \tag{8.77}$$

and

$$\chi_l(\rho)\equiv-\sqrt{\frac{\pi\rho}{2}}N_{l+\frac{1}{2}}(\rho). \tag{8.78}$$

The general solution to (8.55) is then

$$rR=c_l\psi_l\left(\sqrt{\epsilon}\,k_0 r\right)+d_l\chi_l\left(\sqrt{\epsilon}\,k_0 r\right) \tag{8.79}$$

where c_1, and d_1 are unknown constants to be determined by satisfying the boundary conditions.

$\psi_l(\rho)$ is finite for finite ρ and can be used to represent fields within the sphere. $\chi_l(\rho)$ is singular at $\rho = 0$ and cannot be used to represent the fields inside the sphere. Moreover, both ψ_l and χ_l oscillate as $R \rightarrow \infty$, so they cannot be used individually to represent the field in the outer incident medium. However, the Riccati-Bessel function

$$\zeta_l^{(1)}(\rho)\equiv\psi_l(\rho)+i\chi_l(\rho)=\sqrt{\frac{\pi\rho}{2}}H_{l+\frac{1}{2}}^{(2)}(\rho) \tag{8.80}$$

where $H^{(2)}_{l+\frac{1}{2}}(\rho)$ is a Hankel function of the first kind vanishes as $\rho \to \infty$ and can be used in a field expansion for the scattered field in the outer medium.

Both $\psi_l(\rho)$ and $\chi_l(\rho)$ functions are required to satisfy boundary conditions for the fields within the shells. The general solution of the wave equation (8.52) is then

$$r\Pi = \sum_{l=0}^{\infty}\sum_{m=-l}^{+l}[c_l\psi_l(\sqrt{\epsilon}\,k_0 r) + d_l\chi_l(\sqrt{\epsilon}\,k_0 r)]\cdot P_l^{(m)}(\cos\theta)[a_m\cos(m\phi) + b_m\sin(m\phi)]. \tag{8.81}$$

where a_m and b_m are also unknown constants to be determined by satisfying the boundary conditions.

8.5 INCIDENT WAVE POTENTIALS

The incident wave fields in (8.3)–(8.5) must be first expressed in the spherical coordinate system using the transform equations (8.9)–(8.11). Let there be N concentric shells around the inner sphere and the different media be numbered from 0 to $N + 1$, with the 0th medium being inside the sphere, and the $(N + 1)$st medium being the outer medium of the incident wave. The (normalized) fields for the incident wave in spherical coordinates are

$$E_r^{(i)} = \mathrm{Exp}\left[i\sqrt{\epsilon_{N+1}}\,k_0\,r\cos\theta\right]\sin\theta\cos\phi, \tag{8.82}$$

$$E_\theta^{(i)} = \mathrm{Exp}\left[i\sqrt{\epsilon_{N+1}}\,k_0\,r\cos\theta\right]\sin\theta\cos\phi, \tag{8.83}$$

$$E_\phi^{(i)} = -\mathrm{Exp}\left[i\sqrt{\epsilon_{N+1}}\,k_0\,r\cos\theta\right]\sin\phi, \tag{8.84}$$

$$H_r^{(i)} = i\epsilon_{N+1}\mathrm{Exp}\left[i\sqrt{\epsilon_{N+1}}\,k_0\,r\cos\theta\right]\sin\theta\sin\phi, \tag{8.85}$$

$$H_\theta^{(i)} = i\epsilon_{N+1}\mathrm{Exp}\left[i\sqrt{\epsilon_{N+1}}\,k_0\,r\cos\theta\right]\cos\theta\sin\phi, \tag{8.86}$$

and

$$H_\phi^{(i)} = i\epsilon_{N+1}\mathrm{Exp}\left[i\sqrt{\epsilon_{N+1}}\,k_0\,r\cos\theta\right]\cos\phi. \tag{8.87}$$

From (8.44),

$$\frac{\partial^2}{\partial r^2}\left(r\Pi^{(i,e)}\right) + \epsilon\,k_0^2\,r\Pi^{(i,e)} = \mathrm{Exp}\left[i\sqrt{\epsilon_{N+1}}\,k_0\,r\cos\theta\right]\sin\theta\cos\phi. \tag{8.88}$$

The exponential term can be expanded in Legendre polynomials, P_l,

$$\mathrm{Exp}\left[i\sqrt{\epsilon_{N+1}}\,k_0\,r\cos\theta\right] = \sum_{l=0}^{\infty} i^l(2l+1)\frac{\psi_l\left(\sqrt{\epsilon_{N+1}}\,k_0 r\right)}{\sqrt{\epsilon_{N+1}}\,k_0 r}P_l(\cos\theta). \tag{8.89}$$

Three additional relations are

$$\mathrm{Exp}\left[i\sqrt{\epsilon_{N+1}}\,k_0\,r\cos\theta\right] = \frac{i}{\sqrt{\epsilon_{N+1}}\,k_0\,r\sin\theta}\frac{\partial}{\partial\theta}\mathrm{Exp}\left[i\sqrt{\epsilon_{N+1}}\,k_0\,r\cos\theta\right], \tag{8.90}$$

$$\frac{dP_l(\cos\theta)}{d\theta} = -P_l^{(1)}(\cos\theta), \tag{8.91}$$

and

$$P_0^{(1)}(\cos\theta) = 0. \tag{8.92}$$

From (8.90),

$$\operatorname{Exp}\left[i\sqrt{\epsilon_{N+1}}k_0\, r\cos\theta\right]\sin\theta\cos\phi = -\frac{\cos\phi}{i\sqrt{\epsilon_{N+1}}k_0 r}\frac{\partial}{\partial\theta}\operatorname{Exp}\left[i\sqrt{\epsilon_{N+1}}k_0 r\cos\theta\right]. \tag{8.93}$$

Substituting (8.89),

$$= \frac{i\cos\phi}{\sqrt{\epsilon_{N+1}}k_0 r}\sum_{l=0}^{\infty} i^l(2l+1)\frac{\psi_l\left(\sqrt{\epsilon_{N+1}}k_0 r\right)}{\sqrt{\epsilon_{N+1}}k_0 r}\frac{\partial}{\partial\theta}P_l(\cos\theta), \tag{8.94}$$

and substituting (8.91),

$$= -\frac{i\cos\phi}{\sqrt{\epsilon_{N+1}}k_0 r}\sum_{l=0}^{\infty} i^l(2l+1)\frac{\psi_l\left(\sqrt{\epsilon_{N+1}}k_0 r\right)}{\sqrt{\epsilon_{N+1}}k_0 r}P_l^{(1)}(\cos\theta). \tag{8.95}$$

Explicitly evaluating the $l = 0$ term using (8.91),

$$= -\frac{i\cos\phi}{\sqrt{\epsilon_{N+1}}k_0 r}\left\{\frac{\psi_0\left(\sqrt{\epsilon_{N+1}}k_0 r\right)}{\sqrt{\epsilon_{N+1}}k_0 r}P_0^{(1)}(\cos\theta) + \sum_{l=1}^{\infty} i^l(2l+1)\frac{\psi_l\left(\sqrt{\epsilon_{N+1}}k_0 r\right)}{\sqrt{\epsilon_{N+1}}k_0 r}P_l^{(1)}(\cos\theta)\right\} \tag{8.96}$$

or finally,

$$\operatorname{Exp}\left[i\sqrt{\epsilon_{N+1}}k_0 r\cos\theta\right]\sin\theta\cos\phi = \frac{\cos\phi}{\epsilon_{N+1}k_0^2 r^2}\sum_{l=1}^{\infty} i^{l-1}(2l+1)\psi_l\left(\sqrt{\epsilon_{N+1}}k_0 r\right)P_l^{(1)}(\cos\theta). \tag{8.97}$$

This suggests trying a solution for the potential of the incident wave of the form

$$r\Pi_i = \frac{1}{\epsilon_{N+1}k_0^2}\sum_{l=1}^{\infty}\alpha_l\psi_l\left(\sqrt{\epsilon_{N+1}}k_0 r\right)P_l^{(1)}(\cos\theta)\cos\phi. \tag{8.98}$$

Substituting (8.97) and (8.98) into (8.88) gives

$$\alpha_l P_l^{(1)}(\cos\theta)\frac{\partial^2}{\partial r^2}\left(\sum_{l=1}^{\infty}\psi_l\left(\sqrt{\epsilon_{N+1}}k_0 r\right)\right) + \epsilon_{N+1}k_0^2\sum_{l=1}^{\infty}\alpha_l\psi_l\left(\sqrt{\epsilon_{N+1}}k_0 r\right)P_l^{(1)}(\cos\theta) =$$

$$\frac{1}{r^2}\sum_{l=1}^{\infty} i^{l-1}(2l+1)\psi_l\left(\sqrt{\epsilon_{N+1}}k_0 r\right)P_l^{(1)}(\cos\theta). \tag{8.99}$$

Equating each term in the expansion,

$$\alpha_l\left[\frac{\partial^2}{\partial r^2}\left(\psi_l\left(\sqrt{\epsilon_{N+1}}k_0 r\right)\right)+\epsilon_{N+1}k_0^2\psi_l\left(\sqrt{\epsilon_{N+1}}k_0 r\right)\right]=\frac{i^{l-1}}{r^2}(2l+1)\psi_l\left(\sqrt{\epsilon_{N+1}}k_0 r\right). \tag{8.100}$$

A solution of (8.55) for $\alpha = l\,(l + 1)$ from (8.79) is

$$rR(r)=c_l\psi_l\left(\sqrt{\epsilon}\,k_0 r\right) \tag{8.101}$$

with $c_l = 1$ and $d_l = 0$. Substituting (8.101) into (8.79),

$$\frac{d^2}{dr^2}\psi_l\left(\sqrt{\epsilon}\,k_0 r\right)+\left(\epsilon\,k_0^2-\frac{l(l+1)}{r^2}\right)\psi_l\left(\sqrt{\epsilon}\,k_0 r\right)=0 \tag{8.102}$$

and comparing to (8.100), it must be true that

$$\alpha_l=\frac{i^{l-1}(2l+1)}{l(l+1)}. \tag{8.103}$$

Therefore, (8.98) is finally

$$r\Pi_i^{(e)}=\frac{1}{\epsilon_{N+1}k_0^2}\sum_{l=1}^{\infty}\frac{i^{l-1}(2l+1)}{l(l+1)}\psi_l\left(\sqrt{\epsilon_{N+1}}k_0 r\right)P_l^{(1)}(\cos\theta)\cos\phi. \tag{8.104}$$

Following the same procedure for the magnetic potential of the incident wave, it is found that

$$r\Pi_i^{(m)}=\frac{1}{\sqrt{\epsilon_{N+1}}k_0^2}\sum_{l=1}^{\infty}\frac{i^{l-1}(2l+1)}{l(l+1)}\psi_l\left(\sqrt{\epsilon_{N+1}}k_0 r\right)P_l^{(1)}(\cos\theta)\sin\phi. \tag{8.105}$$

8.6 BOUNDARY CONDITIONS

As previously stated, $\epsilon\,k_0 r\Pi^{(e)}$, $k_0 r\Pi^{(m)}$, $\frac{\partial\left(r\Pi^{(e)}\right)}{\partial r}$, and $\frac{\partial\left(r\Pi^{(m)}\right)}{\partial r}$ must be continuous across each interface. Let r_n be the inner radius of shell n, where the shells are numbered from inside to outside. Therefore, for N shells around the inner sphere, r_{N+1} is the radius of the outer surface. Including the scattered field potentials in the incident medium, at the outer surface

$$\epsilon_{N+1}k_0\left(r_{N+1}\Pi_{N+1}^{(e,i)}+r_{N+1}\Pi_{N+1}^{(e,s)}\right)=\epsilon_N k_0 r_{N+1}\Pi_N^{(e)}\ , \tag{8.106}$$

$$\frac{\partial}{\partial r}\left(r\Pi_{N+1}^{(e,i)}+r\Pi_{N+1}^{(e,s)}\right)\Bigg|_{r_{N+1}}=\frac{\partial}{\partial r}\left(r\Pi_N^{(e)}\right)\Bigg|_{r_{N+1}}, \tag{8.107}$$

$$r_{N+1}\Pi_{N+1}^{(m,i)}+r_{N+1}\Pi_{N+1}^{(m,s)}=r_{N+1}\Pi_N^{(m)}, \tag{8.108}$$

and

$$\frac{\partial}{\partial r}\left(r\Pi_{N+1}^{(m,i)}+r\Pi_{N+1}^{(m,s)}\right)\Bigg|_{r_{N+1}}=\frac{\partial}{\partial r}\left(r\Pi_{N}^{(m)}\right)\Bigg|_{r_{N+1}}. \tag{8.109}$$

At the interface between the $(n + 1)^{st}$ and n^{th} shells,

$$\epsilon_{n+1}k_0\left(r_{n+1}\Pi_{n+1}^{(e)}\right)=\epsilon_n k_0\left(r_{n+1}\Pi_{n}^{(e)}\right), \tag{8.110}$$

$$\frac{\partial}{\partial r}\left(r\Pi_{n+1}^{(e)}\right)\Bigg|_{r_{n+1}}=\frac{\partial}{\partial r}\left(r\Pi_{n}^{(e)}\right)\Bigg|_{r_{n+1}}, \tag{8.111}$$

$$r_{n+1}\Pi_{n+1}^{(m)}=r_{n+1}\Pi_{n}^{(m)}, \tag{8.112}$$

and

$$\frac{\partial}{\partial r}\left(r\Pi_{n+1}^{(m)}\right)\Bigg|_{r_{n+1}}=\frac{\partial}{\partial r}\left(r\Pi_{n}^{(m)}\right)\Bigg|_{r_{n+1}}. \tag{8.113}$$

At the inner shell-sphere boundary,

$$\epsilon_1\left(r_1\Pi_1^{(e)}\right)=\epsilon_0\left(r_1\Pi_0^{(e)}\right), \tag{8.114}$$

$$\frac{\partial}{\partial r}\left(r\Pi_{1}^{(e)}\right)\Bigg|_{r_1}=\frac{\partial}{\partial r}\left(r\Pi_{0}^{(e)}\right)\Bigg|_{r_1} \tag{8.115}$$

$$r_1\Pi_1^{(m)}=r_1\Pi_0^{(m)}, \tag{8.116}$$

and

$$\frac{\partial}{\partial r}\left(r\Pi_{1}^{(m)}\right)\Bigg|_{r_1}=\frac{\partial}{\partial r}\left(r\Pi_{0}^{(m)}\right)\Bigg|_{r_1}. \tag{8.117}$$

Examining the previously obtained expansions for the potentials in (8.80), and comparing to the expressions for the incident wave potentials (8.103) and (8.104), it must be true that $m = 1$ in (8.80) and so $l \geq 1$.

$$r\Pi=\sum_{l=1}^{\infty}\left[c_l\psi_l\left(\sqrt{\epsilon}k_0 r\right)+d_l\chi_l\left(\sqrt{\epsilon}k_0 r\right)\right]P_l^{(1)}\left(\cos\theta\right)\left[a_1\cos\phi+b_1\sin\phi\right]. \tag{8.118}$$

Furthermore, for the electric potential, $b_1 = 0$, and for the magnetic potential, $a_1 = 0$. Within the inner sphere only the ψ_l functions remain finite and physical, so

$$r\Pi_0^{(e)}=\frac{1}{\epsilon_0 k_0^2}\sum_{l=1}^{\infty}A_{l,0}^{(e)}\psi_l\left(\sqrt{\epsilon_0}k_0 r\right)P_l^{(1)}\left(\cos\theta\right)\cos\phi, \tag{8.119}$$

and

$$r\Pi_0^{(m)} = \frac{1}{\sqrt{\epsilon_0}k_0^2}\sum_{l=1}^{\infty} A_{l,0}^{(m)}\psi_l\left(\sqrt{\epsilon_0}k_0 r\right)P_l^{(1)}(\cos\theta)\sin\phi. \tag{8.120}$$

As previously discussed, the scattered wave can be represented by the function ζ_l in (8.79) and the scattered wave potentials in the outer medium are

$$r\Pi_{N+1}^{(e,s)} = \frac{1}{\epsilon_{N+1}k_0^2}\sum_{l=1}^{\infty} B_{l,N+1}^{(e)}\zeta_l^{(1)}\left(\sqrt{\epsilon_{N+1}}k_0 r\right)P_l^{(1)}(\cos\theta)\cos\phi \tag{8.121}$$

and

$$r\Pi_{N+1}^{(m,s)} = \frac{1}{\sqrt{\epsilon_{N+1}}k_0^2}\sum_{l=1}^{\infty} B_{l,N+1}^{(m)}\zeta_l^{(1)}\left(\sqrt{\epsilon_{N+1}}k_0 r\right)P_l^{(1)}(\cos\theta)\sin\phi. \tag{8.122}$$

For shell regions between the outer medium and the inner sphere, both functions ψ_l and χ_l are required.

$$r\Pi_n^{(e)} = \frac{1}{\epsilon_n k_0^2}\sum_{l=1}^{\infty}\left[A_{l,n}^{(e)}\psi_l\left(\sqrt{\epsilon_n}k_0 r\right) + B_{l,n}^{(e)}\chi_l^{(1)}\left(\sqrt{\epsilon_n}k_0 r\right)\right]P_l^{(1)}(\cos\theta)\cos\phi, \tag{8.123}$$

and

$$r\Pi_n^{(m)} = \frac{1}{\sqrt{\epsilon_n}k_0^2}\sum_{l=1}^{\infty}\left[A_{l,n}^{(m)}\psi_l\left(\sqrt{\epsilon_n}k_0 r\right) + B_{l,n}^{(m)}\chi_l^{(1)}\left(\sqrt{\epsilon_n}k_0 r\right)\right]P_l^{(1)}(\cos\theta)\sin\phi. \tag{8.124}$$

(8.104), (8.105), and (8.119)–(8.124) are now substituted into the boundary conditions (8.106)–(8.117). To simplify the notation, let

$$q_n \equiv \sqrt{\epsilon_{n-1}}\, k_0\, r_n, \tag{8.125}$$

$$\rho_n \equiv \sqrt{\epsilon_n}\, k_0\, r_n, \tag{8.126}$$

$$\psi_l'(x_m) \equiv \left.\frac{d\psi_l(x)}{dx}\right|_{x=x_m}, \tag{8.127}$$

$$\chi_l'(x_m) \equiv \left.\frac{d\chi_l(x)}{dx}\right|_{x=x_m} \tag{8.128}$$

and

$$\zeta_l'(x_m) \equiv \left.\frac{d\zeta_l(x)}{dx}\right|_{x=x_m}. \tag{8.129}$$

8.6.1 Outer Surface

Matching boundary conditions at the outer surface,

$$\zeta_l^{(1)}\left(\rho_{N+1}\right)B_{l,N+1}^{(e)}+\frac{i^{l-1}\left(2l+1\right)}{l\left(l+1\right)}\psi_l\left(\rho_{N+1}\right)=A_{l,N}^{(e)}\psi_l\left(q_{N+1}\right)+B_{l,N}^{(e)}\chi_l^{(1)}\left(q_{N+1}\right), \tag{8.130}$$

$$\frac{1}{\sqrt{\epsilon_{N+1}}}\left[\zeta_l'\left(\rho_{N+1}\right)B_{l,N+1}^{(e)}+\frac{i^{l-1}\left(2l+1\right)}{l\left(l+1\right)}\psi_l'\left(\rho_{N+1}\right)\right]=\frac{1}{\sqrt{\epsilon_N}}\left[\psi_l'\left(q_{N+1}\right)A_{l,N}^{(e)}+\chi_l'\left(q_{N+1}\right)B_{l,N}^{(e)}\right], \tag{8.131}$$

$$\frac{1}{\sqrt{\epsilon_{N+1}}}\left[\zeta_l'\left(\rho_{N+1}\right)B_{l,N+1}^{(m)}+\frac{i^{l-1}\left(2l+1\right)}{l\left(l+1\right)}\psi_l'\left(\rho_{N+1}\right)\right]=\frac{1}{\sqrt{\epsilon_N}}\left[\psi_l'\left(q_{N+1}\right)A_{l,N}^{(m)}+\chi_l'\left(q_{N+1}\right)B_{l,N}^{(m)}\right], \tag{8.132}$$

and

$$\zeta_l'\left(\rho_{N+1}\right)B_{l,N+1}^{(m)}+\frac{i^{l-1}\left(2l+1\right)}{l\left(l+1\right)}\psi_l'\left(\rho_{N+1}\right)=\psi_l'\left(q_{N+1}\right)A_{l,N}^{(m)}+\chi_l'\left(q_{N+1}\right)B_{l,N}^{(m)}. \tag{8.133}$$

These equations can be rewritten as matrix equations,

$$\tilde{\boldsymbol{M}}_{N+1}^{(e)}B_{l,N+1}^{(e)}+\boldsymbol{V}_{N+1}^{(e)}=\tilde{\boldsymbol{N}}_N^{(e)}\begin{bmatrix}A_{l,N}^{(e)}\\ B_{l,N}^{(e)}\end{bmatrix}, \tag{8.134}$$

and

$$\tilde{\boldsymbol{M}}_{N+1}^{(m)}B_{l,N+1}^{(m)}+\boldsymbol{V}_{N+1}^{(m)}=\tilde{\boldsymbol{N}}_N^{(m)}\begin{bmatrix}A_{l,N}^{(m)}\\ B_{l,N}^{(m)}\end{bmatrix} \tag{8.135}$$

where

$$m_{N+1,1}^{(e)}\equiv\zeta_l\left(\rho_{N+1}\right), \tag{8.136}$$

$$m_{N+1,2}^{(e)}\equiv\frac{1}{\sqrt{\epsilon_{N+1}}}\zeta_l'\left(\rho_{N+1}\right), \tag{8.137}$$

$$\mathrm{v}_{N+1,1}^{(e)}\equiv\frac{i^{l-1}\left(2l+1\right)}{l\left(l+1\right)}\psi_l\left(\rho_{N+1}\right), \tag{8.138}$$

$$\mathrm{v}_{N+1,2}^{(e)}\equiv\frac{1}{\sqrt{\epsilon_{N+1}}}\frac{i^{l-1}\left(2l+1\right)}{l\left(l+1\right)}\psi_l'\left(\rho_{N+1}\right), \tag{8.139}$$

$$n_{N,11}^{(e)} \equiv \psi_l\left(q_{N+1}\right), \tag{8.140}$$

$$n_{N,12}^{(e)} \equiv \chi_l\left(q_{N+1}\right), \tag{8.141}$$

$$n_{N,21}^{(e)} \equiv \frac{1}{\sqrt{\epsilon_N}}\psi_l'\left(q_{N+1}\right), \tag{8.142}$$

$$n_{N,22}^{(e)} \equiv \frac{1}{\sqrt{\epsilon_N}}\chi_l'\left(q_{N+1}\right), \tag{8.143}$$

$$m_{N+1,1}^{(m)} \equiv \frac{1}{\sqrt{\epsilon_{N+1}}}\zeta_l\left(\rho_{N+1}\right), \tag{8.144}$$

$$m_{N+1,2}^{(m)} \equiv \zeta_l'\left(\rho_{N+1}\right), \tag{8.145}$$

$$\mathrm{v}_{N+1,1}^{(m)} \equiv \frac{1}{\sqrt{\epsilon_{N+1}}}\frac{i^{l-1}(2l+1)}{l(l+1)}\psi\left(\rho_{N+1}\right), \tag{8.146}$$

$$\mathrm{v}_{N+1,2}^{(m)} \equiv \frac{i^{l-1}(2l+1)}{l(l+1)}\psi_l'\left(\rho_{N+1}\right), \tag{8.147}$$

$$n_{N,11}^{(m)} \equiv \frac{1}{\sqrt{\epsilon_N}}\psi_l\left(q_{N+1}\right), \tag{8.148}$$

$$n_{N,12}^{(m)} \equiv \frac{1}{\sqrt{\epsilon_N}}\chi_l'\left(q_{N+1}\right), \tag{8.149}$$

$$n_{N,21}^{(m)} \equiv \psi_l'\left(q_{N+1}\right), \tag{8.150}$$

and

$$n_{N,22}^{(m)} \equiv \chi_l'\left(q_{N+1}\right). \tag{8.151}$$

8.6.2 Interfaces Between Shells

For interfaces between shells, the boundary conditions require

$$A_{l,n+1}^{(e)}\psi_l\left(\sqrt{\epsilon_{n+1}}k_0 r_{n+1}\right) + B_{l,n+1}^{(e)}\chi_l^{(1)}\left(\sqrt{\epsilon_{n+1}}k_0 r_{n+1}\right) =$$

$$A_{l,n}^{(e)}\psi_l\left(\sqrt{\epsilon_n}k_0 r_{n+1}\right) + B_{l,n}^{(e)}\chi_l^{(1)}\left(\sqrt{\epsilon_n}k_0 r_{n+1}\right), \tag{8.152}$$

$$A^{(e)}_{l,n+1}\frac{1}{\epsilon_{n+1}}\frac{\partial}{\partial r}\psi_l\left(\sqrt{\epsilon_{n+1}}k_0r\right)\Big|_{r_{n+1}}+B^{(e)}_{l,n+1}\frac{1}{\epsilon_{n+1}}\frac{\partial}{\partial r}\chi^{(1)}_l\left(\sqrt{\epsilon_{n+1}}k_0r\right)\Big|_{r_{n+1}}=$$

$$A^{(e)}_{l,n}\frac{1}{\epsilon_n}\frac{\partial}{\partial r}\psi_l\left(\sqrt{\epsilon_n}k_0r\right)\Big|_{r_{n+1}}+B^{(e)}_{l,n}\frac{1}{\epsilon_n}\frac{\partial}{\partial r}\chi^{(1)}_l\left(\sqrt{\epsilon_n}k_0r\right)\Big|_{r_{n+1}}, \tag{8.153}$$

$$A^{(m)}_{l,n+1}\frac{1}{\sqrt{\epsilon_{n+1}}}\psi_l\left(\sqrt{\epsilon_{n+1}}k_0r_{n+1}\right)+B^{(m)}_{l,n+1}\frac{1}{\sqrt{\epsilon_{n+1}}}\chi^{(1)}_l\left(\sqrt{\epsilon_{n+1}}k_0r_{n+1}\right)=$$

$$A^{(m)}_{l,n}\frac{1}{\sqrt{\epsilon_n}}\psi_l\left(\sqrt{\epsilon_n}k_0r_{n+1}\right)+B^{(m)}_{l,n}\frac{1}{\sqrt{\epsilon_n}}\chi^{(1)}_l\left(\sqrt{\epsilon_n}k_0r_{n+1}\right), \tag{8.154}$$

and

$$A^{(m)}_{l,n+1}\frac{1}{\sqrt{\epsilon_{n+1}}}\frac{\partial}{\partial r}\psi_l\left(\sqrt{\epsilon_{n+1}}k_0r\right)\Big|_{r_{n+1}}+B^{(m)}_{l,n+1}\frac{1}{\sqrt{\epsilon_{n+1}}}\frac{\partial}{\partial r}\chi^{(1)}_l\left(\sqrt{\epsilon_{n+1}}k_0r\right)\Big|_{r_{n+1}}=$$

$$A^{(m)}_{l,n}\frac{1}{\sqrt{\epsilon_n}}\frac{\partial}{\partial r}\psi_l\left(\sqrt{\epsilon_n}k_0r\right)\Big|_{r_{n+1}}+B^{(m)}_{l,n}\frac{1}{\sqrt{\epsilon_n}}\frac{\partial}{\partial r}\chi^{(1)}_l\left(\sqrt{\epsilon_n}k_0r\right)\Big|_{r_{n+1}}. \tag{8.155}$$

Again, there are two matrix equations,

$$\tilde{\boldsymbol{M}}^{(e)}_{n+1}\begin{bmatrix}A^{(e)}_{l,n+1}\\ B^{(e)}_{l,n+1}\end{bmatrix}=\tilde{\boldsymbol{N}}^{(e)}_{n}\begin{bmatrix}A^{(e)}_{l,n}\\ B^{(e)}_{l,n}\end{bmatrix}, \tag{8.156}$$

and

$$\tilde{\boldsymbol{M}}^{(m)}_{n+1}\begin{bmatrix}A^{(e)}_{l,n+1}\\ B^{(e)}_{l,n+1}\end{bmatrix}=\tilde{\boldsymbol{N}}^{(m)}_{n}\begin{bmatrix}A^{(e)}_{l,n}\\ B^{(e)}_{l,n}\end{bmatrix} \tag{8.157}$$

where

$$m^{(e)}_{n+1,11}\equiv\psi_l\left(\sqrt{\epsilon_{n+1}}k_0r_{n+1}\right), \tag{8.158}$$

$$m^{(e)}_{n+1,12}\equiv\chi^{(1)}_l\left(\sqrt{\epsilon_{n+1}}k_0r_{n+1}\right), \tag{8.159}$$

$$m^{(e)}_{n+1,21}\equiv\frac{k_0}{\sqrt{\epsilon_{n+1}}}\frac{\partial\psi_l\left(\sqrt{\epsilon_{n+1}}k_0r\right)}{\partial\left(\sqrt{\epsilon_{n+1}}k_0r\right)}\Bigg|_{r_{n+1}}, \tag{8.160}$$

$$m^{(e)}_{n+1,22}\equiv\frac{k_0}{\sqrt{\epsilon_{n+1}}}\frac{\partial\chi^{(1)}_l\left(\sqrt{\epsilon_{n+1}}k_0r\right)}{\partial\left(\sqrt{\epsilon_{n+1}}k_0r\right)}\Bigg|_{r_{n+1}}, \tag{8.161}$$

$$n_{n,11}^{(e)} \equiv \psi_l\left(\sqrt{\epsilon_n}k_0 r_{n+1}\right), \tag{8.162}$$

$$n_{n,12}^{(e)} \equiv \chi_l^{(1)}\left(\sqrt{\epsilon_n}k_0 r_{n+1}\right), \tag{8.163}$$

$$n_{n,21}^{(e)} \equiv \frac{k_0}{\sqrt{\epsilon_n}} \left.\frac{\partial\psi_l\left(\sqrt{\epsilon_n}k_0 r\right)}{\partial\left(\sqrt{\epsilon_n}k_0 r\right)}\right|_{r_{n+1}}, \tag{8.164}$$

$$n_{n,22}^{(e)} \equiv \frac{k_0}{\sqrt{\epsilon_n}} \left.\frac{\partial\chi_l^{(1)}\left(\sqrt{\epsilon_n}k_0 r\right)}{\partial\left(\sqrt{\epsilon_n}k_0 r\right)}\right|_{r_{n+1}}, \tag{8.165}$$

$$m_{n+1,11}^{(m)} \equiv \frac{1}{\sqrt{\epsilon_{n+1}}}\psi_l\left(\sqrt{\epsilon_{n+1}}k_0 r_{n+1}\right), \tag{8.166}$$

$$m_{n+1,12}^{(m)} \equiv \frac{1}{\sqrt{\epsilon_{n+1}}}\chi_l^{(1)}\left(\sqrt{\epsilon_{n+1}}k_0 r_{n+1}\right), \tag{8.167}$$

$$m_{n+1,21}^{(m)} \equiv k_0 \left.\frac{\partial\psi_l\left(\sqrt{\epsilon_{n+1}}k_0 r\right)}{\partial\left(\sqrt{\epsilon_{n+1}}k_0 r\right)}\right|_{r_{n+1}}, \tag{8.168}$$

$$m_{n+1,22}^{(m)} \equiv k_0 \left.\frac{\partial\chi_l^{(1)}\left(\sqrt{\epsilon_{n+1}}k_0 r\right)}{\partial\left(\sqrt{\epsilon_{n+1}}k_0 r\right)}\right|_{r_{n+1}}, \tag{8.169}$$

$$n_{n,11}^{(m)} \equiv \frac{1}{\sqrt{\epsilon_n}}\psi_l\left(\sqrt{\epsilon_n}k_0 r_{n+1}\right), \tag{8.170}$$

$$n_{n,12}^{(m)} \equiv \frac{1}{\sqrt{\epsilon_n}}\chi_l^{(1)}\left(\sqrt{\epsilon_n}k_0 r_{n+1}\right) \tag{8.171}$$

$$n_{n,21}^{(m)} \equiv k_0 \left.\frac{\partial\psi_l\left(\sqrt{\epsilon_n}k_0 r\right)}{\partial\left(\sqrt{\epsilon_n}k_0 r\right)}\right|_{r_{N+1}}, \tag{8.172}$$

and

$$n_{n,22}^{(m)} \equiv k_0 \left.\frac{\partial\chi_l^{(1)}\left(\sqrt{\epsilon_n}k_0 r\right)}{\partial\left(\sqrt{\epsilon_n}k_0 r\right)}\right|_{r_{n+1}}. \tag{8.173}$$

8.6.3 Inner Surface

Finally, at the innermost interface the boundary conditions give

$$\psi_l(\rho_1)A_{l,1}^{(e)}+\chi_l(\rho_1)B_{l,1}^{(e)}=\psi_l(q_1)A_{l,0}^{(e)}, \tag{8.174}$$

$$\frac{1}{\sqrt{\epsilon_1}}\psi_l'(\rho_1)A_{l,1}^{(e)}+\frac{1}{\sqrt{\epsilon_1}}\chi_l'(\rho_1)B_{l,1}^{(e)}=\frac{1}{\sqrt{\epsilon_0}}\psi_l'(q_1)A_{l,0}^{(e)}, \tag{8.175}$$

$$\frac{1}{\sqrt{\epsilon_1}}\psi_l(\rho_1)A_{l,1}^{(m)}+\frac{1}{\sqrt{\epsilon_1}}\chi_l(\rho_1)B_{l,1}^{(m)}=\frac{1}{\sqrt{\epsilon_0}}\psi_l\left(\sqrt{\epsilon_0}k_0r_1\right)A_{l,0}^{(m)}, \tag{8.176}$$

and

$$\psi_l'(\rho_1)A_{l,1}^{(m)}+\chi_l'(\rho_1)B_{l,1}^{(m)}=\psi_l'(q_1)A_{l,0}^{(m)}. \tag{8.177}$$

The two matrix equations are

$$\tilde{\boldsymbol{M}}_1^{(e)}\begin{bmatrix}A_{l,1}^{(e)}\\ B_{l,1}^{(e)}\end{bmatrix}=\tilde{\boldsymbol{N}}_0^{(e)}A_{l,0}^{(e)}, \tag{8.178}$$

and

$$\tilde{\boldsymbol{M}}_1^{(m)}\begin{bmatrix}A_{l,1}^{(m)}\\ B_{l,1}^{(m)}\end{bmatrix}=\tilde{\boldsymbol{N}}_0^{(m)}A_{l,0}^{(m)} \tag{8.179}$$

where

$$m_{1,11}^{(e)}\equiv\psi_l(\rho_1), \tag{8.180}$$

$$m_{1,12}^{(e)}\equiv\chi_l(\rho_1), \tag{8.181}$$

$$m_{1,21}^{(e)}\equiv\frac{1}{\sqrt{\epsilon_1}}\psi_l'(\rho_1), \tag{8.182}$$

$$m_{1,22}^{(e)}\equiv\frac{1}{\sqrt{\epsilon_1}}\chi_l'(\rho_1), \tag{8.183}$$

$$n_{0,1}^{(e)}\equiv\psi_l(q_1), \tag{8.184}$$

$$n_{0,2}^{(e)}\equiv\frac{1}{\sqrt{\epsilon_0}}\psi_l'(q_1), \tag{8.185}$$

$$m_{1,11}^{(m)} \equiv \frac{1}{\sqrt{\epsilon_1}} \psi_l(\rho_1), \tag{8.186}$$

$$m_{1,12}^{(m)} \equiv \frac{1}{\sqrt{\epsilon_1}} \chi_l(\rho_1), \tag{8.187}$$

$$m_{1,21}^{(m)} \equiv \psi_l'(\rho_1), \tag{8.188}$$

$$m_{1,22}^{(m)} \equiv \chi_l'(\rho_1), \tag{8.189}$$

$$n_{0,1}^{(m)} \equiv \frac{1}{\sqrt{\epsilon_0}} \psi_l(q_1), \tag{8.190}$$

and

$$n_{0,2}^{(m)} \equiv \psi_l'(q_1). \tag{8.191}$$

8.7 GENERAL SOLUTION OF THE POTENTIALS AND FIELDS

(8.177) and (8.178) can be rearranged,

$$\begin{bmatrix} A_{l,1}^{(e)} \\ B_{l,1}^{(e)} \end{bmatrix} = \left(\tilde{\boldsymbol{M}}_1^{(e)}\right)^{-1} \tilde{\boldsymbol{N}}_0^{(e)} A_{l,0}^{(e)}, \tag{8.192}$$

and

$$\begin{bmatrix} A_{l,1}^{(m)} \\ B_{l,1}^{(m)} \end{bmatrix} = \left(\tilde{\boldsymbol{M}}_1^{(m)}\right)^{-1} \tilde{\boldsymbol{N}}_0^{(m)} A_{l,0}^{(m)}. \tag{8.193}$$

From (8.155) and (8.156), the next shell equations are

$$\begin{bmatrix} A_{l,2}^{(e)} \\ B_{l,2}^{(e)} \end{bmatrix} = \left(\tilde{\boldsymbol{M}}_2^{(e)}\right)^{-1} \tilde{\boldsymbol{N}}_1^{(e)} \begin{bmatrix} A_{l,1}^{(e)} \\ B_{l,1}^{(e)} \end{bmatrix} = \left(\tilde{\boldsymbol{M}}_2^{(e)}\right)^{-1} \tilde{\boldsymbol{N}}_1^{(e)} \left(\tilde{\boldsymbol{M}}_1^{(e)}\right)^{-1} \tilde{\boldsymbol{N}}_0^{(e)} A_{l,0}^{(e)}, \tag{8.194}$$

and

$$\begin{bmatrix} A_{l,2}^{(e)} \\ B_{l,2}^{(e)} \end{bmatrix} = \left(\tilde{\boldsymbol{M}}_2^{(m)}\right)^{-1} \tilde{\boldsymbol{N}}_1^{(m)} \begin{bmatrix} A_{l,1}^{(e)} \\ B_{l,1}^{(e)} \end{bmatrix} = \left(\tilde{\boldsymbol{M}}_2^{(m)}\right)^{-1} \tilde{\boldsymbol{N}}_1^{(m)} \left(\tilde{\boldsymbol{M}}_1^{(m)}\right)^{-1} \tilde{\boldsymbol{N}}_0^{(m)} A_{l,0}^{(m)}. \tag{8.195}$$

This sequence is continued until finally the outer medium is reached. (8.134) and (8.135) then become

$$\tilde{\boldsymbol{M}}_{N+1}^{(e)} \boldsymbol{B}_{l,N+1}^{(e)} + \boldsymbol{V}_{N+1}^{(e)} = \tilde{\boldsymbol{N}}_N^{(e)} \begin{bmatrix} A_{l,N}^{(e)} \\ B_{l,N}^{(e)} \end{bmatrix} = \left[\tilde{\boldsymbol{N}}_N^{(e)} \cdots \left(\tilde{\boldsymbol{M}}_2^{(e)} \right)^{-1} \tilde{\boldsymbol{N}}_1^{(e)} \left(\tilde{\boldsymbol{M}}_1^{(e)} \right)^{-1} \tilde{\boldsymbol{N}}_0^{(e)} \right] A_{l,0}^{(e)}$$

$$\equiv \tilde{\boldsymbol{P}}_N^{(e)} A_{l,0}^{(e)}, \tag{8.196}$$

and

$$\tilde{\boldsymbol{M}}_{N+1}^{(m)} \boldsymbol{B}_{l,N+1}^{(m)} + \boldsymbol{V}_{N+1}^{(m)} = \tilde{\boldsymbol{N}}_N^{(m)} \begin{bmatrix} A_{l,N}^{(m)} \\ B_{l,N}^{(m)} \end{bmatrix} = \left[\tilde{\boldsymbol{N}}_N^{(m)} \cdots \left(\tilde{\boldsymbol{M}}_2^{(m)} \right)^{-1} \tilde{\boldsymbol{N}}_1^{(m)} \left(\tilde{\boldsymbol{M}}_1^{(m)} \right)^{-1} \tilde{\boldsymbol{N}}_0^{(m)} \right] A_{l,0}^{(m)}$$

$$\equiv \tilde{\boldsymbol{P}}_N^{(m)} A_{l,0}^{(m)}. \tag{8.197}$$

Rearranging,

$$\tilde{\boldsymbol{P}}_N^{(e)} A_{l,0}^{(e)} - \tilde{\boldsymbol{M}}_{N+1}^{(e)} \boldsymbol{B}_{l,N+1}^{(e)} = \boldsymbol{V}_{N+1}^{(e)} \tag{8.198}$$

and

$$\tilde{\boldsymbol{P}}_N^{(m)} A_{l,0}^{(m)} - \tilde{\boldsymbol{M}}_{N+1}^{(m)} \boldsymbol{B}_{l,N+1}^{(m)} = \boldsymbol{V}_{N+1}^{(m)}. \tag{8.199}$$

The vectors on the right hand side of (8.198) and (8.199) are known quantities for each value of l. $\tilde{\boldsymbol{P}}^{(e)}$ and $\tilde{\boldsymbol{M}}_{N+1}^{(e)}$ are matrices of rank 2, so there are two equations for the electric wave and two for the magnetic wave, and four unknown parameters, $A_{l,0}^{(e)}$, $A_{l,0}^{(m)}$, $B_{l,N+1}^{(e)}$, and $B_{l,N+1}^{(m)}$ for each value of l. Once the values for these parameters have been calculated, the values for each shell may also be iteratively calculated from (8.191), (8.192), (8.155), and (8.156). After calculating the coefficients for the electric and magnetic potentials, the fields are found from (8.44)–(8.49).

$$E_r = E_r^{(e)} + E_r^{(m)} = \frac{\partial^2}{\partial r^2}\left(r\Pi^{(e)}\right) + \epsilon k_0^2 \left(r\Pi^{(e)}\right). \tag{8.200}$$

$$E_\theta = E_\theta^{(e)} + E_\theta^{(m)} = \frac{1}{r}\frac{\partial^2\left(r\Pi^{(e)}\right)}{\partial r\,\partial\theta} + \frac{ik_0}{r\sin\theta}\frac{\partial\left(r\Pi^{(m)}\right)}{\partial\phi}. \tag{8.201}$$

$$E_\phi = E_\phi^{(e)} + E_\phi^{(m)} = \frac{1}{r\sin\theta}\frac{\partial^2\left(r\Pi^{(e)}\right)}{\partial r\,\partial\phi} - \frac{ik_0}{r}\frac{\partial\left(r\Pi^{(m)}\right)}{\partial\theta}. \tag{8.202}$$

$$H_r = H_r^{(e)} + H_r^{(m)} = \frac{\partial^2}{\partial r^2}\left(r\Pi^{(m)}\right) + \epsilon k_0^2 \left(r\Pi^{(m)}\right). \tag{8.203}$$

$$H_\theta = H_\theta^{(e)} + H_\theta^{(m)} = -\frac{i\epsilon k_0}{r\sin\theta}\frac{\partial\left(r\Pi^{(e)}\right)}{\partial\phi} + \frac{1}{r}\frac{\partial^2\left(r\Pi^{(m)}\right)}{\partial r\,\partial\theta}. \tag{8.204}$$

$$H_\phi = H_\phi^{(e)} + H_\phi^{(m)} = \frac{i\epsilon k_0}{r}\frac{\partial\left(r\Pi^{(e)}\right)}{\partial\theta} + \frac{1}{r\sin\theta}\frac{\partial^2\left(r\Pi^{(m)}\right)}{\partial r\,\partial\phi}. \tag{8.205}$$

The derivatives of the potentials are found from (8.123) and (8.124),

$$r\Pi_n^{(e)} = \frac{1}{\epsilon_n k_0^2}\sum_{l=1}^{\infty}\left[A_{l,n}^{(e)}\psi_l\left(\sqrt{\epsilon_n}k_0 r\right) + B_{l,n}^{(e)}\chi_l^{(1)}\left(\sqrt{\epsilon_n}k_0 r\right)\right]P_l^{(1)}(\cos\theta)\cos\phi\,, \tag{8.206}$$

$$\frac{\partial\left(r\Pi_n^{(e)}\right)}{\partial r} = \frac{1}{\sqrt{\epsilon_n}k_0}\sum_{l=1}^{\infty}\left[A_{l,n}^{(e)}\frac{\partial\psi_l\left(\sqrt{\epsilon_n}k_0 r\right)}{\partial\left(\sqrt{\epsilon_n}k_0 r\right)} + B_{l,n}^{(e)}\frac{\partial\chi_l^{(1)}\left(\sqrt{\epsilon_n}k_0 r\right)}{\partial\left(\sqrt{\epsilon_n}k_0 r\right)}\right]P_l^{(1)}(\cos\theta)\cos\phi\,, \tag{8.207}$$

$$\frac{\partial^2\left(r\Pi_n^{(e)}\right)}{\partial r^2} = \sum_{l=1}^{\infty}\left[A_{l,n}^{(e)}\frac{\partial^2\psi_l\left(\sqrt{\epsilon_n}k_0 r\right)}{\partial\left(\sqrt{\epsilon_n}k_0 r\right)^2} + B_{l,n}^{(e)}\frac{\partial^2\chi_l^{(1)}\left(\sqrt{\epsilon_n}k_0 r\right)}{\partial\left(\sqrt{\epsilon_n}k_0 r\right)^2}\right]P_l^{(1)}(\cos\theta)\cos\phi\,, \tag{8.208}$$

$$\frac{\partial^2\left(r\Pi_n^{(e)}\right)}{\partial r\partial\theta} = \frac{1}{\sqrt{\epsilon_n}k_0}\sum_{l=1}^{\infty}\left[A_{l,n}^{(e)}\frac{\partial\psi_l\left(\sqrt{\epsilon_n}k_0 r\right)}{\partial\left(\sqrt{\epsilon_n}k_0 r\right)} + B_{l,n}^{(e)}\frac{\partial\chi_l^{(1)}\left(\sqrt{\epsilon_n}k_0 r\right)}{\partial\left(\sqrt{\epsilon_n}k_0 r\right)}\right]\frac{\partial P_l^{(1)}(\cos\theta)}{\partial\theta}\cos\phi\,, \tag{8.209}$$

$$\frac{\partial^2\left(r\Pi_n^{(e)}\right)}{\partial r\partial\phi} = -\frac{1}{\sqrt{\epsilon_n}k_0}\sum_{l=1}^{\infty}\left[A_{l,n}^{(e)}\frac{\partial\psi_l\left(\sqrt{\epsilon_n}k_0 r\right)}{\partial\left(\sqrt{\epsilon_n}k_0 r\right)} + B_{l,n}^{(e)}\frac{\partial\chi_l^{(1)}\left(\sqrt{\epsilon_n}k_0 r\right)}{\partial\left(\sqrt{\epsilon_n}k_0 r\right)}\right]P_l^{(1)}(\cos\theta)\sin\phi\,, \tag{8.210}$$

$$\frac{\partial\left(r\Pi_n^{(e)}\right)}{\partial\theta} = \frac{1}{\epsilon_n k_0^2}\sum_{l=1}^{\infty}\left[A_{l,n}^{(e)}\psi_l\left(\sqrt{\epsilon_n}k_0 r\right) + B_{l,n}^{(e)}\chi_l^{(1)}\left(\sqrt{\epsilon_n}k_0 r\right)\right]\frac{\partial P_l^{(1)}(\cos\theta)}{\partial\theta}\cos\phi\,, \tag{8.211}$$

$$\frac{\partial\left(r\Pi_n^{(e)}\right)}{\partial\phi} = -\frac{1}{\epsilon_n k_0^2}\sum_{l=1}^{\infty}\left[A_{l,n}^{(e)}\psi_l\left(\sqrt{\epsilon_n}k_0 r\right) + B_{l,n}^{(e)}\chi_l^{(1)}\left(\sqrt{\epsilon_n}k_0 r\right)\right]P_l^{(1)}(\cos\theta)\sin\phi\,, \tag{8.212}$$

$$r\Pi_n^{(m)} = \frac{1}{\sqrt{\epsilon_n}k_0^2}\sum_{l=1}^{\infty}\left[A_{l,n}^{(m)}\psi_l\left(\sqrt{\epsilon_n}k_0 r\right) + B_{l,n}^{(m)}\chi_l^{(1)}\left(\sqrt{\epsilon_n}k_0 r\right)\right]P_l^{(1)}(\cos\theta)\sin\phi\,, \tag{8.213}$$

$$\frac{\partial\left(r\Pi_n^{(m)}\right)}{\partial r} = \frac{1}{k_0}\sum_{l=1}^{\infty}\left[A_{l,n}^{(m)}\frac{\partial\psi_l\left(\sqrt{\epsilon_n}k_0 r\right)}{\partial\left(\sqrt{\epsilon_n}k_0 r\right)} + B_{l,n}^{(m)}\frac{\partial\chi_l^{(1)}\left(\sqrt{\epsilon_n}k_0 r\right)}{\partial\left(\sqrt{\epsilon_n}k_0 r\right)}\right]P_l^{(1)}(\cos\theta)\sin\phi\,, \tag{8.214}$$

$$\frac{\partial^2\left(r\Pi_n^{(m)}\right)}{\partial r^2} = \sqrt{\epsilon_n}\sum_{l=1}^{\infty}\left[A_{l,n}^{(m)}\frac{\partial^2\psi_l\left(\sqrt{\epsilon_n}k_0 r\right)}{\partial\left(\sqrt{\epsilon_n}k_0 r\right)^2} + B_{l,n}^{(m)}\frac{\partial^2\chi_l^{(1)}\left(\sqrt{\epsilon_n}k_0 r\right)}{\partial\left(\sqrt{\epsilon_n}k_0 r\right)^2}\right]P_l^{(1)}(\cos\theta)\sin\phi\,, \tag{8.215}$$

$$\frac{\partial^2\left(r\Pi_n^{(m)}\right)}{\partial r\,\partial\theta}=\frac{1}{k_0}\sum_{l=1}^{\infty}\left[A_{l,n}^{(m)}\frac{\partial\psi_l\left(\sqrt{\epsilon_n}k_0r\right)}{\partial\left(\sqrt{\epsilon_n}k_0r\right)}+B_{l,n}^{(m)}\frac{\partial\chi_l^{(1)}\left(\sqrt{\epsilon_n}k_0r\right)}{\partial\left(\sqrt{\epsilon_n}k_0r\right)}\right]\frac{\partial P_l^{(1)}(\cos\theta)}{\partial\theta}\sin\phi\,, \tag{8.216}$$

$$\frac{\partial^2\left(r\Pi_n^{(m)}\right)}{\partial r\partial\phi}=\frac{1}{k_0}\sum_{l=1}^{\infty}\left[A_{l,n}^{(m)}\frac{\partial\psi_l\left(\sqrt{\epsilon_n}k_0r\right)}{\partial\left(\sqrt{\epsilon_n}k_0r\right)}+B_{l,n}^{(m)}\frac{\partial\chi_l^{(1)}\left(\sqrt{\epsilon_n}k_0r\right)}{\partial\left(\sqrt{\epsilon_n}k_0r\right)}\right]P_l^{(1)}(\cos\theta)\cos\phi\,, \tag{8.217}$$

$$\frac{\partial\left(r\Pi_n^{(m)}\right)}{\partial\theta}=\frac{1}{\sqrt{\epsilon_n}k_0^2}\sum_{l=1}^{\infty}\left[A_{l,n}^{(m)}\psi_l\left(\sqrt{\epsilon_n}k_0r\right)+B_{l,n}^{(m)}\chi_l^{(1)}\left(\sqrt{\epsilon_n}k_0r\right)\right]\frac{\partial P_l^{(1)}(\cos\theta)}{\partial\theta}\sin\phi\,, \tag{8.218}$$

and

$$\frac{\partial\left(r\Pi_n^{(m)}\right)}{\partial\phi}=\frac{1}{\sqrt{\epsilon_n}k_0^2}\sum_{l=1}^{\infty}\left[A_{l,n}^{(m)}\psi_l\left(\sqrt{\epsilon_n}k_0r\right)+B_{l,n}^{(m)}\chi_l^{(1)}\left(\sqrt{\epsilon_n}k_0r\right)\right]P_l^{(1)}(\cos\theta)\cos\phi. \tag{8.219}$$

8.7.1 Fields Inside the Sphere

Inside the sphere all $B_{l,0}^{(e)}=B_{l,0}^{(m)}=0$. The field equations can correspondingly be simplified.

$$E_r=\sum_{l=1}^{\infty}\left(\frac{\partial^2\psi_l(x_0)}{\partial(x_0)^2}+\psi_l(x_0)\right)A_{l,0}^{(e)}P_l^{(1)}(\cos\theta)\cos\phi\,, \tag{8.220}$$

$$E_\theta=\frac{\cos\phi}{x_0}\sum_{l=1}^{\infty}\left(A_{l,0}^{(e)}\frac{\partial\psi_l(x_0)}{\partial x_0}\frac{\partial P_l^{(1)}(\cos\theta)}{\partial\theta}+\mathrm{i}A_{l,0}^{(m)}\psi_l(x_0)\frac{P_l^{(1)}(\cos\theta)}{\sin\theta}\right), \tag{8.221}$$

$$E_\phi=-\frac{\sin\phi}{x_0}\sum_{l=1}^{\infty}\left(A_{l,0}^{(e)}\frac{\partial\psi_l(x_0)}{\partial x_0}\frac{P_l^{(1)}(\cos\theta)}{\sin\theta}+iA_{l,0}^{(m)}\psi_l(x_0)\frac{\partial P_l^{(1)}(\cos\theta)}{\partial\theta}\right), \tag{8.222}$$

$$H_r=\sqrt{\epsilon_0}\sin\phi\sum_{l=1}^{\infty}A_{l,0}^{(m)}\left(\frac{\partial^2\psi_l(x_0)}{(\partial x_0)^2}P_l^{(1)}(\cos\theta)+\psi_l(x_0)P_l^{(1)}(\cos\theta)\right), \tag{8.223}$$

$$H_\theta=\frac{\sqrt{\epsilon_0}\sin\phi}{x_0}\sum_{l=1}^{\infty}\left(iA_{l,0}^{(e)}\psi_l(x_0)\frac{P_l^{(1)}(\cos\theta)}{\sin\theta}+A_{l,0}^{(m)}\frac{\partial\psi_l(x_0)}{\partial x_0}\frac{\partial P_l^{(1)}(\cos\theta)}{\partial\theta}\right), \tag{8.224}$$

and

$$H_\phi=\frac{\sqrt{\epsilon_0}\cos\phi}{x_0}\sum_{l=1}^{\infty}\left(iA_{l,0}^{(e)}\psi_l(x_0)\frac{\partial P_l^{(1)}(\cos\theta)}{\partial\theta}+A_{l,0}^{(m)}\frac{\partial\psi_l(x_0)}{\partial x_0}\frac{P_l^{(1)}(\cos\theta)}{\sin\theta}\right) \tag{8.225}$$

where

$$x_n \equiv \sqrt{\epsilon_n} k_0 r. \tag{8.226}$$

8.7.2 Scattered Fields Outside the Sphere (No Shells)

Similarly, in the incident medium outside the sphere, $A_{l,i}^{(e)} = A_{l,i}^{(m)} = 0$, and the scattered field equations are again simplified.

$$E_{r,N+1}^{(s)} = \cos\phi \sum_{l=1}^{\infty} \left(\frac{\partial^2 \zeta_l^{(1)}(x_i)}{\partial (x_i)^2} + \zeta_l^{(1)}(x_i) \right) B_{l,i}^{(e)} P_l^{(1)}(\cos\theta), \tag{8.227}$$

$$E_{\theta,N+1}^{(s)} = \frac{\cos\phi}{x_i} \sum_{l=1}^{\infty} \left[B_{l,i}^{(e)} \frac{\partial \zeta_l^{(1)}(x_i)}{\partial x_i} \frac{\partial P_l^{(1)}(\cos\theta)}{\partial\theta} + iB_{l,i}^{(m)} \zeta_l^{(1)}(x_i) \frac{P_l^{(1)}(\cos\theta)}{\sin\theta} \right], \tag{8.228}$$

$$E_{\phi,N+1}^{(s)} = -\frac{\sin\phi}{x_i} \sum_{l=1}^{\infty} \left[B_{l,i}^{(e)} \frac{\partial \zeta_l^{(1)}(x_i)}{\partial x_i} \frac{P_l^{(1)}(\cos\theta)}{\sin\theta} + iB_{l,i}^{(m)} \zeta_l^{(1)}(x_i) \frac{\partial P_l^{(1)}(\cos\theta)}{\partial\theta} \right], \tag{8.229}$$

$$H_{r,N+1}^{(s)} = \sqrt{\epsilon_i} \sin\phi \sum_{l=1}^{\infty} \left(B_{l,i}^{(m)} \frac{\partial^2 \zeta_l^{(1)}(x)}{(\partial x_i)^2} + B_{l,i}^{(m)} \zeta_l^{(1)}(x_1) \right) P_l^{(1)}(\cos\theta), \tag{8.230}$$

$$H_{\theta,N+1}^{(s)} = \frac{\sqrt{\epsilon_i} \sin\phi}{x_i} \sum_{l=1}^{\infty} \left[iB_{l,i}^{(e)} \zeta_l^{(1)}(x_i) \frac{P_l^{(1)}(\cos\theta)}{\sin\theta} + B_{l,i}^{(m)} \frac{\partial \zeta_l^{(1)}(x_i)}{\partial x_i} \frac{\partial P_l^{(1)}(\cos\theta)}{\partial\theta} \right], \tag{8.231}$$

and

$$H_{\phi,N+1}^{(s)} = \frac{\sqrt{\epsilon_i} \cos\phi}{x_i} \sum_{l=1}^{\infty} \left[iB_{l,i}^{(e)} \zeta_l^{(1)}(x_i) \frac{\partial P_l^{(1)}(\cos\theta)}{\partial\theta} + B_{l,i}^{(m)} \frac{\partial \zeta_l^{(1)}(x_i)}{\partial x_i} \frac{P_l^{(1)}(\cos\theta)}{\sin\theta} \right]. \tag{8.232}$$

In the incident medium, in the far field, the spherical harmonics can be approximated by the relations

$$\zeta_l^{(1)}(x) \simeq (-i)^{l+1} e^{ix}, \tag{8.233}$$

$$\frac{d\zeta_l^{(1)}(x)}{dx} \simeq (-i)^l e^{ix}, \tag{8.234}$$

and

$$\frac{d^2\zeta_l^{(1)}(x)}{dx^2} \simeq (-i)^{l-1} e^{ix}. \tag{8.235}$$

Inserting these relations into the scattered field equations (8.227)–(8.232) further simplifies them.

$$E^{(s)}_{r,N+1} = 0\,. \tag{8.236}$$

$$E^{(s)}_{\theta,N+1} = \frac{\cos\phi}{x_i} e^{ix_i} \sum_{l=1}^{\infty} (-i)^l \left[B^{(e)}_{l,i} \frac{\partial P^{(1)}_l(\cos\theta)}{\partial\theta} + B^{(m)}_{l,i} \frac{P^{(1)}_l(\cos\theta)}{\sin\theta} \right]. \tag{8.237}$$

$$E^{(s)}_{\phi,N+1} = -\frac{\sin\phi}{x_i} e^{ix_i} \sum_{l=1}^{\infty} (-i)^l \left[B^{(e)}_{l,i} \frac{P^{(1)}_l(\cos\theta)}{\sin\theta} + B^{(m)}_{l,i} \frac{\partial P^{(1)}_l(\cos\theta)}{\partial\theta} \right]. \tag{8.238}$$

$$H^{(s)}_{r,N+1} = 0\,. \tag{8.239}$$

$$H^{(s)}_{\theta,N+1} = \frac{\sqrt{\epsilon_i}\,\sin\phi}{x_i} e^{ix_i} \sum_{l=1}^{\infty} (-i)^l \left[B^{(e)}_{l,i} \frac{P^{(1)}_l(\cos\theta)}{\sin\theta} + B^{(m)}_{l,i} \frac{\partial P^{(1)}_l(\cos\theta)}{\partial\theta} \right]. \tag{8.240}$$

$$H^{(s)}_{\phi,N+1} = \frac{\sqrt{\epsilon_i}\,\cos\phi}{x_i} e^{ix_i} \sum_{l=1}^{\infty} (-i)^l \left[B^{(e)}_{l,i} \frac{\partial P^{(1)}_l(\cos\theta)}{\partial\theta} + B^{(m)}_{l,i} \frac{P^{(1)}_l(\cos\theta)}{\sin\theta} \right]. \tag{8.241}$$

To calculate the fields within and without the sphere, then, the four parameters $A^{(e)}_{l,0}$, $A^{(m)}_{l,0}$, $B^{(e)}_{l,N+1}$, $B^{(m)}_{l,N+1}$ must be calculated. At the inner boundary, if there are no shells, (8.178) and (8.179) are

$$\tilde{\boldsymbol{M}}^{(e)}_1 \begin{bmatrix} A^{(e)}_{l,1} \\ B^{(e)}_{l,1} \end{bmatrix} = \tilde{\boldsymbol{N}}^{(e)}_0 A^{(e)}_{l,0}\,, \tag{8.242}$$

and

$$\tilde{\boldsymbol{M}}^{(m)}_1 \begin{bmatrix} A^{(m)}_{l,1} \\ B^{(m)}_{l,1} \end{bmatrix} = \boldsymbol{N}^{(m)}_0 A^{(m)}_{l,0} \tag{8.243}$$

where $\boldsymbol{N}^{(e,m)}_0$ is a two-element vector. At the outer boundary, (8.134) and (8.135) are

$$\boldsymbol{M}^{(e,m)}_{N+1} B^{(e,m)}_{l,N+1} + \boldsymbol{V}^{(e,m)}_{N+1} = \tilde{\boldsymbol{N}}^{(e,m)}_N \begin{bmatrix} A^{(e,m)}_{l,N} \\ B^{(e,m)}_{l,N} \end{bmatrix} \tag{8.244}$$

where $\boldsymbol{M}^{(e,m)}_{N+1}$ is also a two-element vector. Combining these equations when there are no shells,

$$\tilde{\boldsymbol{M}}^{(e,m)}_1 B^{(e,m)}_{l,1} + \boldsymbol{V}^{(e,m)}_1 = \boldsymbol{N}^{(e,m)}_0 A^{(e,m)}_{l,0} \tag{8.245}$$

and rearranging,

$$\boldsymbol{N}_0^{(e,m)} A_{l,0}^{(e,m)} - \tilde{\boldsymbol{M}}_1^{(e,m)} B_{l,1}^{(e,m)} = \boldsymbol{V}_1^{(e,m)} \tag{8.246}$$

or

$$\tilde{\boldsymbol{Q}}^{(e,m)} \begin{bmatrix} A_{l,0}^{(e,m)} \\ B_{l,1}^{(e,m)} \end{bmatrix} = \boldsymbol{V}_1^{(e,m)} \tag{8.247}$$

where

$$\tilde{\boldsymbol{Q}}^{(e,m)} = \begin{bmatrix} Q_{11}^{(e,m)} & Q_{12}^{(e,m)} \\ Q_{21}^{(e,m)} & Q_{22}^{(e,m)} \end{bmatrix} \equiv \begin{bmatrix} N_{0,1}^{(e,m)} & -M_{1,1}^{(e,m)} \\ N_{0,2}^{(e,m)} & -M_{1,2}^{(e,m)} \end{bmatrix}. \tag{8.248}$$

Then the unknown potential amplitudes are simply

$$\begin{bmatrix} A_{l,0}^{(e,m)} \\ B_{l,1}^{(e,m)} \end{bmatrix} = \left(\tilde{\boldsymbol{Q}}^{(e,m)}\right)^{-1} \boldsymbol{V}_1^{(e,m)}. \tag{8.249}$$

The spherical harmonic expansion must be truncated after a finite number of terms. Debye, as quoted in [1], apparently suggested that only the first q terms are required where

$$q \equiv n_i k_0 r_{N+1} \tag{8.250}$$

and r_{N+1} is the outer radius of the object. However, a better relation will be subsequently described.

8.8 FAR FIELD SCATTERING OF SMALL SPHERES

The far field approximations for ψ_1 and ζ_1 are

$$\psi_l(x) \simeq \frac{1}{2}\left(i^{l+1} e^{-ix} + (-i)^{l+1} e^{ix}\right), \tag{8.251}$$

$$\frac{d\psi_l(x)}{dx} \simeq \frac{1}{2}\left(i^l e^{-ix} + (-i)^l e^{ix}\right), \tag{8.252}$$

$$\zeta_l^{(1)}(x) \simeq (-i)^{l+1} e^{ix}, \tag{8.253}$$

and

$$\frac{d\zeta_l^{(1)}(x)}{dx} \simeq (-i)^l e^{ix}. \tag{8.254}$$

If x is real then

$$\psi_l(x) \simeq \sin\left(x - \frac{l\pi}{2}\right), \tag{8.255}$$

and

$$\psi_l'(x) \simeq \cos\left(x - \frac{l\pi}{2}\right). \tag{8.256}$$

The scattered field is found from (8.45), (8.46), (8.48), and (8.49) by substituting from (8.121), (8.122), (8.233), and (8.234). From (8.45),

$$E_\theta^{(s)} = E_\theta^{(s,e)} + E_\theta^{(s,m)} = \frac{1}{r}\frac{\partial^2\left(r\Pi^{(e)}\right)}{\partial r\,\partial\theta} + \frac{ik_0}{r\sin\theta}\frac{\partial\left(r\Pi^{(m)}\right)}{\partial\phi}, \tag{8.257}$$

and

$$E_\theta^{(s)} = \frac{e^{i\left(\sqrt{\epsilon_{N+1}}k_0 r\right)}}{\sqrt{\epsilon_{N+1}}k_0 r}\cos\phi\sum_{l=1}^{\infty}(-i)^l\left(B_{l,1}^{(e)}\frac{\partial P_l^{(1)}(\cos\theta)}{\partial\theta} + \frac{1}{\sin\theta}B_{l,1}^{(m)}P_l^{(1)}(\cos\theta)\right). \tag{8.258}$$

In the limit of a sphere diameter much smaller than the wavelength, only the $l = 1$ term needs to be considered and

$$E_\theta^{(s)} \cong \frac{-ie^{i\left(\sqrt{\epsilon_{N+1}}k_0 r\right)}\cos\phi}{\sqrt{\epsilon_{N+1}}k_0 r}\left(B_{1,1}^{(e)}\cos\theta + B_{1,1}^{(m)}\right). \tag{8.259}$$

Furthermore, it can be shown that $B_{1,1}^{(e)} \propto \frac{r_1}{\lambda}$ while $B_{1,1}^{(m)} \propto \left(\frac{r_1}{\lambda}\right)^2$ and so the latter may also be neglected.

$$E_\theta^{(s)} \cong \frac{-ie^{i\left(\sqrt{\epsilon_{N+1}}k_0 r\right)}}{\sqrt{\epsilon_{N+1}}k_0 r}\cos\theta\cos\phi B_{1,1}^{(e)}. \tag{8.260}$$

Following the same procedure for E_ϕ,

$$E_\phi = -\frac{e^{i\sqrt{\epsilon_{N+1}}k_0 r}}{\sqrt{\epsilon_{N+1}}k_0 r}\sin\phi\sum_{l=1}^{\infty}(-i)^l\left[B_{l,N+1}^{(e)}\frac{P_l^{(1)}(\cos\theta)}{\sin\theta} + B_{l,N+1}^{(m)}\frac{\partial P_l^{(1)}(\cos\theta)}{\partial\theta}\right]. \tag{8.261}$$

Again, in the limit of a sphere diameter much smaller than the wavelength, only the $l = 1$ term of the $B_{l,N+1}^{(e)}$ term needs to be considered, so

$$E_\phi \cong E_\phi^{(e)} = \frac{1}{r\sin\theta}\frac{\partial^2\left(r\Pi^{(e)}\right)}{\partial r\,\partial\phi} \cong \frac{ie^{i\left(\sqrt{\epsilon_{N+1}}k_0 r\right)}}{\sqrt{\epsilon_{N+1}}k_0 r}\sin\phi B_{1,1}^{(e)}. \tag{8.262}$$

The magnetic field is found from (8.48),

$$H_\theta = \frac{e^{i\sqrt{\epsilon_{N+1}}k_0 r}}{\sqrt{\epsilon_{N+1}}k_0 r}\sin\phi\sum_{l=1}^{\infty}(-i)^l\left[\sqrt{\epsilon_{N+1}}B_{l,N+1}^{(e)}\frac{P_l^{(1)}(\cos\theta)}{\sin\theta}+B_{l,N+1}^{(m)}\frac{\partial P_l^{(1)}(\cos\theta)}{\partial\theta}\right]. \tag{8.263}$$

In the limit of a sphere diameter much smaller than the wavelength,

$$H_\theta \cong H_\theta^{(e)} = -\frac{i\epsilon\, k_0}{r\sin\theta}\frac{\partial\left(r\Pi^{(e)}\right)}{\partial\phi} \cong -\frac{ie^{i\left(\sqrt{\epsilon_{N+1}}k_0 r\right)}}{\sqrt{\epsilon_{N+1}}\,k_0 r}\sin\phi B_{1,N+1}^{(e)}. \tag{8.264}$$

From (8.49),

$$H_\phi = \frac{e^{i\sqrt{\epsilon_{N+1}}k_0 r}}{\sqrt{\epsilon_{N+1}}k_0 r}\cos\phi\sum_{l=1}^{\infty}(-i)^l\left[\sqrt{\epsilon_{N+1}}\,B_{l,N+1}^{(e)}\frac{\partial P_l^{(1)}(\cos\theta)}{\partial\theta}+B_{l,N+1}^{(m)}\frac{P_l^{(1)}(\cos\theta)}{\sin\theta}\right]. \tag{8.265}$$

In the limit of a sphere diameter much smaller than the wavelength,

$$H_\phi \cong H_\phi^{(e)} = \frac{i\epsilon\, k_0}{r}\frac{\partial\left(r\Pi^{(e)}\right)}{\partial\theta} \cong -\frac{ie^{i\left(\sqrt{\epsilon_{N+1}}k_0 r\right)}}{\sqrt{\epsilon_{N+1}}\,k_0 r}\cos\theta\cos\phi B_{1,N+1}^{(e)}. \tag{8.266}$$

The far field intensity pattern in the limit of a small sphere is then

$$I_\theta^{(s)} \propto \frac{\cos^2\theta\cos^2\phi}{\epsilon_1 k_0^2 r^2}, \tag{8.267}$$

and

$$I_\phi^{(s)} \propto \frac{\sin^2\phi}{\epsilon_1 k_0^2 r^2}. \tag{8.268}$$

In the first example at the end of the chapter, the far field polarization of scattered light from a gold sphere is considered. The quantities $I_\parallel^{(s)}$ and $I_\perp^{(s)}$ are defined in terms of the magnitude of the square of the scattered electric field,

$$\left|E_\theta^{(s)}\right|^2 = I_\parallel^{(s)}\cos^2\phi, \tag{8.269}$$

and

$$\left|E_\phi^{(s)}\right|^2 = I_\perp^{(s)}\sin^2\phi. \tag{8.270}$$

The plane of observation is the plane that contains both the direction of propagation, that is, the z-axis, and the point of observation. ϕ is the angle between this plane and the polarization of the incident wave. Therefore, when the incident wave is polarized in this plane, $\phi = 0$ and $I_\perp^{(s)} = 0$. At 90° to this plane, $I_\parallel^{(s)} = 0$. At these two points, the scattered light is linearly polarized. For other angles it is generally elliptically polarized, though for very small particles (Rayleigh limit) the light is linearly polarized in all directions. The degree of polarization of the scattered light is defined as

$$P = \left| \frac{I_{\perp}^{(s)} - I_{\parallel}^{(s)}}{I_{\perp}^{(s)} + I_{\parallel}^{(s)}} \right|. \tag{8.271}$$

8.9 SCATTERING CROSS SECTION

In other common Mie scattering derivations, such as [6] (p. 101), alternate parameters are defined,

$$a_l = \frac{\kappa \psi_l(\kappa x)\psi_l'(x) - \psi_l(x)\psi_l'(\kappa x)}{\kappa \psi_l(\kappa x)\xi_l'(x) - \xi_l(x)\psi_l'(\kappa x)}, \tag{8.272}$$

and

$$b_l = \frac{\psi_l(\kappa x)\psi_l'(x) - \kappa \psi_l(x)\psi_l'(\kappa x)}{\psi_l(\kappa x)\xi_l'(\kappa x) - \kappa \xi_l(x)\psi_l'(\kappa x)} \tag{8.273}$$

where

$$\kappa \equiv \sqrt{\frac{\epsilon_s}{\epsilon_i}} = \frac{n_s}{n_i} \tag{8.274}$$

is the ratio of the refractive indices of the sphere to those of the incident medium, and

$$x \equiv \sqrt{\epsilon_i}\, k_0 R = n_i k_0 R \tag{8.275}$$

so

$$\kappa x = n_s k_0 R \tag{8.276}$$

where R is the radius of the sphere. The scattering cross section is then

$$Q_{\text{sc}} = \frac{2\pi}{\epsilon_i k_0^2} \sum\nolimits_{l=1}^{\infty} (2l+1)\left(|a_l|^2 + |b_l|^2\right). \tag{8.277}$$

The extinction cross section is

$$Q_{\text{ex}} = \frac{2\pi}{\epsilon_i k_0^2} \sum\nolimits_{l=1}^{\infty} (2l+1)\,\text{Re}(a_l + b_l). \tag{8.278}$$

From [4], we also have

$$B_l^{(e)} = i^{l+1} \frac{2l+1}{l(l+1)} \frac{ik_0 n_s k_0 \psi_l'(n_i k_0 R)\psi_l(n_s k_0 R) - ik_0 n_i k_0 \psi_l'(n_s k_0 R)\psi_l(n_i k_0 R)}{ik_0 n_s k_0 \zeta_l'(n_i k_0 R)\psi_l(n_s k_0 R) - ik_0 n_i k_0 \psi_l'(n_s k_0 R)\zeta_l(n_i k_0 R)}, \tag{8.279}$$

and

$$B_l^{(m)} = i^{l+1} \frac{2l+1}{l(l+1)} \frac{ik_0 n_s k_0 \psi_l(n_i k_0 R)\psi_l'(n_s k_0 R) - ik_0 n_i k_0 \psi_l'(n_i k_0 R)\psi_l(n_s k_0 R)}{ik_0 n_s k_0 \zeta_l(n_i k_0 R)\psi_l'(n_s k_0 R) - ik_0 n_i k_0 \zeta_l'(n_i k_0 R)\psi_l(n_s k_0 R)}. \tag{8.280}$$

These simplify to

$$B_l^{(e)} = i^{l+1} \frac{2l+1}{l(l+1)} \frac{\kappa\psi_l'(x)\psi_l(\kappa x) - \psi_l'(\kappa x)\psi_l(x)}{\kappa\zeta_l'(x)\psi_l(\kappa x) - \psi_l'(\kappa x)\zeta_l(x)}. \tag{8.281}$$

and

$$B_l^{(m)} = i^{l+1} \frac{2l+1}{l(l+1)} \frac{\kappa\psi_l(x)\psi_l'(\kappa x) - \psi_l'(x)\psi_l(\kappa x)}{\kappa\zeta_l(x)\psi_l'(\kappa x) - \zeta_l'(x)\psi_l(\kappa x)}. \tag{8.282}$$

Comparing (8.281) and (8.282) to (8.272) and (8.273), the relationships between the parameters are

$$B_l^{(e)} = i^{l+1} \frac{2l+1}{l(l+1)} a_l, \tag{8.283}$$

and

$$B_l^{(m)} = i^{l+1} \frac{2l+1}{l(l+1)} b_l \tag{8.284}$$

or

$$a_l = (-i)^{l+1} \frac{l(l+1)}{2l+1} B_l^{(e)}, \tag{8.285}$$

and

$$b_l = (-i)^{l+1} \frac{l(l+1)}{2l+1} B_l^{(m)}. \tag{8.286}$$

Then, from (8.277),

$$Q_{sc} = \frac{2\pi}{\epsilon_i k_0^2} \sum_{l=1}^{\infty} \frac{l^2(l+1)^2}{2l+1} \left(\left|(-i)^{l+1} B_l^{(e)}\right|^2 + \left|(-i)^{l-1} B_l^{(m)}\right|^2 \right) \tag{8.287}$$

$$= \frac{2\pi}{\epsilon_i k_0^2} \sum_{l=1}^{\infty} \left(\frac{l^2(l+1)^2}{2l+1} \right) \left(\left|B_l^{(e)}\right|^2 + \left|B_l^{(m)}\right|^2 \right). \tag{8.288}$$

From (8.278),

$$Q_{ex} = \frac{2\pi}{\epsilon_i k_0^2} Re\left\{ \sum_{l=1}^{\infty} (-i)^{l+1} l(l+1)\left(B_l^{(e)} + B_l^{(m)}\right) \right\}. \tag{8.289}$$

The absorption cross section is simply the difference between the extinction cross section and the scattering cross section,

$$Q_{\text{ab}} = Q_{\text{ex}} - Q_{\text{sc}}. \tag{8.290}$$

When the sphere has no imaginary part to its refractive index, it cannot absorb any radiation, so $Q_{\text{ab}} = 0$. In that case, $Q_{\text{ex}} = Q_{\text{sc}}$. Because the extinction and scattering cross sections are calculated independently, their equality in this situation is a good check of the internal consistency of the computer algorithm.

The backscattering cross section is defined in Bohren and Huffman as

$$Q_{\text{bk}} = \frac{\pi}{\epsilon_i k_0^2} \left| \sum\nolimits_{l=1}^{\infty} (2l+1)(-1)^l (a_l - b_l) \right|^2 \tag{8.291}$$

$$= \frac{\pi}{\epsilon_i k_0^2} \left| \sum\nolimits_{l=1}^{\infty} (i)^{l+1} l(l+1) \left(B_l^{(e)} - B_l^{(m)} \right) \right|^2. \tag{8.292}$$

The definition of the backscattering cross section is not obvious and is considered in detail in [6]. Normalizing the scattering cross section by the cross-sectional area of the sphere, πR^2, gives the cross-section *efficiencies.*

8.10 COMPUTER ALGORITHM IMPLEMENTATION

The infinite series must obviously be truncated at a number of terms that achieves the desired accuracy of the calculation. Although one relation for the number of terms was given by Debye in (8.250), a careful numerical study was carried out by Wiscombe [9] who concluded that the appropriate number of terms, the cutoff order, is better given by the relations

$$N_c = \begin{cases} x + 4x^{1/3} + 1 & 0.02 \le x \le 8 \\ x + 4.05x^{1/3} + 2 & 8 < x < 4200 \\ x + 4x^{1/3} + 2 & 4200 \le x \le 20000 \end{cases}. \tag{8.293}$$

Further increasing N_c by an additional six terms seems to provide a significant improvement in accuracy for at least some calculations.

Unfortunately, in many situations of interest for larger spheres, the number of terms required is in the tens of thousands and even 64-bit numerical precision is insufficient to ensure convergence to the proper results using the formulation (8.281) and (8.282) in [4], or (8.272) and (8.273) in [6]. This issue is explicitly considered in [6], and a better approach to calculating the scattering coefficients using the equations rearranged from (8.272) and (8.273) is described,

$$a_l = \frac{\left(\dfrac{D_l(\kappa x)}{\kappa} + \dfrac{l}{x} \right) \psi_l(x) - \psi_{l-1}(x)}{\left(\dfrac{D_l(\kappa x)}{\kappa} + \dfrac{l}{x} \right) \zeta_l(x) - \zeta_{l-1}(x)}, \tag{8.294}$$

and

$$b_l = \frac{\left(\kappa D_l(\kappa x) + \dfrac{l}{x} \right) \psi_l(x) - \psi_{l-1}(x)}{\left(\kappa D_l(\kappa x) + \dfrac{l}{x} \right) \zeta_l(x) - \zeta_{l-1}(x)} \tag{8.295}$$

where

$$D_l(\rho) \equiv \frac{d}{d\rho} ln(\psi_l(\rho)) = \frac{\frac{d}{d\rho}\psi_l(\rho)}{\psi_l(\rho)} = \frac{\psi_l'(\rho)}{\psi_l(\rho)}. \tag{8.296}$$

(8.294) and (8.295) make use of the recurrence relations

$$\psi_l'(x) = \psi_{l-1}(x) - \frac{l}{x}\psi_l(x), \tag{8.297}$$

and

$$\zeta_l'(x) = \zeta_{l-1}(x) - \frac{l}{x}\zeta_l(x). \tag{8.298}$$

The logarithmic derivative (8.296) also satisfies a recurrence relation,

$$D_{l-1}(\rho) = \frac{l}{\rho} - \frac{1}{D_l(\rho) + \frac{l}{\rho}}. \tag{8.299}$$

The spherical Bessel functions can be calculated by either upward recurrence or downward recurrence. The downward recurrence algorithm is stable as first pointed by Aden [18]. However, to obtain accurate results, the recurrence must be started at a high order $l = N$ that is sufficiently larger than the smaller required cutoff order N_c that enough precision is obtained when N_c is finally reached by downward recurrence. Sometimes $N \gg N_c$ so it can be much faster to use upward recurrence if it is sufficiently precise.

As pointed out by Du, [17] ψ will overflow the double precision computations when its operand $n_s k_0 R$ has an imaginary component if $l > 720$. However, the ratio ψ' / ψ is well-behaved for both small and large particles for $|Im(n_s k_0 R)|$ as large as 100,000. Similarly, the ratio ψ_{n-1} / ψ_n is stable. Using relations (8.297) and (8.298), (8.272), and (8.273) can also be put into the form

$$a_l = \frac{\left(\frac{D(\kappa x)}{\kappa} + \frac{l}{x}\right)\psi_l(x) - \psi_{l-1}(x)}{\left(\frac{D(\kappa x)}{\kappa} + \frac{l}{x}\right)\xi_l(x) - \xi_{l-1}(\kappa x)}, \tag{8.300}$$

and

$$b_l = \frac{\left(\kappa D(\kappa x) + \frac{l}{x}\right)\psi_l(x) - \psi_{l-1}(x)}{\left(\kappa D(\kappa x) + \frac{l}{x}\right)\xi_l(x) - \xi_{l-1}(\kappa x)}. \tag{8.301}$$

These equations are discussed by Du [18]. They are put into the form

$$a_l = \frac{\left[\frac{r(\kappa x)}{\kappa} + \frac{l}{x}\left(1-\frac{1}{\kappa^2}\right)\right]\psi_l(x) - \psi_{l-1}(x)}{\left[\frac{r(\kappa x)}{\kappa} + \frac{l}{x}\left(1-\frac{1}{\kappa^2}\right)\right]\zeta_l(x) - \zeta_{l-1}(x)}, \tag{8.302}$$

and

$$b_l = \frac{\kappa r_l(\kappa x)\psi_l(x) - \psi_{l-1}(x)}{\kappa r_l(\kappa x)\zeta_l(x) - \zeta_{l-1}(x)} \tag{8.303}$$

where $r_l(\kappa x)$ is a ratio given by

$$r_l(\kappa x) \equiv \frac{\psi_{l-1}(\kappa x)}{\psi_l(\kappa x)}. \tag{8.304}$$

The upward recurrence relation for this ratio is found directly from (8.297),

$$r_{l+1}(\kappa x) = \frac{1}{\left(\frac{2l+1}{\kappa x}\right) - r_l(\kappa x)}. \tag{8.305}$$

Du shows that the number of significant digits that are lost using the upward recurrence relation is given approximately by

$$n_l(\kappa x) \approx \frac{\left|\mathrm{Im}(\kappa x)\right| - \ln(2) - l\left\{\mathrm{Re}\left[\ln\left(\frac{\kappa x}{l}\right) + \sqrt{1-\left(\frac{\kappa x}{l}\right)^2} - \ln\left(1+\sqrt{1-\left(\frac{\kappa x}{l}\right)^2}\right)\right]\right\}}{\ln(10)}. \tag{8.306}$$

Prior to calculating ψ, n_l is calculated and if the number of significant digits that are lost is acceptable, then the upward recurrence relation,

$$\psi_{l+1}(x) = \frac{(2l+1)}{x}\psi_l(x) - \psi_{l-1}(x), \tag{8.307}$$

is used to find ψ for all orders up to N_c. Du uses upward recurrence if $n_l < 4$, though for 64-bit precision this requirement can be considerably relaxed while still retaining at least six digits of precision. The upward recurrence relation is (8.305). The initial value in the recurrence relation for r_0 is cot($\kappa\, x$).

When $n_l \geq 8$, downward recurrence can be used. An initial value N for the order l is chosen, and the number of lost significant digits is calculated from (8.306). The difference, $n_N - n_{N_c}$, is the approximate number of significant digits in the solution. If this is less than the desired number of significant digits, then following the procedure of Du, the trial value for N is increased by either

$4|\kappa x|^{1/3}$ or 5, whichever is larger, until a difference, $n_N - n_{N_c}$, is achieved with the desired number of significant digits. Once a starting value N_1 for the order is determined, the downward recurrence is calculated by rearranging (8.305),

$$r_l(\kappa x) \equiv \frac{\psi_{l-1}(x)}{\psi_l(x)} = \left(\frac{2l+1}{\kappa x}\right) - \frac{1}{r_{l+1}(\kappa x)} \tag{8.308}$$

for all orders. r_{l+1} is set to zero to obtain the initial value for the ratio. Because this downward recurrence algorithm can be a time-consuming process when $N \gg N_c$, upward recurrence should be used whenever it is sufficiently accurate.

8.11 SINGLE SHELL EQUATIONS

In Appendix B of [6] explicit equations for the parameters a_l and b_l are given in a manner more suitable for computation with finite precision. Unlike the no shell case, however, there is no way in principle of arranging the equations to guarantee convergence for computation of high order spherical Bessel functions for every example.

In order to calculate the electric and magnetic fields throughout the computation space one must calculate the eight parameters, $A_{l,0}^{(e)}$, $A_{l,0}^{(m)}$, $A_{l,1}^{(e)}$, $A_{l,1}^{(m)}$, $B_{l,1}^{(e)}$, $B_{l,1}^{(m)}$, $B_{l,2}^{(e)}$, and $B_{l,2}^{(m)}$. The required equations are obtained by combining results in [6] and [19],

$$a_n = \frac{\psi_n(y)(D_n y + m_2 n) - m_2 y \psi_{n-1}(x)}{\zeta_n(y)(D_n y + m_2 n) - m_2 y \zeta_{n-1}(x)}, \tag{8.309}$$

$$b_n = \frac{\psi_n(x)(G_n m_2 x + n) - x\psi_{n-1}(x)}{\zeta_n(x)(G_n m_2 x + n) - x\zeta_{n-1}(x)}, \tag{8.310}$$

$$c_n = \frac{y\{m_2 x R_n(x_2)\psi_n(x_2) - m_2 x B_n \chi_{n-1}(x_2) + n B_n \chi_n(x_2)\}\{\psi_{n-1}(y)\zeta_n(y) - \psi_n(y)\zeta_{n-1}(y)\}}{x R_n(x_1)\psi_n(x_1)[\psi_n(y_2) - B_n \chi_n(y_2)][m_2 y G_n \zeta_n(y) - y\zeta_{n-1}(y) + n\zeta_n(y)]}, \tag{8.311}$$

$$d_n = \frac{m_2 y[\psi_n(x_2) - A_n \chi_n(x_2)][\psi_{n-1}(y)\zeta_n(y) - \psi_n(y)\zeta_{n-1}(y)]}{\psi_n(x_1)[\psi_n(y_2) - A_n \chi_n(y_2)][y D_n \zeta_n(y) - m_2 y \zeta_{n-1}(y) + m_2 n \zeta_n(y)]}, \tag{8.312}$$

$$f_n = \frac{m_2 y[\psi_{n-1}(y)\zeta_n(y) - \psi_n(y)\zeta_{n-1}(y)]}{[\psi_n(y_2) - B_n \chi_n(y_2)][m_2 y G_n \zeta_n(y) - y\zeta_{n-1}(y) + n\zeta_n(y)]} \tag{8.313}$$

$$g_n = \frac{m_2 y[\psi_{n-1}(y)\zeta_n(y) - \psi_n(y)\zeta_{n-1}(y)]}{[\psi_n(y_2) - A_n \chi_n(y_2)][y D_n \zeta_n(y) - m_2 y \zeta_{n-1}(y) + m_2 n \zeta_n(y)]}, \tag{8.314}$$

$$\nu_n = \frac{m_2 y B_n[\psi_{n-1}(y)\zeta_n(y) - \psi_n(y)\zeta_{n-1}(y)]}{[\psi_n(y_2) - B_n \chi_n(y_2)][m_2 y G_n \zeta_n(y) - y\zeta_{n-1}(y) + n\zeta_n(y)]}, \tag{8.315}$$

and

$$\omega_n = \frac{m_2 y A_n \left[\psi_{n-1}(y)\zeta_n(y) - \psi_n(y)\zeta_{n-1}(y)\right]}{\left[\psi_n(y_2) - A_n\chi_n(y_2)\right]\left[yD_n\zeta_n(y) - m_2 y\zeta_{n-1}(y) + m_2 n\zeta_n(y)\right]} \quad (8.316)$$

where

$$x \equiv \frac{2\pi r_1 n_2}{\lambda} = q_1, \quad (8.317)$$

$$y \equiv \frac{2\pi r_2 n_2}{\lambda} = \rho_2, \quad (8.318)$$

$$x_1 = m_1 x = \frac{n_0}{n_2}x = \frac{2\pi r_1 n_0}{\lambda} = q_1, \quad (8.319)$$

$$x_2 = m_2 x = \frac{n_1}{n_2}x = \frac{2\pi r_1 n_1}{\lambda} = \rho_1, \quad (8.320)$$

$$y_2 = \frac{n_1}{n_2}y = \frac{2\pi r_2 n_1}{\lambda} = q_2, \quad (8.321)$$

$$R_n \equiv \frac{\psi_n'}{\psi_n}, \quad (8.322)$$

$$D_n = \frac{R_n(q_2) - A_l\dfrac{\chi_n'(q_2)}{\psi_n(q_2)}}{1 - A_n\dfrac{\chi_n(q_2)}{\psi_n(q_2)}}, \quad (8.323)$$

$$G_l = \frac{R_n(q_2) - B_n\dfrac{\chi_n'(q_2)}{\psi_n(q_2)}}{1 - B_n\dfrac{\chi_n(q_2)}{\psi_n(q_2)}}, \quad (8.324)$$

$$A_n = \psi_n(\rho_1)\frac{n_1 R_n(q_1) - n_0 R_n(\rho_1)}{n_1 R_n(q_1)\chi_n(\rho_1) - n_0\chi_n'(\rho_1)}, \quad (8.325)$$

and

$$B_l = \psi_n(\rho_1)\frac{n_1 R_n(\rho_1) - n_0 R_n(q_1)}{n_1\chi_n'(q_1) - n_0 R_n(q_1)\chi_n(\rho_1)}. \quad (8.326)$$

To tie these results back into the notation of this chapter, one must use the following equations to calculate the fields. For inside the sphere,

$$A_{0,n}^{(e)} = \Gamma_n d_n, \tag{8.327}$$

and

$$A_{0,n}^{(m)} = \Gamma_n c_n. \tag{8.328}$$

Inside the shell,

$$A_{1,n}^{(e)} = \Gamma_n g_n, \tag{8.329}$$

$$A_{1,n}^{(m)} = \Gamma_n f_n, \tag{8.330}$$

$$B_{1,n}^{(e)} = -\Gamma_n w_n, \tag{8.331}$$

and

$$B_{1,n}^{(m)} = -\Gamma_n \nu_n. \tag{8.332}$$

In the outer incident medium,

$$B_{2,n}^{(e)} = -\Gamma_n a_n \tag{8.333}$$

and

$$B_{2,n}^{(m)} = -\Gamma_n b_n \tag{8.334}$$

where

$$\Gamma_n \equiv i^{n-1} \frac{2n+1}{n(l+1)}. \tag{8.335}$$

8.12 USEFUL MATHEMATICAL RELATIONS

When programming these equations, there are a number of useful mathematical relations for calculating Legendre polynomials and Riccati-Bessel functions. Legendre polynomials are defined by the equation

$$P_l(\cos\theta) = \sum_{m=0}^{l/2} (-1)^m \frac{(2l-2m)!}{2^l m!(l-m)!(l-2m)!} (\cos\theta)^{l-2m} \tag{8.336}$$

so

$$\frac{dP_l(\cos\theta)}{d(\cos\theta)} = \frac{(l+1)\left[\cos\theta P_l(\cos\theta) - P_{l+1}(\cos\theta)\right]}{\sin^2\theta}, \tag{8.337}$$

$$\frac{dP_l(\cos\theta)}{d\theta} = -\frac{(l+1)\left[\cos\theta P_l(\cos\theta) - P_{l+1}(\cos\theta)\right]}{\sin\theta}, \tag{8.338}$$

$$P_0(\cos\theta) = 1, \tag{8.339}$$

and

$$P_1(\cos\theta) = \cos\theta. \tag{8.340}$$

Associated Legendre functions of the first kind and derivatives are given by

$$P_l^{(1)}(\cos\theta) = -\sin\theta \frac{dP_l(\cos\theta)}{d(\cos\theta)} = \frac{(l+1)\left[P_{l+1}(\cos\theta) - \cos\theta P_l(\cos\theta)\right]}{\sin\theta}, \tag{8.341}$$

$$\frac{\partial P_l^{(1)}(\cos\theta)}{\partial(\cos\theta)} = -\frac{lP_{l+1}^{(1)}(\cos\theta) - (l+1)\cos\theta P_l^{(1)}(\cos\theta)}{\sin^2\theta} \tag{8.342}$$

and

$$\frac{\partial P_l^{(1)}(\cos\theta)}{\partial\theta} = \frac{lP_{l+1}^{(1)}(\cos\theta) - (l+1)\cos\theta P_l^{(1)}(\cos\theta)}{\sin\theta}. \tag{8.343}$$

Note that $\frac{\partial P_l^{(1)}(\cos\theta)}{\partial(\cos\theta)}$ goes to infinity as $\theta \to 0$ while $\frac{\partial P_l^{(1)}(\cos\theta)}{\partial\theta}$ remains finite. Some specific values are

$$P_1^{(1)}(\cos\theta) = \sin\theta \tag{8.344}$$

and

$$\frac{\partial P_1^{(1)}(\cos\theta)}{\partial\theta} = -\cos\theta. \tag{8.345}$$

Some useful relations for the Riccati-Bessel functions are

$$\psi_l(x) = \sqrt{\frac{\pi x}{2}} J_{l+\frac{1}{2}}(x) \tag{8.346}$$

where $J_{l+\frac{1}{2}}(x)$ is a Bessel function of the first kind,

$$\psi_0(x) = \sin x, \tag{8.347}$$

$$\psi_1(x) = -\cos x + \frac{\sin x}{x}, \tag{8.348}$$

$$\psi_{l-1}(x)=\left(\frac{2l+1}{x}\right)\psi_l(x)-\psi_{l+1}(x), \tag{8.349}$$

$$\psi_l'(x)\equiv\frac{\partial\psi_l(x)}{\partial x}=\psi_{l-1}(x)-\frac{l}{x}\psi_l(x)=\left(\frac{l+1}{x}\right)\psi_l(x)-\psi_{l+1}(x), \tag{8.350}$$

$$\psi_l''(x)=\left(\frac{l^2+l-x^2}{x^2}\right)\psi_l(x), \tag{8.351}$$

and

$$\psi_l(0)=\psi_l'(0)=\psi_l''(0)=0 . \tag{8.352}$$

Also,

$$\chi_l(x)=\sqrt{\frac{\pi x}{2}}Y_{l+\frac{1}{2}}(x) \tag{8.353}$$

where $Y_{l+\frac{1}{2}}(x)$ is a Bessel function of the second kind.

$$\chi_0(x)=-\cos x , \tag{8.354}$$

$$\chi_1(x)=-\sin x-\frac{\cos x}{x}, \tag{8.355}$$

$$\chi_{l-1}(x)=\left(\frac{2l+1}{x}\right)\chi_l(x)-\chi_{l+1}(x), \tag{8.356}$$

$$\chi_l'(x)\equiv\frac{\partial\chi_l(x)}{\partial x}=\chi_{l-1}(x)-\frac{l}{x}\chi_l(x)=\left(\frac{l+1}{x}\right)\chi_l(x)-\chi_{l+1}(x), \tag{8.357}$$

$$\chi_l''(x)=\left(\frac{l^2+l-x^2}{x^2}\right)\chi_l(x), \tag{8.358}$$

and

$$\zeta_l(x)\equiv\psi_l(x)+i\chi_l(x). \tag{8.359}$$

For Bessel functions of order −1/2,

$$J_{-\frac{1}{2}}(x)=\sqrt{\frac{2}{\pi x}}\cos x , \tag{8.360}$$

and

$$Y_{-\frac{1}{2}}(x) = J_{\frac{1}{2}}(x). \tag{8.361}$$

8.13 EXAMPLES

8.13.1 Polarization of Light in the Far Field Scattered From Gold Spheres in Water

The $|E|^2$ scattered light intensity from gold spheres in water is shown in Figure 8.2. It is calculated for polarization both parallel and perpendicular to the observation plane and plotted as a function of the polar scattering angle for four different gold sphere radii in water. The intensity is multiplied by the inverse square of the size parameter, $1/q^2 = [(2\pi\ r)/\lambda]^2$, making the plots independent of wavelength. The refractive index at 550 nm for gold is $0.57 + i\ (2.45)$, and for water is 1.33.

8.13.2 Polarization of Light in the Far Field Scattered From Spheres in Air

The $|E|^2$ intensity of the scattered light from a dielectric sphere with $n = 1.25$ in air is calculated for polarization parallel and perpendicular to the observation plane and plotted in Figure 8.3 as a function of polar scattering angle after Figure 13.11 of [4] on p. 655. The intensity is multiplied by the inverse square of the size parameter, $1/q^2 = [\lambda/(2\pi R)]^2$, for comparison between plots.

8.13.3 Extinction Coefficient of a Water Drop

The extinction coefficient of a water drop in air is plotted in Figure 8.4 as a function of the size parameter, $q = 2\pi R / \lambda$ for comparison to Figure 11.5 in [6]. In (a) the wavelength and refractive

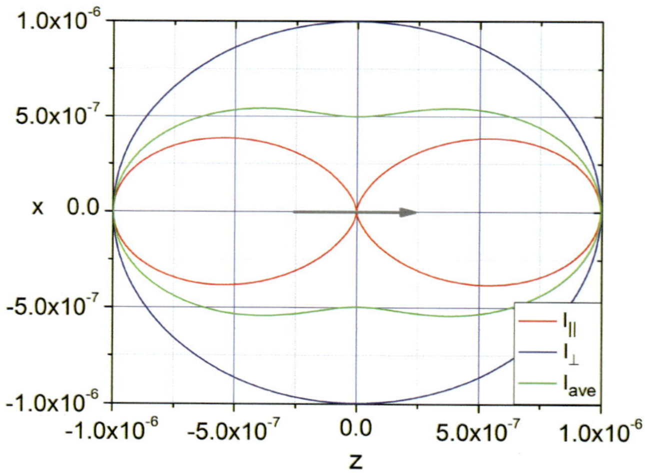

FIGURE 8.2 Polarization scattering diagrams for light propagating along the z-axis in the direction of the arrow, exhibiting I_{par}, I_{prp}, and I_{ave} for a 10 nm diameter gold sphere in water at 550 nm wavelength. The intensity plots are each normalized by the inverse square of the size parameter, $[\lambda_i/(2\pi R)]^2$. After Figure 13.10 in [4].

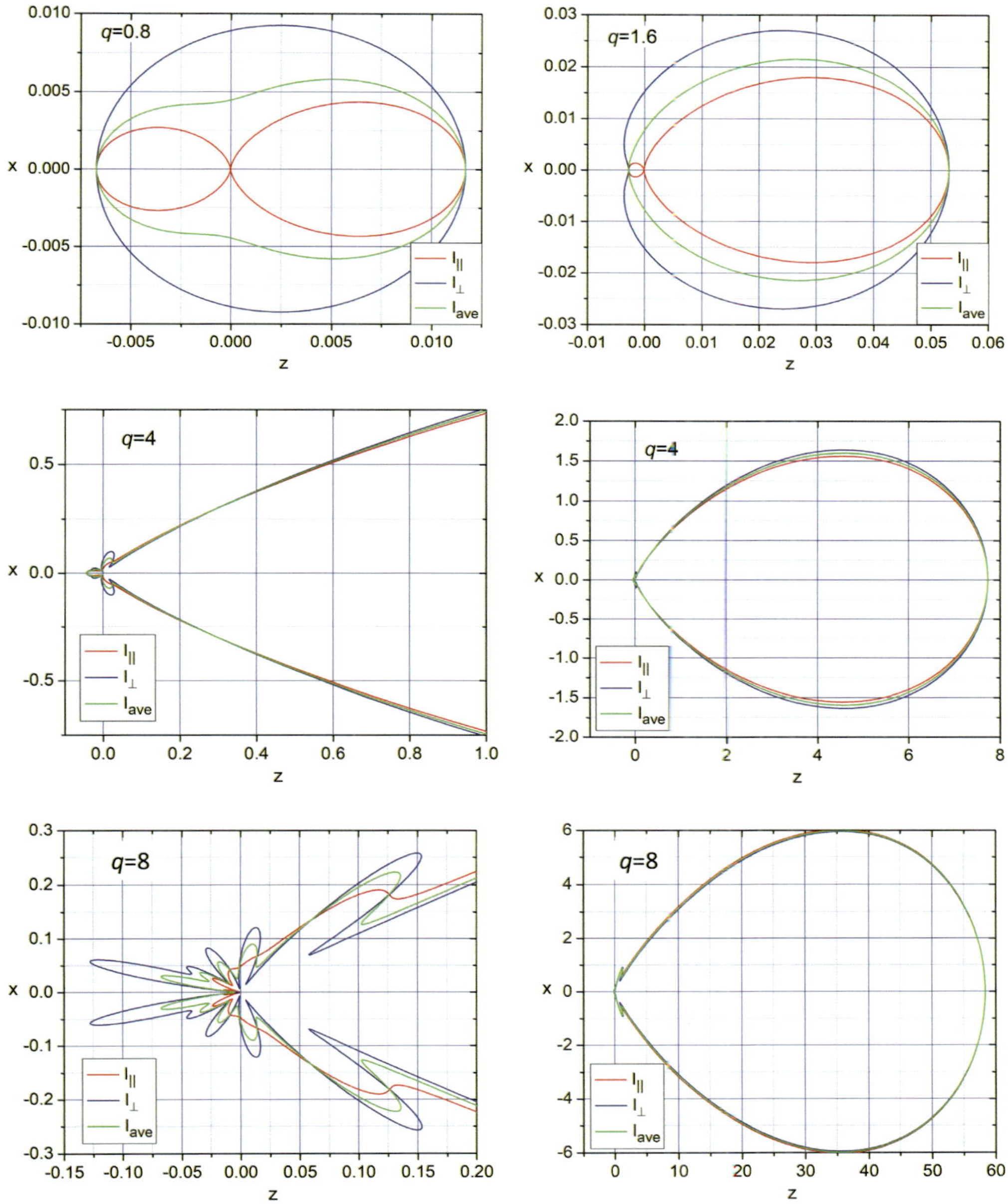

FIGURE 8.3 Polarization scattering diagrams for four different dielectric sphere size parameters. $n_i = 1$. $n_{sp} = 1.25$. $\lambda = 1$ µm. The intensity plots are each multiplied by the inverse square of the size parameter, $\lambda_i/(4\pi R^2)$.

After Figure 13.11 in [4]. Note that (e) for $q = 8$ is somewhat different than the figure in [4].

index are fixed while the radius is varied. In (b) the radius is fixed and the wavelength is varied with the appropriate refractive index for each wavelength using the refractive index data in [20]. New features appear in plot (b) not found in plot (a) for size parameters less than 5 due to infrared absorption bands in water.

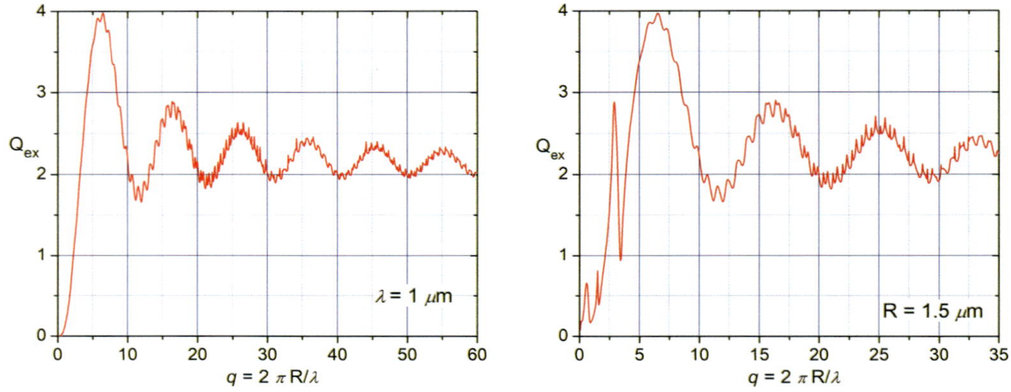

FIGURE 8.4 (a) Extinction efficiency as a function of the size parameter, q, for a water drop with refractive index $1.33 + i\ (10^{-8})$ in air. The wavelength is fixed while the radius is varied. (b) Extinction coefficient as a function of the size parameter. The wavelength is varied and the radius is fixed. The refractive index of the water also varies with wavelength. After Figure 11.5 in [6].

8.13.4 Angle-Dependent Polarized Scattering of Water Drop

Consider a water drop at a wavelength of 0.55 μm with a refractive index of $1.33 + i\ (10^{-8})$ after Figure 4.9 in [6]. The size parameter is 3, giving a radius of 0.2626 μm. The perpendicular and parallel scattered polarization intensities normalized by the inverse square of the size parameter are shown in Figure 8.5(a) in the xz plane and in (b) as a function of scattering angle (unnormalized). The polarization defined by (8.268) is plotted in (c).

8.13.4.1 Intensity Profile of Light Scattered by Water Drop in the Near Field

Consider a 1 μm water drop at a wavelength of 200 nm with a refractive index of 1.33. The water drop focuses the incident light wave in the near field, enhancing the local field intensity as shown in Figure 8.6.

8.13.5 Scattering Cross Sections for Iron Sphere

The extinction, scattering, and absorption cross sections of an iron sphere are plotted as a function of size parameter in Figure 8.7 after Figure 13.15 in [4].

8.13.6 Plasmonic Resonances in Highly Conducting Metals

The extinction cross-section efficiencies for 60 nm silver, gold, copper, and aluminum spherical nanoparticles are plotted as a function of wavelength in Figure 8.8 for particles embedded in both air and glass [21].

8.13.7 Plasmonic Resonances of Silver and Gold Particles in Glass

Stained glass windows in the Middle Ages were made by adding silver and gold nanoparticles to the glass. Specifically, silver nanoparticles of ~20–30 nm size produce a yellow glass while gold nanoparticles of ~10–15 nm produce a red glass when viewed in transmission. In Figures 8.9 and 8.10 the extinction efficiency is plotted as a function of wavelength for various silver and gold nanosphere

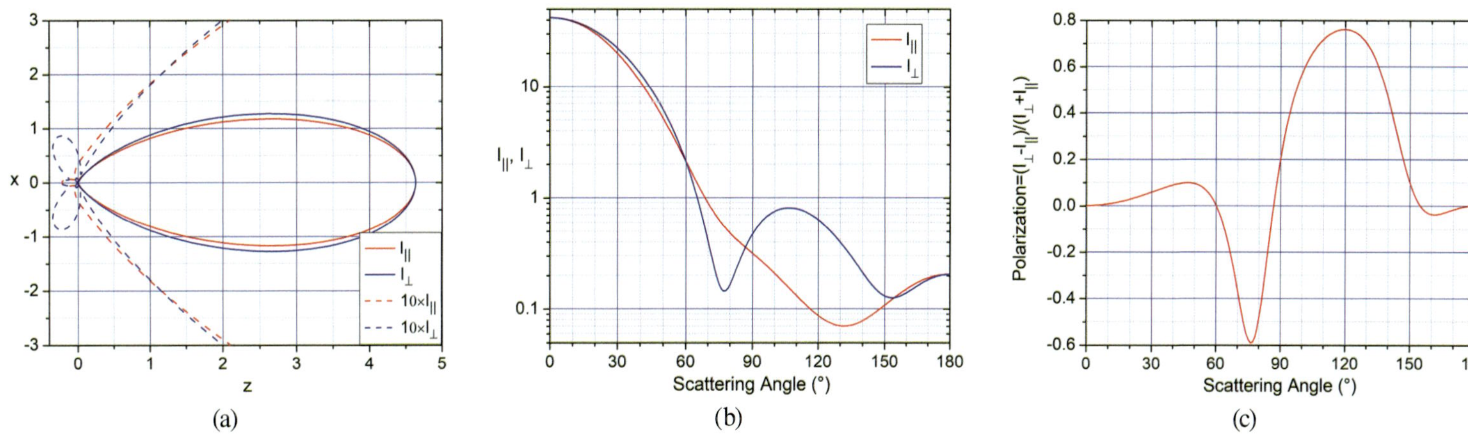

FIGURE 8.5 (a) Polarized scattering by 0.263 μm water drop with refractive index $1.33 + i\,(10^{-8})$ at $\lambda = 0.55$ μm normalized by the inverse square of the size parameter. (b) Same as (a) but plotted as a function of polar scattering angle and unnormalized. Note that $I_{\parallel}$ and $I_{\perp}$ are incorrectly reversed in Figure 4.9(b) of [6]. (c) Polarization of scattered light.

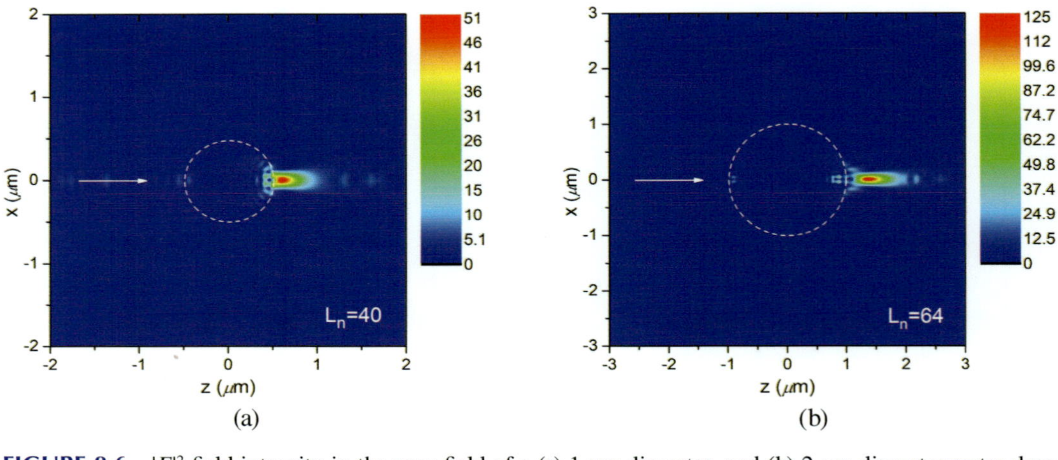

FIGURE 8.6 $|E|^2$ field intensity in the near field of a (a) 1 μm diameter, and (b) 2 μm diameter water drop with $n = 1.33$ at $\lambda = 200$ nm. Both drops exhibit a nanojet of light at the output side of the drops.

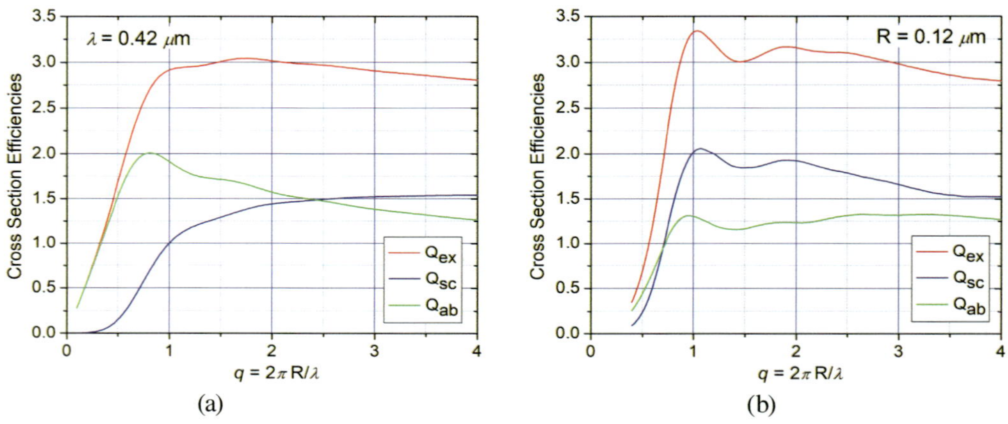

FIGURE 8.7 Cross-section efficiencies of iron sphere with refractive index $1.27 + i\,(1.37)$ as a function of the size parameter. (a) Fixed wavelength at 420 nm for comparison to Figure 13.15 in [4]. (b) Fixed radius of 120 nm with varying wavelength and refractive index for iron.

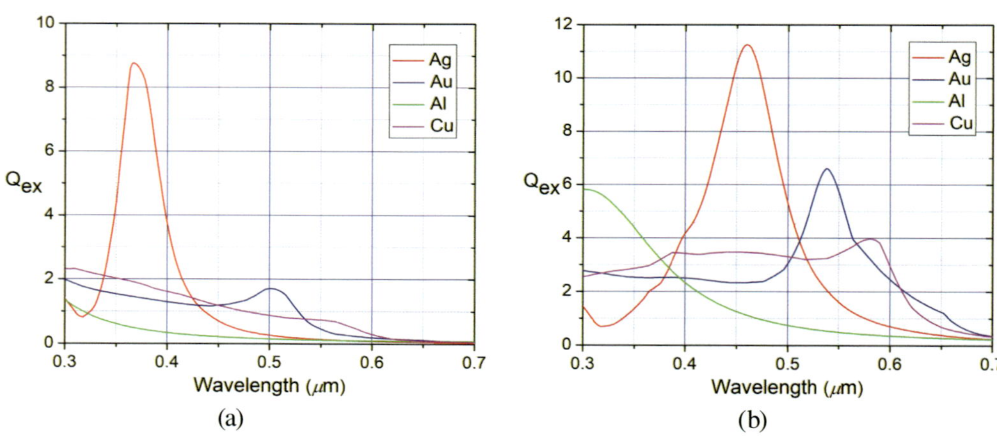

FIGURE 8.8 Extinction cross-section efficiency as a function of wavelength for several highly conducting metal nanospheres with 60 nm diameters. Plasmonic resonances are clearly visible for silver and gold. (a) In air medium, after Figure 9.2 of [21]. (b) In glass medium. Resonances shift to longer wavelengths when particles are embedded in a medium of higher refractive index.

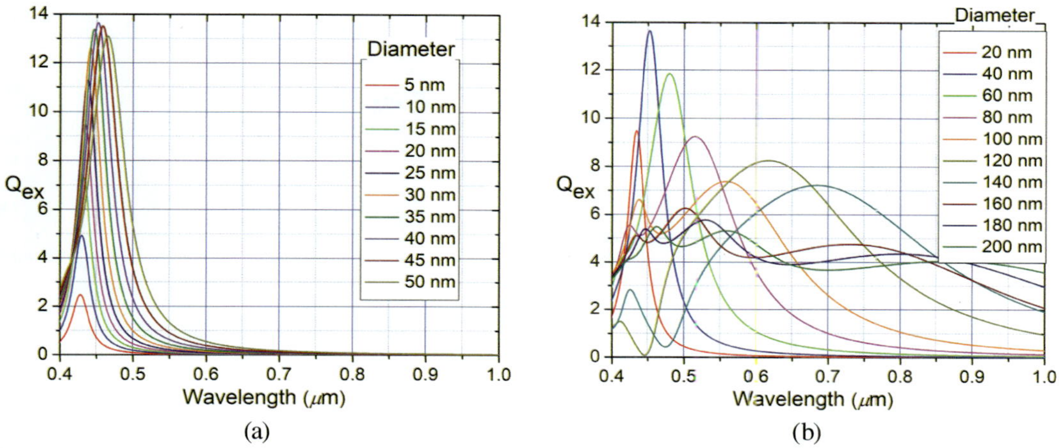

FIGURE 8.9 Extinction efficiency of silver spheres in glass as a function of wavelength for different sphere diameters.

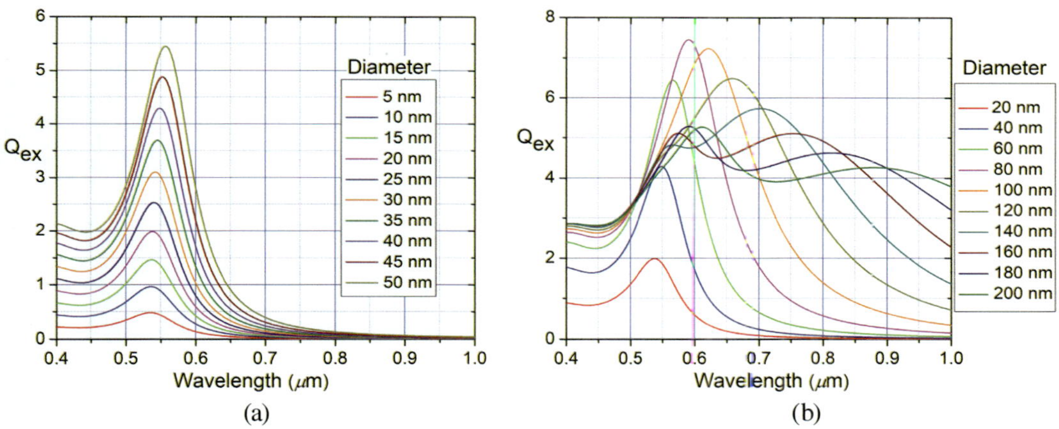

FIGURE 8.10 Extinction efficiency of gold spheres in glass as a function of wavelength for different sphere diameters.

sizes, respectively, exhibiting a very strong absorption at short wavelengths and small diameters due to excitation of localized surface plasmons [21].

The field intensity profile of a 40 nm silver sphere excited by 450 nm light at its dipole resonance, and a 120 nm silver sphere excited at 360 nm at its quadrupole resonance are shown in Figure 8.11.

8.13.8 Extinction Efficiency of Silver Particles in Glass as a Function of Radius

It is possible to calculate cross-section efficiencies for quite large particles using the algorithm of Du [17]. In Figure 8.12 the extinction efficiency is calculated for silver particles in glass up to a radius of 10 cm. The extinction efficiency goes to a value of 2 for large radii. This somewhat counter-intuitive result is discussed in detail in [4] and [6].

8.13.9 Absorption Efficiency of Water Drop in Air as a Function of Radius

The absorption efficiency of a water drop in air as a function of radius is plotted for two different wavelengths in Figure 8.13. At the longer wavelength where water is more absorbing, the resonant absorption effects are greatly damped.

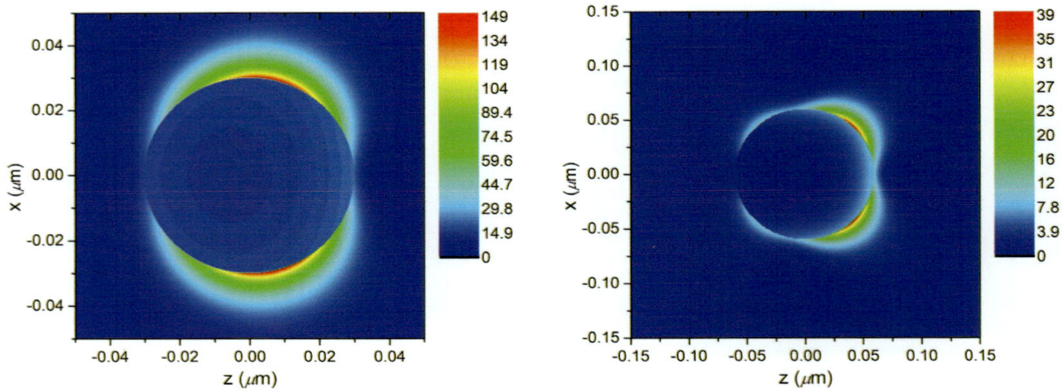

(a) (b)

FIGURE 8.11 $|E|^2$ field profile for (a) a 60 nm Ag sphere, n = 0.1332 + i (2.526), in glass, n = 1.464, at $\lambda = 472$ nm for which Q_{ex} exhibits a surface plasmon dipole resonance peak, and (b) a 120 nm silver sphere, n = 0.196 + i (1.533) in air, which exhibits a surface plasmon quadrupole resonance at $\lambda = 360$ nm.

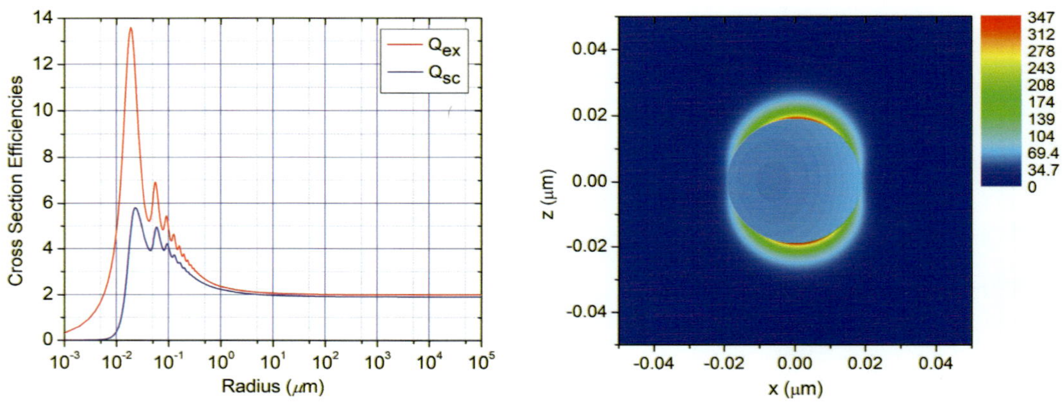

(a) (b)

FIGURE 8.12 (a) Extinction and scattering efficiencies as a function of radius for a silver nanoparticle (n_{Ag} = 0.13611 + i (2.32836)) in glass (n = 1.5) at a wavelength of 450 nm. The largest radius in the plot corresponds to a 20 cm diameter sphere. This calculation requires ~2.1 × 10^6 terms in the Riccati-Bessel function expansion. The peak in the extinction efficiency occurs at a 19 nm radius which only requires 19 terms in the expansion. The $|E|^2$ intensity contour plot at this radius is shown in (b).

8.13.10 Rainbows

The rainbow such as that in Figure 8.14 has always been a beautiful phenomenon begging for an explanation. Mie theory can be used to study rainbows in detail.

A simple ray tracing diagram as in Figure 8.15 is often presented as a means to grasp the essential idea behind a rainbow, though the details are sometimes glossed over. The relevant angular equations are

$$n_i \sin\theta_1 = n_s \sin\theta_2 \,, \tag{8.362}$$

$$\theta_3 = \sin^{-1} h \,, \tag{8.363}$$

$$\theta_1 = \theta_3 - \theta_i \,, \tag{8.364}$$

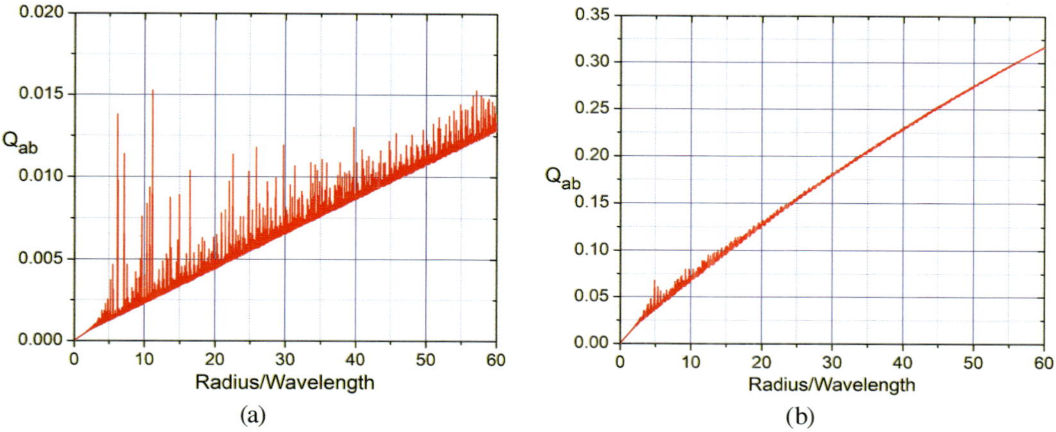

FIGURE 8.13 Absorption efficiency for a water drop in air at two wavelengths as a function of radius. (a) At $\lambda = 1.20\ \mu m$, $n_{water} = 1.323 + i\ (9.74 \times 10^{-6})$. (b) At $\lambda = 1.45\ \mu m$, $n_{water} = 1.319 + i\ (3.0 \times 10^{-4})$. After Figures 7.2 and 7.3 in [6].

FIGURE 8.14 Double rainbow photographed from the author's backyard. The region inside the primary rainbow is much brighter than that between the primary and secondary rainbows.

$$\theta_4 = \pi - 2\theta_2 , \tag{8.365}$$

$$\theta_5 = 2\pi - \theta_3 - 2\theta_4 , \tag{8.366}$$

and

$$\theta_6 = \theta_5 - \theta_1 . \tag{8.367}$$

The included angle of the reflected ray is

$$\theta_r = \theta_i + \theta_6 = 2\theta_i - 2\sin^{-1}h + 4\sin^{-1}\left[\frac{n_i}{n_s}\sin\left(\sin^{-1}h - \theta_i\right)\right]. \tag{8.368}$$

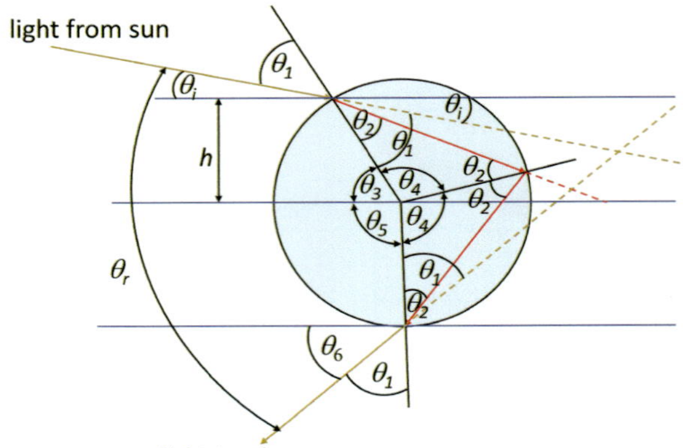

FIGURE 8.15 Ray tracing diagram for rainbow formation. Light enters a spherical drop of unit radius at an angle θ_1 to the horizontal and at a height h above the center of the drop. After two refractions and one reflection, the ray exits the drop at angle θ_6 below the horizontal. The total included angle between the reflected ray and the incident ray is $\theta_r = \theta_1 + \theta_6$.

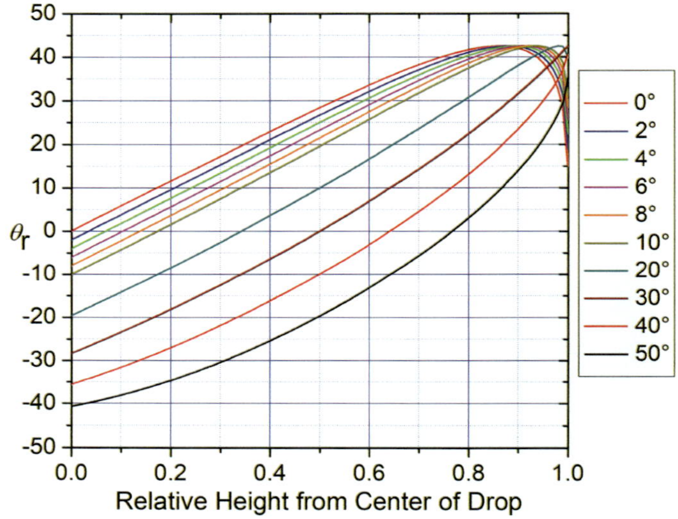

FIGURE 8.16 The total reflected included angle is plotted as a function of light entering a water drop with refractive index 1.33 at a relative height compared to the radius of the drop for several different angles of incidence from the horizontal.

(8.308) indicates that the reflected angle is a function of both the angle of the incident ray and the height at which it strikes the drop. Yet clearly rainbows reflect brightly at one specific angle and not at others. The subtlety is that there is a narrow range of included angles over which many of the incident rays are reflected, while the other incident rays get spread over a large angle range. As shown in Figure 8.16, for light rays incident upon the drop between its equator ($h = 0$) and its top ($h = 1$), the rays emerge from the drop after two refractions and a reflection spread over an 80° range. However, light rays incident at relative heights between ~0.8 and 0.95 all get backscattered at angles near to 42°. This pile up of light rays at approximately one angle is what causes the bright color of the rainbow.

Solid state physicists recognize this as a density of states phenomenon. Where the slope of the curves in Figure 8.16 goes to zero, a large fraction of the incident light rays all reflect at nearly the same included angle, giving a bright color at that angle. Moreover, it is apparent from this figure that this angle is the same regardless of the angle of the incident light ray, at least until the angle of incidence from the horizontal is slightly greater than 20° for which there is no longer a point of zero slope. This surprising result can be expressed mathematically. The derivative of θ_r in (8.368) with respect to the relative height, h, is

$$\frac{\partial \theta_r}{\partial h} = -\frac{2}{\sqrt{1-h^2}} + \frac{4r\cos\left(\theta_i - \sin^{-1}h\right)}{\sqrt{1-h^2}\sqrt{1-r^2\sin^2\left(\theta_i - \sin^{-1}h\right)}} \tag{8.369}$$

where r is the ratio of refractive indices,

$$r = \frac{n_i}{n_s}. \tag{8.370}$$

Setting the slope to zero gives the relation

$$\sin^{-1}h = \theta_i + \cos^{-1}\sqrt{\frac{\left(\frac{n_s}{n_i}\right)^2 - 1}{3}}. \tag{8.371}$$

Substituting this into (8.369) gives finally

$$\theta_r = 4\sin^{-1}\left(\frac{n_i}{n_s}\sin(x)\right) - 2x \tag{8.372}$$

where

$$x \equiv \cos^{-1}\sqrt{\frac{\left(\frac{n_s}{n_i}\right)^2 - 1}{3}}. \tag{8.373}$$

It should be noted that θ_r in (8.372) is independent of the angle of incidence θ_i and the relative height, h! For $n_i = 1$ and $n_s = 1.33$, $\theta_r = 42.52°$, which is the primary rainbow angle. The color/angle spread in a real rainbow is due to the dispersion of the refractive index with wavelength.

Although this analysis provides a qualitative description of the rainbow, it does not provide any information about the intensity of the reflected light at the rainbow angle. However, it is here that Mie theory comes to the rescue. The unpolarized relative intensity, $4\pi r^2\left(\left|E_\parallel\right|^2 + \left|E_\perp\right|^2\right)$, can be calculated as a function of polar scattering angle at a single wavelength for different water drop radii [22] as shown in Figure 8.17. Note that the scattering angles in Figure 8.17 must be subtracted from 180° to obtain the reflected included angle previously discussed. Refractive index data for water are taken from [20]. The primary and secondary rainbows begin to form for droplets with radii on the order of 100 μm, and are clearly visible in the scattering of drops with 1 mm radii. A well-developed dark band, called Alexander's dark band, [23] between the primary and secondary rainbow is also present for the larger drops.

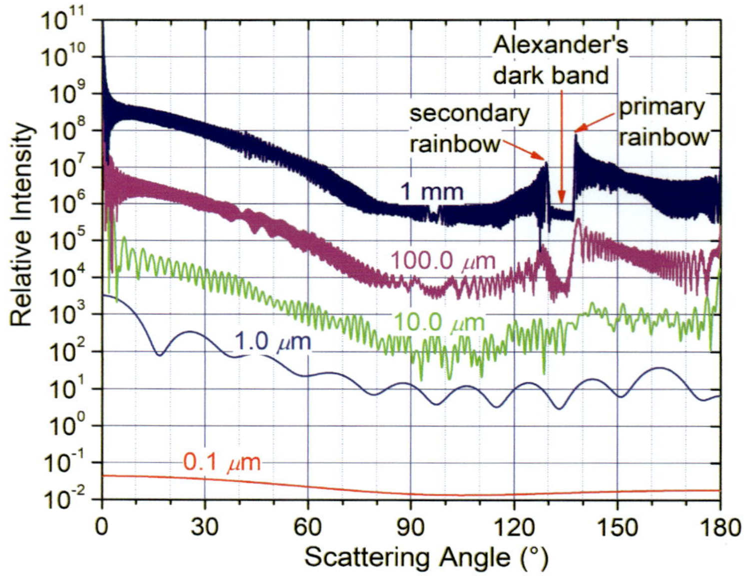

FIGURE 8.17 Unpolarized relative scattering intensity as a function of polar scattering angle for water drop radii between 0.1 mm and 1 mm. The refractive index of water is 1.331 at a wavelength of 650 nm. The drops with larger radii were calculated at a 0.01° angular resolution, but it is likely that even this resolution was too large to capture all the spectral features. After Figure 1 in [22].

Another way to analyze the Mie scattering of water drops is to calculate the unpolarized relative intensity as a function of polar scattering angle over a narrow range of angles for several different wavelengths to see how the rainbow angles shift with wavelength. That is shown in Figure 8.18.

8.13.11 Dielectric Nanoparticles

Dielectric nanoparticles exhibit multipole peaks in their scattering spectra [24]. Figure 8.19 shows the scattering cross section for a nanosphere with index $n = 4$, and a silicon nanosphere with refractive index as tabulated in [25].

8.13.12 Coated Particles

The extinction efficiency for an absorbing hematite particle coated with yttrium carbonate suspended in water has been calculated as shown in Figure 8.20 [26].

Similarly, the extinction efficiency of silica particles coated with titania dispersed in water have also been calculated as shown in Figure 8.21.

8.13.13 Electric Field for Water Drop and the Effect of the Number of Terms in the Riccati-Bessel Function Expansion

The electric field intensity is calculated for water drops ($n = 1.33$) in air at a wavelength of 500 nm. The results are shown in Figure 8.22 for comparison to Figure 3 in [27].

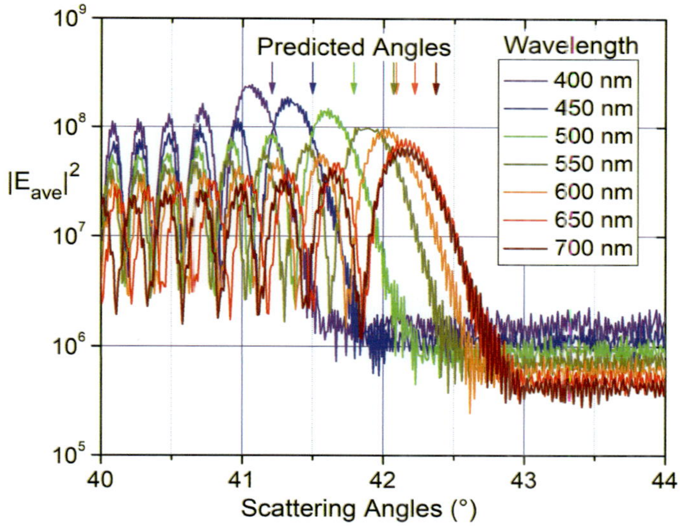

FIGURE 8.18 $|E_{ave}|^2$ calculated as a function of polar scattering angle (subtracted from 180° for back scattering) for 1 mm water drops at several different wavelengths using the tabulated values in [20] for the refractive index of water. The angles calculated from (8.313) are shown by arrows. Note the nearly two orders of magnitude difference between peak brightness and the angular region corresponding to Alexander's dark band.

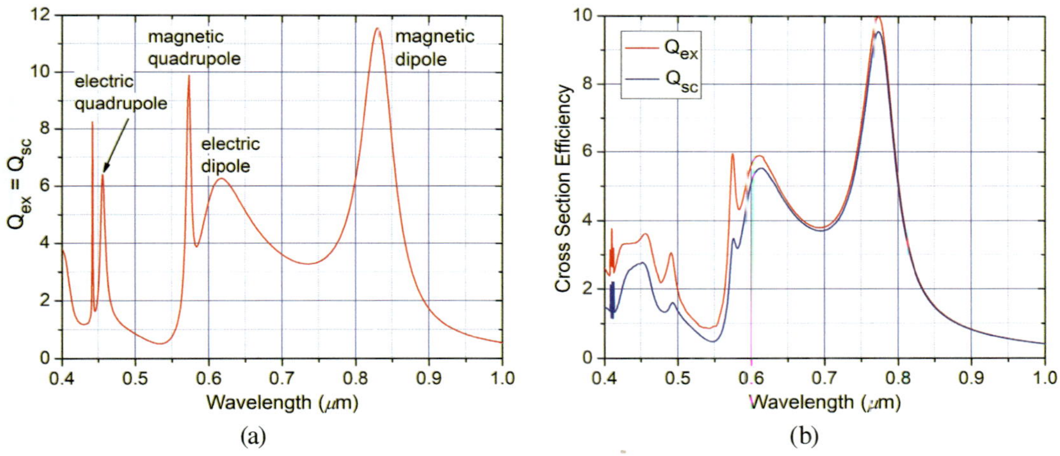

FIGURE 8.19 Scattering efficiencies of 100 nm radii dielectric nanospheres in air. (a) $n = 4$. (b) Silicon particle. After figures in [24].

The equations provided by Wiscombe in (8.291) are used to determine the number of terms required for the Riccati-Bessel function expansion. In the case of a water drop with a refractive index of 1.33 in air when the wavelength equals the drop radius, the number of terms calculated from these equations is 19. However, for these figures six additional terms are used in the expansion, that is, 25 total. In Figure 8.23 the $|E|^2$ field intensity is calculated for expansions with 17, 19, 25, and 30 terms, respectively. In all four plots, there is a high intensity light "jet" emanating from the right side of the drop, but in the first two plots the jet exhibits a minimum before $z = 1.5$ μm. There are slight differences between the final two plots, but the minimum has disappeared in both and the results more closely match that of Figure 3(e) in [27].

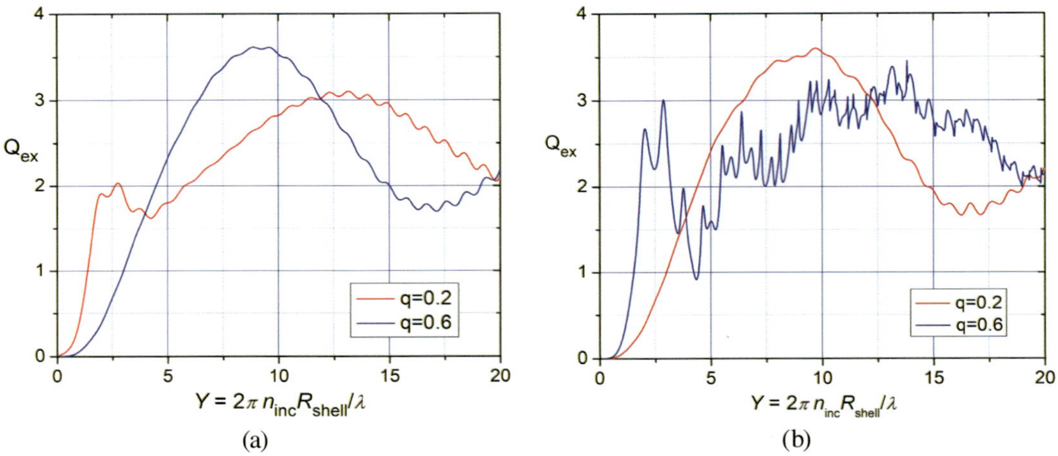

FIGURE 8.20 (a) Extinction coefficient as a function of the size parameter $Y = 2\pi n_w R/\lambda_0$ where R is the coating radius for a hematite particle coated with Y_2CO_3 suspended in water. $n_H = 3.09 + i(0.49)$, $n_{YCO} = 1.65$, and $n_w = 1.3368$. The ratio of the sphere radius to the coating radius is $q = 0.2, 0.6$. After Figures 2 and 3 in [26]. Note that the size parameter Y is calculated from the incident medium, not the core, as incorrectly stated in the reference. (b) Same as (a) but with $n_H = 3.09$. After Figures 6 and 7 in [26].

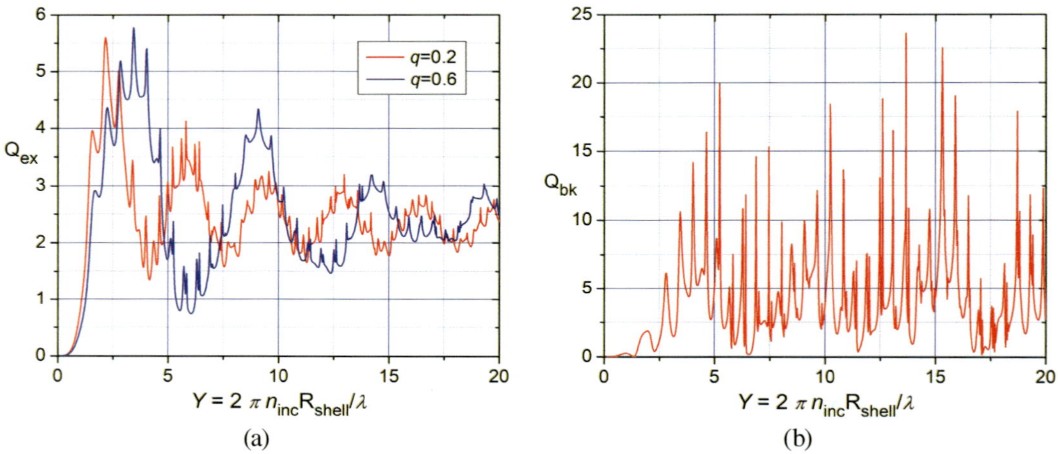

FIGURE 8.21 (a) Extinction coefficient as a function of the size parameter $Y = 2\pi n_w R/\lambda_0$ where R is the coating radius for a silica particle coated with YO_x suspended in water. $n_{SiO2} = 1.48$, $n_{YOx} = 2.6$, and $n_w = 1.3368$. The ratio of the sphere radius to the coating radius is $q = 0.2, 0.6$. After Figures 10 and 11 in [26]. (b) Backscattering efficiency for $q = 0.6$.

8.13.14 Electric Field for Silver Nanoshell

A good example of the difficulties involved in Mie calculations is given in Figure 2 of [28] for a thin silver nanoshell coated on a glass sphere in air. The example is shown in Figure 8.24(a).

Consider the factor in the denominator of (8.310), (8.312), and (8.314),

$$c_n \equiv \psi_n(q_2) - B_n\chi_n(q_2). \tag{8.374}$$

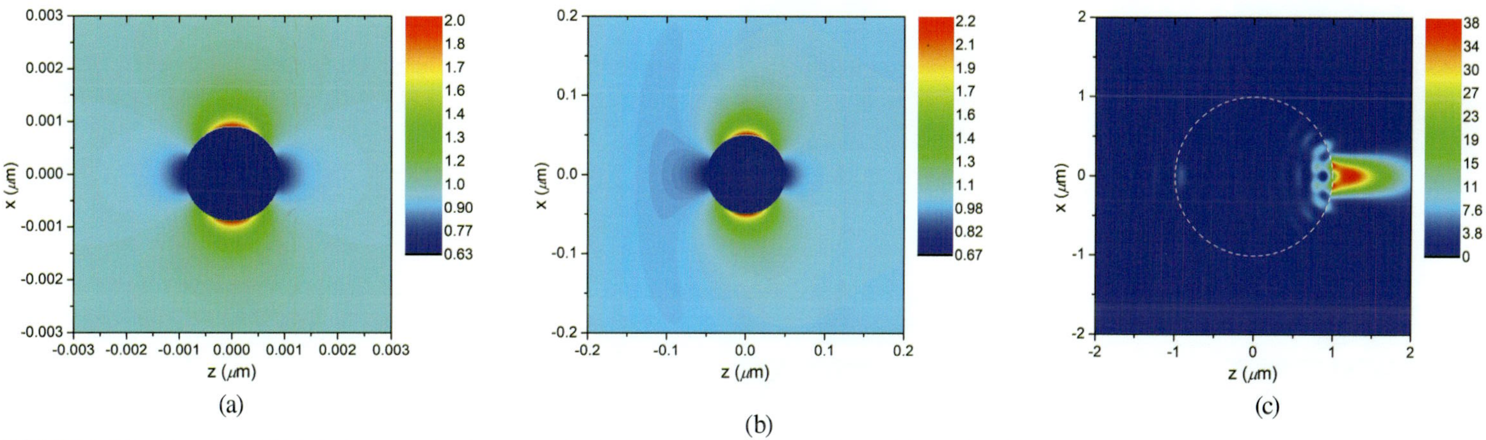

FIGURE 8.22 Electric field intensity for water drops ($n = 1.33$) of various radii in air. (a) $r = 0.9$ nm, $\lambda = 496$ nm. (b) $r = 50$ nm, $\lambda = 500$ nm. (c) $r = 1$ μm, $\lambda = 500$ nm. After Figure 3 in [27].

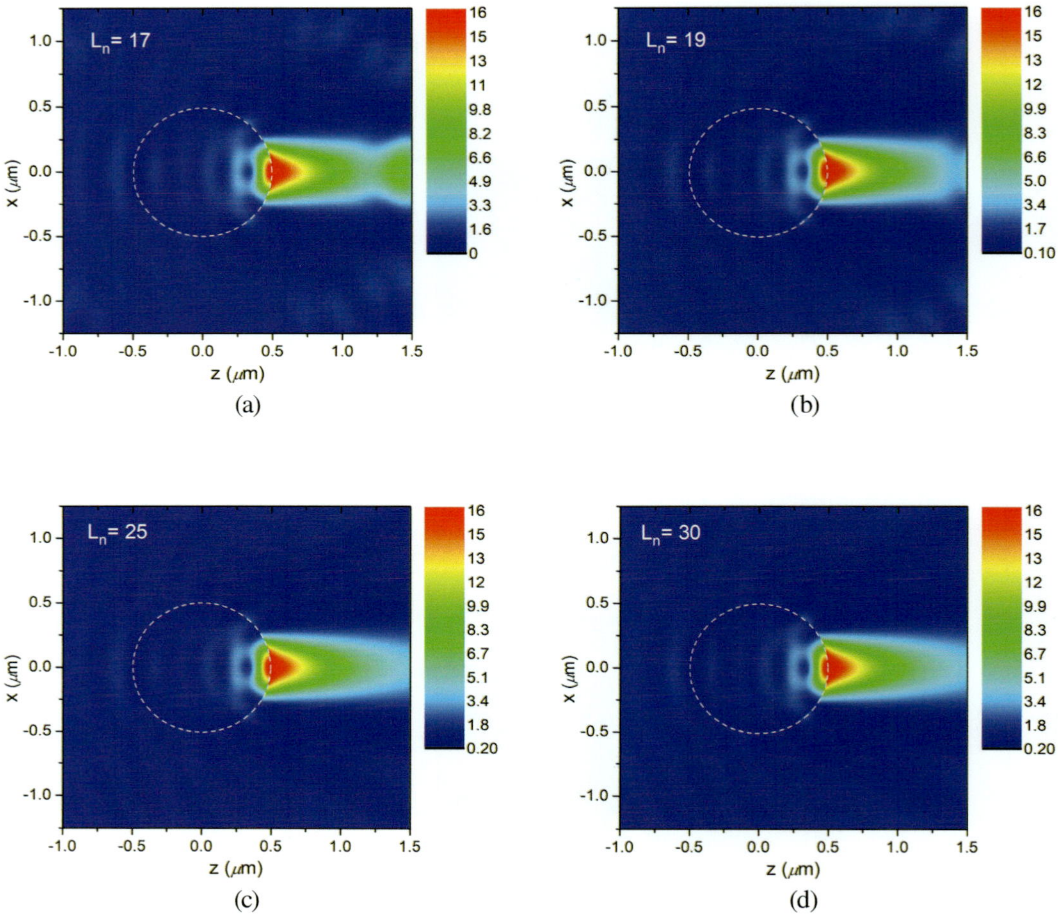

FIGURE 8.23 Effect of number of terms in the Riccati-Bessel function expansion. (a) 17 terms, (b) 19 terms, (c) 25 terms, and (d) 30 terms. The final two plots are very similar to each other and to Figure 3(e) of [27].

where

$$B_n = \psi_n(\rho_1)\frac{\frac{n_1}{n_0}R_n(\rho_1) - R_n(q_1)}{\frac{n_1}{n_0}\chi_n'(\rho_1) - R_n(q_1)\chi_n(\rho_1)} = \psi_n(\rho_1)\frac{n_1 R_n(\rho_1) - n_0 R_n(q_1)}{n_1\chi_n'(\rho_1) - n_0 R_n(q_1)\chi_n(\rho_1)} \tag{8.375}$$

and

$$R_n(x) \equiv \frac{\psi_n'(x)}{\psi_n(x)}. \tag{8.376}$$

Therefore,

$$c_n \equiv \psi_n(q_2) - \frac{n_1\psi_n'(\rho_1)\psi_n(q_1) - n_0\psi_n'(q_1)\psi_n(\rho_1)}{n_1\chi_n'(\rho_1)\psi_n(q_1) - n_0\psi_n'(q_1)\chi_n(\rho_1)}\chi_n(q_2). \tag{8.377}$$

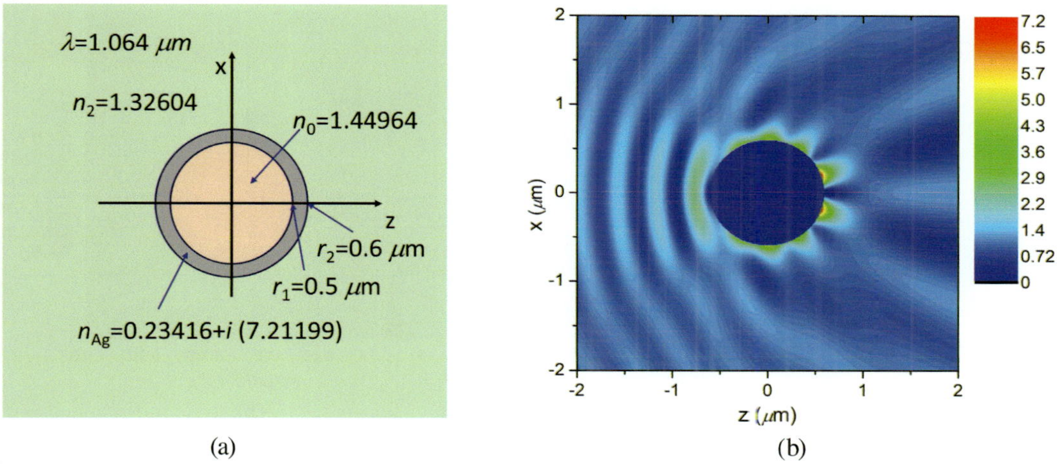

FIGURE 8.24 (a) 100 nm thick silver nanoshell coating a 1 μ*m* diameter glass sphere in water at $\lambda = 1064$ nm. (b) $|E|^2$ intensity distribution.

TABLE 8.1
Calculated Values in Extended Precision for Terms in (8.377)

c_n	$y_n(q_2)$	$\frac{n_1 y'_n(r_1) y_n(q_1) - n_0 y'_n(q_1) y_n(r_1)}{n_1 c'_n(r_1) y_n(q_1) - n_0 y'_n(q_1) c_n(r_1)} c_n(q_2)$
[−17.08720389 − *i* (2.608943979)] × 10^{-10}	[−4.055119701 + *i* (4.442779105)] × 10^{10}	[−4.055119701 + *i* (4.442779105)] × 10^{10}
[−2.220124325 + *i* (3.582277678)] × 10^{-8}	[−4.112341595 − *i* (3.733698946)] × 10^{10}	[−4.112341595 − *i* (3.733698946)] × 10^{10}
[5.297872283 + *i* (3.377046814)] × 10^{-8}	[3.299215794 − *i* (3.662657894)] × 10^{10}	[3.299215794 − *i* (3.662657894)] × 10^{10}

To achieve numerical convergence, the first 45 terms of the Riccati-Bessel function expansion are required. For illustration, the numerical values of the first three terms of the expansion in (8.377) are listed in Table 8.1.

It can be clearly seen that the two terms on the righthand side of (8.377), the second and third columns of the table, are nearly equal – to about twenty decimal places! Unfortunately, REAL*8 precision for floating point numbers involving 64 bits only has a precision of about 16 digits and a maximum exponent of 307. Therefore, it has insufficient precision to solve this example. However, quadruple precision (128 bits) has about 33 digits of precision and an exponent up to 4931. Therefore, to solve this problem, the higher precision is required. Extended precision calculations can be performed in software on a 64-bit machine, but they run substantially more slowly. The result is shown in Figure 8.24(b).

Figure 8.1 of [28] is another example of an electric field calculation for a 100 nm glass sphere with only a 10 nm thick silver film as shown in Figure 8.25(a). The calculated electric field in (b) is rather different than that in [28] in which the dielectric constant for silver has been adjusted to correct for electron scattering within the very thin silver film. No such adjustment has been made for the calculation in Figure 8.25(b) though the difference between the two plots seems larger than can be accounted for by a small change in the refractive index of silver.

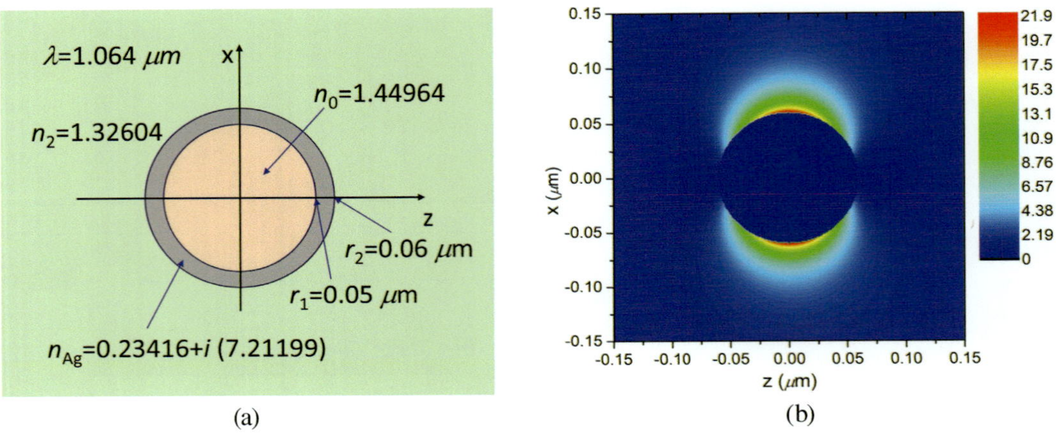

FIGURE 8.25 (a)100 nm thick silver nanoshell coating a 1 μm diameter glass sphere in water at λ = 1064 nm. (b) $|E|^2$ intensity distribution.

REFERENCES

1 G. Mie, "Beiträge zur Optik trüber Medien, speziell kolloidaler Metallösungen," [Contributions to the optics of turbid media, particularly of colloidal metal solutions]. *Annalen der Phys. (in German).* **25** (1908) 377–445.

2 L. Lorenz, "Lysbevaegelsen i og uden for en af plane Lysbolger belyst Kugle," *Det Kongelige Danske Videnskabernes Selskabs Skrifter* **6** Raekke. **6** Bind (1890) 1, 1–62.

3 P. Debye, "Der Lichtdruck auf Kugeln von beliebigem material," *Annalen der Phys., Vierte Folge*, Band **30** (1909) 57–136.

4 M. Born and E. Wolf, *Principles of Optics*, 5th ed. (New York: Pergamon Press) chapter 13.

5 H. C. van de Hulst, *Light Scattering by Small Particles*, (New York: Dover Pub., 1981).

6 C. F. Bohren and D. R. Huffman, *Absorption and Scattering of Light by Small Particles* (New York: John Wiley & Sons, Inc., 1983).

7 J. V. Dave, "Subroutines for computing the parameters of the electromagnetic radiation scattered by a sphere," *Technical Report.* Rep. No. 320–3237. IBM Scientific Center, Palo Alto, Calif. (1968).

8 C. D. Cantrell, "Numerical methods for the accurate calculation of spherical Bessel functions and the location of Mie resonances," *Technical Report,* Center for Applied Optics University of Texas at Dallas Richardson,TX75083-0688(1988)Availableathttps://utd-ir.tdl.org/server/api/core/bitstreams/7a7995ee-a70c-4582-82e8-97cc7463b876/content.

9 W. J. Wiscombe, "Mie scattering calculations: Advances in technique and fast, vector-speed computer codes," *National Center for Atmospheric Research Technical Note.* NCAR/TN-140+STR. (June, 1979). https://www.google.com/url?sa=t&rct=j&q=&esrc=s&source=web&cd=&ved=2ahUKEwjYheLA2fz_AhUEkIkEHfOcC6sQFnoECBMQAQ&url=https%3A%2F%2Fopensky.ucar.edu%2Fislandora%2Fobject%2Ftechnotes%253A232%2Fdatastream%2FPDF%2Fdownload%2FMie_Scattering_Calculations__Advances_in_Technique_and_Fast__Vector-speed_Computer_Codes.citation&usg=AOvVaw30CROA4B7DcxBYHA7n5I5g&opi=89978449

10 W. J. Wiscombe, "Improved Mie scattering algorithms," *Appl. Opt.* **19** (1980) 1505.

11 G. Grehan and G. Gouesbet, "Mie theory calculations: New progress, with emphasis on particle sizing," *Appl. Opt.* **18** (1979) 3489.

12 A. Gogoi, A. Choudhury, and G. Ahmed, "Mie scattering computation of spherical particles with very large size parameters using an improved program with variable speed and accuracy," *J. Mod. Opt.* **57** (2010) 2192.

13 https://miepython.readthedocs.io/en/latest/02_efficiencies.html

14 http://www.philiplaven.com/mieplot.htm

15 G. W. Kattawar and G. N. Plass, "Electromagnetic scattering from absorbing spheres," *Appl. Opt.* **6** (1967) 1377–1382.

16 W. J. Wiscombe, "Improved Mie scattering algorithms," *Appl. Opt.* **19** (1980) 1505–1509.

17 H. Du, "Mie-scattering calculation," *Appl. Opt.* **43** (2004) 1951–1956.

18 A. L. Aden, "Electromagnetic scattering from spheres with sizes comparable to the wavelength," *J. Appl. Phys.* **22** (1951) 601–605.
19 H. Suzuki and I.-Y. S. Lee, "Calculation of the Mie scattering field inside and outside a coated spherical particle," *Intl. J. Phys. Sci.* **3** (2008) 38–41. Note that there is a typographical error in Eq. (9).
20 G. M. Hale and M. R. Querry, "Optical constants of water in the 200-nm to 200-μm wavelength region," *Appl. Opt.* **12** (1973) 555–563.
21 D. Sarid and W. Challener, *Modern Introduction to Surface Plasmons* (New York: Cambridge University Press, 2010).
22 P. Laven, "Simulation of rainbows, coronas and glories using Mie theory and the Debye series," *J. Quant. Spectrosc. Rad. Transf.* **89** (2004) 257–269.
23 According to Wikipedia, this is named after Alexander of Aphrodisias who first described this phenomenon in Aphrodisias, *Commentary on Book IV of Aristotle's* Meteorology.
24 Wikipedia article on Mie scattering. https://en.wikipedia.org/wiki/Mie_scattering
25 D. Palik, ed., *Handbook of Optical Constants of Solids* (New York: Academic Press, Inc., 1985) 547–552.
26 A. Quirantes and A. V. Delgado, "The scattering of light by a suspension of coated spherical particles: Effects of polydispersity on cross-sections," *J. Phys. D: Appl. Phys.* **30** (1997) 2123–2131.
27 X. Fan, W. Zheng, and D. J. Singh, "Light scattering and surface plasmons on small spherical particles," *Light: Sci. Applic.* **3** (2014) e179. https://doi.org/10.1038/lsa.2014.60
28 H. Suzuki and I.-Y. S. Lee, "Calculation of the Mie scattering field inside and outside a coated spherical particle," *Intl. J. Phys. Sci.* **3** (2008) 38–41.

9 Scattered Field Finite Difference Time Domain

9.1 INTRODUCTION

The previous chapters have been primarily focused on optical models that can be solved analytically or semi-analytically. The partial exception to this is the method of lines for solving arbitrary waveguide geometries. In contrast to these models, the finite difference time domain (FDTD) technique is a relatively straightforward numerical approach for solving electromagnetic problems that lack the high symmetry that enables an analytical approach. As a result, it is more general and more powerful in the range of optical problems that it can address. On the other hand, it is also usually orders of magnitude slower in solving them than an analytical technique. FDTD also can only handle a small computational volume relative to the wavelength of the incident light, usually only on the order of a few wavelengths. As the volume of the computation space increases, the time required to find a steady state solution scales proportionally. As the grid size decreases, the time for a solution scales as the inverse fourth power of the grid dimension.

There are two main approaches to FDTD. The most common and simplest is the total field approach. In this technique, the incident magnetic and electric fields are specified on the borders of the computation space (often just one face of the computation space), and the discretized time difference equations of Maxwell are employed to calculate the resulting electric and magnetic fields at each time step sequentially as the incident field propagates through the computation space. Although this technique is mathematically the simplest, it suffers from some important disadvantages that make it unsuitable for some of the complex problems that the author has encountered. In particular, it is difficult or impossible to model the interaction of a tightly focused laser beam on a particle of arbitrary shape. The total field that is calculated includes both the incident field, which may be very intense, and the scattered field, which may be very small. The FDTD algorithm must then employ sophisticated procedures to suppress artificial reflected fields from the boundaries of the computation space.

An alternative technique is that of scattered field FDTD. In this case, the incident electric field is calculated analytically throughout the computation space and only the scattered field is calculated by FDTD at each time step. The boundary condition algorithm is usually much simpler because it only needs to prevent scattering of the typically much smaller scattered field. Another advantage is that the incident field can be calculated analytically throughout the computation space to greater precision than it would be calculated numerically in the total field FDTD technique. This is particularly true for an incident plane wave, but is also the case for a focused field or a waveguide field, for instance. The scattered field FDTD algorithm only calculates the scattered field at each time step. The scattered field FDTD algorithm described in this chapter follows the development of [1].

It will be noted that many of the examples in this chapter are of spherical particles of various materials embedded in various dielectrics. The advantage of modeling spherical particles is that analytical solutions from Mie theory (Chapter 8) can be calculated for comparison to the FDTD results. In the author's experience, when first learning to use the FDTD technique, it is extremely useful to know the correct answer to an example problem so that one can experiment with different cell sizes, computation space volumes, number of time steps, and even size of the time step until one gets a feel for the proper values for these FDTD parameters for different types of materials and scattering object shapes. The other example for which an analytical solution can be obtained is the infinite diffraction grating, which is also considered in this chapter.

DOI: 10.1201/9781003714231-9

9.2 MAXWELL'S EQUATIONS DISCRETIZED

Maxwell's curl equations are

$$\frac{\partial \boldsymbol{H}^t}{\partial t} = -\frac{1}{\mu_0}\left(\nabla \times \boldsymbol{E}^t\right) \tag{9.1}$$

and

$$\varepsilon_s \frac{\partial \boldsymbol{E}^t}{\partial t} = -\sigma_s \boldsymbol{E}^t + \nabla \times \boldsymbol{H}^t. \tag{9.2}$$

The permeability of all materials at optical frequencies is assumed to be identical to that of free space, μ_0. The permittivity of the computation space is a function of position which may be different from free space in the regions where there are scattering objects and is denoted by ε_s. The superscript t in these equations indicates that the fields are the total (incident + scattered) fields within the computation space.

The total field is separated into incident and scattered field components.

$$\boldsymbol{E}^t = \boldsymbol{E}^i + \boldsymbol{E}^s. \tag{9.3}$$

$$\boldsymbol{H}^t = \boldsymbol{H}^i + \boldsymbol{H}^s. \tag{9.4}$$

The incident fields within the computation space in the absence of the scattering object satisfy

$$\frac{\partial \boldsymbol{H}^i}{\partial t} = -\frac{1}{\mu_0}\left(\nabla \times \boldsymbol{E}^i\right) \tag{9.5}$$

and

$$\varepsilon_m \frac{\partial \boldsymbol{E}^i}{\partial t} = \nabla \times \boldsymbol{H}^i. \tag{9.6}$$

ε_m is the permittivity of the computation space surrounding the scattering object. It is assumed that this medium is non-absorbing so that the $\sigma_s \boldsymbol{E}^i$ term in (9.2) is absent in the computation space for the incident field. Substituting (9.3) and (9.4) into (9.1) and (9.2) gives

$$\nabla \times \left(\boldsymbol{E}^i + \boldsymbol{E}^s\right) = -\mu_0 \frac{\partial\left(\boldsymbol{H}^i + \boldsymbol{H}^s\right)}{\partial t} \tag{9.7}$$

and

$$\nabla \times \left(\boldsymbol{H}^i + \boldsymbol{H}^s\right) = \varepsilon_s \frac{\partial\left(\boldsymbol{E}^i + \boldsymbol{E}^s\right)}{\partial t} + \sigma_s\left(\boldsymbol{E}^i + \boldsymbol{E}^s\right). \tag{9.8}$$

The incident fields defined by (9.5) and (9.6) are subtracted from (9.7) and (9.8).

$$\nabla \times \boldsymbol{E}^s = -\mu_0 \frac{\partial \boldsymbol{H}^s}{\partial t} \tag{9.9}$$

and

$$\nabla \times \boldsymbol{H}^s = \varepsilon_s \frac{\partial \boldsymbol{E}^s}{\partial t} + \sigma_s \boldsymbol{E}^s + (\varepsilon_s - \varepsilon_m)\frac{\partial \boldsymbol{E}^i}{\partial t} + \sigma_s \boldsymbol{E}^i. \tag{9.10}$$

Outside of the scattering object within the computation space, (9.8) reduces to

$$\nabla \times \boldsymbol{H}^s = \varepsilon_m \frac{\partial \boldsymbol{E}^s}{\partial t}. \tag{9.11}$$

(9.9) and (9.11) are the necessary equations for the FDTD algorithm. These are solved for the time-dependent term.

$$\frac{\partial \boldsymbol{H}^s}{\partial t} = -\frac{1}{\mu_0} \nabla \times \boldsymbol{E}^s \tag{9.12}$$

and

$$\frac{\partial \boldsymbol{E}^s}{\partial t} = -\frac{\sigma_s}{\varepsilon_s}\boldsymbol{E}^s - \frac{\sigma_s}{\varepsilon_s}\boldsymbol{E}^i - \left(1 - \frac{\varepsilon_m}{\varepsilon_s}\right)\frac{\partial \boldsymbol{E}^i}{\partial t} + \frac{1}{\varepsilon_s}\nabla \times \boldsymbol{H}^s. \tag{9.13}$$

9.3 FINITE DIFFERENCE EQUATIONS FOR A PERFECT CONDUCTOR

When the scattering object is a perfect conductor ($\sigma_s \rightarrow \infty$), the difference equations are simplified. It should be stressed that at optical frequencies there are no perfect conductors and the perfect conductor approximation should only be used when testing the software for bugs. At microwave and lower frequencies, the perfect conductor approximation may be suitable. In this present formulation, the terms for lossy magnetic materials ($\mu \neq \mu_0$) have not been included which may also be necessary at much lower frequencies.

(9.13) can be rewritten,

$$\frac{\varepsilon_s}{\sigma_s}\frac{\partial \boldsymbol{E}^s}{\partial t} = -\boldsymbol{E}^s - \boldsymbol{E}^i - \left(\frac{\varepsilon_s - \varepsilon_m}{\sigma_s}\right)\frac{\partial \boldsymbol{E}^i}{\partial t} + \frac{1}{\sigma_s}\nabla \times \boldsymbol{H}^s \tag{9.14}$$

which in the limit $\sigma_s \rightarrow \infty$ reduces to just

$$\boldsymbol{E}^s = -\boldsymbol{E}^i. \tag{9.15}$$

In other words, inside a perfect conductor the total field is zero as expected.

In FDTD, the partial derivatives in (9.12) and (9.13) are replaced by differences. In particular,

$$\frac{\partial F}{\partial t} \cong \frac{F(t + \Delta t) - F(t)}{\Delta t} \tag{9.16}$$

and

$$\frac{\partial F}{\partial x} \cong \frac{F(x + \Delta x) - F(x)}{\Delta x}. \tag{9.17}$$

The vector curl equations (9.12), (9.13), and (9.14) are separated into scalar equations. Within the perfect conductor,

$$E_x^s = -E_x^i, \tag{9.18}$$

$$E_y^s = -E_y^i, \tag{9.19}$$

and

$$E_z^s = -E_z^i. \tag{9.20}$$

Outside of the perfect conductor,

$$\frac{\partial E_x^s}{\partial t} = \frac{1}{\varepsilon_m}\left(\frac{\partial H_z^s}{\partial y} - \frac{\partial H_y^s}{\partial z}\right), \tag{9.21}$$

$$\frac{\partial E_y^s}{\partial t} = \frac{1}{\varepsilon_m}\left(\frac{\partial H_x^s}{\partial z} - \frac{\partial H_z^s}{\partial x}\right), \tag{9.22}$$

and

$$\frac{\partial E_x^s}{\partial t} = \frac{1}{\varepsilon_m}\left(\frac{\partial H_y^s}{\partial x} - \frac{\partial H_x^s}{\partial y}\right). \tag{9.23}$$

Everywhere within the computation space,

$$\frac{\partial H_x^s}{\partial t} = -\frac{1}{\mu_0}\left(\frac{\partial E_z^s}{\partial y} - \frac{\partial E_y^s}{\partial z}\right), \tag{9.24}$$

$$\frac{\partial H_y^s}{\partial t} = -\frac{1}{\mu_0}\left(\frac{\partial E_x^s}{\partial z} - \frac{\partial E_z^s}{\partial x}\right), \tag{9.25}$$

and

$$\frac{\partial H_z^s}{\partial t} = -\frac{1}{\mu_0}\left(\frac{\partial E_y^s}{\partial x} - \frac{\partial E_x^s}{\partial y}\right). \tag{9.26}$$

The derivatives are replaced by differences. The E_x^s component outside the perfect conductor is given by

$$\frac{E_x^{s,n} - E_x^{s,n-1}}{\Delta t} = \frac{1}{\varepsilon_m}\left(\frac{\Delta H_z^{s,n-\frac{1}{2}}}{\Delta y} - \frac{\Delta H_y^{s,n-\frac{1}{2}}}{\Delta z}\right) \tag{9.27}$$

and the H_y^s component everywhere within the computation space is given by

$$\frac{H_y^{s,n+\frac{1}{2}} - H_y^{s,n-\frac{1}{2}}}{\Delta t} = \frac{1}{\mu_0}\left(\frac{\Delta E_z^{s,n}}{\Delta x} - \frac{\Delta E_z^{s,n}}{\Delta z}\right). \tag{9.28}$$

The other Cartesian field components are found similarly. The n superscripts represent time steps. Note that the time steps are interleaved. The electric field components are calculated on integral time steps and the magnetic field components are calculated at half integral time steps. This contributes to the stability of the calculation.

9.4 FINITE DIFFERENCE EQUATIONS FOR ARBITRARY MATERIALS INCLUDING ABSORPTION

The same procedure is followed to generate the equations for general materials. Returning to (9.13),

$$\frac{E_x^{s,n}-E_x^{s,n-1}}{\Delta t}=-\frac{\sigma_s}{\varepsilon_s}E_x^{s,n}-\frac{\sigma_s}{\varepsilon_s}E_x^{i,n}-\left(1-\frac{\varepsilon_m}{\varepsilon_s}\right)\frac{E_x^{i,n}-E_x^{i,n-1}}{\Delta t}+\frac{1}{\varepsilon_s}\left(\frac{\Delta H_z^{s,n-\frac{1}{2}}}{\Delta y}-\frac{\Delta H_y^{s,n-\frac{1}{2}}}{\Delta z}\right). \tag{9.29}$$

Rearranging,

$$\left(\varepsilon_s+\sigma_s\Delta t\right)E_x^{s,n}=\varepsilon_s E_x^{s,n-1}-\sigma_s\Delta t\,E_x^{i,n}-\left(\varepsilon_s-\varepsilon_m\right)\Delta t\,\dot{E}_x^{i,n}+\Delta t\left(\frac{\Delta H_z^{s,n-\frac{1}{2}}}{\Delta y}-\frac{\Delta H_y^{s,n-\frac{1}{2}}}{\Delta z}\right) \tag{9.30}$$

or

$$E_x^{s,n}=\frac{\varepsilon_s}{\left(\varepsilon_s+\sigma_s\Delta t\right)}E_x^{s,n-1}-\frac{\sigma_s\Delta t}{\left(\varepsilon_s+\varepsilon_s\Delta t\right)}E_x^{i,n}-\frac{\left(\varepsilon_s-\varepsilon_m\right)\Delta t}{\left(\varepsilon_s+\sigma_s\Delta t\right)}\dot{E}_x^{i,n}$$
$$+\frac{\Delta t}{\left(\varepsilon_s+\sigma_s\Delta t\right)}\left(\frac{\Delta H_z^{s,n-\frac{1}{2}}}{\Delta y}-\frac{\Delta H_y^{s,n-\frac{1}{2}}}{\Delta z}\right). \tag{9.31}$$

The derivative $\dot{E}_x^{i,n}$ is kept in the update equation for the incident field since it is calculated separately analytically. (9.28) is still the appropriate update equation for the magnetic field. The update equations for each component of the electric and magnetic fields are independent. Therefore, the programming can be easily parallelized at this point to gain a factor of two or more in calculation speed.

9.5 YEE CELL

As previously mentioned, electric and magnetic fields are calculated on alternating time steps. The fields are also calculated at points offset from each other in the Yee cell, [2] which makes the finite difference derivative calculation straightforward. The Yee cell is shown in Figure 9.1. The incident field is calculated analytically at the appropriate offset locations of each field component within the cell.

Parameters like the Poynting vector and dissipated power require knowing all field components at *the same point*, so after completing the FDTD calculation it is necessary to interpolate the magnetic and electric fields to a point within each cell. Usually, the center of the Yee cell is chosen for the interpolation as shown in Figure 9.2 using the following equations.

$$E_{x,c}\left(i,j,k\right)=\left[E_{x,s}\left(i,j,k\right)+E_{x,s}\left(i,j+1,k\right)+E_{x,s}\left(i,j,k+1\right)+E_{x,s}\left(i,j+1,k+1\right)\right]/4. \tag{9.32}$$

$$E_{y,c}\left(i,j,k\right)=\left[E_{y,s}\left(i,j,k\right)+E_{y,s}\left(i+1,j,k\right)+E_{y,s}\left(i,j,k+1\right)+E_{y,s}\left(i+1,j,k+1\right)\right]/4. \tag{9.33}$$

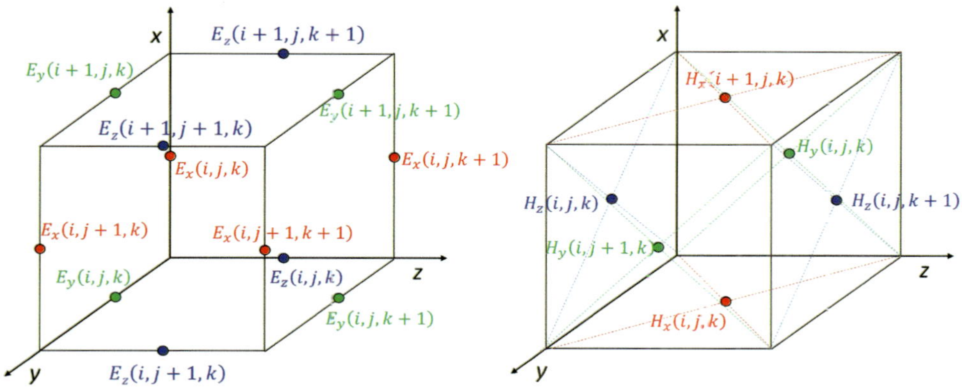

(a) (b)

FIGURE 9.1 (a) Locations of electric field components in the Yee cell in the centers of the edges of the cell. (b) Locations of magnetic field components in the Yee cell in the centers of the faces of the cell. Each cell is labelled by the Cartesian integers (*i*, *j*, *k*). Note that the Yee cell does not need to be a cube but can also be a rectangular parallelepiped.

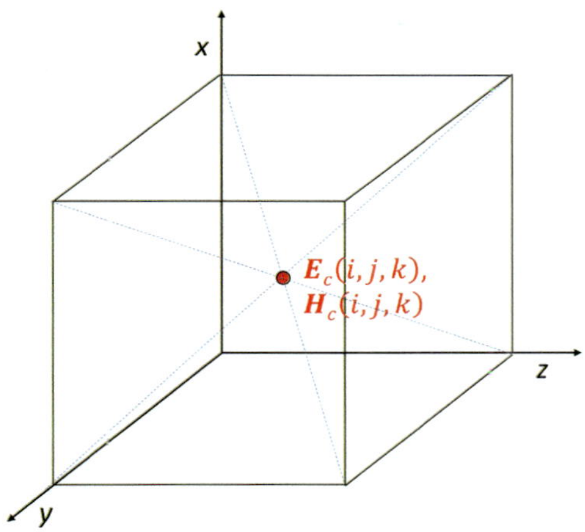

FIGURE 9.2 ***E*** and ***H*** fields at center of the Yee cell.

$$E_{z,c}(i,j,k)=\left[E_{z,s}(i,j,k)+E_{z,s}(i+1,j,k)+E_{z,s}(i,j+1,k)+E_{z,s}(i+1,j+1,k)\right]/4. \quad (9.34)$$

$$H_{x,c}(i,j,k)=\left[H_{x,s}(i,j,k)+H_{x,s}(i+1,j,k)\right]/2 \quad (9.35)$$

$$H_{y,c}(i,j,k)=\left[H_{y,s}(i,j,k)+H_{y,s}(i,j+1,k)\right]/2. \quad (9.36)$$

$$H_{z,c}(i,j,k)=\left[H_{z,s}(i,j,k)+H_{z,s}(i,j,k+1)\right]/2. \quad (9.37)$$

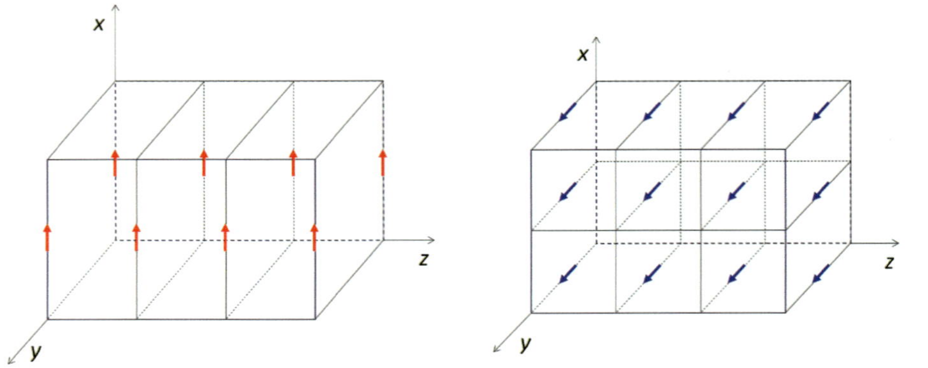

(a) (b)

FIGURE 9.3 (a) Computation space in which there are three complete Yee cells along the z-axis. The locations of the x-components of the electric field are shown in red. (b) Computation space in which there are three complete Yee cells along the z-axis. The locations of the y-components of the electric field are shown in blue.

It is also essential to compute the field quantities at the same point in *time*. The magnetic field in the FDTD algorithm is computed at half a time step between the electric field calculation. So magnetic field quantities can be averaged for two time steps, one a half time step on either side of the electric field calculation.

The effect of this averaging of the fields to the center of the cell should be considered on the size of the computation space. Consider the computation space consisting of three complete Yee cells along the z-axis as shown in Figure 9.3(a). There are E_x components from four Yee cells along the z-axis which are averaged to the centers of three Yee cells. There are also E_x components from two Yee cells along the y-axis shown which are averaged to the center of one Yee cell. In Figure 9.3(b) the E_y field components are shown in blue for a 3×2 computation space. Again, when averaging the field components to the center of each Yee cell, three Yee cells along the x-axis provide an average E_y field component in the centers of two Yee cells, and four Yee cells along the z-axis provide an average E_y field component in the centers of three Yee cells. In general, when calculations require averaging of fields to the centers of the Yee cells, the resulting average fields will represent one fewer cell layer on x, y, and z faces of the FDTD calculation. If the cell space consists of $30 \times 40 \times 50$ cells, then the averaged fields will be calculated in the centers of a $29 \times 39 \times 49$ cell space. In this example, if the center of a symmetric scattering object (a sphere) is located at the center of cell (15, 20, 25), then the average field at the center of this cell should represent the field components at the center of the object, and the FDTD field calculation should also give symmetric results around the object.

9.6 CELL SIZE AND TIME STEP

For accuracy, the largest dimension of the cell size must be much smaller than a wavelength. The appropriate wavelength to consider is the wavelength within the densest medium, that is, the shortest wavelength within the computation space. If a material has a refractive index of 2, for example, then the wavelength within that material is only half that of free space and the cell dimension must be made correspondingly smaller. For scattering calculations that involve only dielectrics and perfect conductors, a maximum cell size dimension of a tenth of a wavelength is often satisfactory, though smaller cell sizes may be needed for absorptive and metallic materials. For example, for calculations performed at 830 nm with gold scattering objects, a maximum cell size of 2.5 nm was sometimes required for accurate results, which is much, much less than the free space wavelength. This is a strong constraint on the maximum size of the computation space and associated scattering objects. A typical calculation may have a cell space composed of $250 \times 250 \times 250 = 15{,}625{,}000$ cells and may take hours to compute on a modern PC.

As previously mentioned, the electric and magnetic fields are offset from each other by half the cell dimensions for easier calculation of the derivatives and calculation stability. The maximum cell dimension is much smaller than the wavelength, and electric and magnetic fields are calculated in alternate time steps. There is one other parameter, the size of the time step, which is crucial in obtaining a stable and accurate result. The Courant condition for stability is that a plane wave propagating through a cell must not pass through more than one cell in one time step. For a rectangular parallelepiped Yee cell with edge lengths Δx, Δy, and Δz, the Courant condition is

$$\Delta t \leq \frac{1}{\mathrm{v}\sqrt{\dfrac{1}{\Delta x^2 + \Delta y^2 + \Delta z^2}}} \tag{9.38}$$

where v is the speed of light in the medium. If the computation space is embedded in a transparent dielectric, v is the speed of light divided by the refractive index of the medium. In actual practice, when there is a metallic object in the computation space, the calculation becomes unstable for the equality in (9.38). In such a case, usually a time step that is ~90% of the equality works well.

For a steady state calculation, the number of time steps must be sufficient to allow high frequency transients in the calculation to die out. This generally requires at least five complete periods of the calculation wavelength, though it is wise to examine the results of a calculation as the number of time steps is increased until there is no significant change in the result.

9.7 BOUNDARY CONDITIONS

Obviously, the computation space for FDTD is finite. In the scattered field FDTD formulation, the incident field is computed exactly for each cell of the computation space, so the boundaries are not an issue. This is an advantage over the total field FDTD technique. However, there is the issue of what happens when the scattered waves reach the boundaries. If the scattered waves are not handled carefully, they can be artificially reflected back into the computation space, thereby invalidating the computation. The scattered field reflections at the boundaries are generally much smaller than the incident field amplitude. The boundary conditions, therefore, generally do not need to be as efficient in the scattered field FDTD approach as in the total field FDTD approach to produce accurate results.

A variety of techniques for handling the boundaries have been invented. Perhaps the most popular is the so-called "perfectly matched layer" or PML boundary condition invented by Berenger [3, 4]. Although this boundary condition is excellent, it is difficult to program and must be adjusted properly for the materials at the boundary. Moreover, the update equations are complex and time-consuming.

9.7.1 Huygens Boundary Conditions

A very simple boundary that is generally not good enough to be used by itself is the Huygens boundary condition [5]. The basic idea is that at the edge of the computation space, if the outermost field is an electric field, then in order to update the electric field the algorithm requires a value for the magnetic field that is half a cell outside of the computation space. In this boundary condition, a value for this missing magnetic field is estimated from the value of the magnetic field one half cell inside of the electric field position by essentially assuming a plane wave propagating at normal incidence. In particular, on the xy faces,

$$E_x^{(s)}(i,j,1) = -\sqrt{\frac{\mu_0}{\epsilon_r \epsilon_0 + \sigma \Delta t}} H_y^{(s)}(i,j,1), \tag{9.39}$$

$$E_x^{(s)}(i,j,n_z)=+\sqrt{\frac{\mu_0}{\epsilon_r\epsilon_0+\sigma\Delta t}}H_y^{(s)}(i,j,n_z), \tag{9.40}$$

$$E_y^{(s)}(i,j,1)=+\sqrt{\frac{\mu_0}{\epsilon_r\epsilon_0+\sigma\Delta t}}H_x^{(s)}(i,j,1), \tag{9.41}$$

and

$$E_y^{(s)}(i,j,n_z)=-\sqrt{\frac{\mu_0}{\epsilon_r\epsilon_0+\sigma\Delta t}}H_x^{(s)}(i,j,n_z). \tag{9.42}$$

On the *xz* faces,

$$E_x^{(s)}(i,1,k)=+\sqrt{\frac{\mu_0}{\epsilon_r\epsilon_0+\sigma\Delta t}}H_z^{(s)}(i,1,k), \tag{9.43}$$

$$E_x^{(s)}(i,n_y,k)=-\sqrt{\frac{\mu_0}{\epsilon_r\epsilon_0+\sigma\Delta t}}H_z^{(s)}(i,n_y-1,k), \tag{9.44}$$

$$E_z^{(s)}(i,1,k)=-\sqrt{\frac{\mu_0}{\epsilon_r\epsilon_0+\sigma\Delta t}}H_x^{(s)}(i,1,k), \tag{9.45}$$

and

$$E_z^{(s)}(i,n_y,k)=+\sqrt{\frac{\mu_0}{\epsilon_r\epsilon_0+\sigma\Delta t}}H_x^{(s)}(i,n_y-1,k). \tag{9.46}$$

On the *yz* faces,

$$E_y^{(s)}(1,j,k)=-\sqrt{\frac{\mu_0}{\epsilon_r\epsilon_0+\sigma\Delta t}}H_z^{(s)}(1,j,k), \tag{9.47}$$

$$E_y^{(s)}(n_x,j,k)=+\sqrt{\frac{\mu_0}{\epsilon_r\epsilon_0+\sigma\Delta t}}H_z^{(s)}(n_x-1,j,k), \tag{9.48}$$

$$E_z^{(s)}(1,j,k)=+\sqrt{\frac{\mu_0}{\epsilon_r\epsilon_0+\sigma\Delta t}}H_x^{(s)}(1,j,k), \tag{9.49}$$

and

$$E_z^{(s)}(n_x,j,k)=-\sqrt{\frac{\mu_0}{\epsilon_r\epsilon_0+\sigma\Delta t}}H_x^{(s)}(n_x-1,j,k). \tag{9.50}$$

In these equations, *i*, *j*, and *k* are indices that run across the entire faces of the computation space.

9.7.2 Mur Boundary Conditions

Mur boundary conditions, a form of absorbing boundary condition, are described in [1] for scattered field FDTD [6]. Although this is an improvement on the simple Huygens boundary conditions, for the most part Mur boundary conditions are no longer used because better techniques have since been developed. The first order Mur update to the scattered field is applied to edges of the computation space and is given, for example, for the E_z component on an $x=0$ edge by

$$E_z^{(n+1)}(1,j,k)=E_z^{(n)}(2,j,k)+\left(\frac{c\Delta t-\Delta x}{c\Delta t+\Delta x}\right)\left\{E_z^{(n+1)}(2,j,k)-E_z^{(n)}(1,j,k)\right\}. \tag{9.51}$$

The first order Mur boundary condition uses the field from one time step earlier at one spatial cell inward from the boundary. On faces other than edges of the computation space, a second order Mur update can be applied which is based on the two previous time steps and two inward cell locations. For example, on the $x=0$ face (aside from the edges),

$$\begin{aligned}E_z^{(n+1)}(1,j,k)=&-E_z^{(n-1)}(2,j,k)+\left(\frac{c\Delta t-\Delta x}{c\Delta t+\Delta x}\right)\left\{E_z^{(n+1)}(2,j,k)+E_z^{(n-1)}(1,j,k)\right\}\\&+\left(\frac{2\Delta x}{c\Delta t+\Delta x}\right)\left\{E_z^{(n)}(2,j,k)+E_z^{(n)}(1,j,k)\right\}\\&+\left(\frac{\Delta x(c\Delta t)^2}{2(\Delta y)^2(c\Delta t+\Delta x)}\right)\left\{E_z^{(n)}(1,j+1,k)-2E_z^{(n)}(1,j,k)+E_z^{(n)}(1,j-1,k)\right.\\&\left.+E_z^{(n)}(2,j+1,k)-2E_z^{(n)}(2,j,k)+E_z^{(n)}(2,j-1,k)\right\}\\&+\frac{\Delta x(c\Delta t)^2}{2(\Delta z)^2(c\Delta t+\Delta x)}\left\{E_z^{(n)}(1,j,k+1)-2E_z^{(n)}(1,j,k)+E_z^{(n)}(1,j,k-1)\right.\\&\left.+E_z^{(n)}(2,j,k+1)-2E_z^{(n)}(2,j,k)+E_z^{(n)}(2,j,k-1)\right\}.\end{aligned} \tag{9.52}$$

9.7.3 Reradiating Boundary Conditions

Reradiating boundary conditions (rRBCs) were first proposed by Díaz and Scherbatko [5]. The idea behind the reradiating boundary condition is simple. The field amplitude of the wave propagating towards the boundary is measured several cells in front of the boundary and then reintroduced with opposite sign one time step later and one cell space closer to the boundary to cancel the propagating wave. Multiple rRBCs may be stacked before the boundary to provide even greater reflected field cancellation. A disadvantage of rRBCs is that they do "use up" three cells around the edges of the computation space, or eight cells with a dual rRBC. At the faces in the computation space, Huygens boundary conditions are also applied.

In order to "teleport" a field to a new location, the surface current of the electric field calculated from one cell must be added into the update equation for the new cell. First, the tangential magnetic fields on the six faces indented by a few cells are stored. For example, on the inner and outer xy faces of the computation space,

$$H_{x,in}^{(1)}(i,j)=H_{x,in}^{(s)}(i,j,k_{\text{off}}), \tag{9.53}$$

$$H_{y,in}^{(1)}(i,j)=H_{y,in}^{(s)}(i,j,k_{\text{off}}), \tag{9.54}$$

$$H_{x,out}^{(1)}(i,j)=H_{x,out}^{(s)}(i,j,N_z-k_{\text{off}}), \tag{9.55}$$

and

$$H_{y,out}^{(1)}(i,j)=H_{y,out}^{(s)}(i,j,N_z-k_{\text{off}}). \tag{9.56}$$

$k_{\text{off}} = 4$ for the rRBC closest to the face and is increased by 5 Yee cells for each successive inner reradiating layer in the FDTD program developed to perform the calculations in this chapter.

For stability of the calculation, it is necessary to build in a one time step delay for updating the electric fields. This is easily done by copying the fields in (9.53)–(9.56) into new arrays with superscript (2) at the next time step. After copying the fields into new arrays, the new calculated tangential fields at this time step are copied into the arrays with superscript (1). The arrays with superscript (2) are then applied to update the calculated scattered fields. For example, on the *xy* faces,

$$E_x^{(s)}(i,j,k_{\text{off}}) = E_x^{(s)}(i,j,k_{\text{off}}) + \left(\frac{0.99\,\Delta t}{\epsilon_m \epsilon_0 \Delta z}\right) H_{y,in}^{(2)}(i,j,\text{bin } 2), \tag{9.57}$$

$$E_y^{(s)}(i,j,k_{\text{off}}) = E_y^{(s)}(i,j,k_{\text{off}}) - \left(\frac{0.99\,\Delta t}{\epsilon_m \epsilon_0 \Delta z}\right) H_{x,in}^{(2)}(i,j,\text{bin } 2), \tag{9.58}$$

$$E_x^{(s)}(i,j,N_z - k_{\text{off}} + 1) = E_x^{(s)}(i,j,N_z - k_{\text{off}} + 1) - \left(\frac{0.99\,\Delta t}{\epsilon_m \epsilon_0 \Delta z}\right) H_{y,out}^{(2)}(i,j,\text{bin } 2), \tag{9.59}$$

and

$$E_y^{(s)}(i,j,N_z - k_{\text{off}} + 1) = E_y^{(s)}(i,j,N_z - k_{\text{off}} + 1) + \left(\frac{0.99\,\Delta t}{\epsilon_m \epsilon_0 \Delta z}\right) H_{x,out}^{(2)}(i,j,\text{bin } 2). \tag{9.60}$$

As before, the i, j indices range over then entire *xy* faces. The factor of 0.99 in (9.57)–(9.60) is inserted for stability of the algorithm to attenuate any back-reflected fields. Similarly, the corresponding magnetic field re-radiating boundary conditions on the *xy* faces are:

$$H_x^{(s)}(i,j,k_{\text{off}} - 1) = H_x^{(s)}(i,j,k_{\text{off}} - 1) - \left(\frac{0.99\,\Delta t}{\mu_0 \Delta z}\right) E_{y,in}^{(2)}(i,j,\text{bin } 2), \tag{9.61}$$

$$H_y^{(s)}(i,j,k_{\text{off}} - 1) = H_y^{(s)}(i,j,k_{\text{off}} - 1) + \left(\frac{0.99\,\Delta t}{\mu_0 \Delta z}\right) E_{x,in}^{(2)}(i,j,\text{bin } 2), \tag{9.62}$$

$$H_x^{(s)}(i,j,N_z - k_{\text{off}} + 1) = H_x^{(s)}(i,j,N_z - k_{\text{off}} + 1) + \left(\frac{0.99\,\Delta t}{\mu_0 \Delta z}\right) E_{y,out}^{(2)}(i,j,\text{bin } 2), \tag{9.63}$$

and

$$H_y^{(s)}(i,j,N_z - k_{\text{off}} + 1) = H_y^{(s)}(i,j,N_z - k_{\text{off}} + 1) - \left(\frac{0.99\,\Delta t}{\mu_0 \Delta z}\right) E_{x,out}^{(2)}(i,j,\text{bin } 2). \tag{9.64}$$

Figure 9.4(a) gives a visual representation of which magnetic fields are used to update which electric fields in a two-dimensional FDTD space. Figure 9.4(b) gives a visual representation of which electric fields are used to update which magnetic fields. The proper sequence of operations is as follows:

1) Update the three components of the electric field using the standard FDTD equations.
2) Shift the stored tangential magnetic fields on the six surfaces from bin 1 to bin 2.
3) Update the tangential magnetic fields in bin 1 with the current magnetic field values.
4) Apply the re-radiating boundary conditions to update the electric fields from the tangential magnetic fields in bin 2 using (9.53)–(9.56).

5) Apply the Huygens boundary conditions to all faces containing electric fields.
6) Update the three components of the magnetic field using the standard FDTD equations.
7) Shift the stored tangential electric fields on the six surfaces from bin 1 to bin 2.
8) Update the tangential electric fields in bin 1 with the current electric field values.
9) Apply the re-radiating boundary conditions to update the magnetic fields from the tangential electric fields in bin 2.
10) Apply the Huygens boundary conditions to all faces containing magnetic fields.

As shown in Figure 9.4(a) the computation space can be designed so the outer faces are all electric field faces in which case step (10), applying the Huygens boundary condition after updating the magnetic fields, is eliminated.

Re-radiating boundary conditions are also stackable. By placing multiple layers of re-radiating boundaries, reflections from the outer surface can be further reduced at the expense of using more of the computational cell space for the boundary conditions. If reradiating boundary conditions are stacked, there should be at least a three-cell spacing between them [7].

It is interesting to compare the various boundary conditions with a specific example. The scattered fields from a 0.2 μm diameter dielectric sphere with a refractive index of 3 in air are calculated at a wavelength of 1 μm and the results along three planes cutting through the center of the sphere are shown in Figure 9.5. The scattered fields are also calculated analytically from Mie theory. It is interesting to note that the Mie calculation is not much faster than FDTD in this example simply because quad precision is required for the Mie calculation. Although all FDTD boundary conditions in this figure appear to be good, the re-radiating + Huygens boundary conditions are closest to the Mie result.

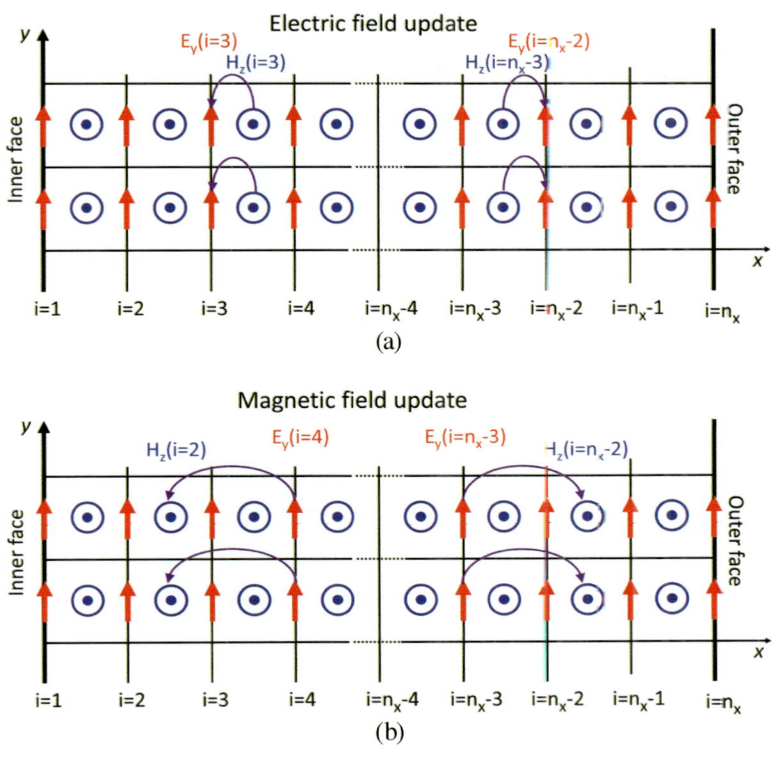

FIGURE 9.4 Illustration of implementation of reradiating boundary conditions. (a)The electric field at cells $i = 3$ and $n_x - 2$ is updated from the stored value of the magnetic field at cells $i = 3$ and $n_x - 3$, respectively. (b) The magnetic field at $i = 2$ and $n_x - 2$ is updated from the stored value of the electric field at cells $i = 4$ and $n_x - 3$, respectively.

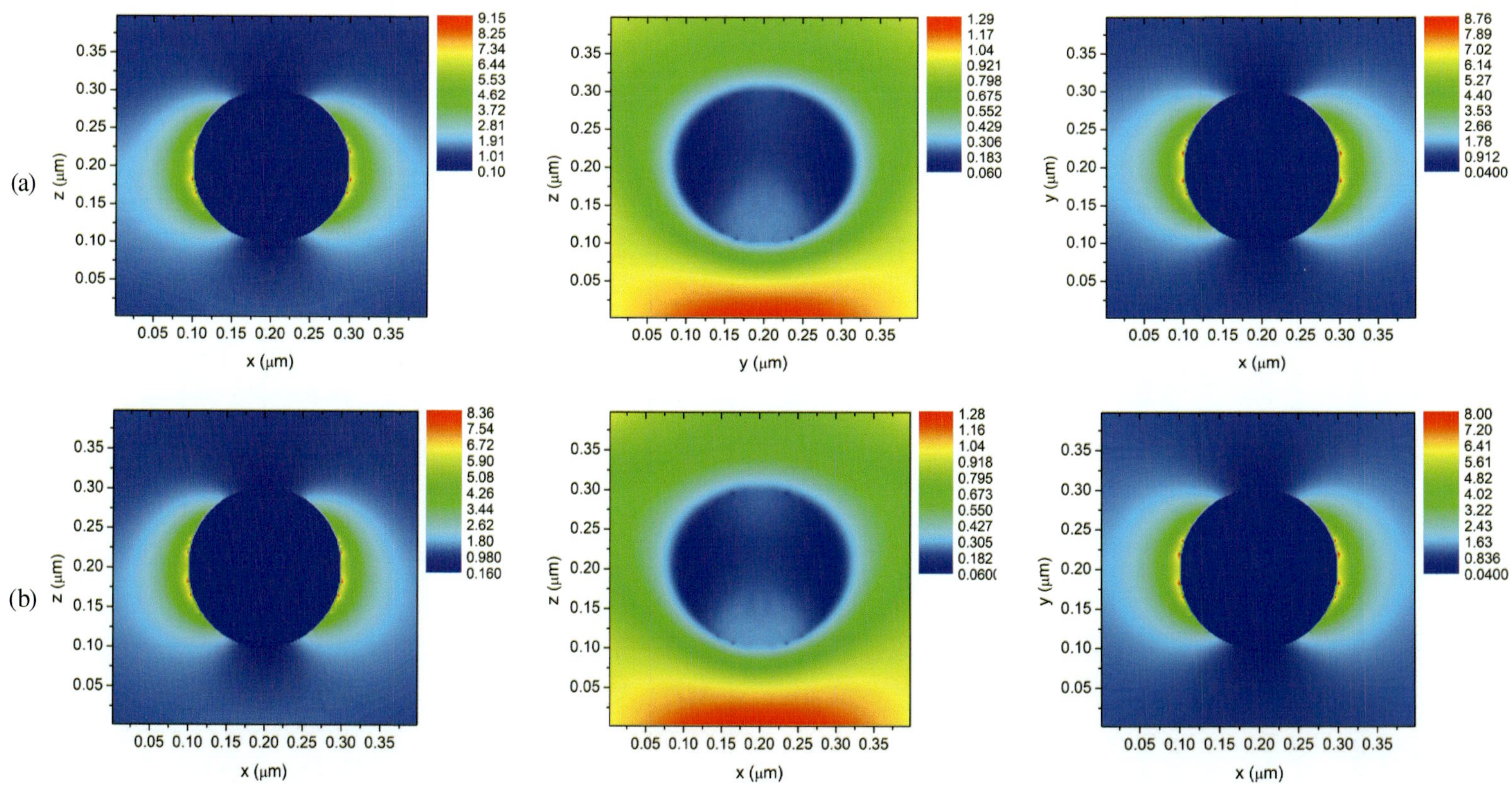

FIGURE 9.5 Total incident + scattered field intensity for a plane wave at 1 μm wavelength from a 0.2 μm diameter sphere with a refractive index of 3. The FDTD simulations involve a computation space with 200 cells on an edge, 0.002 μm cell size, and 5,000 time steps. For each boundary condition, the field intensity is plotted in the *xz*, *yz*, and *xy* planes through the center of the sphere. (a) Huygens boundary conditions alone. (b) Mur boundary conditions. (c) One-layer reradiating and Huygens boundary conditions. (d) Two-layer reradiating and Huygens boundary conditions. (e) Mie theory. Although the field is much smaller in the *yz* plane, the Huygens and Mur boundary conditions place the higher field region on the incident side of the plane wave unlike the reradiating boundary condition and Mie theory.

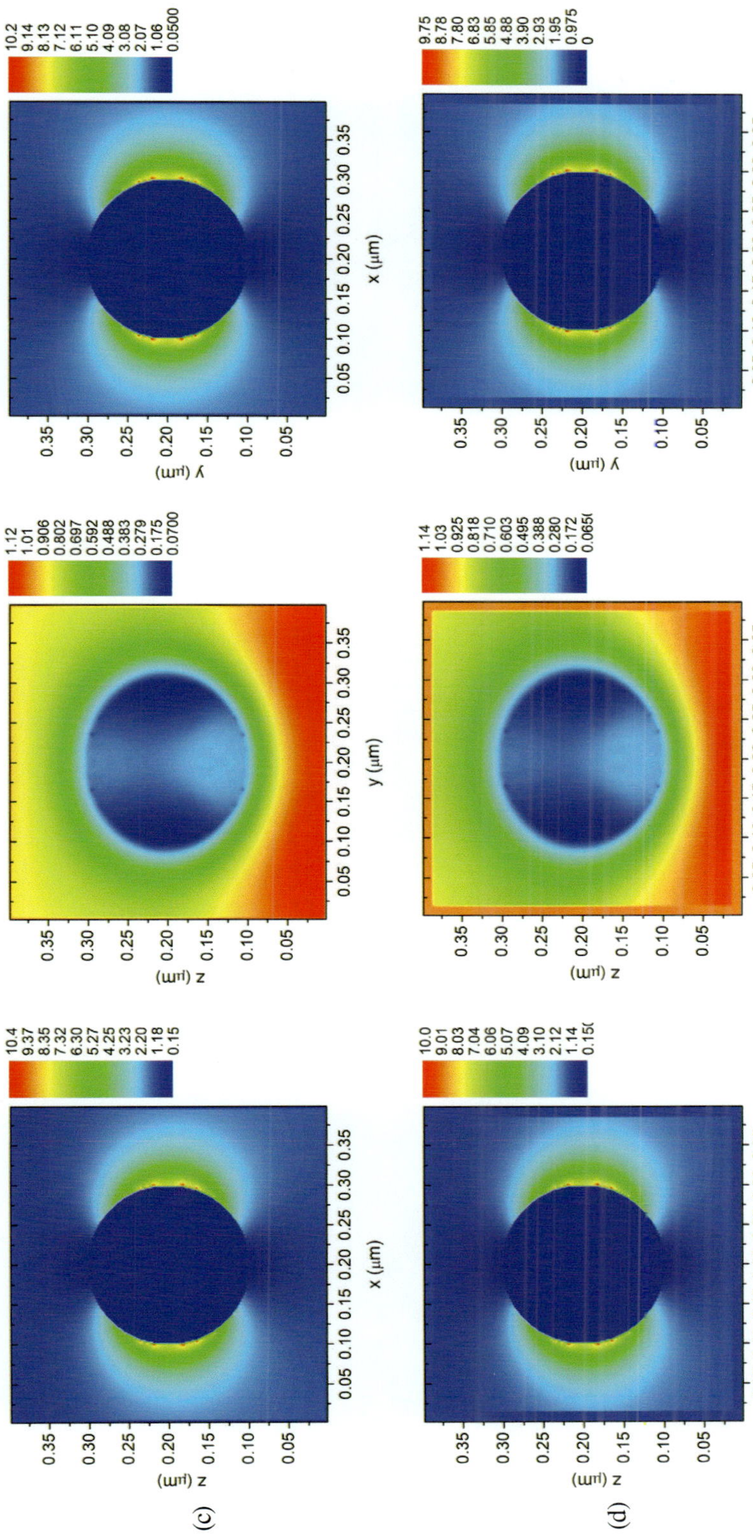

FIGURE 9.5 (Continued)

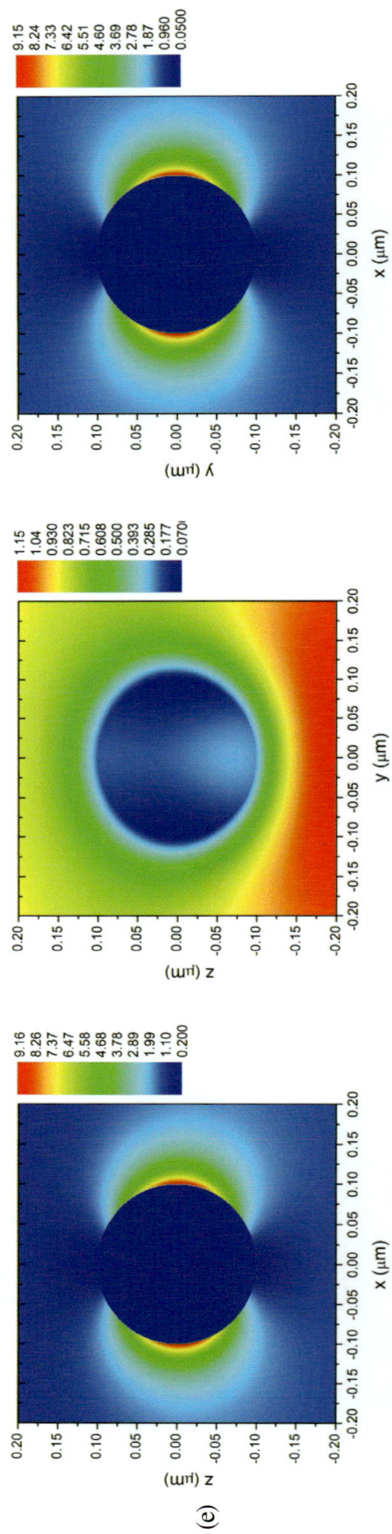

FIGURE 9.5 (Continued)

9.7.4 Periodic Boundary Conditions

Sometimes it is claimed that implementing periodic boundary conditions in FDTD, useful for modeling periodic structures like diffraction gratings, is straightforward. While this may be true, it is not necessarily simple. It is certainly true, however, that developing all the software in complex number format makes it easier to implement periodic boundary conditions for fields that are not normally incident. Obtaining the phase of the scattered field for an incident sinusoidal field requires measurement of the amplitude of the scattered field both at the end of the calculation and at one quarter period earlier. However, when the scattered field is calculated as a complex field amplitude, both amplitude and phase of the scattered field may be directly determined at any time step. Therefore, the scattered field can also be easily phase-shifted by an amount φ for periodic boundary conditions simply by multiplying it by a complex number $e^{i\varphi}$.

The FDTD computation space in [1] has electric fields on all six outer faces. Alternatively, if the computation space is composed of complete Yee cells, then there will be electric fields on three outer faces and magnetic fields on the other three. In this section the procedure for implementing periodic boundary conditions is discussed from the perspective of the former approach while the latter approach is discussed in [8].

Consider a two-dimensional periodic structure upon which a plane wave is incident at an oblique angle as shown in Figure 9.6. Once the calculation reaches a steady-state, the fields within each unit cell of the periodic structure are identical except that they are phase-shifted relative to each other by the time of arrival of the incident wave at each unit cell. Therefore, the calculation can be reduced to calculating just the fields within a single unit cell and then applying the appropriate phase shift, if desired, to visualize the fields in multiple unit cells.

Periodic boundary conditions, therefore, enable one to reduce the calculation of a periodic array of scatterers to a single unit cell. If the wavevector of the incident plane wave lies within the *xz* (or *yz*) plane, then the problem reduces further to two dimensions. The key to applying periodic boundary conditions then is to set the fields at one boundary equal to the fields at the other boundary with an appropriate phase shift after each time step.

Consider the two-dimensional diagram of the FDTD computation space in Figure 9.7 showing the E_y fields in red upward arrows and the H_z fields in blue circles. In this example, the E_y fields are located on the outer surfaces of the computation space and the H_z fields are located on the outer edges of the unit cell. The width of the unit cell is, therefore, two Yee cells shorter than the width of the computation space. The H_z

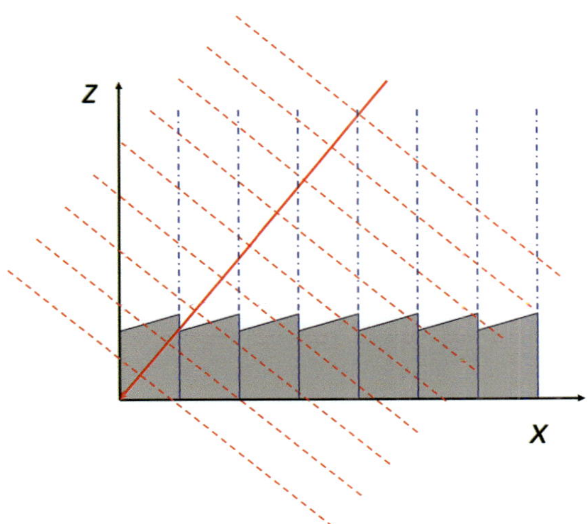

FIGURE 9.6 Periodic structure along the x-axis with a plane wave incident at an oblique angle. The vertical dotted lines divide the computation space into identical unit cells.

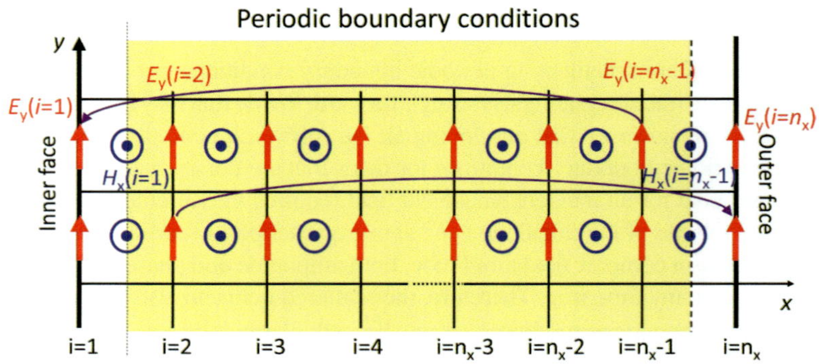

FIGURE 9.7 Technique to implement periodic boundary conditions along the x-axis. After updating the electric field, the second-to-last E_y and E_z values are copied into the first cell, and the second E_y and E_z values are copied into the last cell.

fields at the left edge, $i = 1$, must be equal to the H_z fields at the right edge, $i = n_x - 1$, with the exception of the phase shift caused by the time difference between when an obliquely incident plane wave strikes the two edges of the unit cell. The E_y field at cell $i = n_x - 1$ is used to update the E_y field at cell $i = 1$ with the appropriate phase shift, and the E_y field at cell $i = 2$ is used to update the E_y field at cell $i = n_x$ with the appropriate *negative* phase shift. In this two-dimensional example, the phase shift is

$$\varphi = \left(\frac{2\pi n_m}{\lambda}\right) d_x \sin\theta \tag{9.65}$$

where

$$d_x = (n_x - 2)\Delta x. \tag{9.66}$$

In the three-dimensional case where the plane wave is incident at an angle (θ, ϕ) then the phase shift for the E_y and E_z fields on opposite yz faces is

$$\varphi_x = \left(\frac{2\pi n_m}{\lambda}\right) d_x \sin\theta \cos\phi, \tag{9.67}$$

and for the E_x and E_z fields on opposite xz faces is

$$\varphi_y = \left(\frac{2\pi n_m}{\lambda}\right) d_y \sin\theta \sin\phi \tag{9.68}$$

where

$$d_y = (n_y - 2)\Delta x. \tag{9.69}$$

It should be noted that if a three-dimensional computation space is used to model a two-dimensional example, then the orthogonal dimension can also use periodic boundary conditions with a minimum of four cells along that axis. The standard (Mur, reradiating, or PML) boundary conditions are applied to the top and bottom faces of the computation space as usual. The sequence of operations for each time step is as follows.

1. Update the E_x, E_y, and E_z fields with the standard FDTD equations.
2. Apply the nonperiodic boundary conditions to the top and bottom faces.
3. Apply the periodic boundary conditions to the four side faces.
4. Update the H_x, H_y, and H_z fields with the standard FDTD equations.
5. Apply the nonperiodic boundary conditions to the top and bottom faces.
6. Repeat for the next time step.

The total reflectance is calculated for different angles of incidence in the xz plane from the ratio of the z-component of the Poynting vector of the scattered field summed over the Yee cells in a plane near the top face of the computation space (assuming the incident field comes from the top) divided by the z-component of the Poynting vector of the incident plane wave. The Poynting vector is easily calculated from the complex scattered field amplitudes. Let the average electric and magnetic fields at the center of each Yee cell be

$$E_x^{sct}(i,j,k) \equiv \frac{1}{4}\left[E_x^{sct}(i,j,k+1) + E_x^{sct}(i,j+1,k+1) + E_x^{sct}(i,j,k) + E_x^{sct}(i,j+1,k)\right], \quad (9.70)$$

$$E_y^{sct}(i,j,k) \equiv \frac{1}{4}\left[E_y^{sct}(i,j,k+1) + E_y^{sct}(i+1,j,k+1) + E_y^{sct}(i,j,k) + E_y^{sct}(i+1,j,k)\right], \quad (9.71)$$

$$H_x^{sct}(i,j,k) \equiv \frac{1}{2}\left[H_x^{sct}(i,j,k) + H_x^{sct}(i+1,j,k)\right], \quad (9.72)$$

and

$$H_y^{sct}(i,j,k) \equiv \frac{1}{2}\left[H_y^{sct}(i,j,k) + H_y^{sct}(i,j+1,k)\right]. \quad (9.73)$$

The z-component of the Poynting vector of the scattered field is

$$S_{z,sct} = \frac{1}{2}\mathrm{Re}\left\{\sum_{i=1}^{n_x-2}\sum_{j=1}^{n_y-2}\left[E_x^{sct}(i,j,k_{top})H_y^{*,sct}(i,j,k_{top}) - E_y^{sct}(i,j,k_{top})H_x^{*,sct}(i,j,k_{top})\right]\right\}. \quad (9.74)$$

The z-component of the Poynting vector of the incident field is

$$S_{z,inc} = \frac{1}{2}\sqrt{\frac{\mu_0}{n_m^2\,\epsilon_0}}\cos\theta_i. \quad (9.75)$$

The total reflectance is

$$R_{tot} = \frac{S_{z,sct}}{S_{z,inc}}. \quad (9.76)$$

If there are multiple diffracted orders, R_{tot} is the sum of the reflectances of all reflected propagating orders.

The total transmittance is calculated in a similar manner by integrating the z-component of the Poynting vector across the bottom surface of the unit cell in the computation space. The total absorptance is calculated by a volume summation of the product of the material conductivity with the square of the electric field as will described shortly. The sum of the reflectance, transmittance, and absorptance should be 1. Because the absorptance calculation is independent of the reflectance and transmittance calculations, this sum is a good check of the accuracy of the calculation and that it has reached a steady state.

The reflectance and transmittance of each diffracted order can be calculated using Rayleigh's technique. Rayleigh recognized that the fields outside of the groove region of the grating are quasi-periodic. (Within the groove region, the refractive index varies between the "hill" and "valley" regions, and, hence, the k-vector also varies. Therefore, a simple Fourier expansion between the top and bottom of the grooves is not possible.) In other words, the fields within each unit cell of the grating are identical except for a phase shift that depends on the angle of the incident plane wave.

$$E(x+d,z)=E(x,z)e^{ik_m d\sin\theta_i} \tag{9.77}$$

where d is the period of the grating and θ_i is the angle of incidence. Therefore, the function $E(x,z)e^{-ik_m d\sin\theta_i}$ is periodic with period d and can be expanded in a Fourier series.

$$E(x,z)e^{-ik_m x\sin\theta_i}=\sum_{n=-\infty}^{+\infty}E_n(z)e^{\frac{2\pi inx}{d}}. \tag{9.78}$$

Multiplying each side by $e^{+ik_m x\sin\theta_i}$ gives

$$E(x,z)=e^{+ik_m x\sin\theta_i}\sum_{n=-\infty}^{+\infty}E_n(z)e^{\frac{2\pi inx}{d}}=\sum_{n=-\infty}^{+\infty}E_n(z)e^{i\alpha_n x} \tag{9.79}$$

where

$$\alpha_n\equiv\frac{2\pi n}{d}+k_m\sin\theta_i=k_m\sin\theta_d. \tag{9.80}$$

θ_d is the diffraction angle. The amplitude coefficients of each diffracted order, E_n, are calculated from the inverse Fourier transform.

$$\int_0^d E(x,z_0)e^{-i\alpha_m x}dx=\sum_{n=-\infty}^{+\infty}E_n\int_0^d e^{i\alpha_n x}e^{-i\alpha_m x}dx=\begin{cases}dE_m & m=n\\ 0 & m\neq n\end{cases} \tag{9.81}$$

or

$$E_m=\frac{1}{d}\int_0^d E(x,z_0)e^{-i\alpha_m x}dx. \tag{9.82}$$

$E(x,z_0)$ is the electric field amplitude calculated by FDTD at z-height z_0 outside the grooves of the grating. The electric field amplitude for each vector component of the m'th order is multiplied by the appropriate exponential term $e^{-i\alpha_m x}$ and summed across the $z = z_0$ surface of the unit cell to determine the electric field amplitude of the diffracted order. The power in the incident beam and the diffracted order must be multiplied by the cosine of the angle of incidence or diffraction, respectively, to account for the optical power delivered or reflected in the z-direction. Therefore, the reflectance of order m is

$$R_m=\frac{|E_m|^2\cos\theta_m}{|E_i|^2\cos\theta_i}. \tag{9.83}$$

As an example, consider a silver diffraction grating with a period of 1 μm and a sinusoidal groove profile with a depth of 65 nm and a maximum thickness of 0.31 μm as shown in Figure 9.8. The reflectance of the zero and ±first diffracted orders are plotted in Figure 9.8(b) as calculated by the method of Chandezon described in Chapter 2. A surface plasmon resonance occurs for an infinitely thick grating with p-polarized light at an incident angle of ~19.1° as seen by the narrow, sharp drop in reflectance at this angle of incidence. At the surface plasmon resonance, an evanescent field from the surface plasmon at the top surface of the grating extends a large distance into the region above the grating. This evanescent field then gives spurious contributions to the calculated amplitudes of the diffracted orders. Therefore, accurate calculations of the diffracted order intensities from the fields near the top surface of the computation space require a sufficiently long distance along the *z*-axis away from the grating to minimize these evanescent field effects. The sum of the intensities of the diffracted orders calculated via the Fourier transform should be nearly equal to the total reflected intensity calculated via the Poynting vector. If they are not, then this could be a sign of interference in the diffracted order intensity calculation from the evanescent field.

The flat bottom surface of the grating forms a cavity with the bottom face of the computation space. Suppressing spurious fields in this region may require multiple stacked reradiating boundaries. Four stacked layers of re-radiating boundary conditions were used in this calculation. At the top surface, two reradiating boundaries were sufficient. Calculations near the resonance angle also require more time steps to reach the steady state. The total reflected power is calculated from the scattered field near the top surface of the computation space, while the total absorbed power is calculated from the total field within the grating. Typically, the absorbed power reaches a steady state value before the scattered field at the top surface reaches a steady state. Therefore, the calculation should be run until the sum of the two powers is close to 1 (the thickness of the grating is sufficient to make the transmittance negligible.) For this example, far from resonance, about 10,000 time steps were required to reach steady state values. At resonance, about 25,000 time steps were required.

The scattered electric field calculated by FDTD as a function of the angle of incidence for *p*-polarization (electric field in the plane of incidence) is plotted Figure 9.9 and for incident *s*-polarization in Figure 9.10 for several different angles of incidence. The surface plasmon resonance caused by the periodic surface charge wave is clearly apparent in Figure 9.9 for the 19° and 20° incident angles. No resonance occurs for incident *s*-polarization.

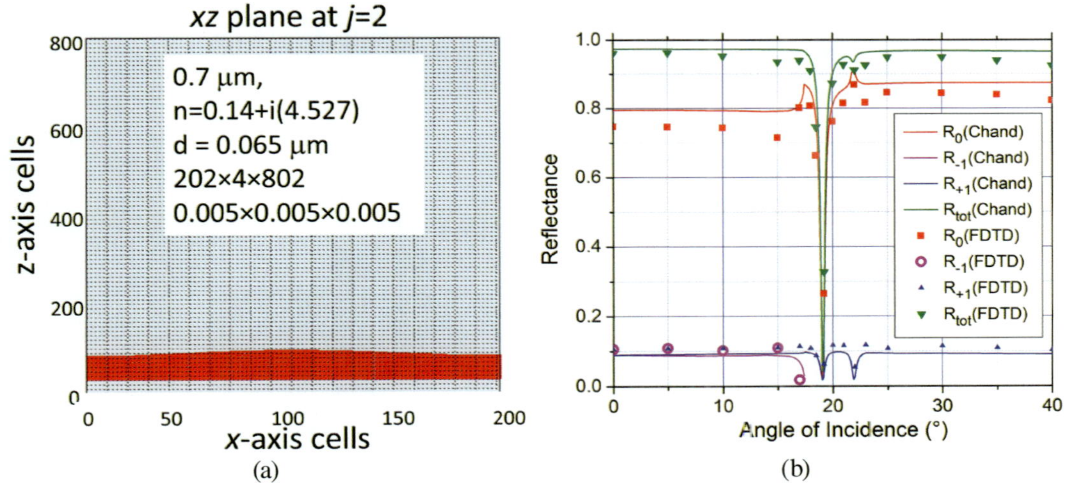

FIGURE 9.8 (a) Unit cell of a sinusoidal diffraction grating made of silver with a refractive index of 0.14 + *i* (4.527) at a wavelength of 0.7 μm. There are 202 cells along the x-axis, 4 cells along the y-axis, and 402 cells along the z-axis. The Yee cell is a cube of 5 nm along each axis. The period of the grating is 1 μm with a groove depth of 65 nm. The maximum thickness of the grating is 0.31 μm. Periodic boundary conditions are applied along the four side faces, and two layers of reradiating + Huygens boundary along the top and bottom faces. FDTD calculations are carried out for 10,000 to 25,000 time steps as described in the text. (b) Zero- and first-order diffraction efficiencies for the surface plasmon resonance calculated by the semi-analytical method of Chandezon (solid lines) for the infinitely thick grating and by FDTD (points) for the grating in (a) as a function of angle of incidence.

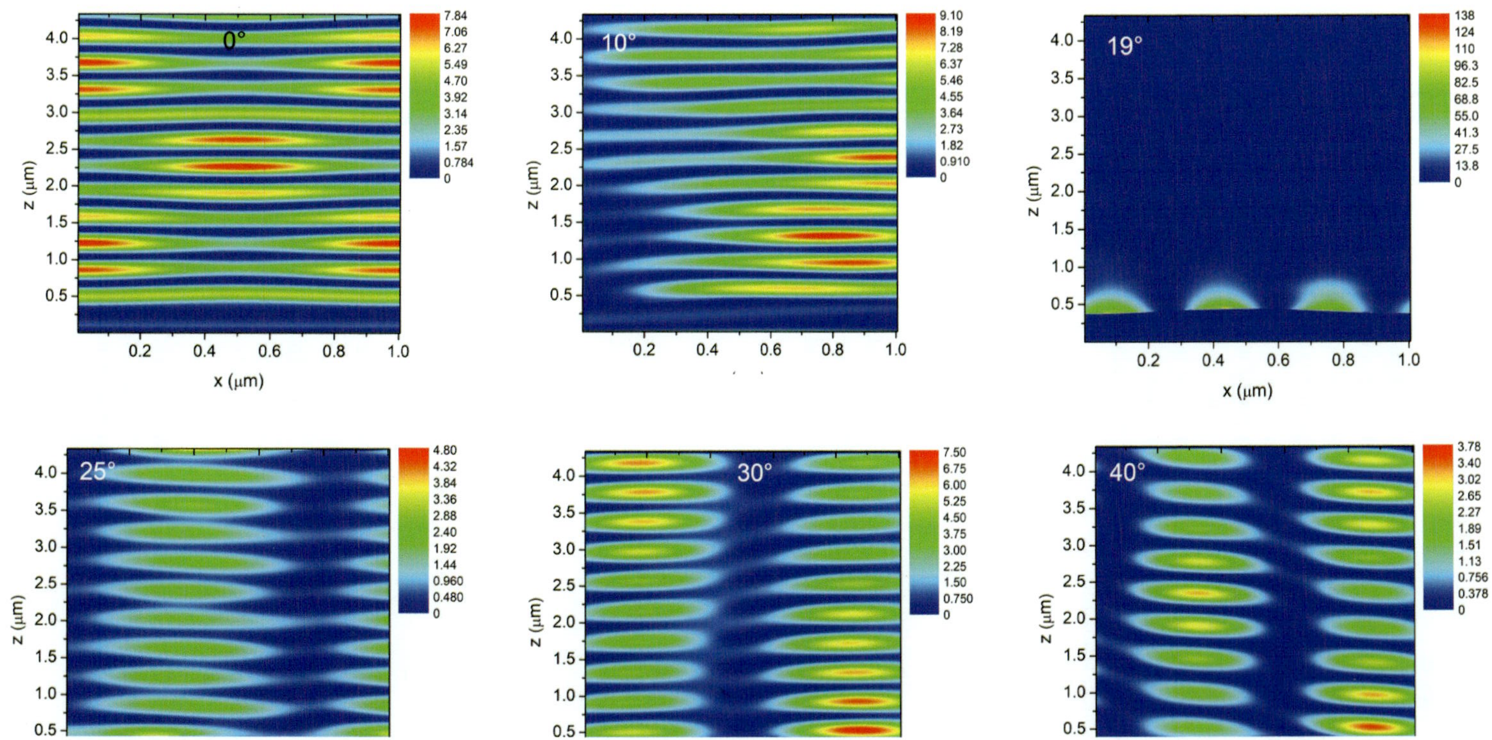

FIGURE 9.9 Electric field intensity in one unit cell of a periodic silver grating for a plane wave incident at several different angles in p-polarization. The anomalous field pattern at 19° is due to excitation of a surface plasmon.

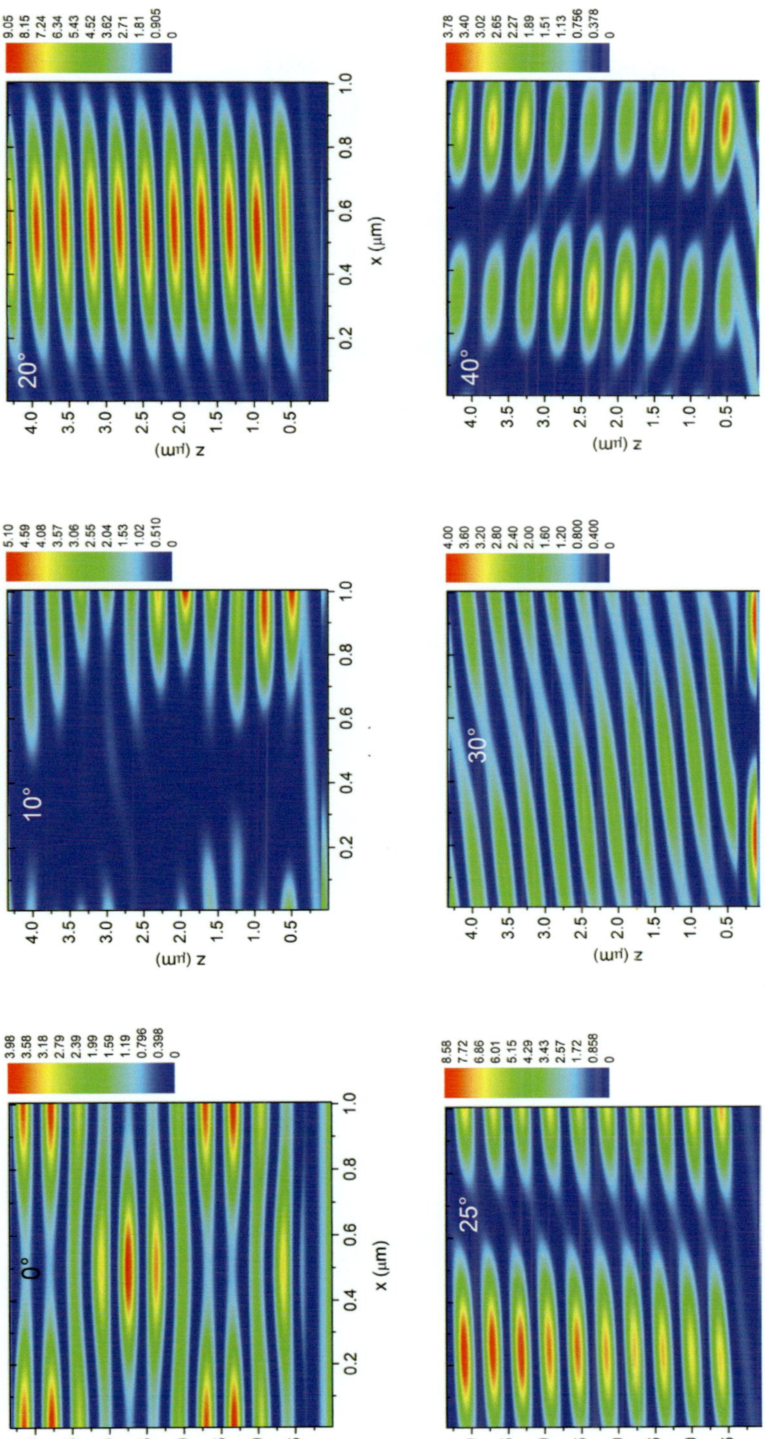

FIGURE 9.10 Electric field intensity in one unit cell of a periodic silver grating for a plane wave incident at several different angles in *s*-polarization. This incident polarization cannot excite a surface plasmon.

9.8 INCIDENT FIELD

Another advantage of the scattered field FDTD technique compared to the total field FDTD technique is that a variety of analytically calculated incident fields can be used. In this section, an incident plane wave and an incident Gaussian pulse are considered.

Let $\hat{\boldsymbol{r}}$ be a unit vector from the origin in the θ, ϕ direction,

$$\hat{\boldsymbol{r}} = \hat{\boldsymbol{x}} \sin\theta \cos\varphi + \hat{\boldsymbol{y}} \sin\theta \sin\varphi + \hat{\boldsymbol{z}} \cos\theta\,, \tag{9.84}$$

and $\hat{\boldsymbol{r}}_{x,\,y,\,z}$ be a vector from the origin to a point in a cell of the FDTD computation space for which the $E_{x,y,z}(i,j,k)$ field component is being calculated as shown in Figure 9.11.

$$\hat{\boldsymbol{r}}_x = (i-0.5)\Delta x\,\hat{\boldsymbol{x}} + (j-1)\Delta y\,\hat{\boldsymbol{y}} + (k-1)\Delta z\,\hat{\boldsymbol{z}}, \tag{9.85}$$

$$\hat{\boldsymbol{r}}_y = (i-1)\Delta x\,\hat{\boldsymbol{x}} + (j-0.5)\Delta y\,\hat{\boldsymbol{y}} + (k-1)\Delta z\,\hat{\boldsymbol{z}}, \tag{9.86}$$

and

$$\hat{\boldsymbol{r}}_z = (i-1)\Delta x\,\hat{\boldsymbol{x}} + (j-1)\Delta y\,\hat{\boldsymbol{y}} + (k-0.5)\Delta z\,\hat{\boldsymbol{z}}\,. \tag{9.87}$$

9.8.1 Plane Wave at Specific Wavelength λ

The simplest incident field is a plane wave. Let the wavevector of the incident field in the background medium be k,

$$k = n_m k_0 \equiv \frac{n_m\,\omega}{c} = \frac{2\pi\,n_m}{\lambda}. \tag{9.88}$$

The phase of the field varies as a function of the time step N as

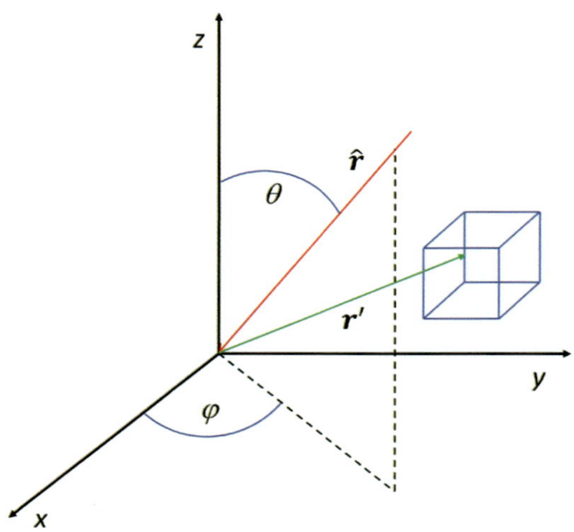

FIGURE 9.11 The plane wave is incident at polar angle (θ, φ) along the opposite direction of the unit vector $\hat{\boldsymbol{r}}$. The field component is calculated at the point $\boldsymbol{r}'$.

$$\phi = \omega t = N(kc\Delta t) = N\left(\frac{2\pi n_m c\ \Delta t}{\lambda}\right). \tag{9.89}$$

For a given point within the computation space r', the field amplitude is zero until time τ where

$$\tau = \frac{n_m d}{c} \tag{9.90}$$

and d is the distance from where the field first enters the computation space to the location of the point along the direction of propagation. For the three Cartesian field components defined in different points of the Yee cell,

$$d_x = \boldsymbol{r}' \cdot \hat{\boldsymbol{r}}_x = (i-0.5)\Delta x \sin\theta\cos\varphi + (j-1)\Delta y \sin\theta\sin\varphi + (k-1)\Delta z \cos\theta\,, \tag{9.91}$$

$$d_y = \boldsymbol{r}' \cdot \hat{\boldsymbol{r}}_y = (i-1)\Delta x \sin\theta\cos\varphi + (j-0.5)\Delta y \sin\theta\sin\varphi + (k-1)\Delta z \cos\theta\,, \tag{9.92}$$

and

$$d_z = \boldsymbol{r}' \cdot \hat{\boldsymbol{r}}_z = (i-1)\Delta x \sin\theta\cos\varphi + (j-1)\Delta y \sin\theta\sin\varphi + (k-0.5)\Delta z \cos\theta\,. \tag{9.93}$$

The time delays are

$$\tau_x = \frac{n_m d_x}{c}, \tag{9.94}$$

$$\tau_y = \frac{n_m d_y}{c}, \tag{9.95}$$

and

$$\tau_z = \frac{n_m d_z}{c}. \tag{9.96}$$

For calculations at the center of the Yee cell,

$$d_c = \boldsymbol{r}' \cdot \hat{\boldsymbol{r}}_c = (i-0.5)\Delta x \sin\theta\cos\varphi + (j-0.5)\Delta y \sin\theta\sin\varphi + (k-0.5)\Delta z \cos\theta\,. \tag{9.97}$$

The electric field at time step n for a plane wave incident upon the edge of the computation space at a polar angle θ and azimuthal angle ϕ is

$$\boldsymbol{E}_i = \begin{cases} \left(E_p\hat{\boldsymbol{\theta}} + E_s\hat{\boldsymbol{\varphi}}\right) \mathrm{Re}\left[e^{ik(\boldsymbol{r}'\cdot\hat{\boldsymbol{r}})}\, e^{i\,kcN\Delta t}\right] & for\ N\Delta t > \tau \\ 0 & otherwise \end{cases}. \tag{9.98}$$

The corresponding Cartesian field components are

$$E_x = \begin{cases} \left(E_p\cos\theta\cos\varphi - E_s\sin\varphi\right)\mathrm{Re}\left[e^{ik(\boldsymbol{r}'\cdot\hat{\boldsymbol{r}})}\, e^{ikcN\Delta t}\right] & for\ N\Delta t > \tau \\ 0 & otherwise \end{cases}, \tag{9.99}$$

$$E_y = \begin{cases} \left(E_p \cos\theta \sin\varphi + E_s \sin\varphi\right) \mathrm{Re}\left[e^{i\,k(r'\cdot\hat{r})}\, e^{i\,k\,c\,N\,\Delta t}\right] & for\, N\,\Delta t > \tau \\ 0 & otherwise \end{cases}, \tag{9.100}$$

and

$$E_z = \begin{cases} \left(-E_p \sin\theta\right) \mathrm{Re}\left[e^{i\,k(r'\cdot\hat{r})}\, e^{i\,k\,c\,N\,\Delta t}\right] & for\, N\,\Delta t > \tau \\ 0 & otherwise \end{cases} \tag{9.101}$$

The incident magnetic field is not needed for the FDTD calculation, but it is given by

$$\boldsymbol{H}_i = \frac{n_m}{\eta_0}\left(E_\varphi \hat{\boldsymbol{\theta}} + E_\theta \hat{\boldsymbol{\varphi}}\right) \mathrm{Re}\left[e^{i\,n_m(r'\cdot\hat{r})k_0}\, e^{i\,k\,c\,N\,\Delta t}\right] \tag{9.102}$$

where n_m is the refractive index of the background medium, and η_0 is the impedance of free space,

$$\eta_0 \equiv \sqrt{\frac{\mu_0}{\varepsilon_0}}. \tag{9.103}$$

The Cartesian components of the magnetic field are

$$H_x = \frac{n_m}{\eta_0}\left(E_\theta \sin\varphi + E_\varphi \cos\theta \cos\varphi\right), \tag{9.104}$$

$$H_y = \frac{n_m}{\eta_0}\left(-E_\theta \cos\varphi + E_\varphi \cos\theta \sin\varphi\right), \tag{9.105}$$

and

$$H_z = -\frac{n_m}{\eta_0} E_\theta \sin\varphi. \tag{9.106}$$

Note that after the field is calculated and stored for the initial time step, at subsequent time steps only the exponential term $e^{i\,\omega\,t} = e^{i\,k\,c\,N\,\Delta t}$ in (9.98) and (9.102) needs to be calculated and multiplied by the stored initial fields.

In the steady state, the amplitude will vary at each point sinusoidally, though with some phase shift relative to the incident field. In order to calculate the amplitude of the scattered electric field at each point in the computation space, the simplest approach is to use the field amplitude calculated for each point at two different times: the final time step, t_f, and the time corresponding to one quarter wavelength before the final time step, t_q. Of course, this approach is only accurate if the FDTD calculation has been carried out for enough time steps that the scattered field has effectively reached the sinusoidal steady state. The number of time steps corresponding to one quarter wavelength before the final time step is ΔN, where

$$\omega\left(t_f - t_q\right) = kc\left(t_f - t_q\right) = \frac{2\pi n_m c\,\Delta N\,\Delta t}{\lambda} = \frac{\pi}{2} \tag{9.107}$$

or

$$\Delta N = \frac{\lambda}{4 n_m c\,\Delta t}. \tag{9.108}$$

Each Cartesian component of the steady state field is characterized by a peak amplitude $\left|E_p\right|$ and phase shift φ_f,

$$E_f = \left|E_p\right| \cos\left(\omega t + \varphi_f\right) \tag{9.109}$$

where

$$\left|E_p\right| = \sqrt{\left(E_{t_q}\right)^2 + \left(E_{t_f}\right)^2} \tag{9.110}$$

and

$$\varphi_f = -\tan^{-1}\left(\frac{E_{t_q}}{E_f}\right) + \frac{\pi}{2}. \tag{9.111}$$

The field amplitudes should be referenced to the center point of each Yee cell using (9.32)–(9.37) in order to calculate total field amplitudes, phases, intensities, Poynting vectors, etc.

The power dissipated within a cell in terms of the peak field amplitude and the local conductivity is

$$P = \frac{1}{2}\sigma' \left|E_p\right|^2. \tag{9.112}$$

σ' is the real part of the complex conductivity at the wavelength of interest. The relationship between conductivity and the complex (relative) dielectric constant is

$$\epsilon(\omega) \equiv \epsilon' + i\epsilon'' = 1 + i\frac{\sigma(\omega)}{\omega\epsilon_0} = 1 + i\frac{Z_0 \lambda_0 \left[\sigma'(\lambda_0) + i\sigma''(\lambda_0)\right]}{2\pi} \tag{9.113}$$

Therefore,

$$\epsilon'(\lambda_0) = 1 - \frac{\sigma''}{\omega\epsilon_0} = 1 - \frac{Z_0 \lambda_0 \sigma''(\lambda_0)}{2\pi}, \tag{9.114}$$

and

$$\epsilon''(\lambda_0) = \frac{\sigma'(\omega)}{\omega\epsilon_0} = \frac{Z_0 \lambda_0 \sigma'(\lambda_0)}{2\pi} \tag{9.115}$$

where ϵ_r is the relative permittivity. Therefore, the real part of the conductivity can be calculated from the imaginary part of the dielectric constant or vice versa.

$$\sigma'(\lambda_0) = \omega\epsilon_0 \epsilon''(\omega) = \frac{2\pi\epsilon''(\lambda_0)}{Z_0 \lambda_0} \tag{9.116}$$

$$\sigma''(\lambda_0) = \omega\epsilon_0\left[1 - \epsilon'(\omega)\right] = \frac{2\pi}{Z_0 \lambda_0}\left(1 - \epsilon'(\lambda_0)\right). \tag{9.117}$$

The complex refractive index n_c is also related to the dielectric constant,

$$\epsilon(\omega) = n_c^2 \equiv (n + iK)^2, \tag{9.118}$$

$$\epsilon'(\lambda_0) = n^2 - K^2, \tag{9.119}$$

$$\epsilon''(\lambda_0) = 2nK, \tag{9.120}$$

$$n = \sqrt{\frac{\epsilon' + \sqrt{\epsilon'^2 + \epsilon''^2}}{2}}, \tag{9.121}$$

and

$$K = \sqrt{\frac{-\epsilon' + \sqrt{\epsilon'^2 + \epsilon''^2}}{2}}. \tag{9.122}$$

(Note: These are the correct equations for the time convention $e^{-i\omega t}$ in a propagating plane wave. The FDTD equations in this chapter, however, use the opposite convention, i.e., $e^{+i\omega t}$. This has the effect of changing the sign of ϵ'' in (9.120), making it always negative, which is generally the opposite convention of that found in optical constant handbooks.) Obviously, the complex conductivity, complex dielectric constant, and the complex refractive index can each be found directly from either of the other two. Sometimes the optical properties of metallic or absorbing materials are also specified in terms of the real part of the dielectric constant and the real part of the conductivity.

Alternatively, the incident field can be initially specified throughout all space and gradually "turned on." This is not physical, but as long as the calculation is run to steady-state, the results appear to be essentially the same as those calculated by a full amplitude plane wave advancing into the computation space from the boundary as previously discussed. Although a gradual turn-on of the field is used to minimize any initial ringing in the calculation that might occur if the scattering object were suddenly immersed in a plane wave at full amplitude, in practice it seems that even turning on the sinusoidal wave at full amplitude at the first time step throughout the computation space gives results consistent with the other two approaches. For an x-polarized incident plane wave,

$$\boldsymbol{E} = E_x \hat{\mathbf{x}} \sin(\varphi_x + \omega t), \tag{9.123}$$

so the values for both E_x and ϕ_x must be stored in a buffer for $t = 0$ at every point (i, j, k) in the computation space. For more complicated incident fields, it may be necessary to also store E_y, E_z, φ_y, and φ_z. The derivative of (9.123) is also required in the FDTD calculation. If the field amplitude is increased linearly from zero to full amplitude at time t_0, then

$$\boldsymbol{E} = \begin{cases} \left(\dfrac{t}{t_0}\right) E_x \hat{\mathbf{x}} \sin(\varphi_x + \omega t) & \text{for } t \le t_0 \\ E_x \hat{\mathbf{x}} \sin(\varphi_x + \omega t) & \text{for } t > t_0 \end{cases}, \tag{9.124}$$

and

$$\frac{d\boldsymbol{E}}{dt} = \begin{cases} \left(\dfrac{E_x}{t_0}\right)\left[\sin(\varphi_x + \omega t) + \omega t \cos(\varphi_x + \omega t)\right]\hat{\mathbf{x}} & \text{for } t \le t_0 \\ \omega E_x \hat{\mathbf{x}} \cos(\varphi_x + \omega t) & \text{for } t > t_0 \end{cases}. \tag{9.125}$$

This approach to turning on the field can be used for other incident fields such as the focused field from a lens in which the steady state field distribution is much simpler to calculate than the time-dependent field after the light is first turned on. However, in many different example calculations, no significant "ringing" or other artifacts have been noticed when the incident field is suddenly turned on, so in general it seems that the additional complexity involved in turning the field on gradually is not required.

9.8.2 Gaussian Pulse

A light pulse is composed of a range of frequencies or wavelengths, so a single FDTD calculation can provide information about the system spectral response rather than having to run the same calculation for multiple different wavelengths. In particular, the Fourier transform of the far zone transient field from an incident pulse ratioed to the frequency-dependent field amplitude in the incident pulse provides the reflectance amplitude as a function of frequency or wavelength from a single FDTD calculation. The length of time over which the FDTD calculation must be run may need to be substantially longer than that of the incident pulse to guarantee that the scattered fields have become negligible.

For a single Gaussian light pulse, the fields in the pulse are defined to enter from the edge of the computation space. Because the Gaussian pulse never goes completely to zero, it must be truncated after a certain time. Let β be the number of time steps from the peak of the pulse to its truncation. The incident field components are

$$E_{xi} = \begin{cases} \left(E_\theta \cos\theta \cos\varphi - E_\varphi \sin\varphi\right) e^{-\alpha\left(\tau_x - \beta\Delta t\right)^2} & for\, 0 \le t \le 2\beta\Delta t \\ 0 & otherwise \end{cases}, \tag{9.126}$$

$$E_{yi} = \begin{cases} \left(E_\theta \cos\theta \sin\varphi + E_\varphi \cos\varphi\right) e^{-\alpha\left(\tau_y - \beta\Delta t\right)^2} & for\, 0 \le t \le 2\beta\Delta t \\ 0 & otherwise \end{cases}, \tag{9.127}$$

and

$$E_{zi} = \begin{cases} \left(-E_\theta \sin\theta\right) e^{-\alpha\left(\tau_z - \beta\,\Delta t\right)^2} & for\ 0 \le t \le 2\beta\Delta t \\ 0 & otherwise \end{cases}. \tag{9.128}$$

τ is the time delay for the pulse to reach a particular field component in cell (i, j, k). α is chosen to be

$$\alpha = \left(\frac{4}{\beta\Delta t}\right)^2. \tag{9.129}$$

With this choice, when $\tau = 2\beta\Delta t$ the field intensity has dropped to e^{-16} or –140 dB.

Although it might seem that a Gaussian light pulse is the preferable approach to FDTD calculations so that multiple calculations at different wavelengths can be avoided, there are two reasons why the author has not generally used pulsed calculations for FDTD optical problems. First of all, any material in the computation space with a wavelength-dependent conductivity must be modeled using an appropriate combination of Drude and/or Lorentz dielectric functions. It can be complicated to determine the appropriate combination of functions that accurately model a given material over a specified wavelength range, and also more complicated to implement the wavelength-dependent FDTD equations to be discussed in the next main section. Secondly, in the author's experience most optical problems involve laser light scattering in which case one is only interested in a single wavelength result.

9.8.3 Focused Beam

Sometimes it is desired to model the effect of a focused beam of light onto a scattering object. The electric and magnetic fields for a focused beam of light for various types of incident polarization states was derived in Chapter 7.

$$\boldsymbol{E} = \mathrm{Re}\left(E_\theta \hat{\theta} + E_\varphi \hat{\varphi}\right) e^{i\omega\left(t + \frac{n_m(\boldsymbol{r}' \cdot \hat{\boldsymbol{r}} + \boldsymbol{R})}{c}\right)}. \tag{9.130}$$

As previously mentioned, the focused fields can also be ramped up linearly in time until they reach the desired amplitude. The total time for the ramp can be adjusted to minimize the time to reach steady state, but a typical ramp time is the amount of time for the wave to move by half a wavelength,

$$t_r = \frac{c}{2n_m f} = \frac{\lambda}{2n_m}. \tag{9.131}$$

The fields are then multiplied by the ramp

$$A = \frac{t}{t_r}. \tag{9.132}$$

9.9 LOSSY MATERIALS

The optical properties of the materials up to now have been described by a dielectric constant and a conductivity. In principle, these two parameters are sufficient to model any material at a specific wavelength with a complex dielectric constant according to (9.116)–(9.122). However, in actuality when the real part of the dielectric constant is negative, the FDTD calculation diverges.

The frequency dependence of the dielectric function can be adequately modeled over a range of wavelengths for at least some materials by simple analytical equations. The Debye dielectric function can be used to model the frequency-dependent dielectric function for some lossy materials if the real part of the dielectric constant is positive. The Drude dielectric function handles lossy materials for which it is negative. The Lorentz dielectric function can also handle materials with a resonant absorption at some wavelength.

9.9.1 Lossy Dielectric

If the material is an absorbing dielectric with a positive real part of its dielectric constant for which the complex refractive index is known only at the wavelength of the calculation, then the previously derived update equation (9.31) can be used. In this case, the complex refractive index is used to determine the real part of the dielectric constant (and permittivity) from (9.119) and the real part of the conductivity from (9.116) and (9.120).

In Figure 9.12 a plane wave is incident upon a lossy dielectric sphere. The real part of the dielectric constant is found from (9.119) which is 6.75, and the real part of the conductivity is found from (9.116), which is $1.7671 \times 10^5 \ (\Omega m)^{-1}$. The calculated electric field intensities for planes through the center of the sphere are plotted in a log contour plot in Figure 9.12(a). The results for the same example calculated from Mie theory are shown in Figure 9.12(b). This example is taken from Figure 2 of [9].

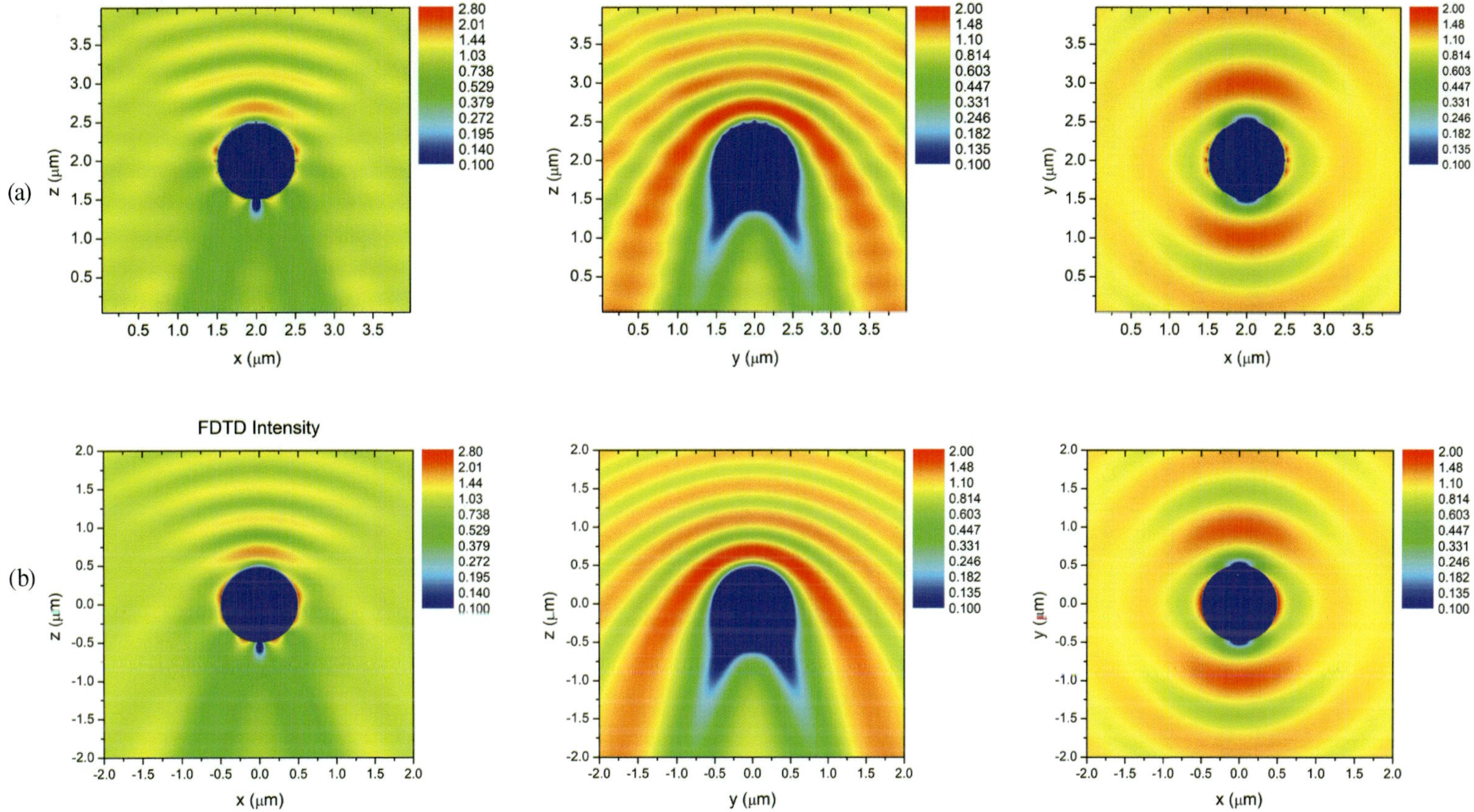

FIGURE 9.12 Log plots of the $|E|^2$ total intensity for a plane wave incident along the z-axis upon a 1 μm diameter absorptive sphere with a complex refractive index of $3.0 + i\,(1.5)$ in vacuum at a wavelength of 850 nm. Plots are in the *xz*, *yz*, and *xy* planes through the center of the sphere. (a) FDTD calculation using a cubic cell space with 200 cells per side and cubic Yee cells of edge length 0.02 μm carried out for 500 time steps. (b) Mie calculation for the same example.

9.9.2 Debye Dielectric Function

A material with a positive real part of its dielectric constant can also be modeled as a Debye material. The complex frequency-dependent dielectric function from the Debye equation is

$$\epsilon \equiv \epsilon' + i\epsilon'' = \epsilon_\infty + \frac{\epsilon_s - \epsilon_\infty}{1 + i\omega t_0} - i\frac{\sigma_0}{\omega\epsilon_0} = \epsilon_\infty + \chi(\omega) - i\frac{\sigma_0}{\omega\epsilon_0}. \tag{9.133}$$

$\chi(\omega)$ is the frequency-dependent susceptibility. The frequency-independent conductivity σ_0 in (9.133) is, strictly speaking, not a part of the Debye dielectric function, but it is easily included in the derivation of the field update equations. As written in (9.133) there are four independent parameters to be specified for the dielectric function, the high frequency dielectric constant ϵ_∞, the static dielectric constant ϵ_s, the relaxation time t_0, and the DC conductivity σ_0 . If the complex refractive index is known at one frequency, then two of these parameters are redundant. In the special case that the user chooses $t_0 = 0$ then the Debye equation reduces to a frequency-independent dielectric constant ϵ_∞ and frequency-independent conductivity σ_0 which are determined by the complex refractive index.

$$\epsilon_\infty = n^2 - K^2, \tag{9.134}$$

and

$$\sigma_0 = 2\omega\epsilon_0 nK. \tag{9.135}$$

The electric field update equation (9.27) is sufficient in this case.

Otherwise, two of the four unknown parameters must be specified by the user together with a complex refractive index to determine the other two parameters in the Debye equation. For example, if ϵ_∞ and σ_0 are given, then t_0 and ϵ_s are found by first separating (9.111) into real and imaginary parts,

$$\epsilon' = \epsilon_\infty + \frac{(\epsilon_s - \epsilon_\infty)}{(1 + \omega^2 t_0^2)}, \tag{9.136}$$

and

$$\epsilon'' = -\frac{\omega t_0 (\epsilon_s - \epsilon_\infty)}{(1 + \omega^2 t_0^2)} - \frac{\sigma_0}{\omega\epsilon_0}. \tag{9.137}$$

The relaxation time and static dielectric constant are then given by

$$t_0 = -\frac{\epsilon'' + \dfrac{\sigma_0}{\omega\epsilon_0}}{\omega(\epsilon' - \epsilon_\infty)}, \tag{9.138}$$

and

$$\epsilon_s = \epsilon_\infty + (\epsilon' - \epsilon_\infty)(1 + \omega^2 t_0^2), \tag{9.139}$$

remembering that ϵ'' is *negative* in (9.138) to ensure that the field in an absorbing medium decays as it propagates because of the $\sin[\omega t - 2\pi(n + iK)r/\lambda]$ time-dependence chosen for the incident field. When the opposite convention for the propagating field, $\sin[2\pi(n + iK)r/\lambda - \omega t]$, is used, the imaginary part of the dielectric function is a positive number as given by (9.123). Clearly, it is possible to also set $\sigma_0 = 0$ in (9.138) and still solve for the two remaining parameters.

If the relaxation time and static dielectric constant are known, then

$$\epsilon_\infty = \frac{\epsilon'\left(1+\omega^2 t_0^2\right)-\epsilon_s}{\omega^2 t_0^2} \tag{9.140}$$

$$\sigma_0 = -\omega\epsilon_0\left[\epsilon'' + \frac{\left(\epsilon_s - \epsilon'\right)}{\omega t_0}\right]. \tag{9.141}$$

If the high frequency and static dielectric constants are specified, then

$$t_0 = \frac{1}{\omega}\sqrt{\frac{\left(\epsilon_s - \epsilon_\infty\right)}{\left(\epsilon' - \epsilon_\infty\right)} - 1} \tag{9.142}$$

and

$$\sigma_0 = \frac{\omega\epsilon_0\epsilon''}{\sqrt{\left(\epsilon_s - \epsilon'\right)\left(\epsilon' - \epsilon_\infty\right)}}. \tag{9.143}$$

The *total* field update equation for a Debye material is derived by Kunz et al. [1] (Sect. 8.2) Their derivation approach will be followed here with certain minor modifications. In the time domain, the displacement field and electric field are related by

$$D(t) = \epsilon_\infty\epsilon_0 E(t) + \epsilon_0\int_0^t E(t-\Lambda)\chi(\Lambda)\,d\Lambda \tag{9.144}$$

where $\chi(t)$ is the time-dependent susceptibility. After discretization, this becomes

$$D(t) \approx D(n\Delta t) = D_n = \epsilon_\infty\epsilon_0 E_n + \epsilon_0\int_0^{n\Delta t} E(n\Delta t - \Lambda)\chi(\Lambda)\,d\Lambda. \tag{9.145}$$

The integration is replaced by a summation. At time step n,

$$D_n = \epsilon_\infty\epsilon_0 E_n + \epsilon_0\sum_{m=0}^{n-1} E_{n-m}\int_{m\Delta t}^{(m+1)\Delta t} \chi(\Lambda)\,d\Lambda. \tag{9.146}$$

Therefore, at time step $n + 1$,

$$D_{n+1} = \epsilon_\infty\epsilon_0 E_{n+1} + \epsilon_0\sum_{m=0}^{n} E_{n+1-m}\int_{m\Delta t}^{(m+1)\Delta t} \chi(\Lambda)\,d\Lambda. \tag{9.147}$$

Subtracting (9.146) from (9.147) gives

$$\begin{aligned} D_{y,n+1}(i) - D_{y,n}(i) = {} & \epsilon_\infty\epsilon_0\left(E_{y,n+1} - E_{y,n}\right) + \epsilon_0\sum_{m=0}^{n} E_{y,n+1-m}(i)\int_{m\Delta t}^{(m+1)\Delta t} \chi(\Lambda)\,d\Lambda \\ & - \epsilon_0\sum_{m=0}^{n-1} E_{y,n-m}(i)\int_{m\Delta t}^{(m+1)\Delta t} \chi(\Lambda)\,d\Lambda \end{aligned} \tag{9.148}$$

which simplifies to

$$D_{y,n+1}(i)-D_{y,n}(i)=\epsilon_{\infty}\epsilon_0\left[E_{y,n+1}(i)-E_{y,n}(i)\right]+\epsilon_0\chi_0 E_{y,n+1}$$
$$+\epsilon_0\sum_{m=0}^{n-1}E_{y,n-m}(i)\left(\chi_{m+1}-\chi_m\right) \tag{9.149}$$

where

$$\int_{m\Delta t}^{(m+1)\Delta t}\chi(\Lambda)\,d\Lambda\equiv\chi_m. \tag{9.150}$$

One of Maxwell's equations is

$$\nabla\times H=\frac{\partial D}{\partial t}+J\,. \tag{9.151}$$

Considering just the y-component of the displacement field,

$$D_{y,n+1}(i)-D_{y,n}(i)=-\left(\frac{\Delta t}{\Delta x}\right)\left[H_{z,n+\frac{1}{2}}\left(i+\frac{1}{2}\right)-H_{z,n+\frac{1}{2}}\left(i-\frac{1}{2}\right)\right]-\sigma\,\Delta t\,E_{y,n+1}(i). \tag{9.152}$$

(9.149) and (9.152) can be equated to eliminate the displacement field.

$$\epsilon_{\infty}\epsilon_0\left[E_{y,n+1}(i)-E_{y,n}(i)\right]+\epsilon_0\chi_0 E_{y,n+1}-\epsilon_0\sum_{m=0}^{n-1}\Delta\varsigma_m E_{y,n-m}(i)$$
$$=-\left(\frac{\Delta t}{\Delta x}\right)\left[H_{z,n+\frac{1}{2}}\left(i+\frac{1}{2}\right)-H_{z,n+\frac{1}{2}}\left(i+\frac{1}{2}\right)\right]-\sigma\,\Delta t\,E_{y,n+1}(i) \tag{9.153}$$

where

$$\Delta\chi_m\equiv\chi_m-\chi_{m-1}. \tag{9.154}$$

Solving (9.153) for $E_{y,n+1}$,

$$E_{y,n+1}(i)=\frac{\epsilon_{\infty}}{\left(\epsilon_{\infty}+\chi_0+\dfrac{\sigma\Delta t}{\epsilon_0}\right)}E_{y,n}(i)$$
$$+\frac{1}{\left(\epsilon_{\infty}+\chi_0+\dfrac{\sigma\Delta t}{\epsilon_0}\right)}\sum_{m=0}^{n-1}\Delta\chi_m E_{y,n-m}(i)-\left(\frac{\Delta t}{\Delta x}\right)\frac{\left[H_{z,n+\frac{1}{2}}\left(i+\frac{1}{2}\right)-H_{z,n+\frac{1}{2}}\left(i+\frac{1}{2}\right)\right]}{\epsilon_0\left(\epsilon_{\infty}+\chi_0+\dfrac{\sigma\Delta t}{\epsilon_0}\right)} \tag{9.155}$$

(9.155) is the update equation at each time step for the *total* field, and it has a new term with a summation that was not present in (9.27). In order to evaluate this term, the time-dependent susceptibility is found from the Fourier transform of the frequency-dependent susceptibility,

$$\chi(t)=\frac{1}{2\pi}\int_{-\infty}^{\infty}e^{i\omega t}\chi(\omega)\,d\omega=\left(\frac{\epsilon_s-\epsilon_\infty}{t_0}\right)e^{-t/t_0}U(t) \tag{9.156}$$

where $U(t)$ is the step function,

$$U(t)=\begin{cases}0 & t\le 0\\ 1 & t>0\end{cases}. \tag{9.157}$$

(9.156) is substituted into (9.150),

$$\chi_m=\int_{m\Delta t}^{(m+1)\Delta t}\left[\left(\frac{\epsilon_0-\epsilon_\infty}{t_0}\right)e^{-t/t_0}U(t)\right]dt=(\epsilon_s-\epsilon_\infty)e^{-m\Delta t/t_0}\left(1-e^{-\Delta t/t_0}\right). \tag{9.158}$$

From this relation it is easily found that

$$\Delta\chi_m=\chi_m-\chi_{m+1}=(\epsilon_s-\epsilon_\infty)\left(1-e^{-\Delta t/t_0}\right)^2e^{-m\Delta t/t_0}. \tag{9.159}$$

Also,

$$\Delta\chi_{m+1}=(\epsilon_s-\epsilon_\infty)\left(1-e^{-\Delta t/t_0}\right)^2e^{-(m+1)\Delta t/t_0}=e^{-\Delta t/t_0}\Delta\chi_m, \tag{9.160}$$

$$\chi_0(t)=(\epsilon_s-\epsilon_\infty)\left(1-\mathrm{e}^{-\Delta t/t_0}\right), \tag{9.161}$$

and

$$\Delta\chi_0=(\epsilon_s-\epsilon_\infty)\left(1-e^{-\Delta t/t_0}\right)^2. \tag{9.162}$$

Let the summation in the update equation (9.156) be denoted by the variable Ψ,

$$\Psi_n(i)\equiv\sum_{m=0}^{n-1}\Delta\chi_m E_{y,n-m}(i). \tag{9.163}$$

The update equation is

$$E_{y,n+1}(i)=\frac{\epsilon_\infty}{\left(\epsilon_\infty+\chi_0+\frac{\sigma\Delta t}{\epsilon_0}\right)}E_{y,n}(i)+\frac{1}{\left(\epsilon_\infty+\chi_0+\frac{\sigma\Delta t}{\epsilon_0}\right)}\Psi_n(i)$$
$$-\left(\frac{\Delta t}{\Delta x}\right)\frac{\left(H_{z,n+\frac{1}{2}}\left(i+\frac{1}{2}\right)-H_{z,n+\frac{1}{2}}\left(i-\frac{1}{2}\right)\right)}{\epsilon_0\left(\epsilon_\infty+\chi_0+\frac{\sigma\Delta t}{\epsilon_0}\right)}. \tag{9.164}$$

$\Psi_n(i)$ can be found iteratively at each time step.

$$\Psi_{n-1}(i)=\sum_{m=0}^{n-2}\Delta\chi_m E_{y,n-1-m}(i) \tag{9.165}$$

and

$$\begin{aligned}\Psi_n(i)&=\sum_{m=-1}^{n-2}\Delta\chi_{m+1}E_{y,n-(m+1)}(i)=\sum_{m=0}^{n-2}\Delta\chi_{m+1}E_{y,n-(m+1)}(i)+\Delta\chi_0 E_{y,n}(i)\\&=e^{-\Delta t/t_0}\Psi_{n-1}(i)+\Delta\chi_0 E_{y,n}(i).\end{aligned} \tag{9.166}$$

The update equation for the magnetic field, (9.28), is unchanged.

Following a similar process using the equations for the *scattered* fields, the update equation for the scattered E_y component of the electric field is

$$\begin{aligned}E^s_{y,n+1}(i)=&\frac{\epsilon_\infty}{\left(\epsilon_\infty+\chi_0+\frac{\sigma\Delta t}{\epsilon_0}\right)}E^s_{y,n}(i)+\frac{1}{\left(\epsilon_\infty+\chi_0+\frac{\sigma\Delta t}{\epsilon_0}\right)}\Psi^s_{n,y}(i)\\&-\frac{\sigma\Delta t}{\epsilon_0\left(\epsilon_\infty+\chi_0+\frac{\sigma\Delta t}{\epsilon_0}\right)}E^i_{y,n+1}(i)-\frac{(\epsilon_\infty-1)\Delta t}{\left(\epsilon_\infty+\chi_0+\frac{\sigma\Delta t}{\epsilon_0}\right)}\frac{\partial E^i_{y,n+1}(i)}{\partial t}\\&-\frac{\Delta t}{\left(\epsilon_\infty+\chi_0+\frac{\sigma\Delta t}{\epsilon_0}\right)}\frac{\partial}{\partial t}\left[E^i_{y,n+1}(i)*\chi(t)\right]\\&+\frac{\Delta t}{\epsilon_0\,\Delta z\left(\epsilon_\infty+\chi_0+\frac{\sigma\Delta t}{\epsilon_0}\right)}\left[H_{x,n+\frac{1}{2}}\left(i+\frac{1}{2}\right)-H_{x,n+\frac{1}{2}}\left(i-\frac{1}{2}\right)\right]\\&-\frac{\Delta t}{\epsilon_0\,\Delta x\left(\epsilon_\infty+\chi_0+\frac{\sigma\Delta t}{\epsilon_0}\right)}\left[H_{z,n+\frac{1}{2}}\left(i+\frac{1}{2}\right)-H_{z,n+\frac{1}{2}}\left(i-\frac{1}{2}\right)\right]\end{aligned} \tag{9.167}$$

where

$$\Psi(i)^s_{n,y}\equiv\sum_{m=0}^{n-1}\Delta\chi_m E^s_{y,n-m}(i). \tag{9.168}$$

The update equations are similar for the other two scattered electric field components with suitable substitutions of $x\rightarrow z$, $z\rightarrow y$, and $y\rightarrow x$ for the E_x component in (9.167), and $x\rightarrow y$ and $z\rightarrow x$, $y\rightarrow z$ for the E_z component.

Comparison of (9.167) to (9.27) show that many of the terms are similar with slightly different prefactors. However, there are two new terms in the update equation. The procedure for calculating Ψ in the first term has already been derived for the total field case in (9.166), and it remains the same for the scattered field case. The second term with the partial derivative is a convolution denoted by the " $\star$ " and depends on the form of the incident field.

9.9.2.1 Incident Sinusoidal Field Advancing From Edge of Computation Space

For an incident sinusoidal field, the convolution is zero for times $\tau \le 0$. For $\tau > 0$,

$$\frac{\partial}{\partial t}\left[E^{i}_{y,n+1}(t) * \chi(t)\right] = \frac{\partial}{\partial t}\int_0^{\tau} E^{i}(\tau - t)\chi(t)\,dt = \frac{\partial}{\partial t}\int_0^{\tau} E^{i}_0 \sin(\omega\tau - \omega t)\left(\frac{\epsilon_0 - \epsilon_\infty}{t_0}\right)e^{-t/t_0}\,dt \quad (9.169)$$

$$= E_0\,\omega\left(\frac{\epsilon_0 - \epsilon_\infty}{1+\omega^2 t_0^2}\right)\left[\cos(\omega\tau) + \omega t_0 \sin(\omega\tau) - e^{-\frac{\tau}{t_0}}\right]. \quad (9.170)$$

The convolution term for the incident field is calculated at each time step from (9.170). At $\tau = 0$, the term in square brackets is zero. As time increases, the exponential term drops out and the function oscillates with a constant amplitude at frequency ω.

It is interesting to investigate the accuracy using the Debye dielectric equation in this manner in FDTD. Consider again the scattering sphere in Figure 9.12. Using the Debye analysis with zero DC conductivity and varying high frequency dielectric constants, the results are shown in Figure 9.13. For a high frequency dielectric constant close to the real part of the dielectric constant at the given wavelength, the field intensity is close to that previously calculated. As the high frequency dielectric constant is reduced towards its minimum value of 1, however, the accuracy of the calculation decreases significantly even though the refractive index at the specified wavelength remains the same. Also shown is the Debye calculation with a nonzero DC conductivity that, not surprisingly, is a close match to the Mie theory results.

9.9.2.2 Sinusoidal Field Throughout the Computation Space That Is Slowly Turned On

Another approach to implementing a plane wave or any other field configuration that varies in time sinusoidally is to calculate the field profile using complex arithmetic throughout the computation space at one instant in time, vary the amplitude at each time step by the factor $e^{i\omega t}$, and finally take the real or imaginary part of the complex amplitude for the FDTD update equation. In this case, to reduce the effects of ringing that might occur if the field is suddenly switched on throughout the computation space, the field amplitude can be linearly ramped from zero to its final value over one period of the incident wave.

Let the complex incident field amplitude at point (i, j, k) be

$$E(i,j,k) = \begin{cases} \left(\dfrac{t}{t_p}\right) E_0(i,j,k)\,e^{i[\omega t + \varphi(i,j,k)]} & 0 \le t < t_p \\ E_0(i,j,k)\,e^{i[\omega t + \varphi(i,j,k)]} & t > t_p \end{cases}. \quad (9.171)$$

Here, t_p can be one period of the incident plane wave. The convolution integral is

$$\frac{\partial}{\partial t}\left[E^{i}_{y,n+1}(t) * \chi(t)\right] =$$

$$\begin{cases} E_0\,e^{i\varphi}\left(\dfrac{\epsilon_0 - \epsilon_\infty}{t_0}\right)\dfrac{\partial}{\partial t}\displaystyle\int_0^{t}\left(\dfrac{t-\tau}{t_p}\right)e^{i(\omega t - \omega\tau)}e^{-\tau/t_0}\,d\tau & 0 \le t < t_p \\ E_0\,e^{i\varphi}\left(\dfrac{\epsilon_0 - \epsilon_\infty}{t_0}\right)\dfrac{\partial}{\partial t}\left[\displaystyle\int_0^{t_p}\left(\dfrac{t-\tau}{t_p}\right)e^{i(\omega t - \omega\tau)}e^{-\tau/t_0}\,d\tau + \int_{t_p}^{t} e^{i(\omega t - \omega\tau)}e^{-\tau/t_0}\,d\tau\right] & t > t_p \end{cases}. \quad (9.172)$$

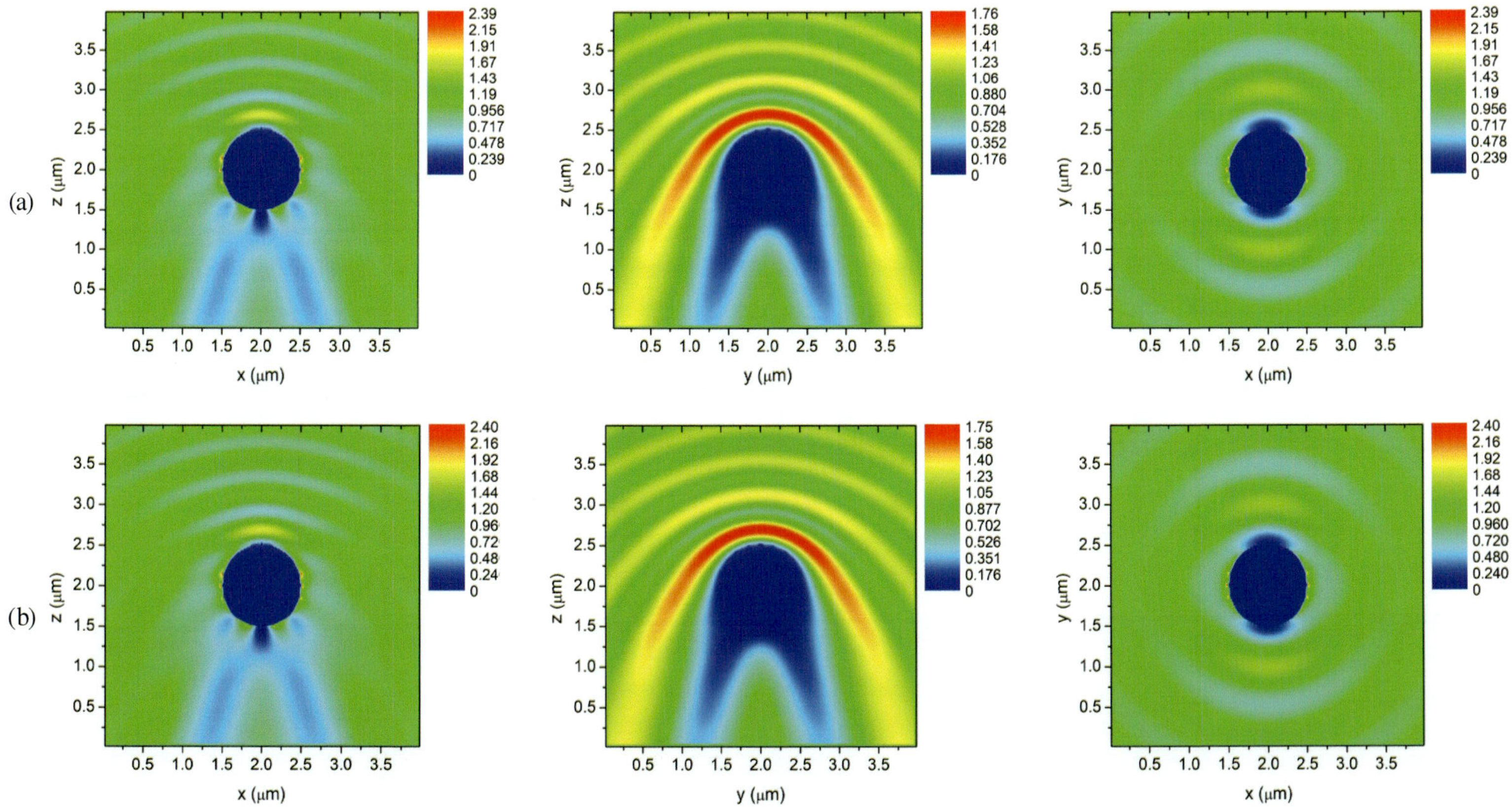

FIGURE 9.13 Linear plot of the $|E|^2$ total intensity for a plane wave incident along the z-axis upon a 1 μm diameter absorptive sphere with a complex refractive index of 3.0 + *i* (1.5) in vacuum at a wavelength of 850 nm. The FDTD cubic cell space is 200 cells per side and cubic Yee cells of edge length 0.02 μm carried out for 800 time steps. (a) Frequency-independent parameters. (b) Debye parameters with $\epsilon_\infty = 10$ and $\sigma = 1.99621 \times 10^5 (\Omega m)^{-1}$.

The assumption is that the calculation is run until the steady state is reached for $t > t_p$. Therefore, the term in the convolution for $t > t_p$ that decays as $exp(-t/t_0)$ can be neglected. The results are

$$\frac{\partial}{\partial t}\left[E^i_{y,n+1}(t)\star\chi(t)\right]\cong$$

$$\begin{cases} E_0 e^{i\phi}\dfrac{(\epsilon_0-\epsilon_\infty)}{t_p(1+i\omega t_0)^2}\left[e^{i\omega t}\left(1+i\omega t-\omega^2 t_0 t\right)-e^{-\frac{t}{t_0}}\right] & 0\le t<t_p \\ E_0 e^{i\omega t+i\varphi}\dfrac{(\epsilon_0-\epsilon_\infty)}{t_p(1+i\omega t_0)^2}\left[\left(1+i\omega t-\omega^2 t_0 t\right)-\left[1+\left(\omega t-2\omega t_p\right)\left(i-\omega t_0\right)\right]e^{-t_p\left(\frac{1}{t_0}+i\omega\right)}\right] & t>t_p \end{cases}. \quad (9.173)$$

For a real incident field, the real or imaginary part of (9.173) is then taken.

9.9.2.3 Incident Cosine Pulse

An incident cosine pulse is discussed by Kunz et al. [1] (pp. 155–161). If τ is the delayed time of the electric field at the cell location, the pulse can be defined by

$$E^i(\tau)=\begin{cases} C_0+\sum_{q=1}^{3}C_q\cos(q\omega_s t) & 0\le\tau\le T \\ 0 & otherwise \end{cases}. \quad (9.174)$$

The time-dependent susceptibility in (9.163) is replaced by the generic form,

$$\chi(t)=A_p e^{-a_p t}U(t). \quad (9.175)$$

The convolution is

$$\frac{\partial}{\partial t}\left[E^i_{y,n+1}(t)\star\chi(t)\right]=\frac{\partial}{\partial\tau}\int_0^\tau E^i(\tau-t)\chi(t)\,dt \quad (9.176)$$

$$=\frac{\partial}{\partial\tau}\left\{C_0A_p\int_0^\tau e^{-a_p t}dt+\sum_{q=1}^{3}\int_0^\tau C_qA_p\cos\left[q\omega_s(\tau-t)\right]e^{-a_p t}\,dt\right\}. \quad (9.177)$$

The partial derivative of the convolution in (9.177) must be evaluated for three different cases. (1) For $\tau\le 0$, the incident field has not yet reached the cell, so the result is zero. (2) The incident field pulse has reached the cell, but has not completely passed by it, so $0<\tau\le T$. Then

$$\frac{\partial}{\partial t}\left[E^i_{y,n+1}(t)\star\chi(t)\right]=\frac{C_0A_p}{a_p}\frac{\partial}{\partial t}\left(1-e^{-a_p t}\right)+A_p\frac{\partial}{\partial t}\int_0^t\sum_{q=1}^{3}C_q\cos\left(q\omega_s t-q\omega_s\tau\right)e^{-a_p\tau}d\tau \quad (9.178)$$

$$=C_0A_pe^{-a_p\tau}+\sum_{q=1}^{3}\frac{A_pC_q}{a_p^2+q^2\omega_s^2}\left[a_p^2e^{-a_p t}+q^2\omega_s^2\cos\left(q\omega_s\tau\right)-a_pq\omega_s\sin\left(q\omega_s\tau\right)\right] \quad (9.179)$$

(3) The pulse has completely passed the cell, $\tau>T$.

$$\frac{\partial}{\partial t}\left[E^{i}_{y,n+1}(t)\star\chi(t)\right]=A_p e^{-a_p\tau}\left(1-e^{-a_p T}\right)\left[C_0+\sum_{q=1}^{3}\frac{C_q a_p^2}{a_p^2+q^2\omega_s^2}\right]. \tag{9.180}$$

9.9.3 Drude Dielectric Function

The Debye dielectric function described in the previous section enables calculations of absorbing dielectrics for which the real part of the dielectric constant is positive. If the real part of the dielectric constant is negative, the FDTD calculation using either the single frequency $\mathrm{Re}(\epsilon)$ and $\mathrm{Re}(\sigma)$ approach, or using the Debye function diverges. In this case, it is necessary to use either the Drude or Lorentz dielectric function. The Drude dielectric function is

$$\epsilon(\omega)\equiv\epsilon'+i\epsilon''\equiv 1-\frac{\omega_p^2}{\omega(\omega-i\nu_c)}=1-\frac{\left(\frac{\omega_p}{\nu_c}\right)^2}{1+i\left(\frac{\omega}{\nu_c}\right)}+\frac{\frac{\omega_p^2}{\nu_c}}{i\omega} \tag{9.181}$$

where ω_p is the plasma frequency and ν_c is the relaxation frequency. A complex permittivity can be defined from Maxwell's equation,

$$\nabla\times\boldsymbol{H}=\sigma\boldsymbol{E}+i\omega\epsilon_r\epsilon_0\boldsymbol{E}=i\omega\epsilon_0\left(\frac{\sigma}{i\omega\epsilon_0}+\epsilon_r\right)\boldsymbol{E}=i\omega\epsilon_0\hat{\epsilon}\,\boldsymbol{E} \tag{9.182}$$

where ϵ_r is the complex relative permittivity, and $\hat{\epsilon}$ is the complex permittivity given by

$$\epsilon_0\hat{\epsilon}\equiv\epsilon_0\epsilon_r-i\frac{\sigma}{\omega}. \tag{9.183}$$

From the Debye equation,

$$\epsilon_0\hat{\epsilon}=1+\chi(\omega)+\frac{\sigma}{i\omega\epsilon_0}. \tag{9.184}$$

A correspondence between the Debye equation and the Drude equation can be obtained by equating (9.181) to (9.184). The two equations are equivalent if

$$\epsilon_\infty=1, \tag{9.185}$$

$$t_0=\frac{1}{\nu_c}, \tag{9.186}$$

$$\chi(\omega)=\frac{\epsilon_s-1}{1+i\omega t_0}=-\frac{\left(\frac{\omega_p}{\nu_c}\right)^2}{1+i\left(\frac{\omega}{\nu_c}\right)}, \tag{9.187}$$

and

$$\sigma = \epsilon_0 \frac{\omega_p^2}{\nu_c}. \tag{9.188}$$

From (9.185)–(9.188),

$$\epsilon_s = 1 - \left(\frac{\omega_p}{\nu_c}\right)^2. \tag{9.189}$$

The previously derived update equation for the Debye function (9.167) can then also be used for the Drude function with a nonzero conductivity. The complex dielectric constant is calculated from the complex refractive index at the frequency of interest from (9.97) and (9.98). The Drude dielectric function (9.181) is separated into real and imaginary parts.

$$\epsilon' = 1 - \frac{\omega_p^2}{\omega^2 + \nu_c^2}, \tag{9.190}$$

and

$$\epsilon'' = -\frac{\omega_p^2 \nu_c}{\omega\left(\omega^2 + \nu_c^2\right)}. \tag{9.191}$$

The static dielectric constant for the Drude function is determined from (9.189).

As mentioned previously, for the $\sin\left[+i\left(\omega t - kr\right)\right]$ time convention for the incident field used to derive the FDTD update equation, $\epsilon'' = -2nK$ which is a negative number, so both plasma and relaxation frequencies in (9.191) are indeed positive. For the opposite incident field time convention, $\sin\left[-i\left(\omega t - kr\right)\right]$, then $\epsilon'' = 2nK$ is a *positive* number, and

$$\epsilon'' = \frac{\omega_p^2 \nu_c}{\omega\left(\omega^2 + \nu_c^2\right)}. \tag{9.192}$$

Once ϵ' and ϵ'' are determined from the complex refractive index, the plasma and relaxation frequencies are determined from (9.190) and (9.192).

$$\nu_c = \frac{1}{t_0} = \frac{\epsilon''\omega}{1 - \epsilon'} \tag{9.193}$$

and

$$\omega_p^2 = \frac{\epsilon''\omega}{\nu_c}\left(\omega^2 + \nu_c^2\right) = \omega^2 \frac{\left(\epsilon' - \epsilon_\infty\right)^2 + \left(\epsilon''\right)^2}{\left(\epsilon' - \epsilon_\infty\right)}. \tag{9.194}$$

The parameters ν_c and ω_p^2 are determined from the refractive index at the frequency of interest from (9.183) and (9.194). ϵ_∞, t_0, $\chi\left(\omega\right)$, σ, ϵ_s are then determined from (9.185)–(9.189). Thus, all parameters for the Debye update equation (9.167) have been found from the given complex refractive index.

An example of the Drude calculation is shown in Figure 9.14. A plane wave is incident upon a silver sphere in vacuum. The calculated electric field intensities in planes through the center of the sphere are plotted in a log contour plot in Figure 9.14(a). The results for the same example calculated from Mie theory are shown in Figure 9.14(b) for comparison.

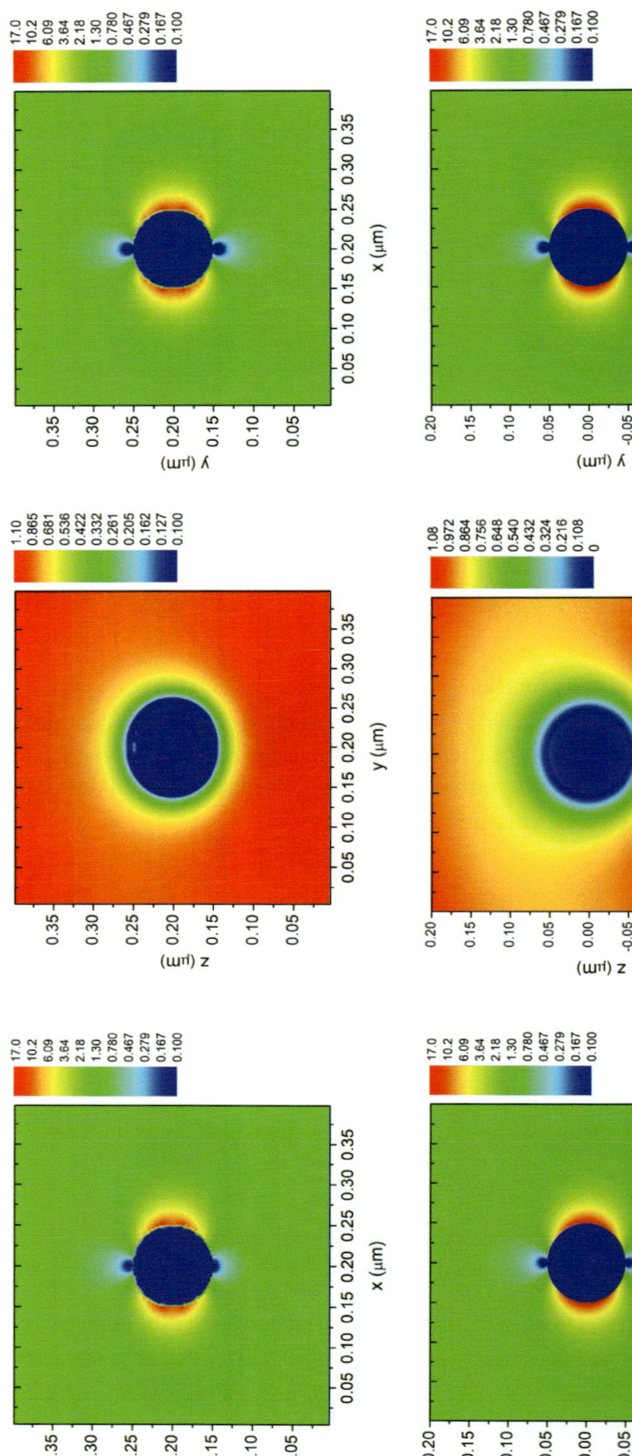

FIGURE 9.14 Log plots of the $|E|^2$ total intensity for a plane wave incident along the *z*-axis upon a 0.1 µm diameter silver sphere in vacuum with a complex refractive index of $0.14 + i\,(4.523)$ at a wavelength of 0.7 µm. Intensity is plotted on the *xz*, *yz*, and *xy* planes through the center of the sphere. (a) FDTD calculation using a cubic cell space with 200 cells per side and cubic Yee cells of edge length 0.002 µm carried out for 3,000 time steps. (b) Mie calculation for the same example.

9.9.4 Lorentz Dielectric Function

A material that exhibits a strong absorption at a specific wavelength due to an electronic or vibrational resonance can be modeled by a Lorentz dielectric function. The complex frequency-dependent Lorentz dielectric function is

$$\epsilon(\omega) = \epsilon_\infty + \frac{\omega_o^2(\epsilon_s - \epsilon_\infty)}{\omega_o^2 - \omega^2 + 2i\omega\delta_o}. \tag{9.195}$$

There are four parameters, the high frequency dielectric constant ϵ_∞, the static dielectric constant ϵ_s, the damping constant δ_o, and the oscillator frequency ω_o. Given a complex refractive index (or complex dielectric constant) at a specific wavelength, as well as ϵ_∞ and ϵ_s, the other two parameters can be determined. Separating the complex dielectric constant into its real and imaginary parts,

$$\epsilon' = \epsilon_\infty + \frac{\omega_p^2(\epsilon_s - \epsilon_\infty)(\omega_o^2 - \omega^2)}{(\omega_o^2 - \omega^2)^2 + 4\omega^2\delta_o^2} \tag{9.196}$$

$$\epsilon'' = -\frac{2\omega\delta_o\omega_o^2(\epsilon_s - \epsilon_\infty)}{(\omega_o^2 - \omega^2)^2 + 4\omega^2\delta_o^2}. \tag{9.197}$$

Solving for the damping constant and oscillator frequency,

$$\omega_o = \frac{\omega}{\sqrt{1 - \frac{(\epsilon_s - \epsilon_\infty)(\epsilon' - \epsilon_\infty)}{\epsilon''^2 + (\epsilon' - \epsilon_\infty)^2}}} \tag{9.198}$$

and

$$\delta_o = \frac{\epsilon''(\omega_o^2 - \omega^2)}{2\omega(\epsilon' - \epsilon_\infty)}. \tag{9.199}$$

(9.199) again assumes a *positive* value for ϵ''.

The frequency-domain susceptibility in (9.199) is

$$\chi(\omega) = \frac{\omega_o^2(\epsilon_s - \epsilon_\infty)}{\omega_o^2 - \omega^2 + 2i\omega\delta_o}. \tag{9.200}$$

The Fourier transform of the susceptibility (Eq. (8.38) in [1]) is

$$\chi(t) = \frac{1}{2\pi}\int_{-\infty}^{\infty} \frac{\omega_o^2(\epsilon_s - \epsilon_\infty)}{\omega_o^2 - \omega^2 + 2i\omega\delta_o} e^{i\omega t} d\omega = \frac{\omega_o^2(\epsilon_s - \epsilon_\infty)}{\sqrt{\omega_o^2 - \delta_p^2}} e^{-\delta_o t} \sin\left(t\sqrt{\omega_o^2 - \delta_o^2}\right) U(t). \tag{9.201}$$

To simplify the notation, let

$$\beta \equiv \sqrt{\omega_o^2 - \delta_o^2}. \tag{9.202}$$

To determine the update equation for χ_m, the susceptibility in (9.201) is put into a complex form,

$$\hat{\chi}(t)=\frac{\omega_o^2}{\beta}(\epsilon_s-\epsilon_\infty)e^{-\delta_o t}e^{i\beta t}U(t) \tag{9.203}$$

so that the susceptibility $\chi(t)$ is the imaginary part of $\hat{\chi}(t)$. Then

$$\hat{\chi}_m=\int_{m\Delta t}^{(m+1)\Delta t}\chi(\Lambda)d\Lambda=\frac{\omega_o^2(\epsilon_s-\epsilon_\infty)}{\beta(\delta_o-i\beta)}e^{-(\delta_o-i\beta)m\Delta t}\left[1-e^{-(\delta_o-i\beta)\Delta t}\right] \tag{9.204}$$

and

$$\hat{\chi}_0=\frac{\omega_o^2(\epsilon_s-\epsilon_\infty)}{\beta(\delta_o-i\beta)}\left[1-e^{-(\delta_o-i\beta)\Delta t}\right]. \tag{9.205}$$

The recursive equations to update these complex parameters are

$$\Delta\hat{\chi}_m=\frac{\omega_o^2(\epsilon_s-\epsilon_\infty)}{\beta(\delta_o-i\beta)}e^{-(\delta_o-i\beta)m\Delta t}\left[1-e^{-(\delta_o-i\beta)\Delta t}\right]^2=\hat{\chi}_m\left[1-e^{-(\delta_o-i\beta)\Delta t}\right], \tag{9.206}$$

$$\Delta\hat{\chi}_{m+1}=e^{-(\delta_o-i\beta)m\Delta t}\,\Delta\hat{\chi}_m, \tag{9.207}$$

$$\Delta\hat{\chi}_0=\frac{\omega_o^2(\epsilon_s-\epsilon_\infty)}{\beta(\delta_o-i\beta)}\left[1-e^{-(\delta_o-i\beta)\Delta t}\right]^2=\hat{\chi}_0\left[1-e^{-(\delta_o-i\beta)\Delta t}\right], \tag{9.208}$$

and

$$\hat{\Psi}_n=E_{inc}\,\Delta\hat{\chi}_0+e^{-(\delta_o-i\beta)\Delta t}\,\hat{\Psi}_{n-1}. \tag{9.209}$$

The imaginary parts of (9.208) and (9.209) provide the values for $\Delta\chi_0$, and $\hat{\Psi}_n$ for the update equation (9.167).

9.9.4.1 Incident Plane Wave

The convolution term in (9.169) must also be calculated for the Lorentz dielectric function.

$$\frac{\partial}{\partial t}\left[E_{inc}(t)*\chi_L(t)\right]=\frac{\partial}{\partial\tau}\int_{-\infty}^{\tau}E_0\,\sin(\omega\tau-\omega t)\frac{\omega_o^2}{\beta}(\epsilon_s-\epsilon_\infty)e^{-\delta_o t}\,\sin(\beta t)U(t)dt \tag{9.210}$$

$$=E_0\,\omega\,\omega_p^2(\epsilon_s-\epsilon_\infty)\frac{\left[(\omega_p^2-\omega^2)\cos(\omega\tau)+2\delta\omega\sin(\omega\tau)\right]-e^{-\delta_o\tau}\left[(\omega_p^2-\omega^2)\cos(\beta\tau)+\frac{\delta}{\beta}(\omega_p^2+\omega^2)\sin(\beta\tau)\right]}{(\omega_p^2-\omega^2)^2+(2\delta\omega)^2}. \tag{9.211}$$

τ is measured from the time at which the incident plane wave first reaches the point in the cell that is being calculated. When $\tau=0$, the convolution term in (9.211) is zero. This means that the contribution to the scattered electric field involving this term turns on slowly after the field reaches this point, reducing any effects of ringing in the calculation. For large times, the exponential part of this term goes to zero. Although it might seem that the exponential term can be neglected if the calculation is run to the steady state, it has been found that the calculation results are more accurate when the exponential part is retained throughout the calculation. However, the steady state is only reached after this term becomes negligible.

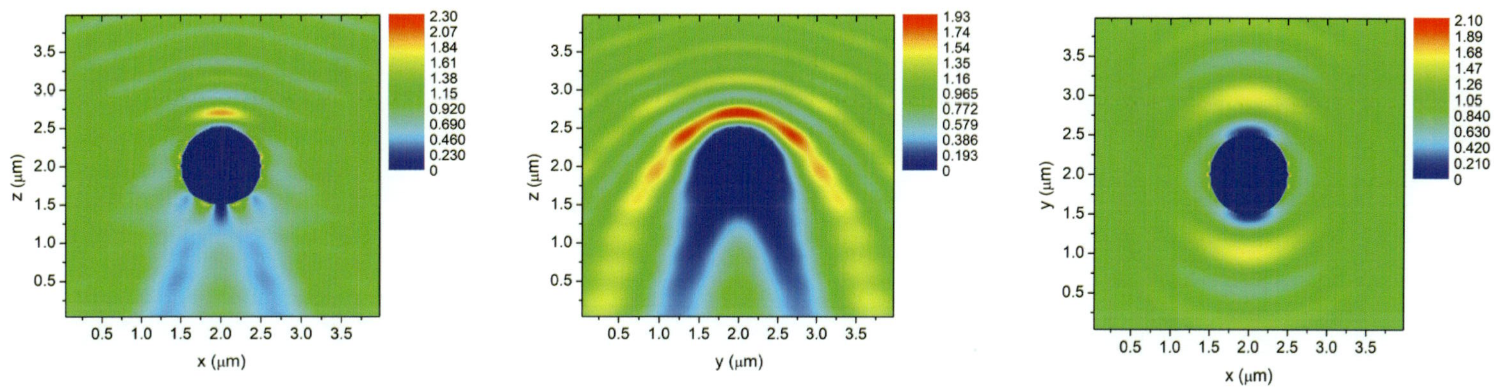

FIGURE 9.15 Scattering from a 1 μm lossy dielectric sphere with refractive index 3.0 + *i*(1.5) embedded in vacuum for an incident plane wave at a wavelength of 0.85 μm. The sphere is modeled with the Lorentz dielectric function using $\epsilon_{\infty} = 1$ and $\epsilon_s = 2$. The cell space is a cube with 200 cells on an edge and each cell is a cube with edge length 0.02 μm. The calculation is carried out for 800 time steps. The intensity is plotted on the *xz*, *yz*, and *xy* planes through the center of the sphere.

As an example, the lossy dielectric sphere in vacuum is again considered. The Lorentz dielectric function is used with $\epsilon_s = 2$ and $\epsilon_\infty = 1$. The resonant frequency is $\omega_0 = 2.2742 \times 10^{15}$ Hz. The damping frequency is $\delta_0 = 9.2071 \times 10^{13}$ Hz. The results are shown in Figure 9.15 to be compared with Figure 9.13.

9.9.4.2 Incident Cosine Pulse

An incident cosine pulse is discussed by Kunz et al. (pp. 155–161 in [1]). Let the susceptibility function be defined in a generic manner,

$$\chi(t) = A_p e^{-a_p t} U(t). \tag{9.212}$$

If τ is the delayed time of the electric field at the cell location, the pulse can be defined by

$$E^i(\tau) = \begin{cases} C_0 + \sum_{q=1}^{3} C_q \cos(q\omega_s t) & 0 \le \tau \le T \\ 0 & otherwise \end{cases}. \tag{9.213}$$

Kunz and Luebbers choose the values $C_0 = \frac{10}{32}, C_1 = -\frac{15}{32}, C_2 = \frac{6}{32}$, and $C_3 = -\frac{1}{32}$. The length of the pulse is T. Also,

$$\omega_s \equiv \frac{2\pi}{T}. \tag{9.214}$$

Following the same procedure to calculate the convolution term, as always for times $t < \tau$ the convolution is zero. For larger times,

$$\frac{\partial}{\partial t}\left[E_{inc}(t) * \chi_L(t)\right] =$$

$$\begin{cases} A_p C_0 e^{-a_p \tau} + \sum_{q=1}^{3} \frac{A_p C_q}{a_p^2 + (q\omega_s)^2}\left[(q\omega_s)^2 \cos(q\omega_s \tau) - a_p q\omega_s \sin(q\omega_s \tau) + a_p^2 e^{-a_p \tau}\right] & 0 \le \tau \le T \\ A_p e^{-a_p \tau}\left(1 - e^{-a_p T}\right)\left(C_0 + \sum_{q=1}^{3} \frac{C_q A_p^2}{a_p^2 + (q\omega_s)^2}\right) & \tau > T \end{cases} \tag{9.215}$$

9.10 SUBSTRATES

Often one wishes to model a scattering object deposited on a substrate. In scattered field FDTD it is straightforward to include substrates. Instead of simply using an incident plane wave, the incident field is that within the combined semi-infinite upper medium and semi-infinite bottom medium as shown in Figure 9.16.

9.10.1 *P*-POLARIZATION

The incident electric field lies in the plane of incidence for *p*-polarization. The electric and magnetic fields throughout space are

$$\boldsymbol{E}_i = E_0\left(\cos\theta_i\,\hat{\boldsymbol{x}} + \sin\theta_i\,\hat{\boldsymbol{z}}\right)e^{-ik_i\left[x\sin\theta_i - (z-z_0)\cos\theta_i\right]}, \tag{9.216}$$

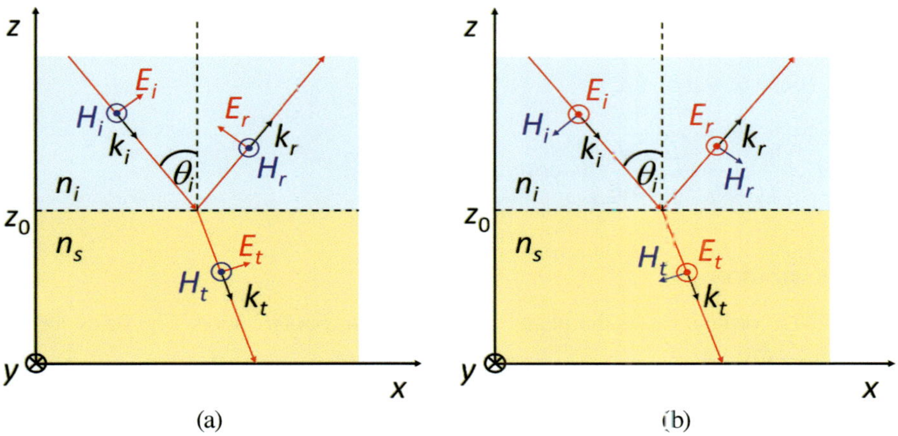

FIGURE 9.16 Plane wave incident at polar angle θ within the medium of refractive index n_i onto a substrate with refractive index n_s in (a) transverse magnetic (TM) or p-polarization, and (b) transverse electric (TE) or s-polarization.

$$\boldsymbol{E}_r = E_r\left(-\cos\theta_i\,\hat{\boldsymbol{x}} + \sin\theta_i\,\hat{\boldsymbol{z}}\right)e^{-ik_i\left[x\sin\theta_i + (z - z_0)\cos\theta_i\right]}, \tag{9.217}$$

$$\boldsymbol{E}_t = E_t\left(\cos\theta_s\,\hat{\boldsymbol{x}} + \sin\theta_s\,\hat{\boldsymbol{z}}\right)e^{-ik_s\left[x\sin\theta_s - (z - z_0)\cos\theta_s\right]}, \tag{9.218}$$

$$\boldsymbol{H}_i = -\frac{n_i E_0}{\eta_0}\,\hat{\boldsymbol{y}}e^{-ik_i\left[x\sin\theta_i - (z - z_0)\cos\theta_i\right]}, \tag{9.219}$$

$$\boldsymbol{H}_r = -\frac{n_i E_r}{\eta_0}\,\hat{\boldsymbol{y}}e^{-ik_i\left[x\sin\theta_i + (z - z_0)\cos\theta_i\right]}, \tag{9.220}$$

and

$$\boldsymbol{H}_t = -\frac{n_s E_t}{\eta_0}\,\hat{\boldsymbol{y}}e^{-ik_s\left[x\sin\theta_s - (z - z_0)\cos\theta_s\right]}. \tag{9.221}$$

According to Snell's law,

$$n_i\,\sin\theta_i = n_s\,\sin\theta_s. \tag{9.222}$$

The tangential components of the fields are continuous at the boundary, $z = z_0$. This provides two equations that can be solved for the unknown transmitted and reflected fields in terms of the incident field amplitude. The equations are simplified using Snell's law. The transmitted electric field is

$$E_t = E_0\left(\frac{2n_i\cos\theta_i}{n_i\cos\theta_s + n_s\cos\theta_i}\right). \tag{9.223}$$

The transmitted reflected field is

$$E_r = E_0\left(\frac{n_s\cos\theta_i - n_i\cos\theta_s}{n_i\cos\theta_s + n_s\cos\theta_i}\right) \tag{9.224}$$

The total electric field in the upper medium for p-polarization is

$$\boldsymbol{E}_{\text{tot}} = E_0\left(\cos\theta_i\,\hat{\boldsymbol{x}} + \sin\theta_i\,\hat{\boldsymbol{z}}\right)e^{-ik_i\left[x\sin\theta_i - (z-z_0)\cos\theta_i\right]}$$

$$+E_0\left(\frac{n_s\cos\theta_i - n_i\cos\theta_s}{n_i\cos\theta_s + n_s\cos\theta_i}\right)\left(-\cos\theta_i\,\hat{\boldsymbol{x}} + \sin\theta_i\,\hat{\boldsymbol{z}}\right)e^{-ik_i\left[x\sin\theta_i + (z-z_0)\cos\theta_i\right]}. \quad (9.225)$$

9.10.2 S-POLARIZATION

The electric field is orthogonal to the plane in incidence for s-polarization. The fields for an incident s-polarized plane wave are

$$\boldsymbol{E}_i = -E_0\hat{\boldsymbol{y}}e^{-ik_i\left[x\sin\theta_i - (z-z_0)\cos\theta_i\right]}, \quad (9.226)$$

$$\boldsymbol{E}_r = -E_r\hat{\boldsymbol{y}}e^{-ik_i\left[x\sin\theta_i + (z-z_0)\cos\theta_i\right]}, \quad (9.227)$$

$$\boldsymbol{E}_t = -E_t\hat{\boldsymbol{y}}e^{-ik_s\left[x\sin\theta_s - (z-z_0)\cos\theta_s\right]}, \quad (9.228)$$

$$\boldsymbol{H}_i = \frac{n_i E_0}{\eta_0}\left(-\cos\theta_i\hat{\boldsymbol{x}} - \sin\theta_i\hat{\boldsymbol{z}}\right)e^{-ik_i\left[x\sin\theta_i - (z-z_0)\cos\theta_i\right]}, \quad (9.229)$$

$$\boldsymbol{H}_r = \frac{n_i E_r}{\eta_0}\left(+\cos\theta_i\hat{\boldsymbol{x}} - \sin\theta_i\hat{\boldsymbol{z}}\right)e^{-ik_i\left[x\sin\theta_i + (z-z_0)\cos\theta_i\right]}, \quad (9.230)$$

and

$$\boldsymbol{H}_t = \frac{n_s E_t}{\eta_0}\left(-\cos\theta_s\hat{\boldsymbol{x}} - \sin\theta_s\hat{\boldsymbol{z}}\right)e^{-ik_s\left[x\sin\theta_s - (z-z_0)\cos\theta_s\right]}. \quad (9.231)$$

Again, after matching boundary conditions at the surface and applying Snell's law, the transmitted electric field is

$$E_t = E_0\left(\frac{2n_i\cos\theta_i}{n_i\cos\theta_i + n_s\cos\theta_s}\right), \quad (9.232)$$

the reflected electric field is

$$E_r = E_0\left(\frac{n_i\cos\theta_i - n_s\cos\theta_s}{n_i\cos\theta_i + n_s\cos\theta_s}\right), \quad (9.233)$$

ands the total electric field in the upper region is

$$\boldsymbol{E}_{\text{tot}} = -E_0\,\hat{\boldsymbol{y}}\left[e^{-ik_i\left[x\sin\theta_i - (z-z_0)\cos\theta_i\right]} + \left(\frac{n_i\cos\theta_i - n_s\cos\theta_s}{n_i\cos\theta_i + n_s\cos\theta_s}\right)e^{-ik_i\left[x\sin\theta_i + (z-z_0)\cos\theta_i\right]}\right]. \quad (9.234)$$

In the FDTD update equations, the incident field is now calculated by (9.223) within the substrate region and (9.225) within the incident medium for p-polarization, and by (9.232) within the substrate region and (9.234) within the incident medium for s-polarization. As previously mentioned,

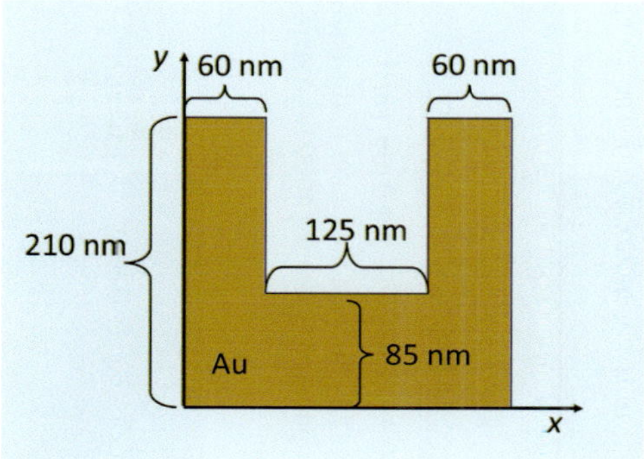

FIGURE 9.17 U-shaped split ring resonator composed of gold. The gold thickness is 50 nm. Values for the refractive index of gold are from [20]. The cell space is 100 × 100 × 100 and each Yee cell is a cube with 5 nm on a side.

an advantage of the scattered field formalism over the total field formalism is that the incident field can be calculated analytically throughout the computation space even for more complicated incident fields than simple plane waves. This section demonstrates the inclusion of a substrate for the incident field. The fields for planar waveguides and strip/rib waveguides derived in previous chapters may also be used as the incident field within the computation space. The FDTD calculation will then model just the scattered fields of a scattering object on the substrate or within the waveguide, thereby improving the accuracy of the result. For a single scattering object on top or within a substrate, reradiating boundary conditions can be applied to all faces of the computation space. For an array of objects on a planar substrate, reradiating boundary conditions can be applied on the top and bottom faces of the computation space and periodic boundary conditions on the other four faces.

As a simple example, consider a gold split ring resonator (SRR) in vacuum shown in Figure 9.17. The electric field for a normally incident plane wave polarized along the x-axis perpendicular to the two prongs is shown for three wavelengths in Figure 9.18 without a substrate and in Figure 9.19 with a substrate of refractive index $n = 2.0$. The effect of the substrate is to shift the resonant mode to longer wavelengths and reduce the maximum scattered $|E|^2$ field intensity. An array of SRR's is considered in the examples at the end of the chapter.

9.11 SCATTERED FAR FIELDS, POWER DISSIPATION, AND CROSS SECTIONS

9.11.1 Scattered Far Fields

The far field radiation from a near field source can be calculated by a procedure due to Shelkunoff [10, 11]. Consider a closed region containing electromagnetic sources. If the fields on the surface are known, fictitious currents on this surface can be calculated that in turn will reproduce the fields outside the surface that would be generated by the internal sources. Although the fields on the surface may have both tangential and normal components, only the tangential components contribute to the far fields.

The electric and magnetic fields at an observation point outside the surface are given in terms of the vector potentials **N** and **L** at that point by (p. 664 in [11])

$$\boldsymbol{E}_{\mathrm{o}} = i\omega \boldsymbol{N}_{\mathrm{o}} + \frac{i\omega}{k^2}\boldsymbol{\nabla}(\boldsymbol{\nabla}\cdot\boldsymbol{N}_{\mathrm{o}}) - \frac{1}{\epsilon}\boldsymbol{\nabla}\times\boldsymbol{L}_{\mathrm{c}}, \tag{9.235}$$

and

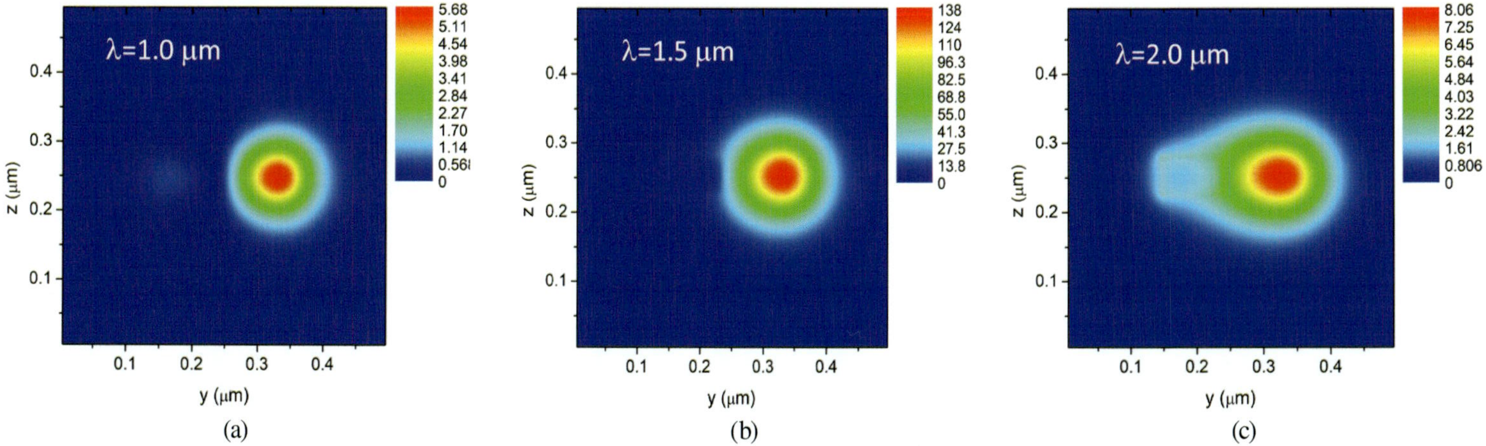

FIGURE 9.18 Total scattered field intensity in the *yz* plane (i = 50) from a gold split ring resonator in air with refractive indices (a) $0.257 + i(6.820)$ at 1.0 μm, (b) $0.53 + i(9.51)$ at 1.5 μm, and (c) $0.85 + i(12.6)$ at 2.0 μm for a normally incident plane wave polarized along the *x*-axis orthogonal to the prongs and through the center of the resonator. The high field location is in the gap between the prongs. The cell space is 100 × 100 × 100 with 0.005 μm cubic cell. The computation is run for 4000 time steps.

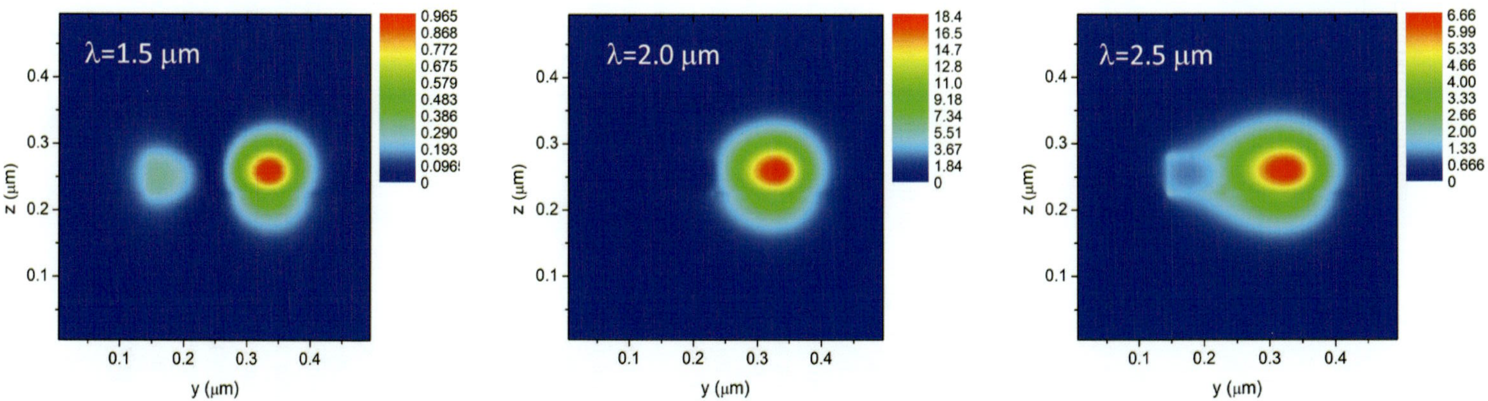

FIGURE 9.19 Same parameters as Figure 9.18 but with the SRR on a substrate with refractive index 2.0.

$$\boldsymbol{H}_{\text{o}} = i\omega \boldsymbol{L}_{\text{o}} + \frac{i\omega}{k^2}\boldsymbol{\nabla}(\boldsymbol{\nabla}\cdot \boldsymbol{L}_{\text{o}}) + \frac{1}{\mu}\boldsymbol{\nabla}\times \boldsymbol{N}_{\text{o}}. \tag{9.236}$$

These equations can be simplified in the far field by keeping only the propagating field terms that decay as $1/r^2$. (p. 664 in [11]) [12] The result in spherical coordinates is

$$E_\theta(\omega) = i\frac{e^{-ikR}}{2\lambda R}\left(-Z_0 N_\theta - L_\phi\right), \tag{9.237}$$

$$E_\phi(\omega) = i\frac{e^{-ikR}}{2\lambda R}\left(-Z_0 N_\phi + L_\theta\right), \tag{9.238}$$

$$H_\theta(\omega) = -\frac{E_\phi(\omega)}{Z_0}, \tag{9.239}$$

and

$$H_\phi(\omega) = \frac{E_\theta(\omega)}{Z_0}. \tag{9.240}$$

The vector potentials are calculated from surface currents which in turn are calculated by FDTD in Cartesian coordinates, so it is helpful to derive the vector potential transform between spherical and Cartesian coordinates. The coordinate transforms for Cartesian to polar coordinates are

$$\hat{\boldsymbol{r}} = \sin\theta(\cos\phi\hat{\boldsymbol{x}} + \sin\phi\hat{\boldsymbol{y}}) + \cos\theta\hat{\boldsymbol{z}}, \tag{9.241}$$

$$\hat{\boldsymbol{\theta}} = \cos\theta(\cos\phi\hat{\boldsymbol{x}} + \sin\phi\hat{\boldsymbol{y}}) - \sin\theta\hat{\boldsymbol{z}}, \tag{9.242}$$

and

$$\hat{\boldsymbol{\phi}} = -\sin\phi\hat{\boldsymbol{x}} + \cos\phi\hat{\boldsymbol{y}}. \tag{9.243}$$

Therefore, the components of the vector potentials in polar coordinates are given by

$$\begin{aligned} N_\theta = \boldsymbol{N}\cdot\hat{\boldsymbol{\theta}}_{\text{o}} &= \boldsymbol{N}\cdot(\hat{\boldsymbol{x}}\cos\theta_{\text{o}}\cos\phi_{\text{o}} + \hat{\boldsymbol{y}}\cos\theta_{\text{o}}\sin\phi_{\text{o}} - \hat{\boldsymbol{z}}\sin\theta_{\text{o}}) \\ &= N_x\cos\theta_{\text{o}}\cos\phi_{\text{o}} + N_y\cos\theta_{\text{o}}\sin\phi_{\text{o}} - N_z\sin\theta_{\text{o}} \end{aligned} \tag{9.244}$$

and

$$N_\phi = \boldsymbol{N}\cdot\hat{\phi}_{\text{o}} = \boldsymbol{N}\cdot(-\hat{\boldsymbol{x}}\sin\phi_{\text{o}} + \hat{\boldsymbol{y}}\cos\phi_{\text{o}}) = -N_x\sin\phi_{\text{c}} + N_y\cos\phi_{\text{c}}, \tag{9.245}$$

and similarly for L_θ and L_ϕ in terms of L_x, L_y, and L_z. The subscript "o" denotes the vector direction towards the point of observation from the center of the computation space. Substituting (9.244) and (9.245) into (9.237) and (9.238),

$$E_\theta(\omega) = i\frac{e^{-ikR}}{2\lambda R}\left[-Z_0\left(N_x\cos\theta_{\text{o}}\cos\phi_{\text{o}} + N_y\cos\theta_{\text{o}}\sin\phi_{\text{o}} - N_z\sin\theta_{\text{o}}\right) - \left(-L_x\sin\phi_{\text{o}} + L_y\cos\phi_{\text{o}}\right)\right], \tag{9.246}$$

and

$$E_\phi(\omega) = i\frac{e^{-ikR}}{2\lambda R}\left[-Z_0\left(-N_x \sin\phi_o + N_y \cos\phi_o\right) + \left(L_x \cos\theta_o \cos\phi_o + L_y \cos\theta_o \sin\phi_o - L_z \sin\theta_o\right)\right].$$

(9.247)

The vector potentials are defined by closed surface integrals over the surface currents (p. 107 in [1]).

$$N(\omega, x_o, y_o, z_o) \equiv \frac{1}{4\pi R}\int_{S'} J_s(\omega, x', y', z') e^{+ik_m \hat{R}\cdot r'} dA', \tag{9.248}$$

and

$$L(\omega, x_o, y_o, z_o) \equiv \frac{1}{4\pi R}\int_{S'} M_s(\omega, x', y') e^{+ik_m \hat{R}\cdot r'} dA'. \tag{9.249}$$

k_m is the wavevector within the background medium and S' is a closed surface around the FDTD computation space which is chosen to be its six outer faces. dA' is the area differential. For the *xy* faces, $dA' = dx\,dy$, for the *xz* faces, $dA' = dx\,dz$, and for the *yz* faces, $dA' = dy\,dz$. r' is the vector from the center of the computation space to the source point on the surface of the computation space. $\hat{R}$ is the unit vector from the center of the computation space along the far field observation angle (θ_o, ϕ_o). The vector from the center of the computation space to the center of cell (i, j, k) is

$$r'(i,j,k) = \Delta x\left(i - \frac{n_x}{2}\right)\hat{x} + \Delta y\left(j - \frac{n_y}{2}\right)\hat{y} + \Delta z\left(k - \frac{n_z}{2}\right)\hat{z}. \tag{9.250}$$

The unit vector from the center of the computation space to the point of observation at polar angles (θ_o, ϕ_o) is

$$\hat{R} = x_o\hat{x} + y_o\hat{y} + z_o\hat{z} = \sin\theta_o\left(\cos\phi_o\hat{x} + \sin\phi_o\hat{y}\right) + \cos\theta_o\hat{z}. \tag{9.251}$$

Therefore, for surface cell (i, j, k),

$$\hat{R}\cdot r' = \left[\Delta x\left(i - \frac{n_x}{2}\right)\cos\phi_o + \Delta y\left(j - \frac{n_y}{2}\right)\sin\phi_o\right]\sin\theta_o + \Delta z\left(k - \frac{n_z}{2}\right)\cos\theta_o. \tag{9.252}$$

In the FDTD calculation, the surface integrals in (9.248) and (9.249) are replaced by summations over all the surface cells on the six faces of the computation space. The *yz* faces occur at the centers of the cells forming the $i = 1$ and $n_x - 1$ faces, respectively. (In practice, the outer few *yz* planes may be used for implementing the boundary condition algorithm, so the actual cell faces used for the summation may need to be moved towards the center of the computation space by a few Yee cells.) The *xz* surface faces occur at the centers of the cells with $j = 1$ and $n_y - 1$, respectively, and the *xy* surface faces occur at cells for which $k = 1$ and $n_z - 1$, respectively.

$$N(\omega, x_o, y_o, z_o) \equiv \frac{1}{4\pi R}\sum_{i=1} J_s(\omega, i, j, k) e^{-ik_m \hat{R}\cdot r'}\Delta y\Delta z + \frac{1}{4\pi R}\sum_{i=n_x} J_s(\omega, i, j, k) e^{-ik_m \hat{R}\cdot r'}\Delta y\Delta z$$

$$+\frac{1}{4\pi R}\sum_{j=1} J_s(\omega, i, j, k) e^{-ik_m \hat{R}\cdot r'}\Delta x\Delta z + \frac{1}{4\pi R}\sum_{j=n_y} J_s(\omega, i, j, k) e^{-k_m \hat{R}\cdot r'}\Delta x\Delta z$$

$$+\frac{1}{4\pi R}\sum_{k=1}\boldsymbol{J}_s(\omega,i,j,k)e^{-ik_m\hat{\boldsymbol{R}}\cdot\boldsymbol{r}'}\Delta x\Delta y+\frac{1}{4\pi R}\sum_{k=n_x}\boldsymbol{J}_s(\omega,i,j,k)e^{-ik_m\hat{\boldsymbol{R}}\cdot\boldsymbol{r}'}\Delta x\Delta y. \quad (9.253)$$

and similarly for $\boldsymbol{L}$ with $\boldsymbol{M}_s$ substituted for $\boldsymbol{J}_s$.

The surface currents in (9.253) at surface cell (*i*, *j*, *k*) are determined by the fields at the center of that cell (p. 663 of [11]).

$$\boldsymbol{J}_s(\omega)=\hat{\boldsymbol{n}}\times\boldsymbol{H}(\omega), \quad (9.254)$$

and

$$\boldsymbol{M}_s(\omega)=-\hat{\boldsymbol{n}}\times\boldsymbol{E}(\omega). \quad (9.255)$$

$\hat{\boldsymbol{n}}$ is the outward pointing normal to the surface. For the *xy* plane at the $k = 0$ face, $\hat{\boldsymbol{n}}=-\hat{\boldsymbol{z}}$. (9.254) and (9.255) reduce to

$$\boldsymbol{J}_s(\omega)=H_y(\omega)\,\hat{\boldsymbol{x}}-H_x(\omega)\,\hat{\boldsymbol{y}}, \quad (9.256)$$

and

$$\boldsymbol{M}_s(\omega)=-E_y(\omega)\,\hat{\boldsymbol{x}}+E_x(\omega)\,\hat{\boldsymbol{y}}. \quad (9.257)$$

For the *xy* plane at $k = k_{max}$ face on the other side of the computation space, the outward normal is $\hat{\boldsymbol{n}}=+\hat{\boldsymbol{z}}$, so the signs of the field components in (9.256) and (9.257) are reversed.

For the $x = 0$ face of the computation space in the *yz* plane, the outward normal is $\hat{\boldsymbol{n}}=-\hat{\boldsymbol{x}}$, so

$$\boldsymbol{J}_s=H_z\,\hat{\boldsymbol{y}}-H_y\,\hat{\boldsymbol{z}}, \quad (9.258)$$

and

$$\boldsymbol{M}_s=-E_z\,\hat{\boldsymbol{y}}+E_y\,\hat{\boldsymbol{z}}. \quad (9.259)$$

The signs of the field components are reversed for the $i = i_{max}$ face of the computation space.

Finally, for the $y = 0$ face of the computation space in the *xz* plane, the outward normal is $\hat{\boldsymbol{n}}=-\hat{\boldsymbol{y}}$, so

$$\boldsymbol{J}_s=-H_z\,\hat{\boldsymbol{x}}+H_x\,\hat{\boldsymbol{z}}, \quad (9.260)$$

and

$$\boldsymbol{M}_s=E_z\,\hat{\boldsymbol{x}}-E_x\,\hat{\boldsymbol{z}}. \quad (9.261)$$

Again, the signs of the field components are reversed for the $j = j_{max}$ face.

The field amplitudes and phases are calculated for each surface cell from the fields at both the end and at one quarter period before the end of the FDTD calculation from (9.75) and (9.79), assuming that the calculation has reached a steady sinusoidal state. Because a quarter period will not generally be equal to an integral number of time steps, it is also appropriate to linearly interpolate between the fields calculated for time steps on either side of the quarter period time, but this is usually just a minor correction. The fields are averaged on the edges and center faces of each surface Yee cell using (9.32)–(9.37) to find both the electric and the magnetic fields at the center of each cell. They must also be found at the same point in time, so the magnetic fields are also averaged for the half time step before and after the electric fields.

Once the far fields are computed in Cartesian coordinates, the more useful form of the far fields in spherical coordinates is easily obtained through the coordinate transforms,

$$E_\theta = -\frac{E_z}{\sin\theta_o}, \tag{9.262}$$

and

$$E_\phi = -\sin\phi_o E_x + \cos\phi_o E_y. \tag{9.263}$$

Of course, it is entirely possible to first compute the vector potentials in spherical coordinates from (9.221) and (9.222) applying the coordinate transforms (9.241) and (9.242), and then find E_θ and E_ϕ directly. In that case, if the Cartesian components of the far field are desired, the inverse relations are

$$E_x = \cos\phi_o \cos\theta_o E_\theta - \sin\phi_o E_\phi, \tag{9.264}$$

$$E_y = \sin\phi_o \cos\theta_o E_\theta + \cos\phi_o E_\phi, \tag{9.265}$$

$$E_z = -\sin\theta_o E_\theta, \tag{9.266}$$

$$H_x = -\frac{1}{Z_0}\left(\cos\phi_o \cos\theta_o E_\phi + \sin\phi_o E_\theta\right), \tag{9.267}$$

$$H_y = -\frac{1}{Z_0}\left(\sin\phi_o \cos\theta_o E_\phi - \cos\phi_o E_\theta\right), \tag{9.268}$$

and

$$H_z = \frac{1}{Z_0}\sin\theta_o E_\theta. \tag{9.269}$$

9.11.2 Scattering Cross Section and Efficiency

The scattering cross section is the ratio of the total scattered power to the incident power per unit area.

$$\sigma_s \equiv \frac{P_s}{S_i}. \tag{9.270}$$

The incident power per unit area of the plane wave is given by the Poynting vector.

$$|\boldsymbol{S}_i| = \frac{1}{2}|\boldsymbol{E}_i \times \boldsymbol{H}_i| = \frac{n_m}{2Z_0}|\boldsymbol{E}_i|^2. \tag{9.271}$$

The scattering efficiency is the scattering cross section normalized by the cross-sectional area of the object to the incident plane wave. For a spherical particle of diameter d,

$$\varepsilon_s \equiv \frac{4\sigma_s}{\pi d^2}. \tag{9.272}$$

The scattered power per unit area can be calculated for each face of the computation space from the scattered electric and magnetic fields in each surface Yee cell. For example, at the $z = 0$ face, the outward normal to the surface is $-\hat{z}$. The Poynting vector integrated over this face becomes a summation over all the Yee cells on this face,

$$\begin{aligned} I_1 &\equiv \iint_{z=0} \left(\frac{1}{2}\boldsymbol{E}_s \times \boldsymbol{H}_s\right)\cdot \hat{\boldsymbol{n}}\, dx\, dy = \frac{\Delta x \cdot \Delta y}{2}\sum_{n,k=1}(\boldsymbol{E}_s \times \boldsymbol{H}_s)\cdot(-\hat{z}) \\ &= \left(\frac{\Delta x \cdot \Delta y}{2}\right)\sum\nolimits_{n,z=0}\left(E_{s,y}H_{s,x} - E_{s,x}H_{s,y}\right) \end{aligned} \tag{9.273}$$

For the $z = z_{\max}$ face of the computation space, the outward normal is $\hat{z}$. The integral of the Poynting vector over this face is

$$I_2 \equiv \frac{\Delta x \cdot \Delta y}{2}\sum\nolimits_{n,k=1}(\boldsymbol{E}_s \times \boldsymbol{H}_s)\cdot \hat{z} = -\left(\frac{\Delta x \cdot \Delta y}{2}\right)\sum\nolimits_{n,z=z_{max}}\left(E_{s,y}H_{s,x} - E_{s,z}H_{s,y}\right). \tag{9.274}$$

Similarly, for the $x = 0$ face,

$$I_3 \equiv \iint_{x=0}\left(\frac{1}{2}\boldsymbol{E}_s \times \boldsymbol{H}_s\right)\cdot \hat{\boldsymbol{n}}\, dy\, dz = \frac{\Delta y \cdot \Delta z}{2}\sum\nolimits_{n,x=0}\left(E_{s,z}H_{s,y} - E_{s,y}H_{s,z}\right), \tag{9.275}$$

for the $x = x_{\max}$ face,

$$I_4 \equiv \iint_{x=x_{max}}\left(\frac{1}{2}\boldsymbol{E}_s \times \boldsymbol{H}_s\right)\cdot \hat{\boldsymbol{n}}\, dy\, dz = -\frac{\Delta y \cdot \Delta z}{2}\sum\nolimits_{n,x=x_{max}}\left(E_{s,z}H_{s,y} - E_{s,y}H_{s,z}\right), \tag{9.276}$$

for the $y = 0$ face,

$$I_5 \equiv \iint_{y=0}\left(\frac{1}{2}\boldsymbol{E}_s \times \boldsymbol{H}_s\right)\cdot \hat{\boldsymbol{n}}\, dx\, dz = \frac{\Delta x \cdot \Delta z}{2}\sum\nolimits_{n,y=0}\left(E_{s,x}H_{s,z} - E_{s,z}H_{s,x}\right), \tag{9.277}$$

and finally for the $y = y_{\max}$ face,

$$I_6 \equiv \iint_{y=y_{max}}\left(\frac{1}{2}\boldsymbol{E}_s \times \boldsymbol{H}_s\right)\cdot \hat{\boldsymbol{n}}\, dx\, dz = -\frac{\Delta x \cdot \Delta z}{2}\sum\nolimits_{n,y_{max}}\left(E_{s,x}H_{s,z} - E_{s,z}H_{s,x}\right). \tag{9.278}$$

Therefore, the total optical power scattered by the object is

$$P_s = I_1 + I_2 + I_3 + I_4 + I_5 + I_6\,. \tag{9.279}$$

As an example, consider a dielectric sphere with a refractive index of 2.0 and a diameter of 0.1 μm, which is much smaller than the incident wavelength of 2.0 μm. If the incident plane wave is polarized along the x-axis, the dominant charge oscillation in the particle is along this axis. There is, therefore, almost no radiation along the x-axis, but almost uniform radiation intensity for all polar angles in the yz plane perpendicular to the x-axis. In the xy plane, the characteristic dipole radiation pattern is observed for E_ϕ with a maximum intensity along the y-axis and zero intensity along the x-axis. The charge oscillation along the x-axis also generates an E_θ field component along the z-axis in the xz-plane. The far field scattering efficiencies for the polar and azimuthal electric field components along the three orthogonal planes through the center of the sphere are plotted in Figure 9.20.

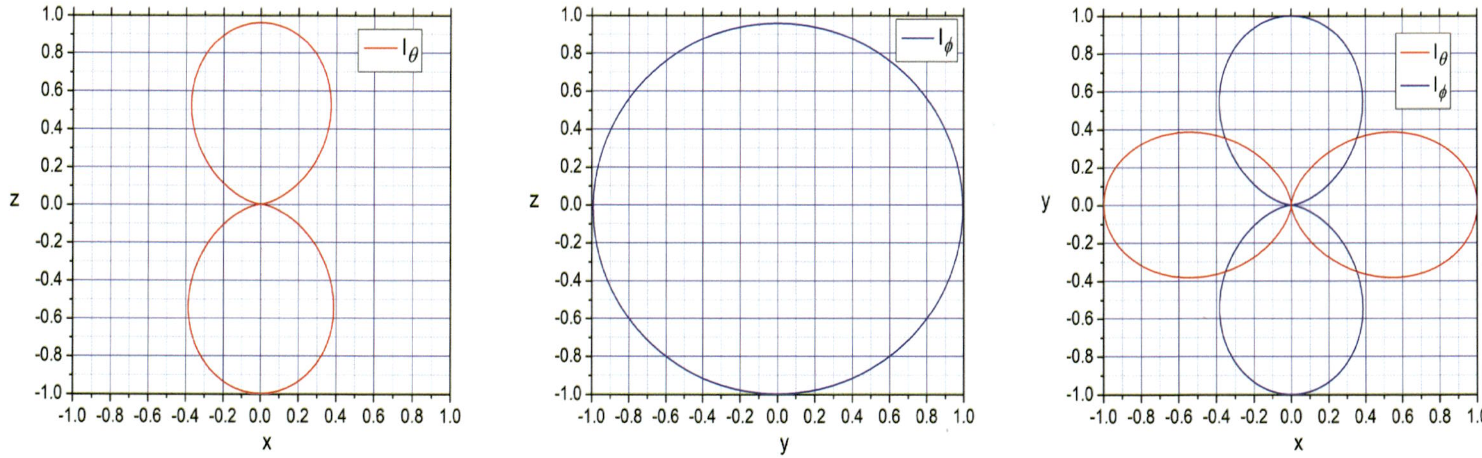

(a) (b) (c)

FIGURE 9.20 Normalized far field scattering efficiencies for (a) *xz* plane at $\phi = 90°$, *yz* plane at $\phi = 0°$, and *xy* plane at $\theta = 90°$. The computation space is 60 cells along each edge with dimensions of 0.01 μm and the calculation is carried out for 5,000 time steps. Note that the scattering efficiencies for the E_θ field component in (a) and the E_θ field component in (b) are too small to be accurately computed by FDTD.

9.11.3 Absorption Cross Section and Dissipated Power

The total power absorbed by an object can be calculated from the electric field amplitude and the object's conductivity. The electric field at the center of each Yee cell within the object is calculated from (9.32)–(9.34) at the final time step and one quarter period before the final time step in order to find the amplitude $|E_{ave}|$ of the sinusoidal wave within that cell from (9.87). The total power absorbed within an object is then the sum over cells within the object,

$$P_a = \frac{1}{2}\sum_n \sigma_n |E_{n,ave}|^2 \tag{9.280}$$

where, as before, σ_n is the real part of the optical conductivity. The factor of $\frac{1}{2}$ comes about due to the time-average of the sinusoidal variation of the electric field squared. Given the complex index of refraction at a specific wavelength, the conductivity at that wavelength is given by (9.116).

The absorption cross section is the area given by the ratio of the total power absorbed by the object to the total power per unit area that is incident upon it. For an incident plane wave,

$$\sigma_a = \frac{P_a}{S_i}. \tag{9.281}$$

The absorption efficiency is the absorption cross section normalized by the cross-sectional area of the object to the incident plane wave.

The extinction cross section is the sum of the scattering and absorption cross sections,

$$\sigma_e = \sigma_a + \sigma_s, \tag{9.282}$$

and the extinction efficiency is likewise the sum of the scattering and absorption efficiencies.

These cross-section efficiencies are calculated for a gold sphere with radius up to 0.4 μm by both FDTD and Mie theory for comparison. The wavelength is 0.55 μm for which the gold index of refraction is 0.497 + *i* (2.37) so the Drude model is used. The cubic cell dimension and number of time steps as well as the calculated cross-section efficiencies are tabulated in Table 9.1 and plotted in Figure 9.21.

9.12 ADDITIONAL EXAMPLES

9.12.1 Scattering in a Dense Medium (1)

The previous examples considered a plane wave in vacuum incident upon a spherical object. This example considers a plane wave in a medium with index of refraction of 2 incident upon a 100 nm air bubble with an index of refraction of 1. The cell space is 200 cubic cells on each edge with cell edge length of 2.5 nm. The vacuum wavelength is 0.7 μm. The FDTD results using the single wavelength dielectric model are shown in Figure 9.22(a) and the corresponding results from Mie theory are shown in Figure 9.22(b).

9.12.2 Scattering in a Dense Medium (2)

As a second example, consider a plane wave in a dense medium of refractive index 2 incident upon a 100 nm silver sphere with complex refractive index 0.14 + *i* (4.523). The vacuum wavelength is 0.7 μm. This example is solved with FDTD using the Debye, Drude, and Lorentz models, as well as by Mie scattering. For the Debye model, the cell space was 100 cells on an edge with cubic cells of edge length 5 nm. $\sigma_0 = 5.0 \times 10^7\ (\Omega\ \text{m})^{-1}$ and $\epsilon_\infty = 4$. For the Drude and Lorentz models, the cell space

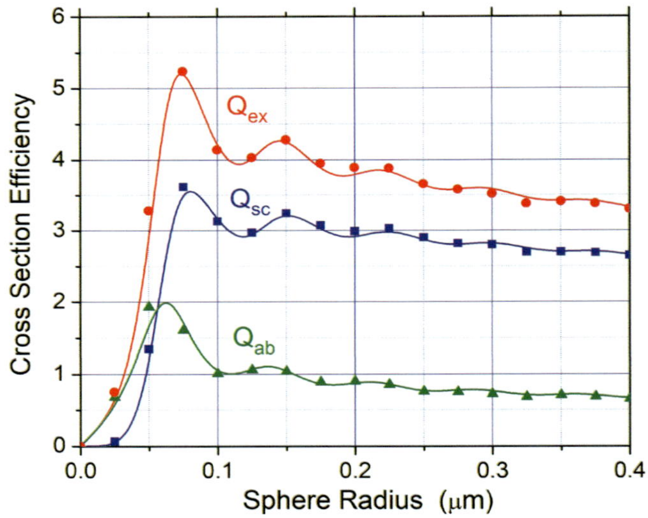

FIGURE 9.21 Cross sections calculated by Mie theory (solid lines) and FDTD (discrete points) for a gold sphere with index 0.497 + i(2.37) at a wavelength of 0.55 μm as a function of sphere radius. The cell space is 200 cells on a side. The other parameters for the FDTD calculation are given in Table 9.1.

TABLE 9.1
Parameters for Figure 9.21

Radius (μm)	Cell size (μm)	# time steps	Q_{sc}	Q_{ab}	Q_{ex}
0	0.0025 μm	3000	0	0	0
0.025	0.0025 μm	3000	.0686	.681	.75
0.05	0.0025 μm	3000	1.35	1.93	3.28
0.075	0.005 μm	2000	3.62	1.61	5.24
0.1	0.005 μm	2000	3.13	1.01	4.14
0.125	0.005 μm	2000	2.97	1.06	4.03
0.15	0.005 μm	2000	3.24	1.04	4.28
0.15	0.003 μm	3000	3.13	1.09	4.22
0.175	0.005 μm	2000	3.07	0.889	3.95
0.2	0.004 μm	3000	2.99	0.902	3.89
0.225	0.005 μm	2000	3.03	0.854	3.88
0.25	0.005 μm	2000	2.90	0.759	3.66
0.275	0.005 μm	2000	2.82	0.751	3.58
0.3	0.005 μm	2000	2.80	0.722	3.52
0.325	0.005 μm	2000	2.70	0.681	3.38
0.35	0.007 μm	1500	2.70	0.704	3.41
0.375	0.0075 μm	1500	2.69	0.686	3.38
0.4	0.008 μm	1500	2.65	0.652	3.30.3

was 200 cells on an edge with 2.5 nm cubic Yee cells, and the calculations were run for 3,000 steps. For the Drude model, $\sigma_0 = 1.1266 \times 10^7 \left(\Omega m\right)^{-1}$ and $\epsilon_\infty = 4$, corresponding to $\omega_p = 1.3321 \times 10^{16}$ Hz, $f_r = 1.3945 \times 10^{14}$ Hz, and $\epsilon_s = -9120.1$. For the Lorentz model, $\epsilon_\infty = 4$, $\epsilon_s = 10$, giving $\omega_0 = 2.4118 \times 10^{15}$ Hz, and $f_r = 1.3715 \times 10^{13}$ Hz. The results are shown in Figure 9.23.

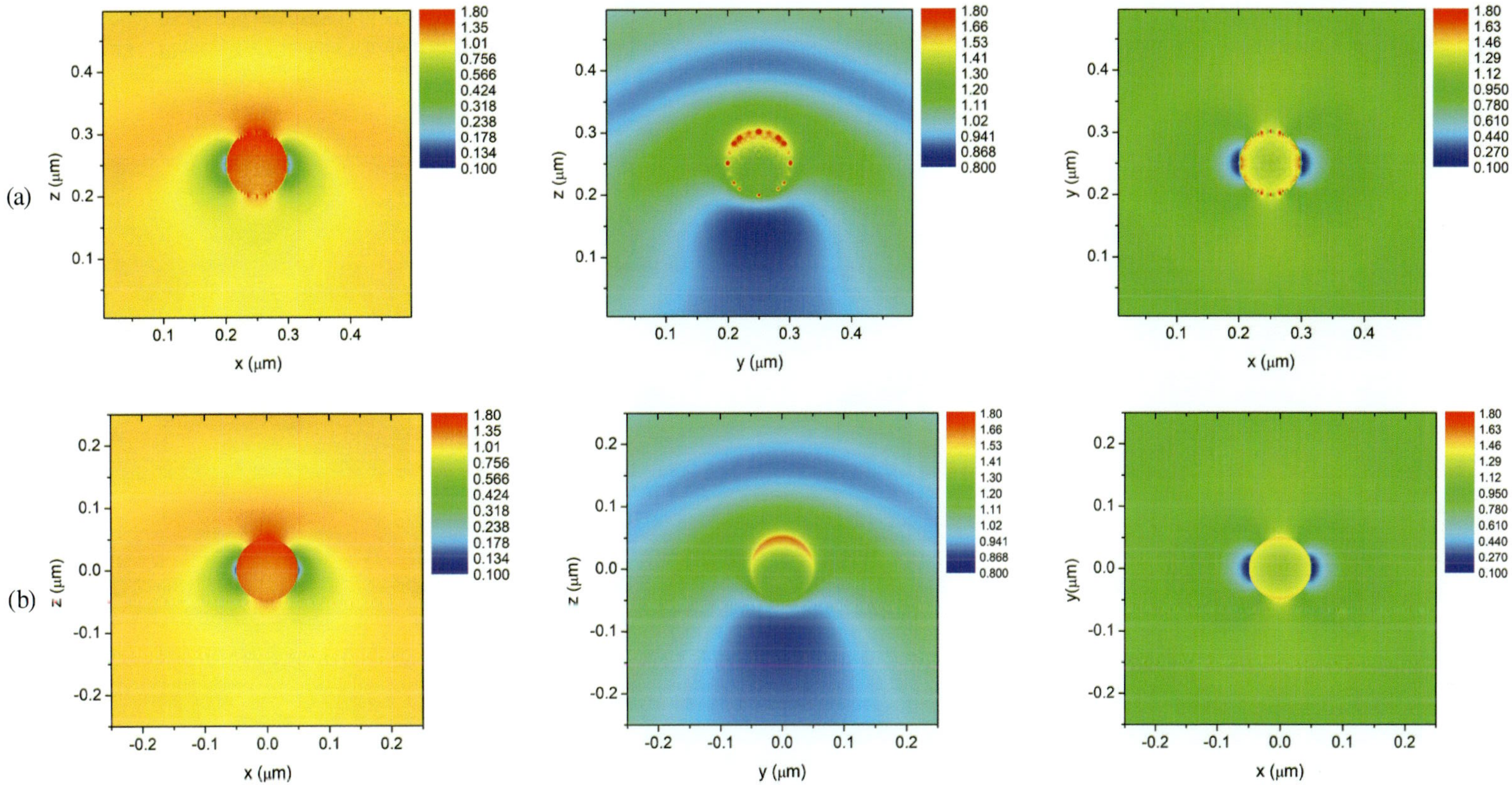

FIGURE 9.22 Total field intensity from a 100 nm spherical air bubble embedded in a dense medium with refractive index = 2 at a wavelength of 0.7 μm. (a) FDTD calculation with 200 cells on each side and 2.5 nm cubic Yee cells after 3,000 time steps. (b) Mie calculation plotted on the same scale. Note that the *xz* plot is a log plot for comparison to Figure 4 in [2], while the *yz* and *xy* plots are scaled linearly.

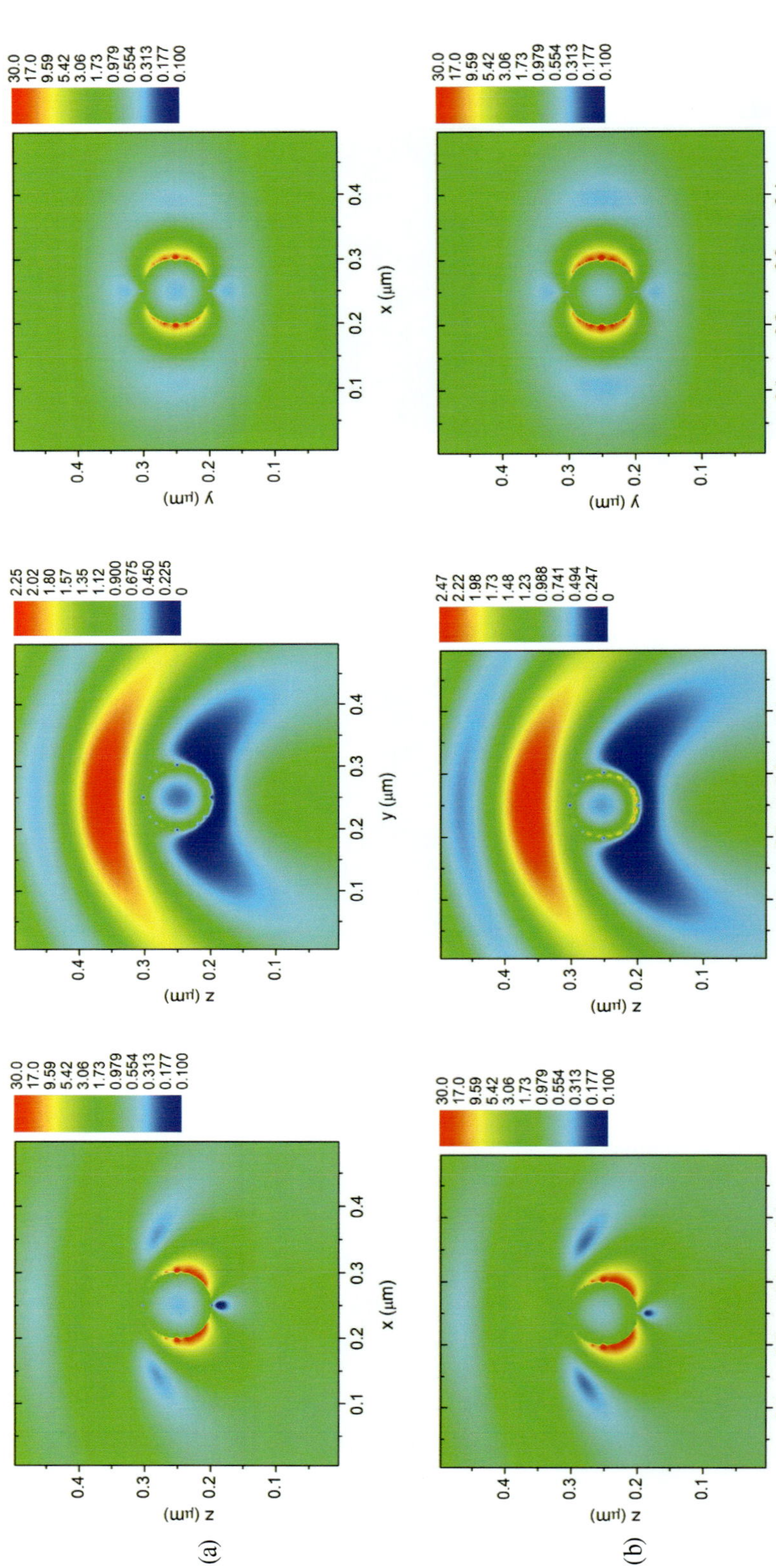

FIGURE 9.23 Plane wave incident within a medium of refractive index 2.0 onto a 100 nm silver sphere with refractive index 0.14 + *i*(4.523) at a wavelength of 0.7 μm. For each model, total field intensity is plotted for the *xz*, *yz*, and *xy* planes through the center of the sphere. (a) FDTD Debye model. (b) FDTD Drude model. (c) Lorentz model. (d) Mie scattering results.

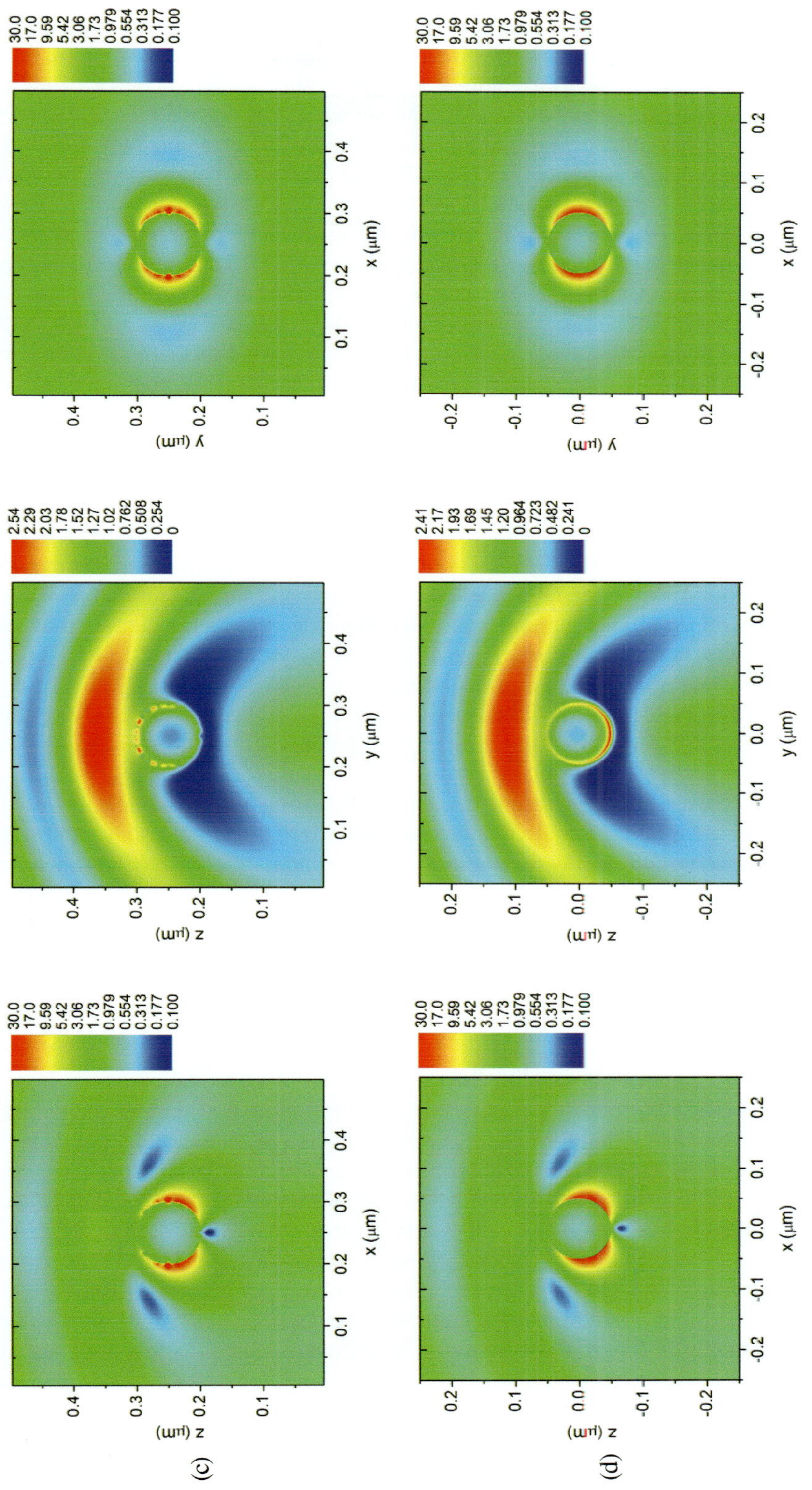

FIGURE 9.23 (Continued)

9.12.3 Far Field Scattering From Sphere

Consider a 0.12 µm diameter silver sphere at a wavelength of 0.36 µm in vacuum. Using the Drude dielectric function results in an electric field intensity distribution as shown in Figure 9.24(a). The far field scattering pattern is shown in Figure 9.24(b). The normalization constant for I_θ in the *xz* far field plane (in arbitrary units) is 3.80×10^{-27}. The normalization constant for I_ϕ in the *yz* far field plane is 3.81×10^{-27}, nearly identical to the previous I_θ value. The normalization constant for I_θ in the *xy* far field plane is 3.82×10^{-28}, and for I_ϕ is 1.73×10^{-27}. The far field cross-section efficiencies are $Q_{sc} = 3.30$, $Q_{ab} = 2.39$, and $Q_{ex} = 5.69$.

Mie theory is used to compute the same example in Figure 9.25 for comparison. The normalization factor for I_θ in the *xz* far field plane (in arbitrary units) is 2.13. The normalization constant for I_ϕ in the *yz* far field plane is also 2.13. The normalization constant for I_θ in the *xy* far field plane is 0.258, and for I_ϕ is 1.125. The far field cross-section efficiencies are $Q_{sc} = 3.11$, $Q_{ab} = 2.07$, and $Q_{ex} = 5.17$. The FDTD and Mie theory far field plots are somewhat different, probably indicating that the computation space was not sufficiently large for evanescent fields near the edges to be negligible in the near- to far-field transform.

9.12.4 Metamaterials

At visible light frequencies the permeability of all simple or homogenous materials is not significantly different from the permeability of vacuum. However, materials composed of arrays of submicron elements, called metamaterials, have been designed and fabricated and shown to exhibit properties not generally found in natural materials. For example, metamaterials may exhibit an effective permeability larger than that of vacuum. They may be designed to exhibit resonance effects in the visible and infrared. An example of an array is shown in Figure 9.26 of the gold split ring resonators in Figure 9.17 similar to that described in [13]. Periodic boundary conditions are applied on the *xz* and *yz* faces, while 2-layer reradiating boundary conditions are applied on the top and bottom *xy* faces.

When an array of these split ring resonators is (somehow) arranged in a plane in free space, the calculated reflectance for light normally incident and polarized along the *x*-axis exhibits a narrow electric quadrupole resonance at a wavelength of ~0.7 µm and a broad magnetic dipole resonance at ~1.4 µm as shown in Figure 9.27(a). In Figure 9.27(b) for the incident light polarized along the *y*-axis, there is a single resonance at ~0.75 µm. When the array is supported on a glass substrate (n = 1.5), the resonances shift towards longer wavelengths as shown in Figure 9.28.

9.12.5 Heat-Assisted Magnetic Recording (HAMR)

The final example is more of an exercise for the interested and ambitious reader. How can one deliver laser power efficiently into a thin metallic film with a spot diameter that is an order of magnitude smaller than the diffraction limit in the presence of a magnetic recording pole? Heat assisted magnetic recording or "HAMR" is the technique of heating a spot on the surface of a spinning hard disc coated with a magnetic alloy and allowing it to cool in an applied magnetic field to record data [14, 15]. The technology has been recently commercialized and is delivering storage densities that were unthinkable a decade ago. In the laboratory, storage densities of 6 TB/in^2 (a terabyte is 1.0×10^{12}) have been demonstrated, [16, 17] corresponding to an area per bit of only 108 nm^2!

The first demonstration of HAMR in the laboratory occurred with a recording head fabricated as shown schematically in Figure 9.29. Many of the software models in this book are obviously applicable for designing this recording head. An efficient coupling grating can be designed using FDTD [18]. The grating is split into halves as shown with a half period offset between them. The grating couples light efficiently into a TE waveguide mode and then into the planar solid immersion mirror (PSIM). The PSIM efficiently concentrates the energy of the mode at the focal

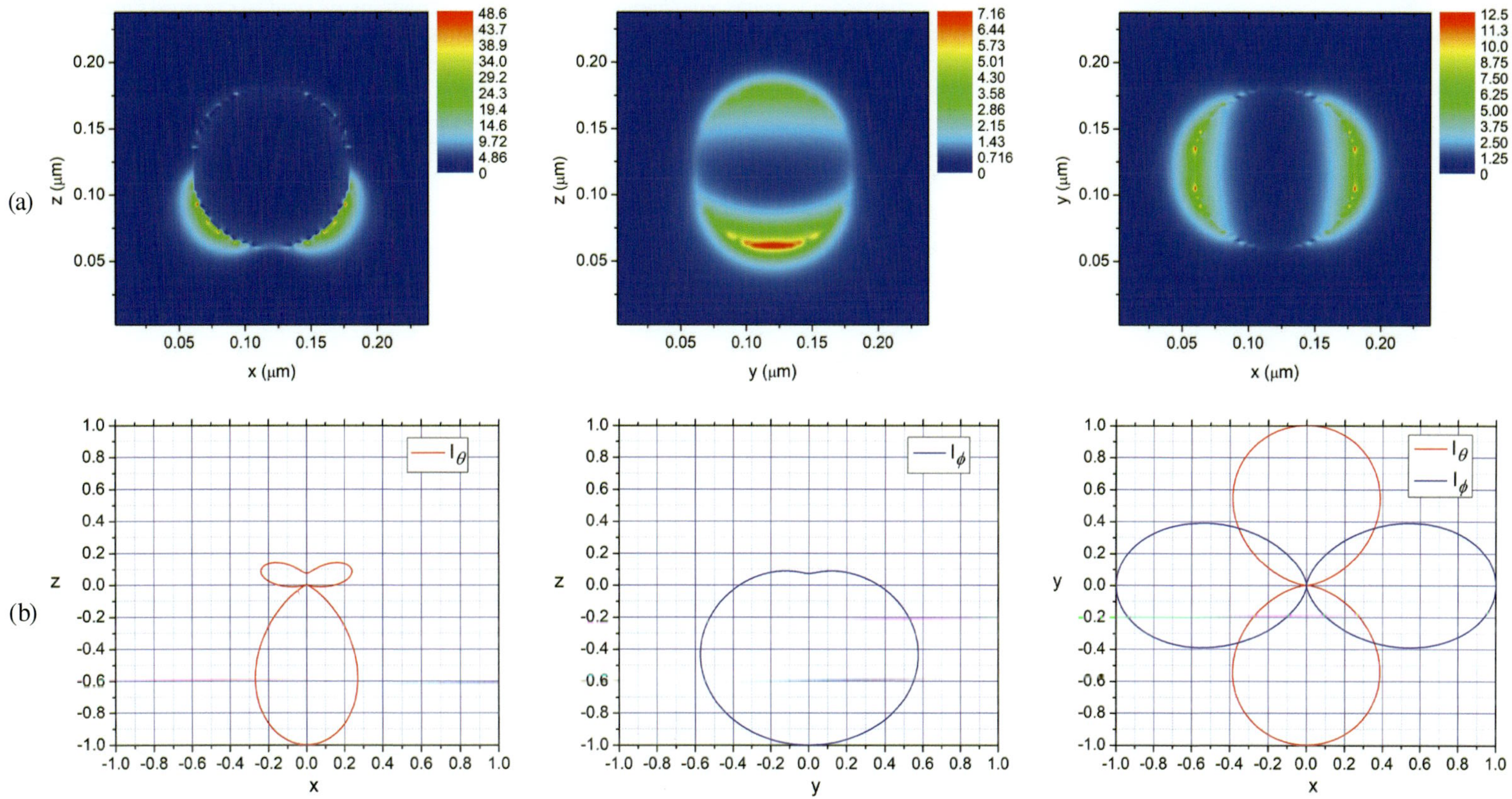

FIGURE 9.24 (a) Scattered near field intensity distribution for a plane wave incident upon a 0.12 μm diameter silver sphere with a refractive index of 0.196 + *i* (1.533) at $l = 0.36$ μm using a Drude dielectric function. The computation space is 120 cells on a side with 0.002 μm cubic Yee cell running the calculation for 4,000 time steps. (b) Normalized far field scattering pattern calculated from the near field.

(a)

(b)

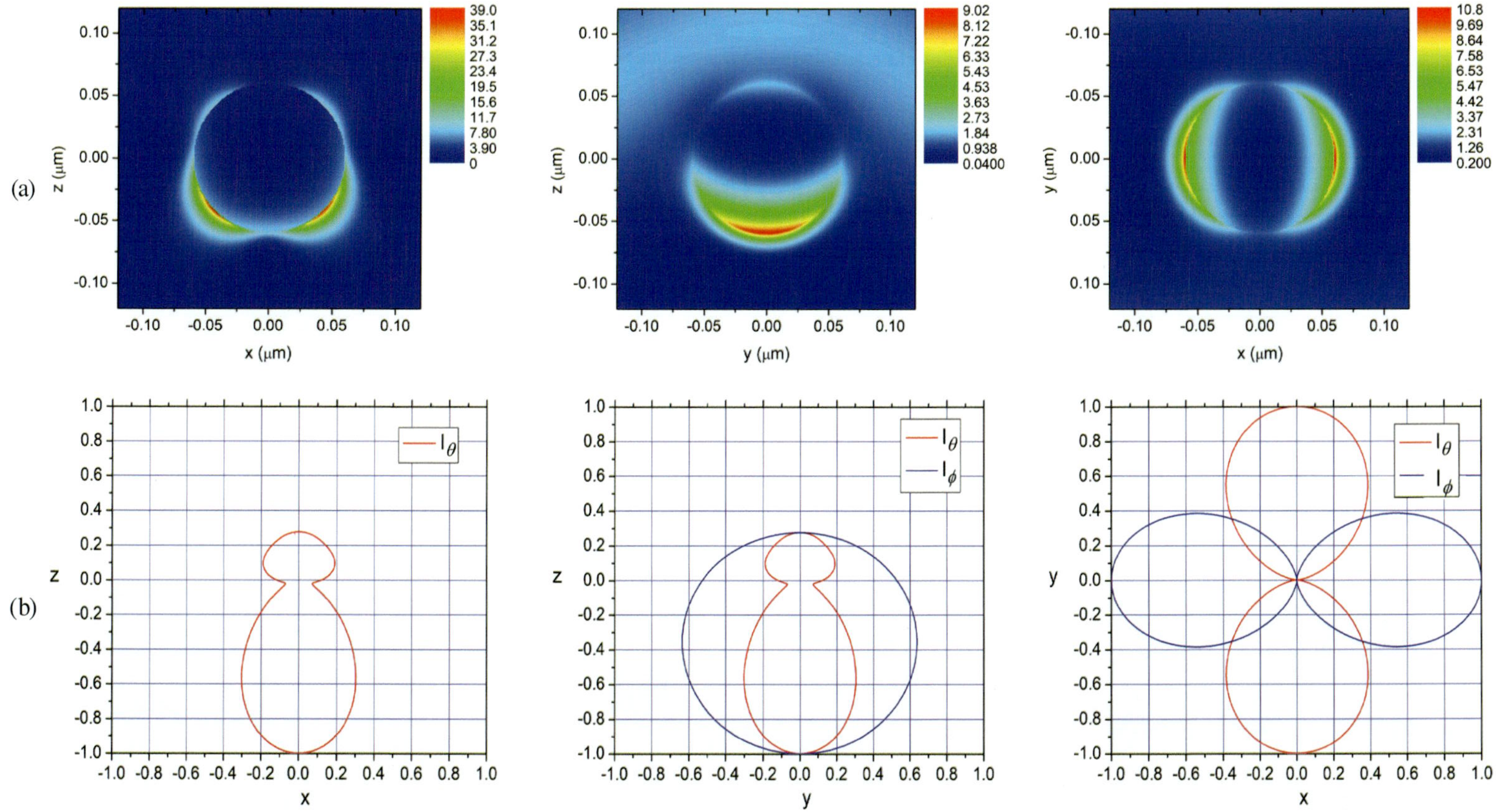

FIGURE 9.25 (a) Near field intensity distribution for a plane wave incident upon a 0.12 μm diameter silver sphere at $\lambda = 0.36$ μm using Mie theory. (b) Normalized far field scattering pattern. The plots are for the *xz*, *yz*, and *xy* planes through the center of the sphere.

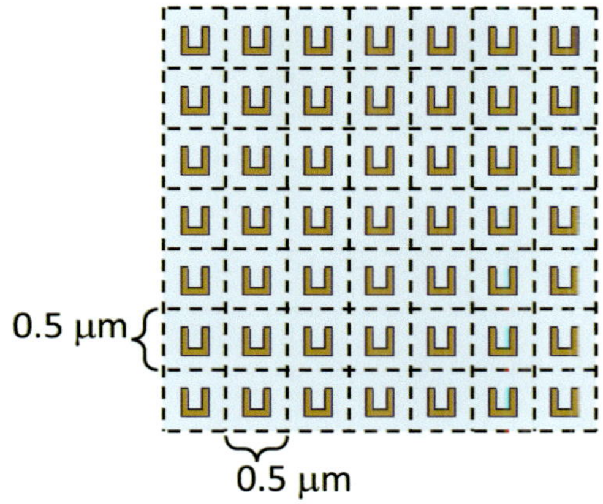

FIGURE 9.26 Array of U-shaped split ring resonator metamaterial composed of gold designed for an electric quadrupole resonance in the visible and a magnetic dipole resonance in the infrared. The SRRs are defined in Figure 9.17. The cell space is 100 × 100 × 100, and each Yee cell is a cube with 5 nm on a side. The array periodicity is 0.5 μm.

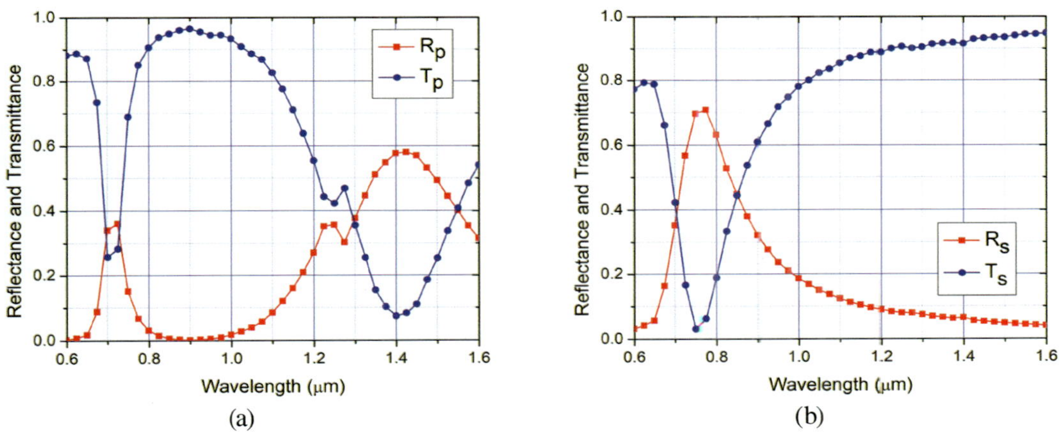

FIGURE 9.27 Reflectance and transmittance for a normally incident plane wave polarized along the (a) y-axis and (b) x-axis on an array of split ring resonators in air. The computation is run for 4,000 time steps.

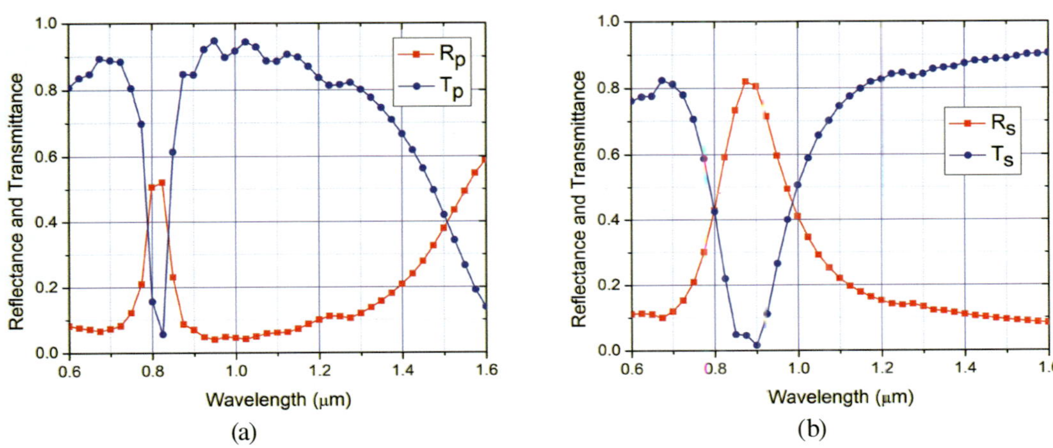

FIGURE 9.28 Reflectance and transmittance for a normally incident plane wave polarized along the (a) y-axis and (b) x-axis on an array of split ring resonators on a glass substrate (n = 1.5). The computation is run for 4,000 time steps.

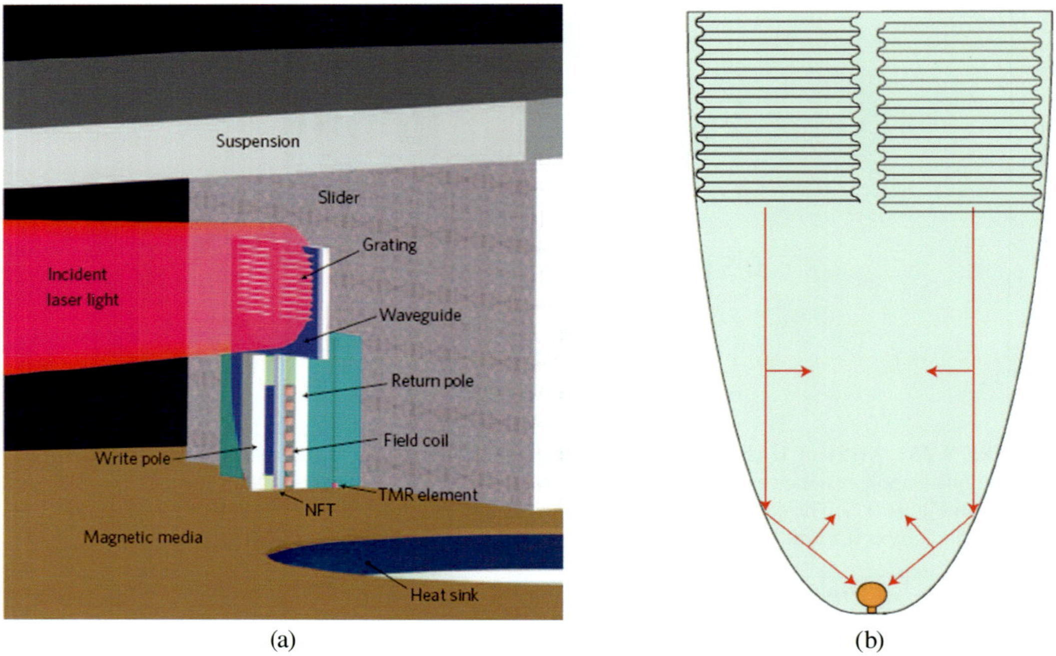

FIGURE 9.29 (a) HAMR recording head design. (b) Laser light at a wavelength of 830 nm is incident upon a split grating that efficiently couples light into the TE mode of a planar waveguide. The waveguide terminates in a planar solid immersion mirror that focuses the waveguide mode onto a "lollipop" near field transducer (NFT). The half wave offset of the two grating halves generates a longitudinal electric field at the focus of the PSIM. The NFT has a plasmonic resonance at the wavelength of the incident light and efficiently couples the optical energy into the magnetic recording medium underneath it, heating the medium above its Curie point in the applied magnetic field.

From W. Challener, Asia Pacific Magnetic Recording Conference, Singapore, 2009.

point where the NFT is located. By analogy to radial polarization, as was discussed in Chapter 7, the split offset grating generates a strong *longitudinal* electric field at the focal point. The gold NFT is a disk with a peg sticking out [19]. The disk has a large absorption cross section at the laser wavelength which excites a localized surface plasmon resonance on the disk as shown in Figure 9.30(a), similar to those discussed for spherical particles from Mie theory. The peg effectively concentrates the field into the recording medium by the lightning rod effect. The NFT design is optimized by scattered field FDTD. The electric field at the NFT is calculated for the PSIM and used as the incident field in the FDTD model to optimize the NFT dimensions and calculate the fraction of the incident optical power absorbed by the magnetic thin film below the NFT and the size of the optical spot. As shown in Figure 9.30(b), an NFT was designed by this process with a theoretical coupling efficiency from the PSIM into the recording medium of ~8% within a FWHM spot size of ~50 nm, an order of magnitude below the diffraction limit of ~415 nm for a lens with a numerical aperture of 1 at this wavelength. The recording head fabricated according to this design actually recorded data with a cross-track width of ~70 nm as shown in Figure 9.31.

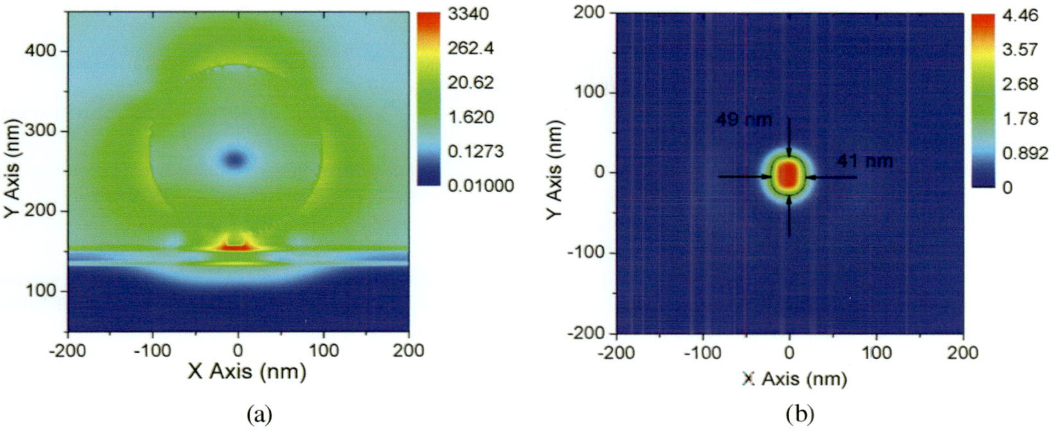

FIGURE 9.30 FDTD calculation of the (a) resonant mode of the lollipop NFT and (b) the optical spot within the magnetic recording medium. From W. Challener, Asia Pacific Magnetic Recording Conference, Singapore, 2009.

FIGURE 9.31 Magnetic force microscopy image of data recorded on a hard disc with a track width of ~70 mm using the lollipop near field transducer. Scale bar is 300 nm. Reprinted from Figure 5 of [15] with permission from Springer Nature.

REFERENCES

1 K. S. Kunz and R. Luebbers, *The Finite Difference Time Domain Method for Electromagnetics* (New York: CRC Press, 1993).

2 K. S. Yee, "Numerical solution of initial boundary value problems involving Maxwell's equations in isotropic media," *IEEE Trans. Ant. Prop.* **14** (1966).

3 J. P. Berenger, "A perfectly matched layer for the absorption of electromagnetic waves," *J. Comp. Phys.* **114** (1994) 185–200.
4 S. D. Gedney, "An anisotropic perfectly matched layer-absorbing medium for the truncation of FDTD lattices," *IEEE Trans. Ant. Prop.* **AP-44** (1996) 1630–1639.
5 R. E. Díaz and I. Scherbatko, "A simple stackable re-radiating boundary condition (rRBC) for FDTD," *IEEE Ant. Prop.* **46** (2004) 124–130.
6 G. Mur, "Absorbing boundary conditions for the finite-difference approximation of the time-domain electromagnetic-field equations," *IEEE Trans. Electromag. Compat.* **EMC-23** (1981) 377–382.
7 R. E. Díaz and I. Scherbatko, "A new multistack radiation boundary condition for FDTD based on self-teleportation of field," *J. Comput. Phys.* **203** (2005) 176–190.
8 A. J. Kogon and C. D. Sarris, "Finite-difference time-domain modeling of periodic structures: A review of constant wavevector techniques," *IEEE Ant. Prop. Mag.* **64** (2022) 59–70. https://doi.org/10.1109/MAP.2021.3061023. See also arXiv:2007.05091v2.
9 W. A. Challener, I. K. Sendur, and C. Peng, "Scattered field formulation of finite difference time domain for a focused light beam in dense media with lossy materials," *Opt. Exp.* **11** (2003) 3160–3170.
10 S. A. Shelkunoff, "A general radiation formula," *Proc. I.R.E.* **27** (1939) 660–666.
11 S. Ramo, J. R. Whinnery, and T. Van Duzer, *Fields and Waves in Communication Electronics* (New York: John Wiley & Sons, 1965) 661 ff.
12 Note a sign error in (7.3) of [1], 108.
13 Y. Ji, B. Wang, L. Fang, Q. Zhao, F. Xiao, and X. Gan, "Exciting magnetic dipole mode of split-ring plasmonic nano-resonator by photonic crystal nanocavity," *Materials.* **14** (2021) 7330.
14 W. A. Challener, "The road to heat assisted magnetic recording," *Asia Pacific Magnetic Recording Conference* (January, 2009).
15 W. A. Challener, C. Peng, A. V. Itagi, D. Karns, W. Peng, Y. Peng, X-M. Yang, X. Zhu, N. J. Gokemeijer, Y.-T. Hsia, G. Ju, R. E. Rottmayer, M. A. Seigler, and E. C. Gage, "Heat-assisted magnetic recording by a near-field transducer with efficient optical energy transfer," *Nat. Phot.* **3** (2009) 220–224.
16 https://www.seagate.com/innovation/hamr/ (accessed 10/25/2025).
17 https://www.seagate.com/blog/the-future-of-data-storage-technology-why-hamr-is-the-new-standard-for-hard-drives/ (accessed 10/25/2025).
18 C. Peng and W. A. Challener, "Input-grating couplers for narrow Gaussian beam: influence of groove depth," *Opt. Exp.* **12** (2004) 6481–6490.
19 W. A. Challener, *Transducer for Heat Assisted Magnetic Recording*. U.S. patent #7,272,079. (2007).
20 E. D. Palik, ed., *Handbook of Optical Constants of Solids* (New York: Academic Press, Inc., 1985) 286–295.

Index

Note: Page numbers in *italics* indicate a figure on the corresponding page.

M

N

O

P

Q

R

S

T

V

W

Y

For Product Safety Concerns and Information please contact our EU representative GPSR@taylorandfrancis.com Taylor & Francis Verlag GmbH, Kaufingerstraße 24, 80331 München, Germany

Batch number: 10706963

Printed by Printforce, the Netherlands